Reactivity and Structure Concepts in Organic Chemistry

Volume 13

G. W. Gokel S. H. Korzeniowski

Macrocyclic Polyether Syntheses

With 89 Tables

Springer-Verlag
Berlin Heidelberg New York 1982

Professor George W. Gokel
Dept. of Chemistry
University of Maryland
College Park, MD 20742
USA

Dr. Stephen H. Korzeniowski
Chemicals and Pigments Department
E. I. Du Pont de Nemours Co., Inc.
Jackson Laboratory
Deepwater, New Jersey 08023
USA

ISBN-13:978-3-642-68453-1 e-ISBN-13:978-3-642-68451-7
DOI: 10.1007/978-3-642-68451-7

Library of Congress Cataloging in Publication Data, Gokel, George W., 1946–. Macrocyclic polyether syntheses. (Reactivity and structure: concepts in organic chemistry; v. 13). Bibliography: p. Includes index. 1. Ethers. 2. Cyclic compounds. I. Korzeniowski, S. H. (Stephen H.), 1949–. II. Title. III. Series. QD341. E7G64. 547'.0350459. 82–654. ISBN-13:978-3-642-68453-1 (U.S.). AACR2

Softcover reprint of the hardcover 1st edition 1982

Typesetting: Elsner & Behrens, Oftersheim

2152/3140-5 4 3 2 1 0

To Our Mothers,
The Memory of Our Fathers
and
The Girls from Montclair State

Preface

It seems likely that no project of this magnitude is undertaken in the absence of several motivations. This book is no exception. There are, from our perspective, at least two rational reasons for conducting this cataloging and organizational effort. The first of these is the obvious fact that a very large number of macrocyclic polyethers have now been prepared and nowhere is there a comprehensive listing of them. A number of excellent reviews have been published dealing with one or more aspects of macrocycle chemistry, but we felt a great need for a single source listing known compounds. We have, therefore, attempted to provide such a listing to assist ourselves in keeping track as well as, we hope, helping others in this field.

We also felt that there was a need for a survey of methods which have been utilized in the syntheses of these cation binders so that the novice could establish some starting point. Wherever possible, we have also attempted to point to the motivations which engendered the various syntheses.

The somewhat irrational reason which propelled us into this endeavor was the questioning of Charlie Morgan who often asked if one or another obscure compound had been synthesized. As often as not, we were uncertain and now at least have a reference which, if not absolutely complete, will at least save face.

We have not attempted to cover all or even most aspects of crown chemistry and some may say that the inclusions are eclectic. We felt that anyone approaching the field would need an appreciation for the jargon currently abounding and for the so-called "template effect" since the latter has a considerable bearing on the synthetic methodology. We have, therefore, included brief discussions of these topics in the first two chapters. In chapters 3–8, we have tried to present an overview of the macrocyclic polyethers which have been prepared. We have taken a decidedly "organic" tack in this attempting to be comprehensive in our inclusion of alkali and alkaline earth cation binders rather than the compounds of use in transition metal chemistry. Nevertheless, many of the latter are included in concert with their overall importance.

Construction of both chapters and tables has required considerable compromise. Is a crown ether containing oxygen, nitrogen, sulfur and a urethane link a crown, an azacrown, a thiacrown, a crown ester or a crown amide? Because of the ambiguity in designating structures there is a certain level of redundancy in the text and perhaps an unfortunate amount of arbitrary classification. In the "notes on table usage" we have made some suggestions which we hope will help locate compounds and make the arbitrariness less onorous.

One may also wonder why a volume with "syntheses" in the title is devoid of detailed preparations. These have been consciously omitted because inclusion of more than just a few would have made the book quite lengthy and the arbitrary choice of procedures would have added to the problem noted above. We have included mention of key experimental details in numerous cases but feel that the reader would be far wiser to refer to the original literature in performing a preparation than following a recipe from a book. The orginial literature would be necessary in any event since it would have been impossible to include spectral data in the preparations.

There are many who deserve our acknowledgement if not our heartfelt thanks. First, and perhaps foremost, are the members of the library staff at E. I. Du Pont de Nemours Co., Chemicals and Pigments Department, Jackson Laboratory, Deepwater, New Jersey, for their remarkable and enthusaistic assistance in culling literature. We particularly appreciate the efforts of Louise Del Signore, Dianne Cowperthwait, Evern Ford, Eleanor Price, Joanne Randolph, and Dot Toulson.

We are also pleased to thank Dennis Dishong and Jim Beadle for their great help in organizing tables after all the compounds had been listed on cards and put through their organizational iteration. It is a special pleasure to acknowledge a number of kind and competent ladies for their assistance in typing and in other aspects of this work. We warmly thank Sandy Alfier, Caren Klarman, Karen Miller, Delores Griffin and Millie Lindenberger for their various contributions.

Any errors which have survived the proofreading process are ours alone and we apologize in advance for any inconvenience these omissions or errors may cause a reader.

George W. Gokel
College Park, Maryland

Stephen H. Korzeniowski
Deepwater, New Jersey

Table of Contents

Listing of Tables

Chapter 3. Tables

Chapter 4. Tables

Chapter 5. Tables

Chapter 6. Tables

Chapter 7. Tables

Chapter 8. Tables

Notes on Table Usage

Because the number of crown compounds thus far prepared is large and because there are numerous variations in structures, many of the compounds are difficult to consign to a single group. For example, a compound with an amide link in it would likely be found in chapter 5 tables, but if it contained an additional nitrogen atom, it might be found in chapter 4. The general scheme for organizing the various compounds is given below, but should be used only as a guide. Regrettably, not all compounds which might be listed in more than one chapter have been duplicated.

Chapter 3: Compounds containing oyxgen in the macroring. Certain heterocyclic subunits are included even if a nonoxygen heteroatom is turned inward and may participate in binding.

Chapter 4: Compounds containing one or more nitrogen atoms in the macroring, whether or not the nitrogen is part of another heterocyclic subunit. Certain pyridine and pyrimidine compounds have been included in chapter 3 as well.

Chapter 5: Compounds which contain either an ester or an amide link in a single macrocyclic ring. This includes lactones, lactams, carbonates, carbamates, ureas, phosphates, etc. including sulfur, nitrogen and other atoms in the ring.

Chapter 6: Miscellaneous crown type compounds, including acetals, and compounds containing sulfur, phosphorus, arsenic, etc. Polymeric species are included here and in chapter three.

Chapter 7: Open-chained materials of all descriptions including nitrogen, sulfur, ester, etc.

Chapter 8: Macrobicyclic, macrotricyclic, or polycyclic compounds including cryptands, spherands, etc.

In using the tables, it would be best to survey the list of tables included in each chapter to determine how many categories might possibly contain the compound of interest. It should be noted that a large number of cyclophanes which contain fewer than three heteroatoms are not included in this book since they are not generally useful as cation binders.

The tables are arranged by increasing carbon number within groups. Within groups of the same number of carbons, the compounds are arranged by increasing ring size of the only or largest ring. Among compounds having the same number of carbons, and the same ring size, the compounds are arranged by increasing number of heteroatoms; increasing numbers of non-oxygen heteroatoms and among virtually identical compounds by increasing order of Cahn-Ingold-Prelog priority of the pendant group, or other substituent.

In counting the size of the ring, only those atoms which actually constitute the ring are counted. The compounds below would all be counted as 18-membered rings by our method.

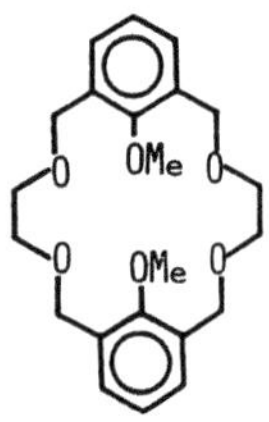

Chapter 3

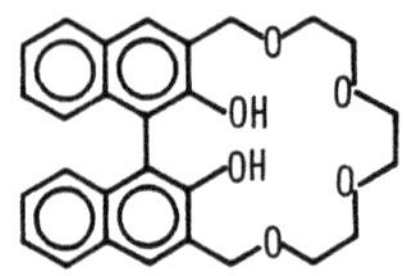

Chapter 3

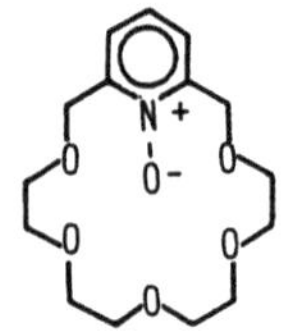

Chapter 4

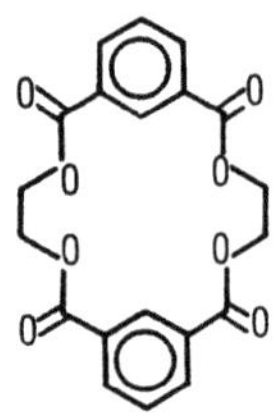

Chapter 5

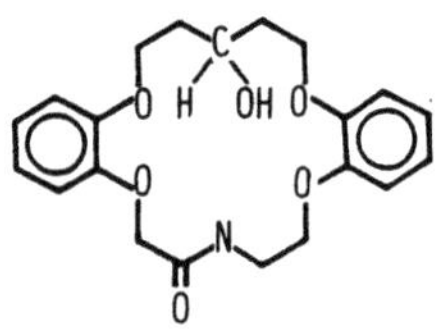

Chapter 5

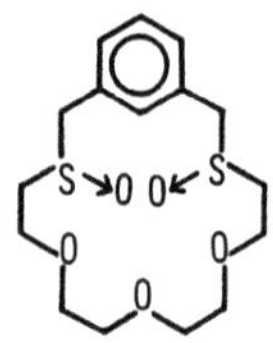

Chapter 6

There are also some abbreviations to which attention should be drawn. Most of these are used primarily in chapter 8 and are given in the legend of the chapter 8 tables. The others are listed below:

Ph = Phenyl = C_6H_5– = 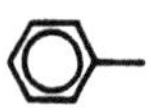

Py = Pyridyl = C_5H_4N– = 

Ms = mesyl = methansulfonyl = CH_3SO_2–

Ts = tosyl = p-toluenesulfonyl = 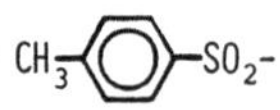

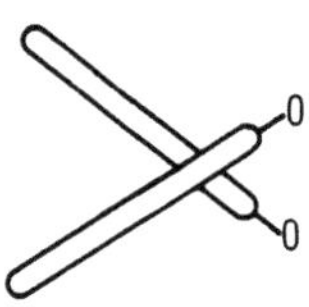

(S)

=

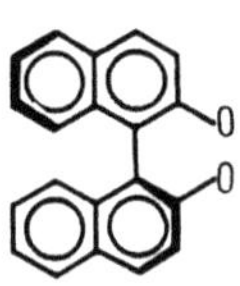

(S)

Note that in many cases, hydrogen atoms have not been included on secondary nitrogen atoms. This has been done for structural clarity.

Chapter 1

Introduction and General Principles

1.1 Background

There are many developments which predate Pedersen's first papers on crown ether chemistry[26,27], and it would be thoughtless indeed to ignore them. The nitrogen macrocycles had been studied for decades prior to 1967[36] and pólyethers and macrocyclic polyethers had been obtained years before from the cycloligomerization of ethylene oxide[34]. At about the same time, cation complexation by naturally occurring polyoxygen containing species was established[31] and the "self-solvating bases" described by Ugelstad[35] bear a striking resemblance to what we now call "crown complexation". Nevertheless, the beginning of crown ether chemistry is generally dated from Pedersen's first disclosures in 1967 because he not only prepared many examples of these compounds, but he recognized clearly their potential as complexing agents. It is largely for the latter reason that Lüttringhaus, who was clearly a pioneer in this area[20–23], is not more widely credited with his early preparations of these compounds.

1.2 Pedersen's Discovery of Crown Compounds

Charles J. Pedersen, a man of Chinese and Norwegian extraction, was employed by E. I. du Pont de Nemours and Company in the Elastomer Chemicals Department. By his own account[28,30] he was seeking a phenolic complexing agent for alkali metal salts. He had prepared the monotetrahydropyranyl ether derivative of catechol for the purpose of converting it into a bidentate ligand. His experimental approach involved the reaction between the phenol ether, diethylene glycol dichloride and sodium hydroxide. The intended reaction sequence is shown in Eq. (1.1). After a double Williamson reaction occurred, it was Pedersen's intention to cleave the THP ethers leaving him with *1*. He was, indeed, able to isolate *1* but he also obtained a small amount of fluffy white crystals which melted at 164 °C, showed no hydroxyl absorption in the IR but which still seemed to interact with alkali metal cations. We now know that the compound he isolated was dibenzo-18-crown-6 (*2*).

Pedersen reasoned that there must have been a small amount of catechol present which had not been protected properly. This must have led to the formation of *1 in situ*

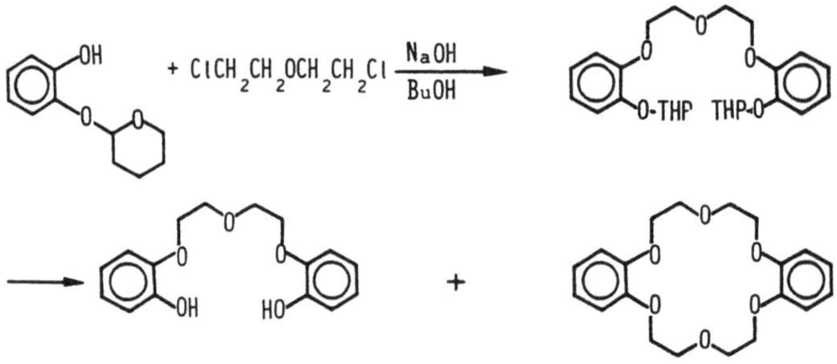

1 *2*

(1.1)

followed by cyclization to form *2*. The amount of catechol in the reaction mixture was small, so the efficiency of the cyclization which led to *2* must have been remarkably high. This efficiency was especially notable since the cyclization was conducted in quite concentrated solution. Furthermore, *2* seemed to interact with, or complex alkali metal cations despite the fact that it contained no hydroxyl groups. These unexpected results led Pedersen to prepare many examples of the compounds we know as "crown ethers"[27].

Extensive and important as Pedersen's efforts were, they might have been even greater had he not prepared the first examples of crown ethers when he was beyond sixty years of age. After giving birth to a remarkable child, he was unable to nurture it because of his retirement in 1969.

1.3 Nomenclature and Jargon

As in any new field, new structures demand novel names. As the variety of structures has increased, the names have kept pace. Although some may be dismayed by this, the different names connote not only different structural types or functional groups, but conceptual differences among the various classes as well. In order to introduce the novitiate and perhaps remind the journeyman, we provide a few definitions and explanations herewith.

1.3.1 Crown Ether

This term was first suggested by Pedersen and has been widely adopted. Although it is assumed by some that the term arises from the appearance of the molecular models as stylized below, Pedersen reported[27] that the term was suggested by the fact that these macrocycles "crowned" cations even as a regal crown adorns the monarch's brow.

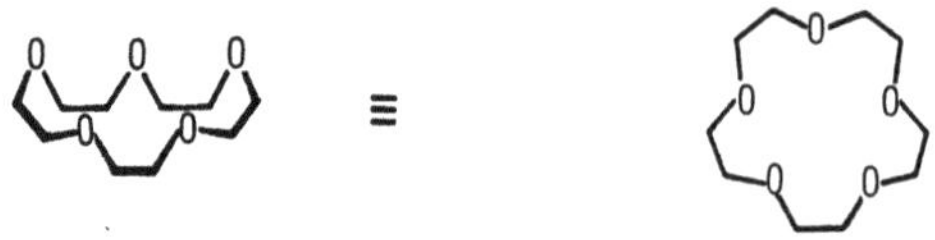

15-crown-5 (*3*)

The term "crown" generally refers to macrocyclic polyethers having the ethyleneoxy unit as the basic repeating structure. That the repeating unit is ethyleneoxy rather than methyleneoxy or propyleneoxy is not an accident. When every third atom is oxygen, binding is obviously higher than when more carbons separate the heteroatoms. In addition, the unfavorable conformational interactions are reduced relative to the carbon analogs. Naturally, these interactions could be reduced even further by using repeating methyleneoxy units, but the hydrolytic stability of such species must necessarily be inferior to crowns, notwithstanding indications to the contrary in certain cases[10].

The nomenclature convention suggested by Pedersen[27] for the simple compounds involves two numbers. The first of these indicates the total number of atoms in the macrocycle. The second number indicates how many heteroatoms are present in the ring. Compound *3* shown above, for example, contains fifteen atoms in the ring, five of which are oxygen. The commonly accepted name for this compound is "15-crown-5". The three compounds shown below can all be named by this simple rule.

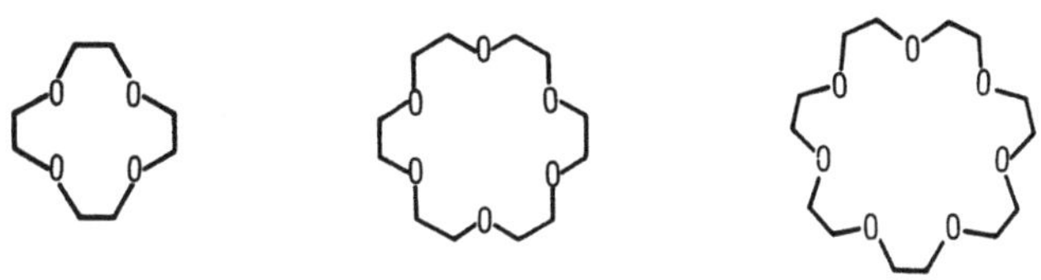

12-crown-4 18-crown-6 21-crown-7

It is often the case that chemists involved or familiar with ethylene oxide (oxiran) chemistry refer to these cyclic oligomers as EO-4, EO-6 and EO-7 respectively. Such designations are informal if not colloquial but, like any name, are useful if they correctly convey an idea. The difficulty with these informalities, indeed with the "crown" nomenclature system is that it cannot adequately deal with complicated structures or even isomers of simple ones.

Compound *2*, below, is essentially 18-crown-6 onto which two benzene rings have been fused. Pedersen suggested the name "dibenzo-18-crown-6" to describe it. The hydrogenated derivative was referred to as dicyclohexyl-18-crown-6, although the slightly more consistent dicyclohexano-18-crown-6 now seems preferred. The name of the latter compound gives no information about the stereochemistry at the cyclohexane ring junctions and the name dibenzo does not adequately describe the relative positions of the benzene rings. Naturally, systematic names may be used for these compounds but such names have generally proved cumbersome. The systematic name for 18-crown-6, for example, is 1,4,7,10,13,16-hexaoxacyclooctadecane. Any substituent on the ring

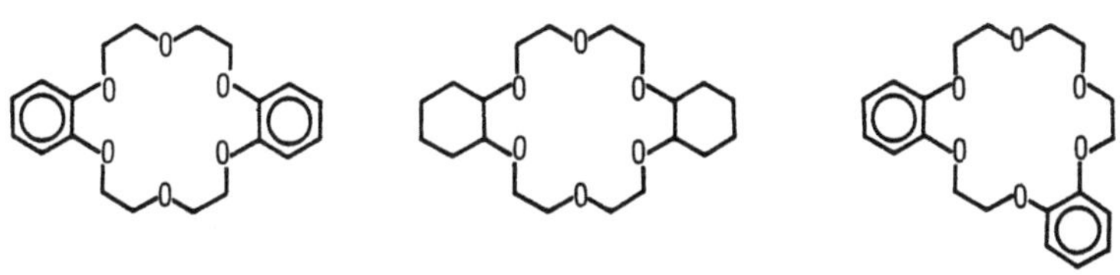

dibenzo-18-crown-6 (*2*) dicyclohexano-18-crown-6 an isomer of *2*

complicates the systematic nomenclature considerably and the less systematic but much more convenient "crown" designations therefore persist. A listing of systematic names for many of Pedersen's first crowns are available as a supplement to his second paper[39].

1.3.2. Azacrown

This terminology is a logical extension of the nonsystematic nomenclature described above. Although either amino-crown or azacrown might connote the presence of a nitrogen atom in place of oxygen in the macroring structure, the latter has more often been used. The structures below are monoaza-18-crown-6 (*7*) and 1,10-diaza-18-crown-6 (*8*). Note in the latter case that the heteroatom positions are designated simply and unambiguously. The inorganic chemists generally refer to polyaza-crowns as "cyclens".

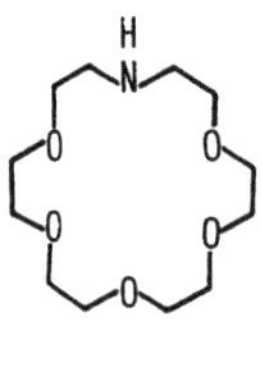

7

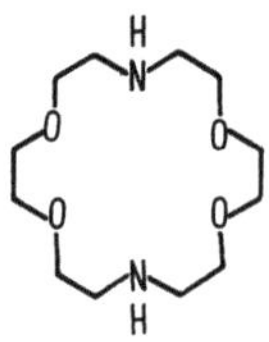

8

1.3.3 In-Out Bicyclic Amines

This class of compounds was developed by Simmons and Park[25,33] at about the same time Lehn and his group were preparing the first cryptands[18]. These compounds are macrobicyclic structures in which both rings contain two tertiary nitrogen atoms. The compounds are illustrated below schematically in their "in-in" and "out-out" forms. These compounds embody the interesting framework of the cryptands but lack the heteroatom cation binding possibilities making them less relevant to the discussion here. Nevertheless, their interesting properties and the timeliness of their discovery have made them of interest to many in the crown field.

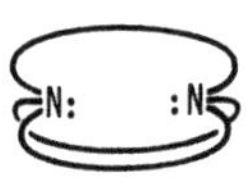

"in-in"

"out-out"

1.3.4 Cryptands and Cryptate Complexes

This remarkable group of compounds was introduced by Lehn and his co-workers in Strasbourg[18]. They are macrobi-, macrotri-, or macropolycyclic compounds which are similar to the "in-out" amines illustrated above, but they have crown-like bridges throughout. Because of their ability not only to complex cations but to encapsulate or entomb them, the name *crypt*and was suggested[7]. The complexes of cryptands are referred to as "cryptates".

For the purposes of nomenclature, the simple cryptands are assumed to be macrobicyclic and nitrogen is assumed to be the bridgehead atom. The different cryptands are designated by assigning numbers according to the number of heteroatoms in each ethylenoxy chain. The three cryptands shown below are designated [1.1.1]-cryptand (*9*), [2.2.1]-cryptand (*10*) and [3.2.2]-cryptand (*11*), respectively.

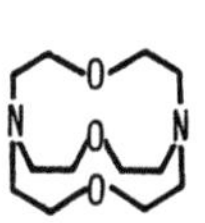

[1.1.1]-cryptand (*9*)

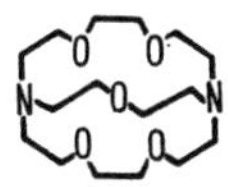

[2.2.1]-cryptand (*10*)

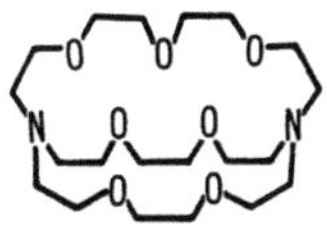

[3.2.2]-cryptand (*11*)

As with the crowns, the situation becomes more complicated when there are other heteroatoms or substituents in one of the cryptand bridges. The symbol "B" is used to designate a benzo- or catechol unit in the bridge and subscripts are used to designate heteroatoms when non-oxygen heteroatoms are present. Examples of this are shown in structures *12* and *13* above.

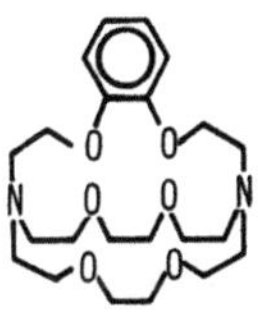

[2.2.B]-cryptand (*12*)

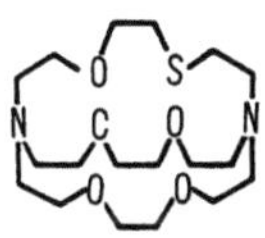

[2_O.2_O.2_S]-cryptand (*13*)

An even more complicated nomenclature problem arises with the closely related all-oxygen cryptands[13]. These compounds do not utilize nitrogen as the three-chain junction. Most examples of this class of compounds have utilized pentaerythritol or glycerol as the junction. This naturally imparts a somewhat lower flexibility to the molecule than would be present in the nitrogen-containing cases. Structures of two such molecules are illustrated below.

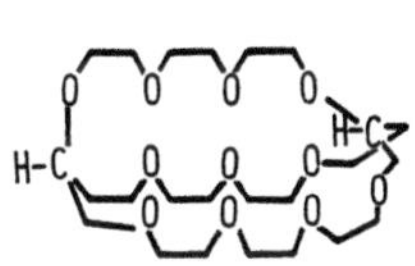

14, d-glycerol junction[13]

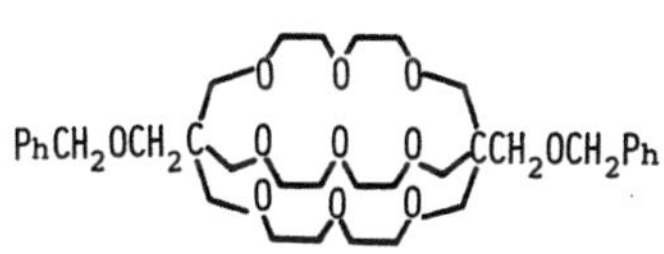

15, pentaerythritol junction[3]

The disadvantages of these names are manifest, but one shudders to think about naming them systematically. It seems likely that for the foreseeable future, nomenclature of crowns, cryptands and related substances will bear a semblance to heiroglyphics.

1.3.5 Lariat Ethers

This is a term introduced by Gokel and co-workers only very recently[9]. The compounds included in this class are those which have single macrorings like crowns but additional pendant donor groups which make them similar in some respects to crytands. The presumption from which the name derives is that a complexed cation will be bound by both the macroring and the sidechain in much the same fashion as a lasso binds an animal. The compounds are named as simple crown derivatives and are illustrated as *16* and *17* below.

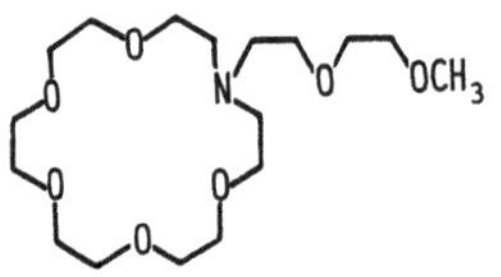

16

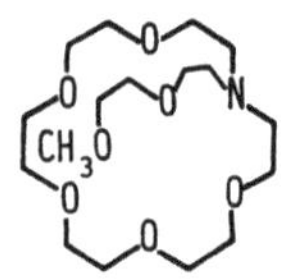

17

Pedersen patented[29] a series of *bis*-crowns related to the above except that the side chain was bound on each side to a crown. Cram[11,14] and Sutherland[16] have both reported similar structures. Pedersen suggested the name "clams" for such species but the name has not found wide acceptance.

1.3.6 Crown Esters

The crown esters are actually macrocyclic lactones, the bulk of which rings are ethyleneoxy units. By far, the largest number of such compounds have been prepared by Bradshaw and his co-workers[1]. Representative examples of such compounds are shown below. The name "crab" ethers has occasionally been applied to such structures, but as with the aquatic species noted above, this name has not found wide acceptance.

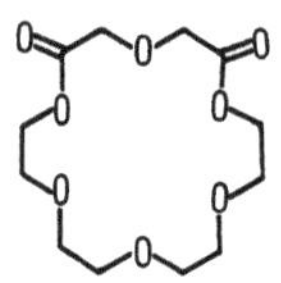

18

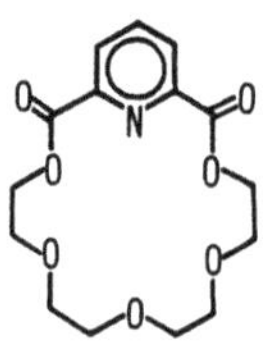

19

1.3.7 Spherands

This is a group of macropolycyclic molecules whose molecular architecture is such that a spherical array of binding sites is presented to an encapsulated ion. The first example of such a structure is illustrated below as *20*; and was prepared by Graf and Lehn[12]. The ten binding sites of this molecule are arranged in an octahedrotetrahedral geometry, the four nitrogens constituting a tetrahedron and the oxygens forming an octahedron. Compound *21* is an all oxygen spherand reported by Cram and coworkers[4]. Note that in *21* no nitrogen atom is present as a "cornerstone"; rather, substituted biphenyls provide the junctions. The cavity of *21* is smaller than that of *20* and shows a selectivity for lithium and sodium cations.

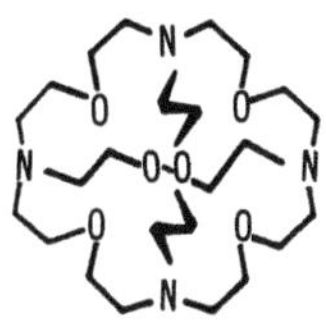

20

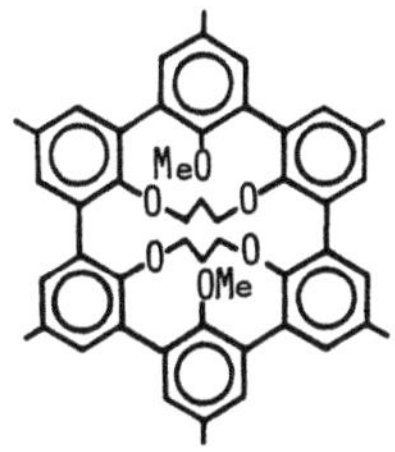

21

1.3.8 Cascades

This is a group of compounds having a ladder-like array of macrorings through which a cation or other species may cascade. Examples of such structures may be found in Ref. 2.

1.3.9 Polypodes and Octopus Molecules

Both of these structures are open-chained compounds corresponding to crown ethers in function if not exactly in structure (see Chap. 7). They have repeating ethyleneoxy side-chains generally terminated in a methyl group. Montanari and co-workers introduced the polypodes *22* as phase transfer catalysts[8]. These compounds were based on the triazine nucleus as illustrated below. The first "octopus molecule (*23*)" was prepared by Vögtle and Weber[37] and is shown below. The implication of the name is that the compound is multiarmed and not specifically that it has eight such side-chains. Related molecules have recently been prepared by Hyatt[15] and the name octopus adopted. For further information on this group of compounds and for examples of structures, refer to the discussion and tables in Chap. 7.

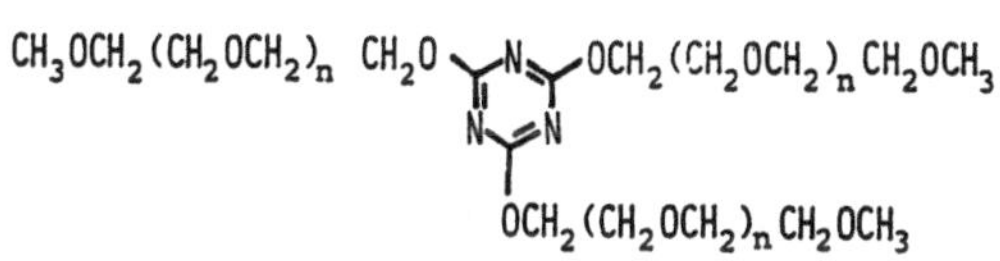

"polypode molecules" *22*

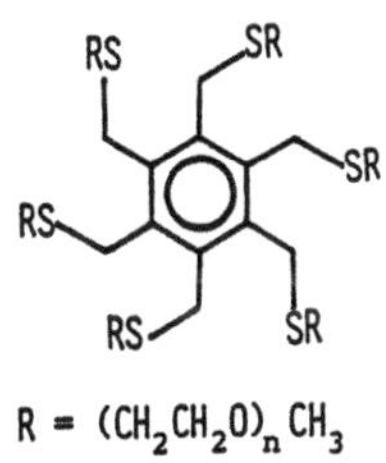

"octopus molecules" *23*

1.3.10 Host-Guest-Chemistry

This useful and general terminology has been applied quite widely in the macrocycle field and in other areas as well. It is obvious from the name that a host is any molecule which can complex, encapsulate or otherwise bind a substrate. The guest is whatever is bound, encapsulated, etc. In the case of the 18-crown-6 complex of potassium ion, the former is the host and the latter is the guest.

1.3.11 Conventions

The terms *crown* and *cryptand* have been universally adopted. A number of other terms have enjoyed less widespread recognition as noted above. Recently, Vögtle and Weber[38] have proposed use of the terms *crown ether, cryptand* and *podand* according to the following scheme. Their suggested definitions are as follows:

Coronand: Any medium sized or macrocyclic system having only one ring and containing any heteroatom(s).
Cryptand: Bi- and polycyclic compounds containing any heteroatoms.
Podand: Open-chained (acyclic) analogs of either *coronands* or *cryptands*.
Crown ether: This term would refer specifically to *coronands* (see above) containing only oxygen heteroatoms in the ring.
Complexes: The term *cryptate* is now accepted to mean the complex formed between a *cryptand* and a substrate. The corresponding complex with a *coronand* would be a *coronate,* a term suggested some years ago by the same authors[37]. Presumably, a complex between a *podand* and some substrate would be a "*podate*".

Of these terms, the names *crown ether, cryptand* and *cryptate* are in general usage. At this writing (summer, 1980), it is not clear whether the system proposed above will receive any official sanction, or indeed, find its way into general usage.

1.4 The Cyclooligomerization of Ethylene Oxide

The cyclooligomerization of ethylene oxide to yield dioxane as well as compounds we now call crowns predates Pedersen's discovery by more than a decade[34]. The full utility of these cyclic oligomers was not recognized, however, and the patent reporting these early efforts remains an interesting historical footnote. The promise of utilizing cyclooligomerization commercially is so important, however, that attention is called to the method and the existence of the patent.

The original cyclooligomerization noted above was conducted using alkyl aluminum, zinc and magnesium compounds[34]. Both open-chained and cyclic polymers were obtained in addition to dioxane. The principal macrocyclic component of the mixture was apparently 12-crown-4 (EO-4) which was said to be "valuable as a high boiling neutral solvent. . . ."[34] The reaction is shown below in Eq. (1.2).

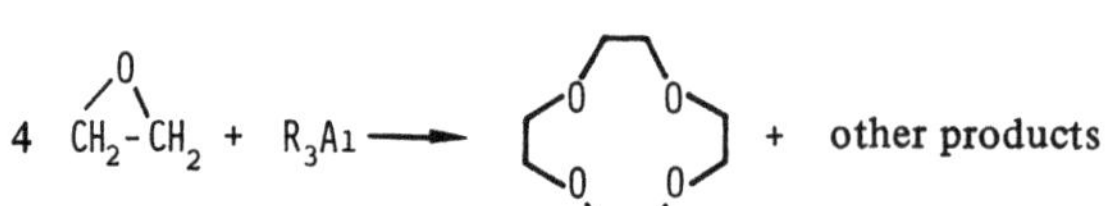

(1.2)

Dale and co-workers examined this reaction in considerable detail some years later and utilized a mixture of HF and BF_3 in dioxane as catalyst[5]. They noted that this catalyst mixture was stable for months at room temperature and did not etch glass. It was useful for initiating the cyclooligomerization reaction which led to a product mixture. The composition of the mixture was apparently independent of the ethylene oxide concentration and the reaction was apparently not kinetically controlled.

Additional empirical observations concerning this reaction were that use of BF_3 as its etherate or addition of small amounts of water (both apparently common practices in certain polymerizations) reduce the overall yield of cyclic products because of chain termination. In a typical reaction mixture, 12-crown-4, 15-crown-5, and 18-crown-6 were formed in 15%, 5% and 4% yields respectively. Dioxane constituted 40% of the product mixture and the remainder was less than 3% each of identified components below the cyclododecamer[5].

Dale improved this procedure by conducting the cyclooligomerization in the presence of certain templating cations paired with non-nucleophilic counterions (e.g., BF_4^-)[6]. The influence of certain cations on the cyclization can be gleaned from an examination of Table 1.1, below. Note that the percentages are distributions and not yields. The only specific statement in the paper concerning yield is "... yields depend on the conditions". From earlier papers[5,34] one can presume that a significant amount of dioxane is formed as are small amounts of other cyclooligomers. The advantage of this method appears to be that the 12, 15 and 18-membered rings separate from the reaction mixture as solid complexes. The non-nucleophilic anion, used because it does not react with the catalyst, exhibits a further advantage here by not initiating crown degradation during pyrolysis of the complex to liberate the macroring.

Recent efforts have also been successful in both the polymerization of ethylene oxide and tetrahydrofuran. Both Saegusa et al.[17], and Robinson and Pruckmayr[32] were successful in oligomerizing ethylene oxide under varying conditions. Of somewhat greater

Table 1.1. Product mixtures from the cycloologomerization of ethylene oxide

Catalyst	Percent product obtained from precipitated complex or volatiles			
	12-crown-4	15-crown-5	18-crown-6	Refs.
1% BF_3/HF	15	5	4	32
CF_3SO_3H (1 h)	16	33	2	35
CF_3SO_3H (2 h)	16	23	6	35
$BF_3/LiBF_4$	30	70	0	33
$BF_3/NaBF_4$	25	50	25	33
BF_3/KBF_4	0	50	50	33
BF_3/KPF_6	20	40	40	33
$BF_3/KSbF_6$	40	20	40	33
$BF_3/RbBF_4$	0	0	100	33
$BF_3/CsBF_4$	0	0	100	33
$BF_3/Cu(BF_4)_2$	5	90	5	33
$BF_3/Zn(BF_4)_2$	5	90	5	33

interest is that the latter workers were able to copolymerize ethylene oxide and tetrahydrofuran using trifluoromethanesulfonic acid as catalyst. When 10 moles of THF and either 5 or 10 moles of ethylene oxide were used in the feedstock, the 2:2 cyclic adduct shown in Eq. (1.3) was isolated in 32% yield. When 10 THF and 2 ethylene oxides were present in the mix, a 1:3 cyclic adduct was isolated in 32% yield[32].

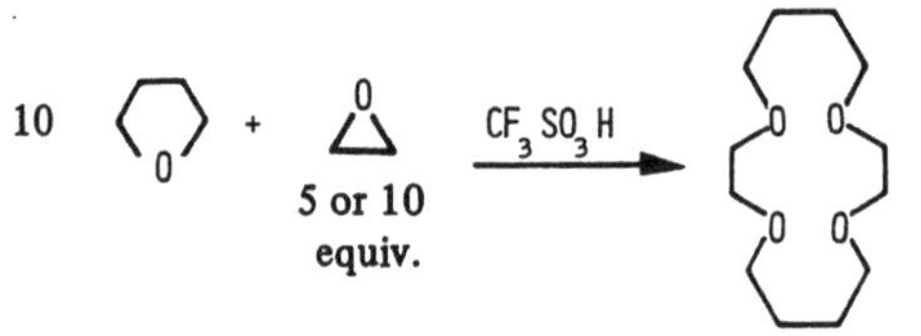

(1.3)

Numerous other methods exist for making the simple crowns. Many of these methods are compared in Sect. 3.1 which deals with variations in the synthesis of 18-crown-6. Further commentary on the influence of templating ions on the synthesis of crown ethers may be found in Chap. 2, which deals with various aspects of the template effect.

1.5 Polyazamacrocycles as Complexing Agents for Transition Metals

Pedersen obviously viewed nitrogen containing macrocycles as derivatives of the crowns and most organic chemists tend to harbor a similar prejudice. Pedersen showed that addition of nitrogen diminished the affinity of these macrocycles for alkali metal ions and increased their affinity for transition metals[27]. This fact was already known by inorganic chemists and numerous organic chemists especially those working with porphyrins and corrins[24]. Nevertheless, we deal in this book with numerous compounds which contain only one or two nitrogen atoms and we feel justified in considering them as crowns.

A great deal of information exists on the syntheses of macrocyclic polynitrogen compounds and it would be redundant to present it here. Outside of the rather specific examples chosen for inclusion here, the reader is directed to the book edited by Melson[24] and to any of the other numerous reviews cited at the end of this chapter[19].

1.6 References

1. Bradshaw, J. S., Maas, G. E., Izatt, R. M., Christensen, J. J.; Chem. Rev. (1979) *79*, 37
2. Buhleier, E., Wehner, W., Vögtle, F.: Synthesis (1978) 155
3. Coxon, A. C. and Stoddart, J. F.: J. Chem. Soc. Chem. Commun. (1974) 537
4. Cram, D. J., Kaneda, T., Lein, G. M. and Helgeson, R. C.: J. Chem. Soc. Chem. Commun. (1979) 948
5. Dale, J., Borgen, G. and Daasvatn, K.: Acta Chem. Scand. (1974) *B28*, 378
6. Dale, J. and Daasvatn, K.: J. Chem. Soc. Chem. Commun. (1976) 295
7. Dietrich, B., Lehn, J. M. and Sauvage, J. P.: Tetrahedron Letters (1969) 2885, 2889
8. Fornasier, R., Montanari, F., Podda, G. and Tundo, P.: Tetrahedron Letters (1976) 1381
9. Gokel, G. W., Dishong, D. M. and Diamond, C. J.: J. Chem. Soc. Chem. Commun. (1980) 1053

10. Gold, V. and Sghibartz, C. M.: J. Chem. Soc. Chem. Commun. (1978) 507
11. Goldberg, I.: Acta. Cryst. (1977) *B33*, 472
12. Graf, E. and Lehn, J. M.: J. Am. Chem. Soc. (1975) *97*, 5022
13. Gregory, B. F., Haines, A. H. and Karntiang, P.: J. Chem. Soc. Chem. Commun. (1977) 918
14. Helgeson, R. C., Tarnowski, T. L., Cram, D. J.: J. Org. Chem. (1979) *44*, 2538
15. Hyatt, J. A.: J. Org. Chem. (1978) *43*, 1808
16. Johnson, M. R., Sutherland, I. O., Newton, R. F.: J. Chem. Soc. Chem. Commun. (1979) 309
17. Kobayashi, S., Morikawa, K. and Saegusa, T.: Macromolecules (1975) *8*, 952
18. Lehn, J. M.: Acc. Chem. Res. (1978) *11*, 49 and references therein
19. Lindoy, L. F.: J. Chem. Soc. Chem. Soc. Rev. (1975) *4*, 421
20. Lüttringhaus, A. and Ziegler K.: Liebigs Ann. Chem. (1937) *528*, 155
21. Lüttringhaus, A.: Liebigs Ann. Chem. (1937) *528*, 181
22. Lüttringhaus, A.: Liebigs Ann. Chem. (1937) *528*, 211
23. Lüttringhaus, A. and Sichert–Modrow, I.: Makromol. Chem. (1956) *18–19*, 511
24. Melson, G. A.: in Melson, G. A. (ed.), Coordination Chemistry of Macrocyclic Compounds. Plenum Press, New York (1979) 1
25. Park, C. H. and Simmons, H. E.: J. Am. Chem. Soc. (1968) *90*, 2429, 2431
26. Pedersen, C. J.: J. Am. Chem. Soc. (1979) *89*
27. Pedersen, C. J.: J. Am. Chem. Soc. (1967) *89*, 7017
28. Pedersen, C. J., Aldrichimica Acta, (1971) *4*, 1
29. Pedersen, C. J. and Bromels, M. H.: U.S. 3,847,949, 12 November 1974
30. Pedersen, C. J.: in Izatt, R. M. and Christensen, J. J. (eds): Synthetic Multidentate Macrocyclic Compounds. Academic Press, New York (1978) 1
31. Pressman, B. C., Harris, E. J., Jaggar, W. S. and Johnson, J. H.: Proc. Natl. Acad. Sci. (1967) *58*, 1949
32. Robinson, I. M., and Pruckmayr, G.: Macromolecules (1979) *12*, 1043
33. Simmons, H. E. and Park, C. H.: J. Am. Chem. Soc. (1968) *90*, 2428
34. Stewart, D. G., Waddan, D. Y. and Borrows, E. T.: British Patent 785,229. Filed 4 February 1955, published 23 October 1957
35. Ugelstad, J., Ellingsen, T. and Berge, A.: Acta Chem. Scand. (1966) *20*, 1593
36. Van Alphen, J.: Rec. Trav. Chim. Pays-Bas. (1936) *55*, 835
37. Vögtle, F. and Weber, E.: Angew. Chem. Int. Ed. Engl. (1974) *13*, 814
38. Vögtle, F. and Weber, E.: Angew. Chem. Int. Ed. Engl. (1979) *18*, 753
39. See footnote 11 in Ref. 27

Chapter 2

The Template Effect

2.1 Introduction

It has been axiomatic for many years that large rings must be prepared under high dilution conditions so that the probability of the two ends meeting exceeds the probability of linear polymer formation[7]. It therefore seems remarkable that many of the syntheses of macrocyclic polyethers which have been reported involve less than dilute if not very concentrated conditions[18]. The success of these large-ring intramolecular S_N2 reactions is generally attributed to the operation of a "template" effect during crown synthesis.

2.2 Operation of the Template Effect

Many of the crown ether syntheses with which we are concerned in this book are one form or another of the Williamson ether synthesis. Although the simplest example of such a reaction would involve an ω-haloethylene glycol oligomer which undergoes intramolecular cyclization, it is more common for two new bonds to be formed in crown syntheses. An early example of the formation of a crown by a double-Williamson can be found in Dale's synthesis of 18-crown-6[5]. The rather obvious chemical steps are shown in Eq. (2.1).

$$HOCH_2(CH_2OCH_2)_2CH_2OH + TsOCH_2(CH_2OCH_2)_2CH_2OTs \xrightarrow{base}$$

$$[HOCH_2(CH_2OCH_2)_5CH_2OTs] \xrightarrow{base}$$

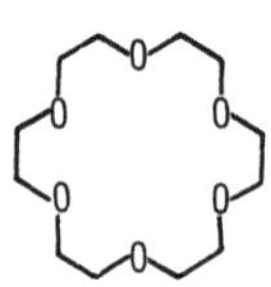

(2.1)

The first C–O bond formation is probably not influenced strongly by the presence of a templating cation. Since it is not crucial for one end of the chain to meet the other rather than reacting with a different molecule, it is not necessary to superimpose either a template or dilution condition on the reaction to prejudice the statistics. In the second step, however, such a prejudicial condition is required. This is available in the form of an alkali metal cation for which the long ethyleneoxy chain has a certain affinity. Presumably, the cation is ion-paired with the alkoxide anion and the remainder of the chain wraps around it. Note that such a picture corresponds to Ugelstad's "self-solvating" bases[22]. The wrapping is illustrated in Eq. (2.2), below.

(2.2)

It is not clear exactly when the association illustrated above actually takes place. It is certainly involved by the final ring closure stage, but it seems reasonable to assume that there is some cation-glyme type interaction taking place from the instant of solution. The fact that wrapping occurs in such a way that the two ends of the molecule are held in proximity, allows the reaction to be conducted at much higher concentrations than might otherwise be practical. The evidence for the operation of such a template effect is presented and considered below.

2.3 Evidence for the Template Effect

A perusal of Pedersen's experimental conditions[18] suggests clearly his understanding that a "template" effect of some sort must be operable, otherwise, he would have surely used high-dilution conditions. In Dale's early synthesis of 18-crown-6 from triethylene glycol and triethylene glycol ditosylate using potassium t-butoxide as base, one finds the comment: "High-dilution techniques are not needed"[5]. There is, however, no mention of a template effect. In this case, one might argue (if only weakly) that considering the low solubility of potassium t-butoxide in benzene, the reaction was conducted, in a sense, under high dilution conditions. This is not likely the case.

One of the prime tenets of the "template effect" is that a cation of appropriate size coordinates with the heteroatom binding sites in the incipient ring system. Evidence to be discussed below indicates that there is a clear relationship between the size of the coordinating cation and the size of the ring formed. Before presenting this evidence, it is useful to note the approximate ring sizes of common crowns and the known ionic diameters of sundry metal cations. This is done in Table 2.1.

The first suggestion of a "template" effect which was offered in the literature was made by Greene in 1972[10]. The illustration of this concept is approximately that shown in Eq. (2.2), above. Greene presented several pieces of evidence in addition to the concept itself. First, he noted that when the final concentration of 18-crown-6 in a reaction mixture (Me_2SO) was increased from 0.04 M to 0.09 M, the yield of crown dropped only slightly (84% to 75%). In a competition experiment, equal amounts of 18-crown-6 and 21-crown-7 were formed when one molar equivalent each of triethylene glycol and tetraethylene glycol were allowed to compete for triethylene glycol ditosylate (KO-t-Bu/THF).

Table 2.1. Relationship between crown hole sizes and cation diameters

Crown Ether (hole size, Å)	Favored Cation (ionic diameter, Å)	
12-Crown-4 (1.2)	Li^+ (1.20)	Mg^{++} (1.30)
14-Crown-4 (1.2–1.5)		
15-Crown-5 (1.7–2.2)	Na^+ (1.90)	Ca^{++} (1.98)
18-Crown-6 (2.6–3.2)	K^+ (2.66)	Sr^{++} (2.26)
21-Crown-7 (3.4–4.3)	Rb^+ (2.96) Cs^+ (3.38)	Ba^{++} (2.70) –
24-Crown-8 (4.5)	–	–

Equal amounts of the two crowns were also formed when Bu_4NOH was used as base, but both rate and yield were considerably reduced. Greene noted that such a template effect would actually involve binding between the open-chained intermediates and cations and these interactions are known to be weak relative to crown-cation interactions. Nevertheless, evidence for such interactions exists (see also Chap. 7)[1].

Chastrette and Chastrette showed that the cyclocooligomerization of furan and acetone was improved by the presence of lithium salts[2]. Liotta and co-workers reported the synthesis of 12-crown-4 by a double Williamson approach in the presence of $LiClO_4$[3]. They mention in a footnote that "A run performed as detailed but without $LiClO_4$ gave no product". Other inferential evidence has accumulated as well.

Reinhoudt, Gray, Smit and Veenstra have studied this question in some detail and offered some additional information[9,20]. They conducted the reaction shown in Eq. (2.3). They noted that at slow addition rates, the ratio of glycolate to dibromide was at least 1,000:1 during the reaction, but the yield of cyclic products and composition remained constant. This suggested to them that the intramolecular cyclization step must be much faster than solvolysis. Further, they suggested that this could be atributed to "a different entropy contribution in the respective transition states, which must be due to the formation of a highly-ordered conformation of the polyether chain around the potassium ion, or to an increased nucleophilicity of the second polyethylene glycol alkoxide due to the formation of a charge-separated ion pair resulting from a similar conformation of the polyether chain"[9].

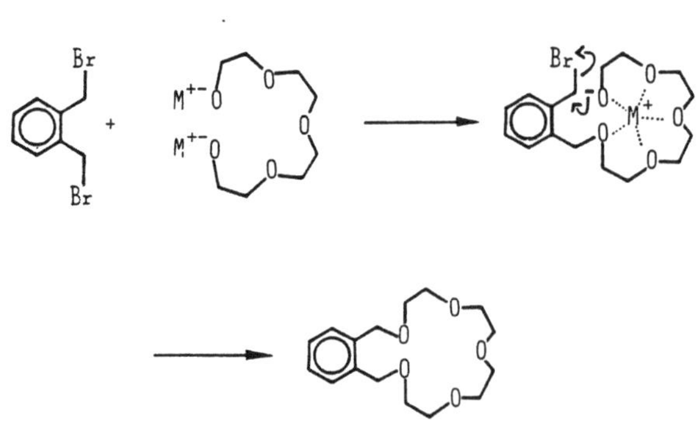

(2.3)

In other work[21], these authors examined the product distribution for Eq. (2.3) as a function of ring size formed. Maximum yields for cyclic product were observed when a five-oxygen, seventeen-membered ring was formed.

In 1977, Mandolini and Masci reported clear kinetic evidence that a template effect was important to the cyclization of *o*-hydroxyphenyl-3,6,9,12-tetraoxa-14-bromotetradecyl ether[5]. The starting material and its subsequent reactions are illustrated in Eq. (2.4).

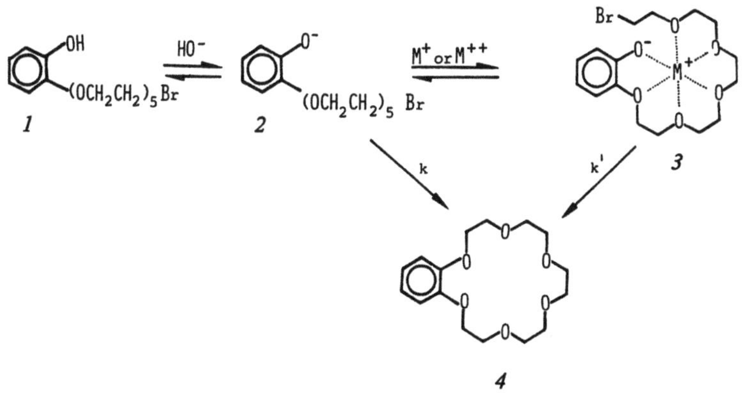

(2.4)

The basis of the analysis is as follows. Phenol *1* is a strong enough acid to be readily deprotonated by hydroxide. If the hydroxide ion is paired with a cation like Et_4N^+ which is clearly too large to fit into the crown arrangement of oxygen atoms shown in *3*, then no template effect should be observed. A base like $Et_4N^+OH^-$ should therefore effect cyclization, but should constitute a baseline against which other reactions can be compared. Lithium cation should be too small to be effective as a templating atom and is, in fact, ineffective in accelerating the reaction. For most of the concentrations examined, sodium and potassium ions did exhibit an enhancement in the rate, indicating that cyclization was occurring by the pathway designated k′ rather than k. The divalent cations Ba^{2+} and Sr^{2+} were even more effective in promoting the cyclization. The authors make the reasonable observation that the "observed effect occurs at reasonably low cation concentration (0.1 *M* or less) so that the operation of significant medium effects on the reaction rates is unlikely". Although, they say, "A medium effect is possibly superimposed into the template effect in the case of K^+ and Na^+, particularly at the highest concentrations used[15]".

For such observations to be compelling, it must be possible to demonstrate an interaction between a metal ion and a polyether as suggested by 3. Indeed, numerous complexes of precisely this type have been prepared by Vögtle and co-workers[23]. Moreover, the conceptualization of "self-solvating bases" provided by Ugelstad is remarkably similar to illustration 3[22]. Finally, assuming that a pseudo-crown conformation is adopted by the reactant in the presence of a templating metal ion, it should be possible to correlate the rates observed by Mandolini and Masci with known cation binding constants between these species and 18-crown-6. The observed rates are $Li^+ < Et_4N^+ < Na^+ < K^+ < Sr^{2+} < Ba^{2+}$. The observed cation binding constants in water are in the order $Na^+ < K^+ < Sr^{2+} < Ba^{2+}$ with constants for Li^+ and Et_4N^+ not available[11].

Mandolini and Masci provided further information on this question quite recently by demonstrating that polyethylene glycol dibromides undergo partial hydrolysis and then cyclization in water containing Ba^{2+} ions[16]. According to the report, yields were higher in this reaction when barium cation was present than when it was absent.

Other indications of the template effect have been offered as well. Fenton, Cook and Nowell reported that "the condensation of pyridine-2,6-dicarbaldehyde with 1,11-diamino-3,6,9-trioxaundecane in 1 : 1 molar ratio in butanol leads only to resinous gums"[8]. In the presence of equimolar amounts of $Pb(SCN)_2$, the macrocycle illustrated in Eq. (2.5) was obtained in "good yield".

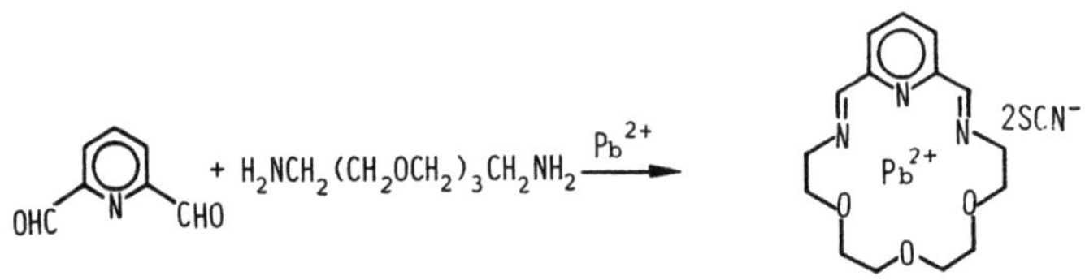

(2.5)

The addition of various alkali and alkaline earth cations to the cyclooligomerization of ethylene oxide by Dale and Daasvatn also provides strong presumptive evidence for the template effect[4]. Recently, Reinhoudt, de Jong and Tomassen utilized several metal fluorides to effect crown formations[20]. The reaction rates were found to be in the order $Cs^+ > Rb^+ \geqslant K^+ \gg Na^+ \sim Li^+$. Such an order would be expected on the basis of binding constants as noted above.

An unusual example of a template effect has been reported by Madan and Cram[14]. Based on an examination of C-P-K molecular models, they presumed that there would be complementary hydrogen bonding between the ether and "guanidinium ion favour the larger rings suggests the ion acts as a template during reaction. . . ." A complex between an appropriately sized crown and guanidinium ion was prepared and recently, Lehn and co-workers have accomplished a similar feat[13].

Desvergne and Bouas-Laurent have shown that photochemical ring closure of a bis-anthracene bridged by a polyether chain is effective only when lithium cation is present[6]. They presume that cyclization is successful because the conformation is "cation locked". The reaction is shown in Eq. (2.6).

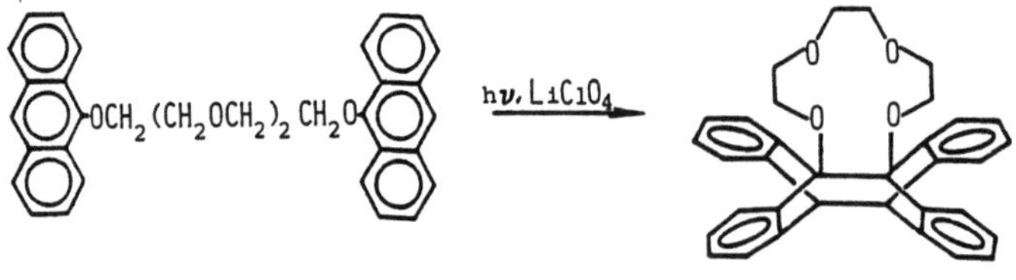

(2.6)

Recently, Okahara and his co-workers have investigated a variety of "one-pot" crown ether syntheses which are referred to in Sects. 3.4, 4.3, 5.4. During the course of these investigations, they examined the temperature dependence of the cyclization yield[12]. Using either sodium hydroxide or potassium hydroxide and forming 15-crown-5, 18-crown-6 and 21-crown-7, an attempt was made to correlate yield and reaction temperature. For most of the reactions, yield was recorded over the range from 20 ° – 120 °C in 20° intervals. The findings were that yield varied from case to case, but generally remained constant over the range 60°–100° and often from 40°–100° with a fall off in yield at either extreme. The reaction for which these data were reported is shown in Eq. (2.7).

$$HOCH_2(CH_2OCH_2)_nCH_2OH \xrightarrow[\text{aprotic solvent}]{\text{TsCl/NaOH or KOH}} \underset{(n = 4)}{\text{15-crown-5}} + \underset{(n = 5)}{\text{18-crown-6}} + \underset{(n = 6)}{\text{21-crown-7}} \quad (2.7)$$

Newkome and co-workers[17] have demonstrated the operation of a template effect in the formation of a pyrido-ester-crown. In the reaction shown in Eq. (2.8), they treated 2-chloronicotinoyl chloride with either the disodium or dipotassium salt of pentaethylene glycol. The two reactions were conducted under identical conditions except for the presence of sodium vs. potassium cations. Since the product is a six-oxygen macrocycle, its formation would be expected to be favored by K^+ rather than Na^+ counter ions for the glycolate. In fact, the yields of crown-lactone were 30% and 48% respectively when Na^+ and K^+ were the templating cations.

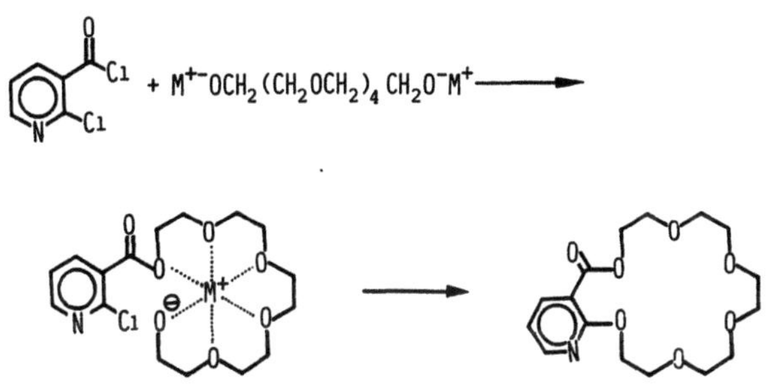

(2.8)

There is probably no better evidence for a template effect than its application directly in the solution of a synthetic problem. Rastetter and Phillion[19] have utilized a substituted 19-crown-6 compound (shown below in Eq. 2.9) in the formation of macrocyclic lactones. Although there were certain experimental variations and the the possibility of intermolecular potassium ion complexation, the overall formation of lactone was favorable.

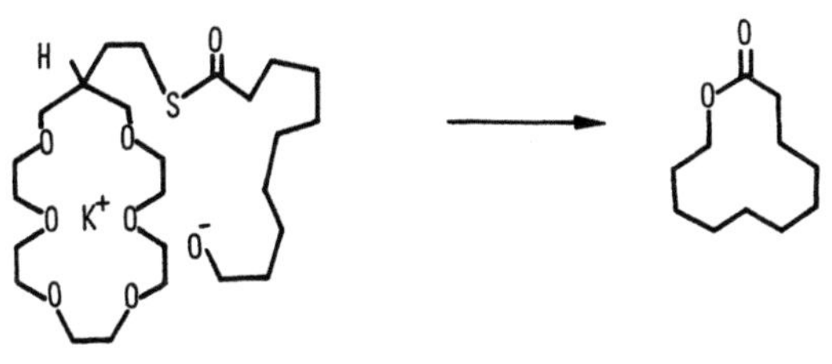

(2.9)

2.4 Complexation Constants and the Template Effect

The template effect has long been accepted *prima facie* by workers in the crown field because of the obvious relationship between ring size and complexation constant. In fact, Cram stated in 1975 that "the templating properties of K^+ for preparing crown ethers is well established. . . ."[14]. In fact, the template effect was widely acknowledged and has always received overwhelming support. Nevertheless, few direct comparisons are available in the literature and we have restricted ourselves in this discussion only to direct rather than the large body of presumptive evidence which is currently available.

2.5 References

1. Chan, L. L., Wong, K. H. and Smid, J.: J. Am. Chem. Soc. (1970) *92*, 1955
2. Chastrette, M. and Chastrette, F.: J. Chem. Soc. Chem. Commun. (1973) 534
3. Cook, F. L., Caruso, T. C., Byrne, M. P., Bowers, C. W., Speck, D. H. and Liotta, C. L.: Tetrahedron Letters (1974) 4092
4. Dale, J. and Daasvatn, K.: J. Chem. Soc. Chem. Commun. (1976) 295
5. Dale, J. and Kristiansen, P. O.: J. Chem. Soc. Chem. Commun. (1971) 670
6. Desvergne, J. P. and Bouas-Laurent, H.: J. Chem. Soc. Commun. (1978) *403*
7. Eliel, E. L.: Stereochemistry of Carbon Compounds. McGraw Hill, New York 1962, 198 and references therein
8. Fenton, D. E., Cook, D. H. and Nowell, I. W.: J. Chem. Soc. Chem. Commun. (1977) 274
9. Gray, R. T., Reinhoudt, D. N., Smit, C. J. and Veenstra, I.: Recl. Trav. Chem. Pays-Bas (1976) *95*, 258
10. Greene, R. N.: Tetrahedron Letters (1972) 1793
11. Izatt, R. M., Terry, R. E., Haymore, B. L., Hansen, L. D., Dalley, N. K., Avondet, A. G. and Christensen, J. J.: J. Am. Chem. Soc. (1976) *98*, 7620
12. Kawamura, N., Miki, M., Ikeda, I. and Okahara, M.: Tetrahedron Letters (1979) 535
13. Lehn, J. M., Vierling, P., Hayward, R. C.: J. Chem. Soc. Chem. Commun. (1979) 296
14. Madan, K. and Cram, D. J.: J. Chem. Soc. Chem. Commun. (1975) 427
15. Mandolini, L. and Masci, B.: J. Am. Chem. Soc. (1977) *99*, 7709
16. Mandolini, L. and Masci, B.: Synth. Commun. (1979) *9*, 851
17. Newkome, G. R., Kawato, T. and Benton, W. H.: J. Org. Chem. (1980) *45*, 626
18. Pedersen, C. J.: J. Am. Chem. Soc. (1967) *89*, 7017
19. Rastetter, W. H., Phillion, D. P.: Tetrahedron Letters (1979) 1469
20. Reinhoudt, D. N., deJong, F. and Tomassen, H. P. M.: Tetrahedron Letters (1979) 2067
21. Reinhoudt, D. H., Gray, R. T., Smit, C. J. and Veenstra, I.: Tetrahedron (1976) *32*, 1161
22. Ugelstad, J., Mork, P. C. and Jensen, B.: Acta Chem. Scand. (1963) *17*, 1455
23. Vögtle, F. and Weber, E.: Angew. Chem. Int. Ed. Engl. (1979) *18*, 753

Chapter 3

Syntheses of Oxygen Macrocycles

3.1 Pedersen's First Crowns

In Pedersen's first full paper on the synthesis of macrocyclic polyethers[169], he reported the synthesis of thirty-three macrocycles having a variety of ring sizes and substituents. These compounds were prepared by a variety of routes and Pedersen offered a scheme in which he generalized the synthetic approaches he used. He designated the four principal methods "V", "W", "X", and "Y". The methods are discussed individually.

The first method is shown in Eq. (3.1). This corresponds to the so-called "one plus one" synthesis of crowns. The notion is that a single diol unit is allowed to react with a single polyethylene glycol having leaving groups at each end. An example of this would be the synthesis of benzo-15-crown-5 from catechol and tetraethylene glycol dichloride. Note that the stoichiometry of this method is identical to that of "method X" which is shown below in Eq. (3.3).

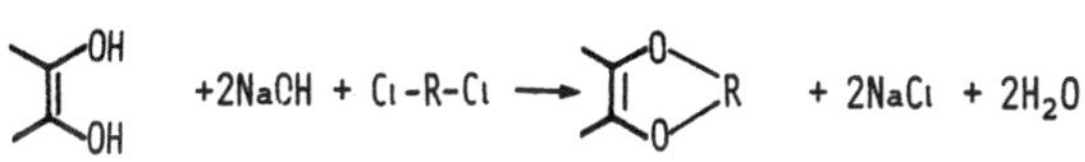

(3.1)

The general approach represented by Eq. (3.1) is much broader than the equation itself indicates. Obviously, the diol component need not be an enediol (see also Sect. 3.10). In addition, bases other than sodium hydroxide may be used and in many cases are preferred. For aliphatic alcohols, for example, potassium *t*-butoxide has been used often, but sodium hydride has been the base of choice. The same is true of the expression "Cl–R–Cl". Although Pedersen successfully utilized chlorides as leaving groups in a number of syntheses and others have likewise done so, the most common choice of leaving group appears to be tosylate or methanesulfonate. Although the conversion of most primary alcohols into the corresponding chlorides (using thionyl chloride and pyridine) is a relatively straightforward reaction, it is somewhat tedious. To our knowledge, all of the polyethylene glycols can readily be converted into the corresponding tosylates using toluenesulfonyl chloride and pyridine at ice-bath temperature. Mono-, di- and triethylene glycol ditosylates are all crystalline and may be stored indefinitely without decomposition. Although the higher homologs are glasses or oils. They also seem to be reasonably stable on storage.

The principal alternative to the chlorides or tosylates is the methansulfonate esters. The preparation of these is similar to the preparation of the tosylates and they generally afford similar yields of products. They are occassionally preferred for manipulative reasons but seem otherwise to offer little advantage.

The second general approach outlined by Pedersen is the so-called "method W" in which the two hydroxyl groups are separated by a portion of the crown chain. A good example of this is the assembly of 18-crown-6 from triethylene glycol and triethylene glycol dichloride.

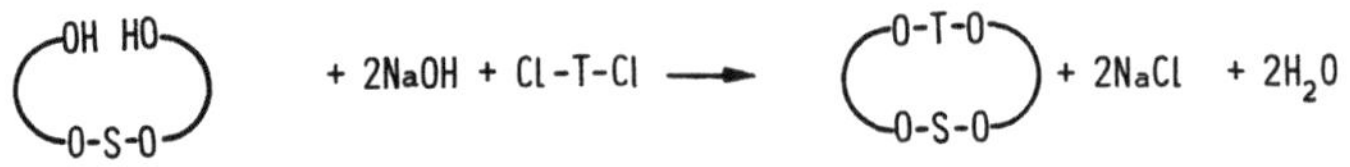

(3.2)

The tactic implied by Eq. 3.2 would be especially important when the two halves of the crown unit are different. We return to this below, after a discussion of Eq. (3.3).

Method X (Eq. 3.3) is almost a corollary to method V since the stoichiometry is identical. In fact, it is not the intention, but the size of the crown which will determine the outcome of the reaction. For example, the synthesis of dibenzo-18-crown-6 (Eq. 3.4) is carried out by treating two equivalents of catechol with two equivalents of diethylene glycol dichloride in the presence of four equivalents of sodium hydroxide. Although the 18-membered ring is formed, division of all reagents by a factor of two might have led to the prediction that the nine-membered ring would predominate. Of course, it is the ring size and not the reagent ratio which determines the outcome.

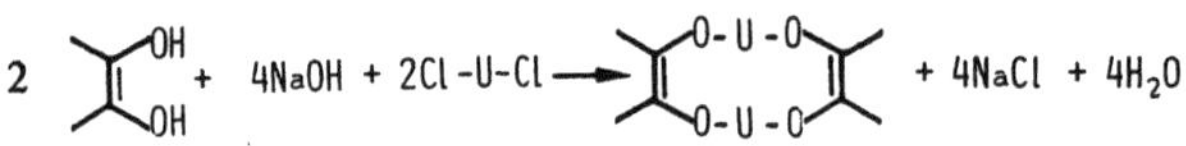

(3.3)

When Kyba, Cram and coworkers[115,116] attempted the synthesis of dibinaphthyl-22-crown-6 from binaphthol and diethylene glycol ditosylate, the reagent ratio was as described either by V or X (Eqs. 3.1 or 3.3). The intent was to make the 22-membered ring. Some of the 11-membered ring was isolated as well (Eq. 3.5). None was apparently observed for the benzo case because the nine-membered ring is less favorable. In short, it is not the intent of the synthesis but the reaction dynamics which determine the product.

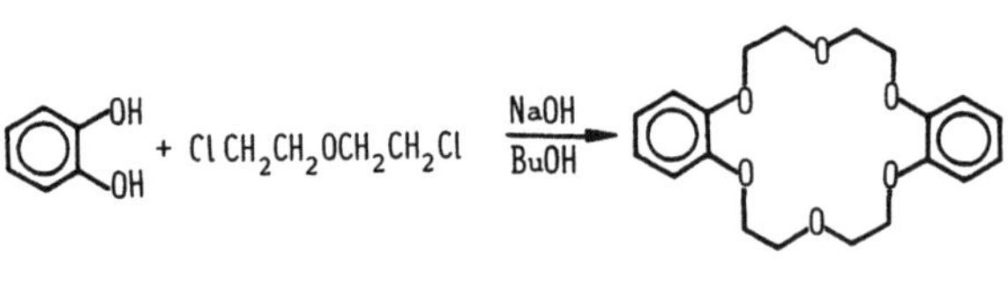

(3.4)

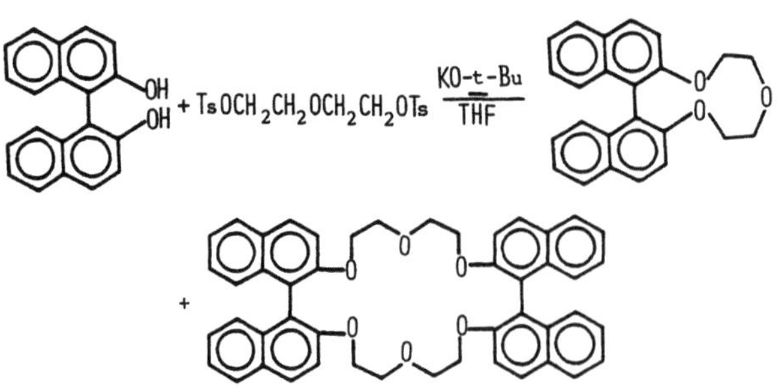

(3.5)

The fourth of Pedersen's general methods is expressed as method Y. In this approach, a single unit may be both nucleophile and electrophile and react with the corresponding portions of its counterpart to yield a macrocycle. This is illustrated in Eq. (3.6). Note that there are really two possibilities here. The first of these is that two units will react as illustrated, but the other possibility is that the single unit will cyclize to afford a crown of half the size. It is precisely this approach which Pedersen used in the first synthesis of 18-crown-6 (see Sect. 3.2).

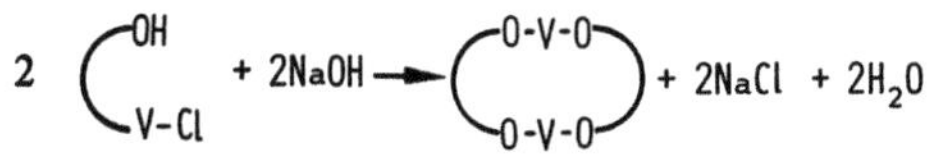

(3.6)

These four methods suggested by Pedersen serve as a useful starting point for the consideration of other approaches and should certainly not be considered an overview of existing methodology. Nevertheless, this work was pioneering and a good deal of what follows is based to a greater or lesser extent upon it.

3.2 Syntheses of 18-Crown-6

Over the years, 18-crown-6 has probably been utilized in more applications than any other crown with the possible exception of dibenzo-18-crown-6. There are several reasons for this. First, simple syntheses of 18-crown-6 have been available for a long time and the molecule may be prepared from very inexpensive starting materials. Equally important, however, is the fact that 18-crown-6 is a very strong binder for a number of alkali metals, especially sodium and potassium cations.

The first synthesis of 18-crown-6 was reported by Pedersen in his first full paper on crowns[169]. The method used was potassium *t*-butoxide catalyzed cyclization of hexaethlyene glycol monochloride in 1,2-dimethoxyethane, as shown in Eq. (3.7), below. Unfortunately, the yield by this approach was only 1.8%[169].

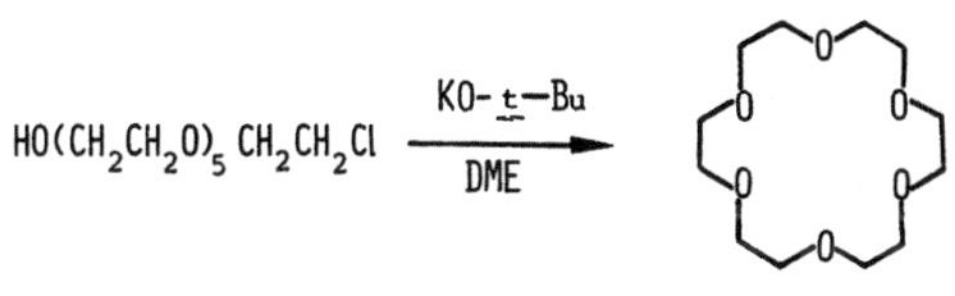

(3.7)

In 1972, Greene, a colleague of Pedersen's at Dupont, offered a much improved approach to 18-crown-6 based on assembly from two-triethylene glycol units[81]. Greene utilized *t*-butoxide as a base as Pedersen had, but triethylene glycol served as nucleophile and the corresponding ditosylate functioned as electrophile. The approach is illustrated in Eq. (3.8). Greene reports that he conducted the reaction in three different solvents and obtained a different yield in each: dimethyl sulfoxide, 84%; tetrahydrofuran, 30–60%; and dimethoxyethane, 93%. He notes that although the yields are poorer in THF than in either of the other two solvents, convenience recommends its use.

$$HO(CH_2CH_2O)_3H + TsO(CH_2CH_2O)_3Ts \xrightarrow[\text{solvent}]{K^{+-}O\text{-}t\text{-}Bu} \text{18-crown-6} \quad (3.8)$$

At the same time, Dale and Kristiansen[52] reported a successful synthesis of 18-crown-6 from the diol and ditosylate just as shown in Eq. (3.2), but using benzene as solvent. They obtained the product as its potassium tosylate complex (mp 164°) in 33% yield. The free macrocycle was liberated by chromatography over a column of alumina, eluting with a benzene-chloroform mixture. Dale and Kristiansen note that "the cyclic ether cannot be liberated from its complex by simple heating"[52].

Kyba's unpublished observation that the cycle could, in fact, be liberated from its complex by heating if conducted *in vacuo* suggested an alternate procedure developed by Gokel[72,73]. In this approach, triethylene glycol is heated with commercially available triethylene glycol dichloride and KOH in 10% aqueous THF solution. Filtration and evaporation of the solvent leave a thick syrup which may be distilled to yield crude 18-crown-6 in 50–60% yield. Unfortunately, the crown is not pure enough for use after a single distillation. Liotta, Harris and Cook noted that crude 18-crown-6 forms a crystalline solvate (complex) with acetonitrile from which pure 18-crown-6 can be obtained. The preparation and purification procedures were combined into a single preparation of 18-crown-6[72,73] which reliably yields 25–30% of the crystalline macrocycle.

$$HOCH_2(CH_2OCH_2)_2CH_2OH + ClCH_2(CH_2OCH_2)_2CH_2Cl \xrightarrow[\text{THF}]{\text{KOH}} \text{crude 18-crown-6} \xrightarrow{\text{MeCN}}$$

$$\text{crystalline solvate (complex)} \xrightarrow{\text{evaporation}} \text{pure 18-crown-6} \quad (3.9)$$

Shortly after the publication of the 18-crown-6 procedure, Liotta and his coworkers applied a similar approach to the synthesis of 12-crown-4 and 15-crown-5[34]. They were able to realize yields of 13% and 15%, respectively, using a modification of the procedure illustrated in Eq. (3.9), above.

In 1976, Johns, Ransom and Reese[98] reported improvements in the previously reported syntheses of 18-crown-6 and 15-crown-5. By using tetraethylene glycol rather than triethylene glycol and the correspondingly shorter dichloride (2.5 equivalents of the latter) in concert with KOH (no water added), they were able to realize a 6% yield improvement in the synthesis of 18-crown-6 over the previously published method[72,73]. The improvement in the yield of 15-crown-5 was of somewhat greater interest, being 38% compared to Liotta's previous report of 15%[34].

Reese and coworkers indicated that "optimum results may not have been obtained"[98] but to our knowledge, no full report of this work has yet appeared.

An entirely different approach was taken by Dale and Daasvatn (see also Sect. 1.4), who isolated 18-crown-6 from a cyclooligomerization reaction of ethylene oxide[49]. These workers found that BF_3 etherate, the traditional catalyst for such reactions, was not effective in this application. Gaseous BF_3 ["or similar acidic fluorine compounds (PF_5, SbF_5)"][49] apparently transformed ethylene oxide into a mixture of cyclic oligomers with no formation of open-chained polymers. Most other Lewis acids "were either ineffective as catalysts ($AlCl_3$, $FeCl_3$) or gave waxy polymers ($SnCl_4$, $SbCl_3$), while oxygen acids ($HClO_4$, *p*-toluenesulfonic acid) are rapidly transformed to half-esters of ethylene glycol"[49]. The reaction is illustrated in Eq. (3.10).

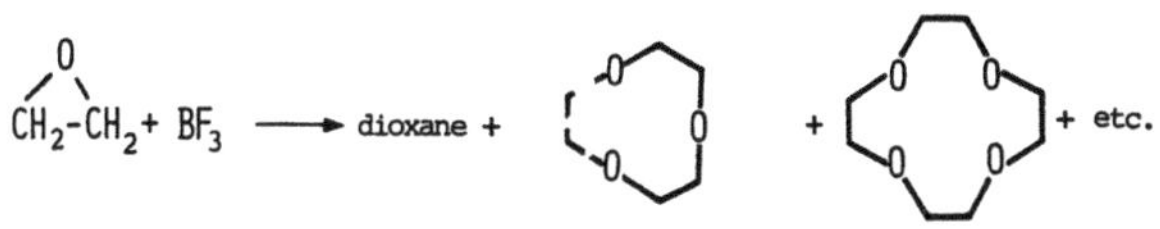

(3.10)

More recently, Dale and Daasvatn have shown that the product composition can be influenced by adding a "templating cation" (see Sect. 2.3)[48,50,51]. Addition of copper or zinc tetrafluoroborate, for example, leads to a product mixture containing greater than 90% of cyclic tetramer (i.e., 15-crown-5)[50]. Although no yields are given in this paper, examples extracted from the patents on this subject[48,51] suggest that yields are not necessarily high for individual products. For example, when ethylene oxide (26 g), and BF_3 (0.03 g) were allowed to react in dioxane (12 ml) in the presence of $CsBF_4$ (2.5 g), 2.5 g of a precipitated complex and 7.4 g of residue are obtained. Pyrolysis of the former affords 1 g of the hexamer, i.e., 18-crown-6[51]. Despite the fact that the yields are not as high as one would like, considering the simplicity of the method, it is quite attractive.

It should probably also be noted before closing this section that 18-crown-6 and some of its aliphatic relatives have been found to form quite a variety of "complexes" or "solvates" with neutral molecules. We have noted the formation of an acetonitrile solvate[72,73]. Although the other complexes are generally beyond the scope of the present work, we wish to call attention to their existence so that the worker attempting to improve purification procedures will check the literature for his particular method. This is especially important since a number of the solvates are found, to our knowledge, only in the patent literature.

3.3 Syntheses of Dibenzo and Dicyclohexano-18-Crown-6

Two of the most widely used crown ethers have been dibenzo-18-crown-6 and dicyclohexano-18-crown-6. (In older literature, the latter is often referred to as "dicyclohexyl-18-crown-6".) A major reason for this is that Pedersen reported complete details of the preparation of both compounds in *Organic Syntheses* in 1972[171]. As a result, both compounds were readily prepared and available.

Pedersen's preparation of dibenzo-18-crown-6 involves catechol and bis(2-chloroethyl) ether. In this procedure, sodium hydroxide is used as base and *n*-butanol as solvent. The reactants are heated overnight and the crude crown is obtained by precipitation from acetone in which it is almost completely insoluble. The yield range specified is 39–48% and is readily realized. The overall preparation is illustrated in Eq. (3.11).

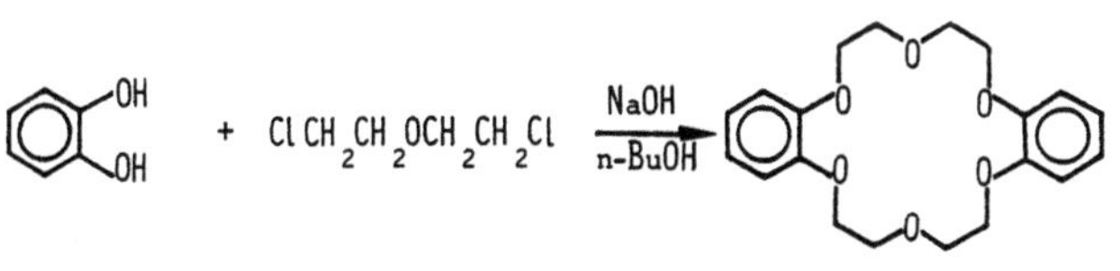

(3.11)

The procedure described above is what Pedersen characterized as a synthesis by "method X"[169]. Pedersen's alphabetical designations (see Sect. 3.1) are rarely used, but the approach is often referred to instead as a "2 + 2" synthesis. The implication is, of course, that two equivalents of nucleophile (in this case, catechol) are allowed to react with two equivalents of electrophile. Obviously, the stoichiometry is identical to the "1 : 1" case; the distinction is based on the expectation of product which, in this case at least, is realized.

Dibenzo-18-crown-6 is obtained by Pedersen's procedure as a tan solid which may be recrystallized to give beautiful, white fibrous needles melting at about 163 °C. The yield range of essentially pure material is, as noted above, 39–48%. Ashby, Hull, Cooper and Ramage[8] have modified the procedure to avoid the "inherent danger of handling"[8] 2,2′-dichlorodiethyl ether. These authors substituted the corresponding diol ditosylates for the dichlorides and modified the procedure only slightly otherwise. Although the yields (32–35%) were lower than those reported by Pedersen, the modification is apparently justified in terms of increased safety. A more important issue also noted by these authors is the question of commercial availability of dichlorides. Whereas many polyethylene glycols are readily available, many of the dichlorides are less accessible. Moreover, the conversion of such diols to the corresponding ditosylates is generally straightforward, the preparation of the dichlorides is usually tedious[169].

The crude, but essentially pure, dibenzo-18-crown-6 obtained from the *Organic Syntheses* preparation described above (Eq. 3.11) may be hydrogenated in *n*-butanol solution over 5% ruthenium on alumina, in a stainless steel autoclave at 1,000 p.s.i.[171]. Dicyclohexano-18-crown-6 is obtained from this treatment in 58–69% yield as a mixture of crystalline diastereomers having a melting range of 38–54°.

The mixture of diastereomers has been separated into its two principal components by Izatt, Haymore, Bradshaw and Christensen[97] who had previously identified the two principal diastereomers as the cis-syn-cis and cis-anti-cis isomers. Their previous separation technique involved a protracted chromatography on alumina but the new method relied upon the difference in water solubility between the lead perchlorate and hydronium perchlorate complexes. The lead perchlorate complex is essentially insoluble in aqueous solution and precipitates from it. Using this method, one may obtain 39% of the high-melting polymorph (mp 83–84°) and 44% of the low-melting compound (mp 62–63°). Note that the former also exists in a second crystalline form, mp 69–70°.

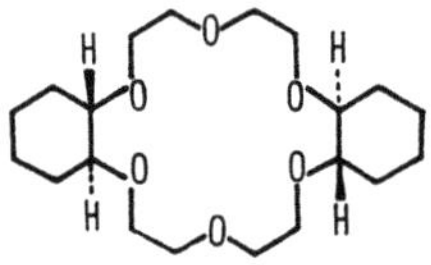

1, trans-anti-trans

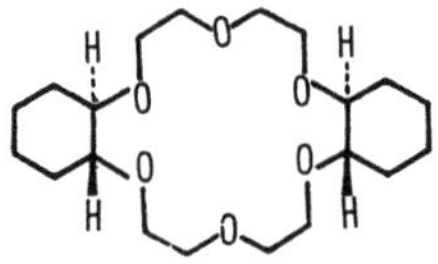

2, trans-syn-trans

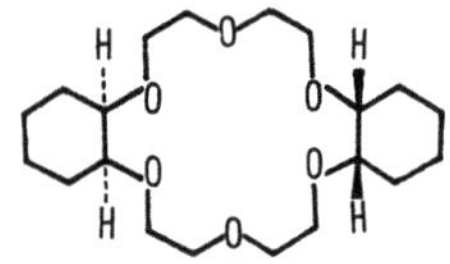

3, cis-anti-cis

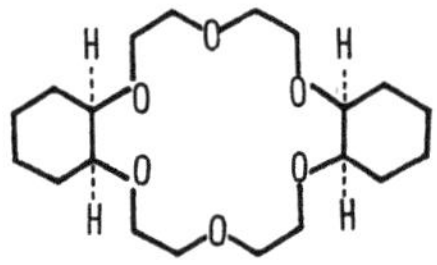

4, cis-syn-cis

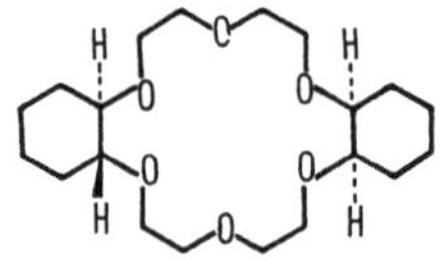

5, trans-cis

Stoddart and his coworkers[30,199] have reported syntheses of the trans-syn-trans and the trans-anti-trans isomers of dicyclohexano-18-crown-6. The synthesis of these two compounds from trans-1,2-cyclohexanediol was accomplished in two stages. First, the diols were temporarily linked on one side by formation of the formal, and this was treated with diethylene glycol ditosylate and sodium hydride to form the hemi-crown formal. Removal of the formal protecting group, followed by a second cyclization completed the synthesis. The synthesis of the trans-anti-trans compound is illustrated below in Eq. (3.12) and the structures of the five possible stereoisomers are shown as structures *1–5*.

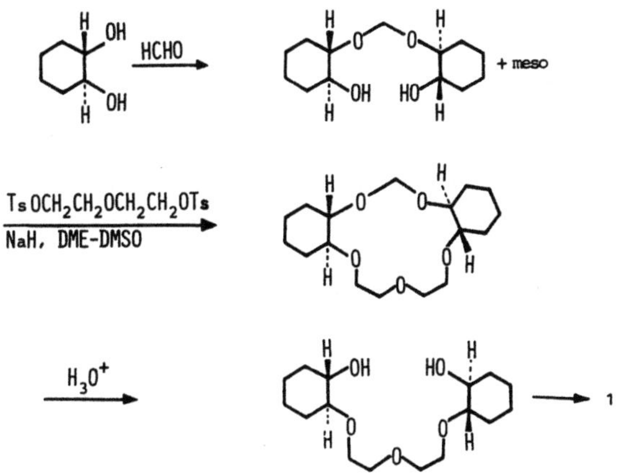

(3.12)

Hayward, Overton and Whitham[87] have prepared the related monocyclohexano-15-crown-5 and 18-crown-6 compounds, also starting with trans-1,2-cyclohexanediol. Construction of the latter was attempted using the diol and diethylene glycol ditosylate in analogy to Pedersen's original synthesis of dibenzo-18-crown-6, but the yield was "unsatisfactory"[87]. Instead, an "indirect approach *via* the 'half-crown' diol" was utilized. Trans-1,2-cyclohexanediol was alkylated with allyl bromide and then cleaved using ozone and sodium borohydride. This material could then be converted into the corresponding tosylate and thus comprised both "halves" of the crown ether. The overall approach to the monocyclohexane crown is illustrated in Eq. (3.13) below which was achieved in 42% yield.

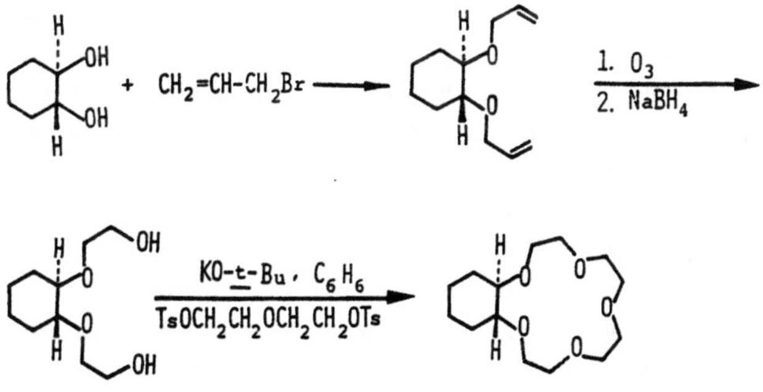

(3.13)

3.4 Synthesis of Substituted Benzocrowns

Because mono and dibenzocrowns were among the first crowns to be synthesized, and because the syntheses of these molecules were relatively straightforward, quite a number of modifications in the basic skeleton have been reported. Some of these modifications have utilized substituted catechols as starting materials and others have involved transformations on the existing crown.

The synthesis of substituted crowns has been accomplished in a straightforward fashion from a number of substituted catechols. Vögtle and Jansen, for example, have prepared a number of crown derivatives of adrenalins, and closely related substances by this method[220]. We have noted (see Sect. 3.8 and Eq. 3.29) that 4-acetylcatechol may serve as a precursor to either the corresponding acetyl crown or, after haloform oxidation, the carboxycrown[22,228]. Likewise, crowns containing the paracyclophane subcyclic unit have been thus prepared (see Sect. 3.7 and Eq. 3.28)[91,101]. Smid and his coworkers have prepared 4-formylbenzo-18-crown-6 from 4-formylcatechol and pentaethylene glycol dichloride (NaOH, BuOH) in 32% yield[196]. This compound was then converted into the 4′-vinylcrown by reaction with methylmagnesium bromide followed by dehydration. An alternative approach to this molecule is described below.

Of the many methods which have been published so far for the substitution of existing crowns, probably the most straightforward are Friedel-Crafts alkylation or acylation reactions. Cygan, Biernat and Chadzynski have reported the successful di-*t*-butylation of dibenzo-24-crown-8 using *t*-butanol as alkylating agent[46]. The crown was heated at 100° for 4 h in the presence of excess *t*-butanol and 85% phosphoric acid. The product was obtained as a crystalline (mp 52–74°) solid in 93% yield. The alkylated crowns are presumably a mixture of isomers substituted once in each ring as illustrated in Eq. (3.14).

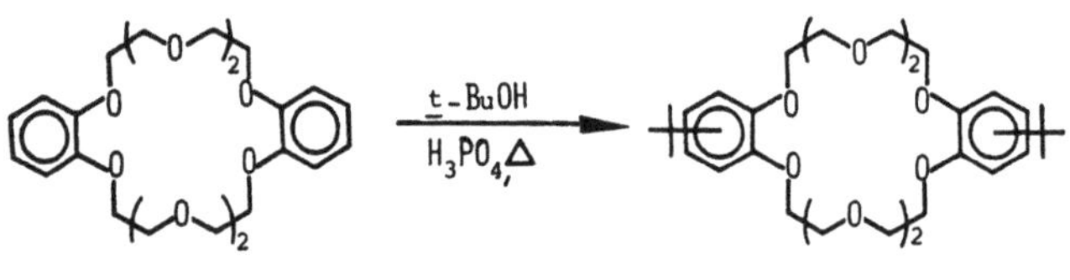

(3.14)

Stott and coworkers have reported the acylation of a number of benzocrown compounds for use as phase transfer catalysts[164]. These workers found that the attempted acylation of dibenzo-18-crown-6 using succinic anhydride and aluminum chloride was unsuccessful. Aluminum chloride proved a generally unsatisfactory catalyst for these reactions and the reactions of acid chlorides were generally catalyzed by boron trifluoride instead. Acylation could be effected using acids when a mixture of phosphorus pentoxide and methanesulfonic acid was used as solvent and catalyst. When relatively long-chain carboxylic acids were used as acylating agents, the mixture required heating to obtain good yields of acylated crowns. Diacylation of most crowns led to isomer mixtures. It was noted that the two isomers generally do not co-crystallize and often undergo partial fractionation. The authors offered the following interesting comments: "The melting ranges of the mixed isomer products were broad. It was noted, however, that samples of the mixed isomers nearly always exhibited two distinct melting points rather than a continual broad melting range"[164]. A synthesis of bis-myristoyldibenzo-18-crown-6 is illustrated in Eq. (3.15).

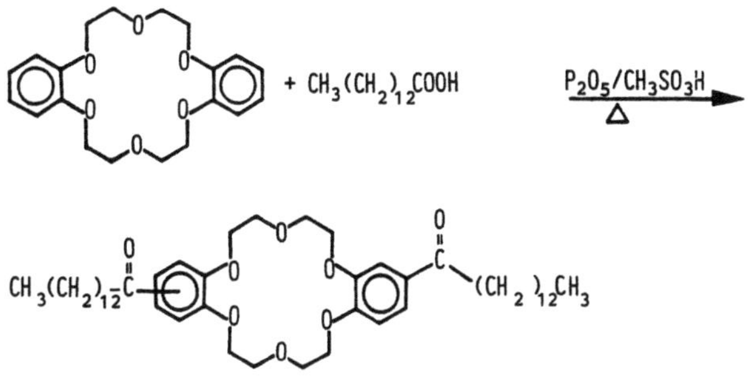

(3.15)

Tusek has reported the Clemmensen reduction of acylated crowns such as shown above[212]. This approach constitutes a useful two-step alkylation procedure, although the yields reported for this sequence are not as high as one might wish. Other examples of closely related acylation approaches can be found in recent work by Tashmukhamedova and coworkers[201–203] and by Kauer[99,100].

We have noted above that benzocrowns may be nitrated quite readily. This approach was used in the formation of a photoresponsive bis-crown (see Sect. 3.8) wherein the nitrobenzo crowns reductively dimerize to the corresponding azobenzene. Kikukawa, Nagira and Matsuda have utilized 4-nitrobenzo-15-crown-5 in a somewhat different way during the synthesis of 4′-vinylbenzo-15-crown-5[103]. Nitration is effected using nitric acid in a mixture of chloroform and acetic acid.

Hydrogenation reduces the nitro group to amino which is then diazotized using sodium nitrite and tetrafluoroboric acid. The diazotized crown was not isolated but the aq. solution was treated directly with sodium acetate and bis(dibenzylideneacetone)-palladium(0) in acetonitrile solution. Ethylene was then introduced to the autoclave and the solution was allowed to stir for 2 days. 4′-Vinylbenzo-15-crown-5 was isolated (30% from 4′-nitrobenzo-15-crown-5) as a colorless solid (mp 43.5–44.2°)[103]. The synthesis is illustrated in Eq. (3.16).

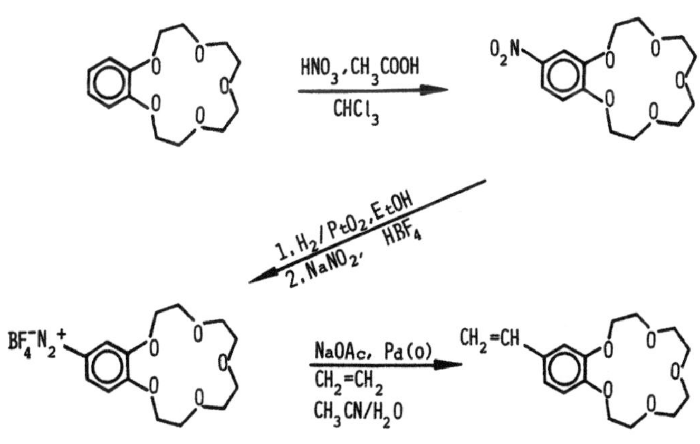

(3.16)

Smid and coworkers have developed a straightforward and effective method for the synthesis of 4'-vinylbenzo-15-crown-5[196]. In this method, 3,4-dihydroxybenzaldehyde is converted into the crown in the usual fashion and then the side chain is transformed by successive treatment with methylmagnesium bromide and then toluenesulfonic acid to effect dehydration. The overall yield is about 20%.

Nakamura, Takagi and Ueno have also utilized 4'-nitrobenzo-15-crown-5 as a starting material[136]. Their goal was the formation of a colored crown which could be utilized in transport studies. They have prepared 4'-picrylaminobenzo-15-crown-5 for this purpose in the following way. 4'-Nitrobenzo-15-crown-5 was hydrogenated and then picryl chloride was added. Nucleophilic aromatic substitution apparently ensued (deep red color) and the product was then isolated by standard techniques as a yellow solid (mp 155°, λ_{max} 395 nm) in 72% yield[136] as shown in Eq. (3.17).

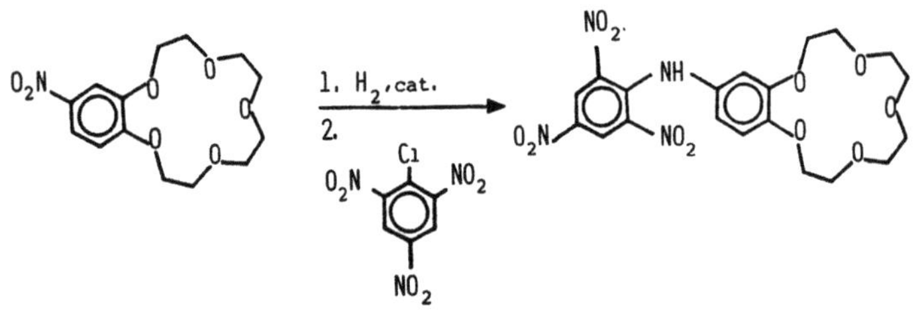

(3.17)

An interesting variant in which nucleophilic aromatic substitution is carried out on the crown nucleus rather than using the crown as nucleophile was reported by Haines[83]. In this approach, hexafluorobenzene was stirred in 1,2-dimethoxyethane at room temperature with pentaethylene glycol and sodium hydride. A double nucleophilic aromatic substitution occurred affording 2',3',4',5'-tetrafluorobenzo-15-crown-5 as an oil in 38% yield. The reaction is illustrated below.

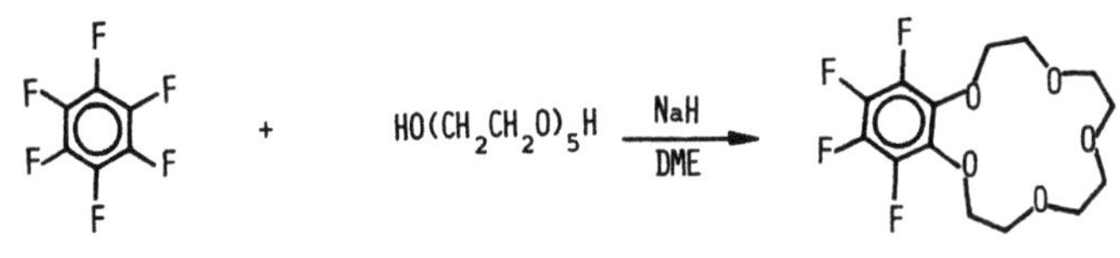

(3.18)

3.5 Crowns Incorporating the Xylylene Unit

The *ortho*-xylyl unit may be considered as the synthetic complement of the catechol unit so often used in crown synthesis. This is especially so for the 1,2-bromomethyl compounds which may be used as electrophiles. These units have been incorporated in numerous crowns over the years and the syntheses are all quite similar.

In 1972, Vögtle and Zuber[224] condensed 1,2-dibromomethylbenzene (xylylene dibromide) with another xylylene unit, 1,2-dihydroxymethylbenzene. The result was formation of the macrocyclic tetraether (40% yield, mp 194°) shown in Eq. (3.19). The reaction of 1,2-dihydrocymethylbenzene with 1,3-dibromomethylbenzene under similar conditions, yielded the isomeric macrocycle containing both *ortho*- and *meta*-xylyl units. Note that at about the same time, Wudl and Gaeta used the *ortho*-xylyl group as an electrophile in the preparation of certain proline-based macrocycles (see Sect. 4.9 and Eq. 4.15).

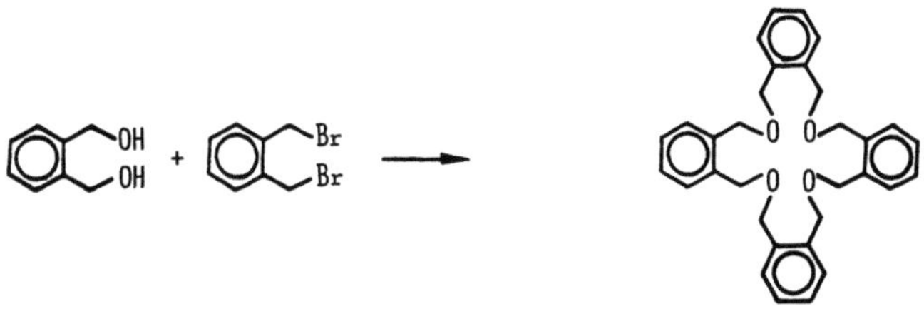

(3.19)

These xylylene groups are typically available commercially[185], by halogenation of xylenes[24], by halomethylation of an aromatic[116,185] or by conversion of a diester into its corresponding diol[155,185] followed by halogen-hydroxyl interconversion[76,155,185].

Reinhoudt, Gray, Smit and Veenstra prepared a number of monomer and dimer crowns based on a variety of substituted xylylene units. They first conducted the reaction of 1,2-dibromomethylbenzene and a polyethylene glycol with sodium hydride or potassium *t*-butoxide in toluene solution. Mixtures of the "1 : 1" and "2 : 2" (monomer and dimer) products were isolated and some polymer was formed[185]. The reaction was conducted at temperatures from 30–60° and appeared to be complete in a maximum of one hour. The authors noted that "the highest yield of 1 : 1 cyclic product was obtained with disodium tetraethylene glycolate instead of dipotassium hexaethylene glycolate" (see also Chap. 2)[185]. Chloromethylation of 1,3-benzodioxole followed by reaction with disodium tetraethylene glycolate afforded the macrocycle (29% yield) illustrated in Eq. (3.20).

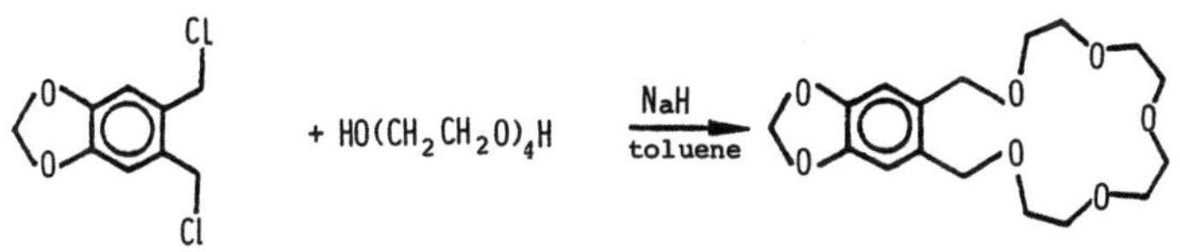

(3.20)

The incorporation of *meta*-xylene units proved somewhat more challenging for Gokel, Timko and Cram[76] who prepared a number of chiral macrocycles incorporating this subunit. In an effort to determine the effects of structural changes on chiral selectivity on molecular complexation, a series of dibinaphthayl crowns were prepared containing aliphatic hydrocarbon, ethyleneoxy, xylyl and pyridyl subunits[76,114] (see also Sect. 3.13). The parent crown of this group may be prepared in the normal 2 + 2 fashion (see Eq. 3.51), but those analogs having different strands bridging the binaphthyl units must be synthesized sequentially. This was accomplished using the benzhydryl unit as a base-stable, acid-labile protecting group for one of the naphthyl-hydroxyl groups. The synthesis of a di-binaphthyl crown having ethyleneoxy and 1,3-xylyl strands in the ring is shown in Eq. (3.21).

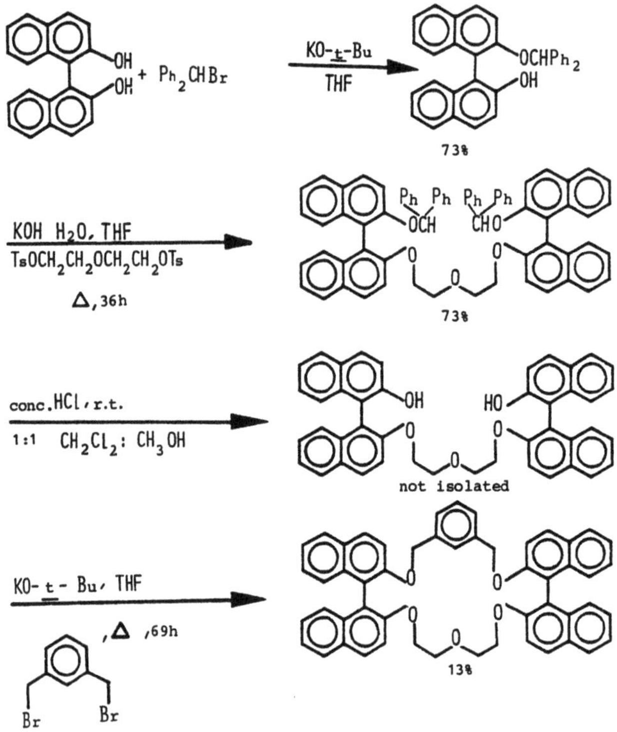

(3.21)

Cram and his coworkers successfully employed the benzhydryl protection strategy in preparing dibinaphthyl crowns containing the nitrogen equivalents of *meta*-xylene, i.e. 2,6-dimethylpyridines[76,114]. Newkome and Robinson[155] had previously utilized 2,6-dihydroxymethylpyridine as a nucleophile in the synthesis of such compounds. This contrasts with Cram's use of 2,6-bromomethylpyridines as electrophiles. Newkome and Robinson[155] obtained the diol by sodium borohydride reduction of methyl pyridine-2,6-dicarboxylate. 2,6-Dihydroxymethylpyridine was treated with sodium hydride in dimethoxyethane and 1,2-dibromomethylbenzene for 25 hours. The 22-membered macrocycle (mp 142–143°) was obtained in 40% yield. The "trimer" or 33-membered ring was isolated in 9% yield and material presumed to be 44- and 55-membered rings were detected. No "monomer" or 11-membered ring was observed. The synthesis is illustrated in Eq. (3.22).

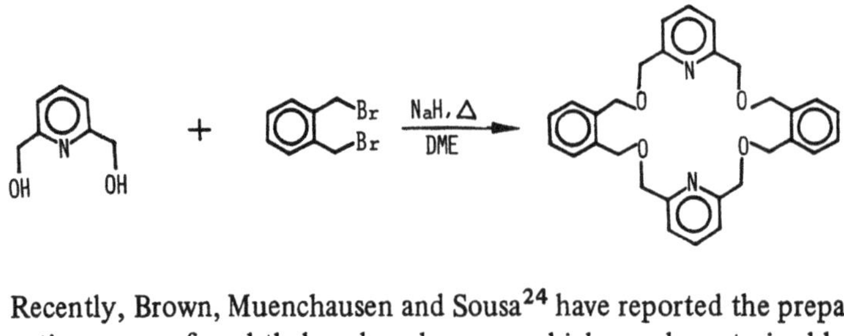

(3.22)

Recently, Brown, Muenchausen and Sousa[24] have reported the preparation of an interesting group of naphthalene-based crowns which are characterized by these authors as "rope-skipping" crown ethers. A discussion of the conformational variations in these structures is beyond the scope of this work, but it is interesting to note that when the ethyleneoxy chain is sufficiently long, a "rope-skipping" phenomenon appears to occur fairly easily (energy barrier 6.3 kcal/mole)[24]. The dibromides required for these syntheses were prepared by free radical halogenation (N-bromosuccinimide in CCl_4) of the appropriate dimethylnaphthalenes. The Williamson reaction was carried out between the dibromonaphthalene and polyethylene glycol using potassium *t*-butoxide as base in THF solution. The "rope-skipping" crown referred to above, 1,5-naphtho-22-crown-6 (mp 55–56°) (see Eq. 3.23) was isolated in 14% yield.

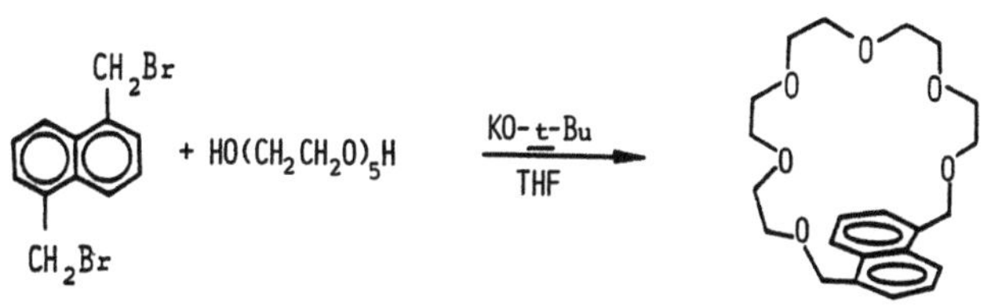

(3.23)

3.6 Furan-containing Macrocycles

Apparently intrigued by the observation that furan and pyrrole do not undergo the Diels-Alder cycloaddition with certain dienes, Ackman, Brown and Wright studied the condensation of furan with certain ketones[1]. They observed that the condensation products, "called anhydrotetramers because the analytical data show that they contain four units each of furan and ketone minus four units of water"[1], can be formed in a hydrochloric acid catalyzed reaction. Varying the molar ratio of furan to acetone, these workers were able to detect up to 18.5% of "anhydrotetramer." In these reaction mixtures, there were also sundry non-cyclic, mixed-condensation products. Compound *6*, shown below and given the suggested name 2,2,7,7,12,12,17,17-octamethyl-21,22,23,24-tetroxaquaterene[1] was actually prepared by condensation of fragments larger than acetone and furan, typically by the reaction shown in Eq. (3.24) which affords *6* (mp 243°) in 37% yield.

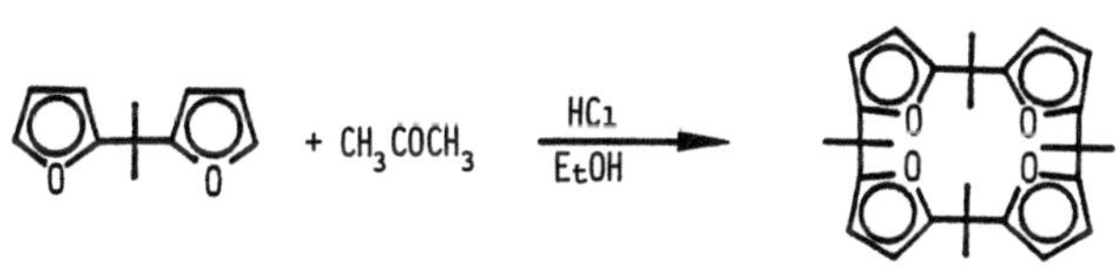

(3.24)

6

Subsequently, Beals and Brown[12] expanded the scope of the earlier work from their laboratory to include the "tetraoxaquaterene" derived from furan and 3-pentanone. Using 3,3-difurylpentane and diethyl ketone in the presence of dry hydrogen chloride gas, the all-ethyl analog of *6* (mp 249°) was obtained in 20% yield[12].

In subsequent work, Brown and his coworkers[25,26] succeeded in extending the "cyclocondensation" reaction which produced a variety of "quaterenes" to molecules containing mixtures of furan and pyrrole nuclei as well as numerous carbonyl-derived residues. Further interest in this class of compounds waned until similar species were investigated for aromaticity. Obviously, the quaterene compounds could not have ring currents since quaternary carbons are interposed between the π-systems, but aromatic properties are observed if the furan residues are connected by an unsaturated carbon.

Badger and coworkers devised a sequential synthesis of [18]annulene-1,4;7,10;13,16-trioxide which is formally the condensation product of three furan molecules and three ethylenes[9,10]. The synthesis is illustrated below in Eq. (3.25). The [18]annulene trioxide was obtained as a red solid (mp 215–216 °d) whose proton nmr spectrum showed two peaks of equal area at 8.66 and 8.68 ppm.

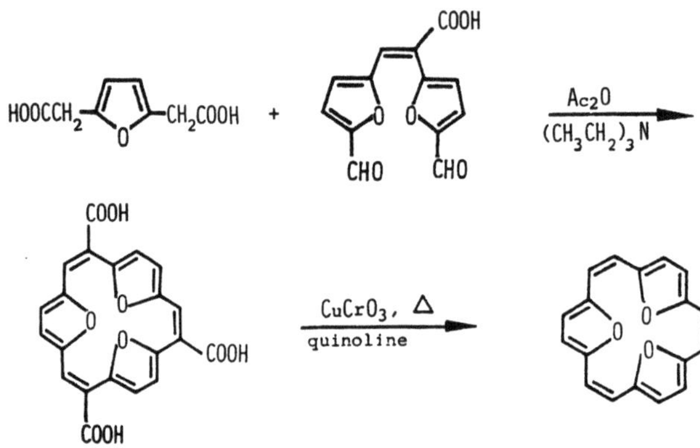

(3.25)

Considerable additional work in this area was reported by Elix[63,64], Brown, Hutchinson and MacKinnon[27], as well as Cresp and Sargent[39–41] but the first real association with crown compounds was reported by Chastrette and Chastrette[32] who effected a direct, template synthesis of *6*. The latter authors found that when acetone (0.2 mol) and furan (0.1 mol) are allowed to react in ethanol (50–60°) in the presence of concentrated HCl, the "anhydrotetramer" (see above) could be isolated in 18–20% yield. When 0.03–0.1 mol of lithium perchlorate is present in the solution, 40–45% yield of *6* is realized. Similar yield enhancements attributed to a template effect were obtained when calcium, zinc and magnesium salts were used but not when tetrabutylammonium perchlorate is substituted for any of the other salts. It should also be noted that certain chloride salts afforded enhanced yields relative to the untemplated reaction. The work of Chastrette and Chastrette was augmented by studies reported by Rest, Smith and Tyler[172] who found that transition metals as well as alkali and alkaline earth metals could also serve as templates, although the yields were either similar or lower.

Timko and Cram were the first to prepare true crown ethers containing the furanyl subcyclic unit[204]. Destructive distillation of sucrose yielded 2-hydroxymethyl-5-formylfuran *7* in 41% yield. This could be reduced to the corresponding diol in 91% yield by treatment with sodium borohydride. Reaction of the diol with tetraethylene glycol ditosylate, and potassium *t*-butoxide in THF solution afforded the crown in 36% yield. The approach is illustrated below as Eq. (3.26).

After conversion of compound *7* into its chloride ($SOCl_2$), followed by treatment with chloroethanol and barium carbonate, a monofuranyl aldehyde could be obtained

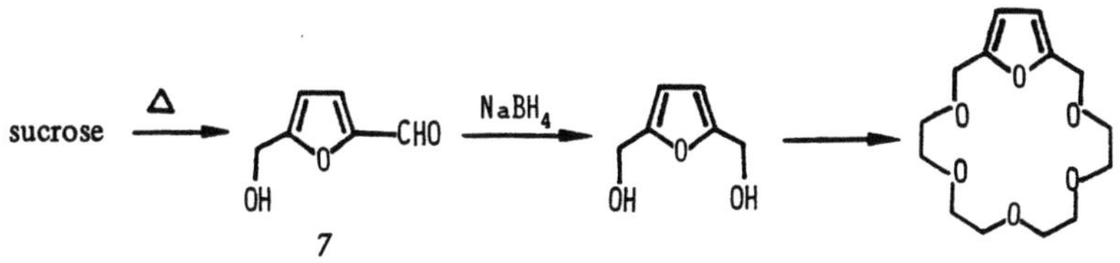

(3.26)

which was further converted to the corresponding chloroalcohol by sodium borohydride reduction. The chloroalcohol was then dimerized by treatment with sodium hydride in DMF solution. The bis-furan crown (mp 109–110°) was isolated in 11% yield as illustrated in Eq. (3.27).

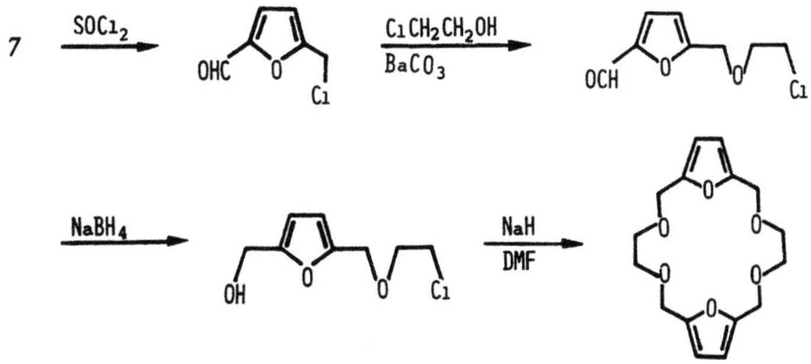

(3.27)

Recently, Gray, Reinhoudt and coworkers have prepared a number of furan-containing crowns by similar methods[79,80,184]. It is worthwhile noting that these workers prepared considerably larger macrocycles than those prepared by Timko and Cram and the complexation profile for these was obviously different.

3.7 Crowns Containing the Paracyclophanyl Unit

Macrocyclic polyethers containing the 2.2-paracyclophane unit are interesting structures and several such compounds have been prepared[89,91,101,205]. Despite the diverse structural possibilities, the syntheses of these molecules have generally been accomplished by straightforward Williamson ether syntheses. The only unusual aspect of the syntheses appears to be a novel approach to certain paracyclophanes developed by Helgeson (see footnote 7a in Ref. 91). The first step of Eq. (3.28) illustrates the formation of the required tetrol, which is then treated with base (KOH or KO-*t*-Bu) and the appropriate diol ditosylate to afford the macrocycle.

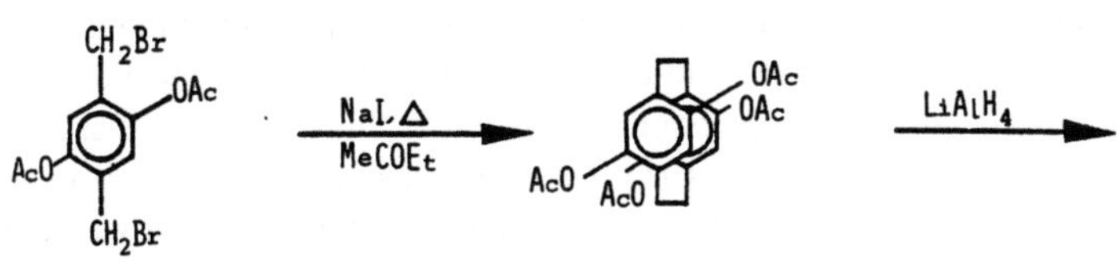

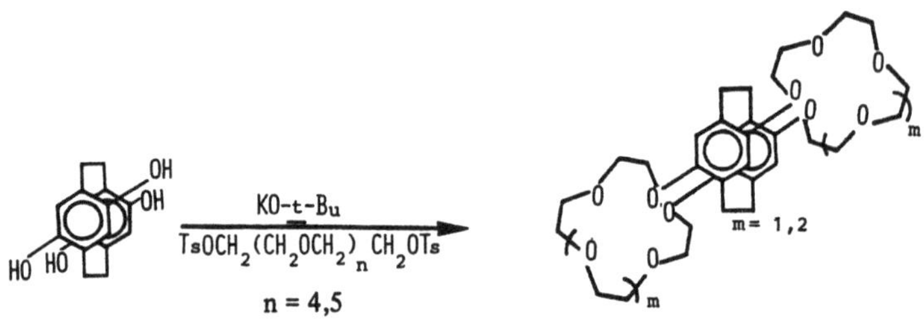

(3.28)

3.8 Bis-Crowns

The first report of a bis-crown appears to be that of Wong, Bourgoin and Smid[228] who observed a 2:1 complex between benzo-15-crown-5 and potassium cation in THF solution. This suggested to them that the appropriate bis-crown might mimic such behavior and bind a cation quite strongly. They set as their target, two benzo-15-crown-5 molecules bridged by a diethylene glycol residue. To this end, 4-acetylcatechol was converted into 4′-acetylbenzo-15-crown-5 in the usual fashion. Haloform reaction afforded the carboxylic acid which was then converted into the acyl chloride. The latter was allowed to react with diethylene glycol in a mixture of chloroform and pyridine to yield the bridged crown (mp 89–91°) in 64% yield. The synthesis is illustrated in Eq. (3.29).

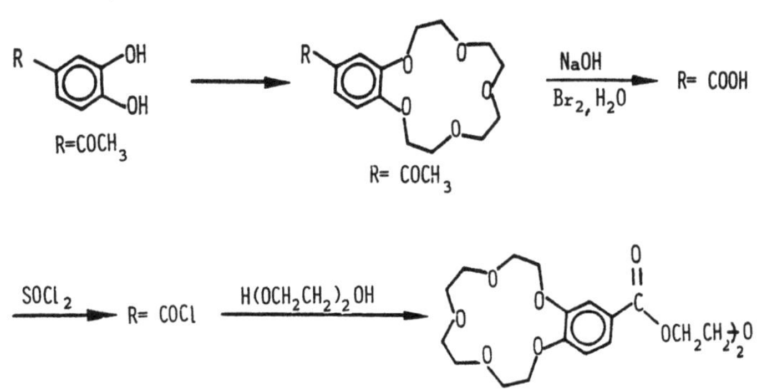

(3.29)

Concerning the efficacy of this approach, the authors state that "compared with the methyl ester of 4′-carboxybenzo-15-crown-5, the macrobicyclic polyethers are considerably more effective in binding K^+ and NH_4^+ cations"[22].

At almost the same time the reports described above appeared, Cram and his coworkers[205] reported a variety of crown ethers and a survey of their binding properties with the *t*-butylammonium cation. One of the compounds reported in this work is a bis-crown quite similar to that shown in Eq. (3.29) above. The most important difference is that instead of the two benzo-15-crown-5 residues being bound through a single strand, it was actually the benzo groups which were held together as part of the [2.2]paracyclophanyl nucleus. Two such compounds are illustrated below as *8* and *9*. For details of the synthesis of such molecules, see Sect. 3.7 (immediately preceding).

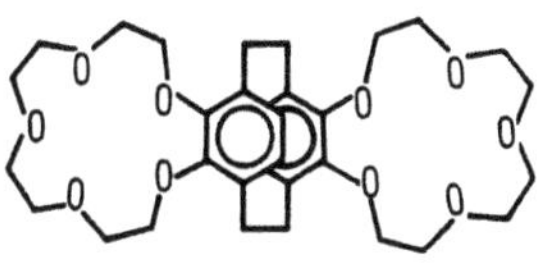

8

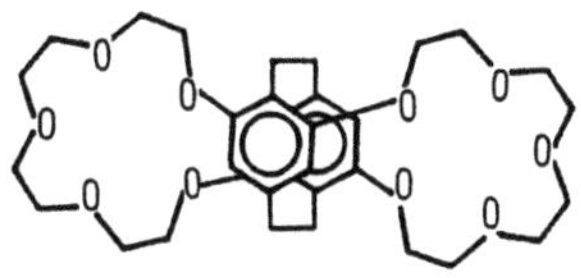

9

Not long thereafter, Tarnowski and Cram[200] reported the first example of a "hinged" bis-crown ether. The compound was prepared in the usual Williamson reaction by heating a mixture of 2,2′,3,3′-tetrahydroxy-1,1′-binaphthyl with pentaethylene glycol ditosylate and KOH in aqueous THF solution. The product (mp 159.5–161°) shown in Eq. (3.30) was obtained in 30% yield. This compound was shown to complex bis-ammonium cations of several varieties[200].

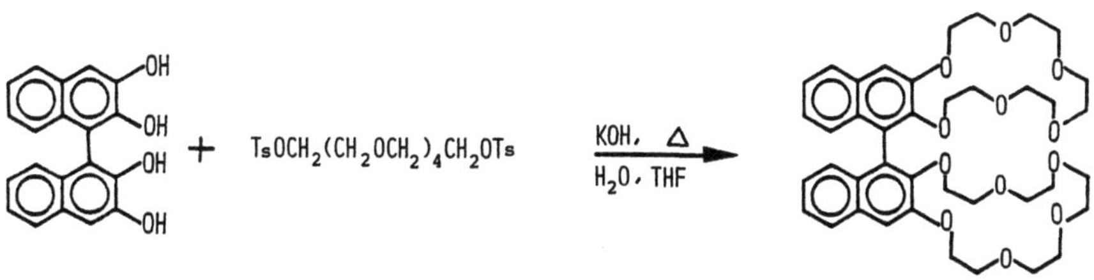

(3.30)

At about the same time, Vögtle and Frensch[219] reported the synthesis of a similar bis-crown based on the papaverine nucleus. This synthesis also was executed in the Williamson fashion using doubly-demethylated papaverine and tetraethylene glycol dichloride. The base, if used, was not specified but it was noted that the high dilution technique was not used. It was presumed that the major product was *10*, shown below, but there was also evidence of other components, presumably *11* and *12*.

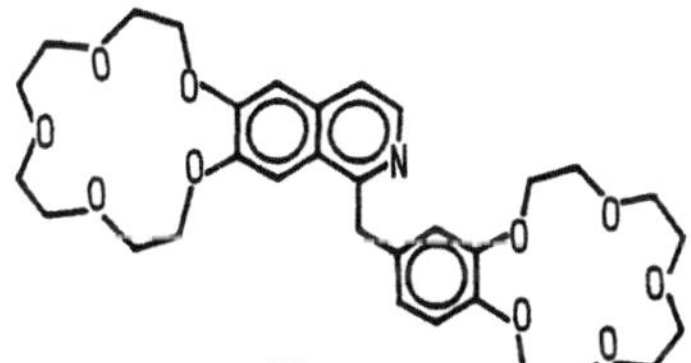

10

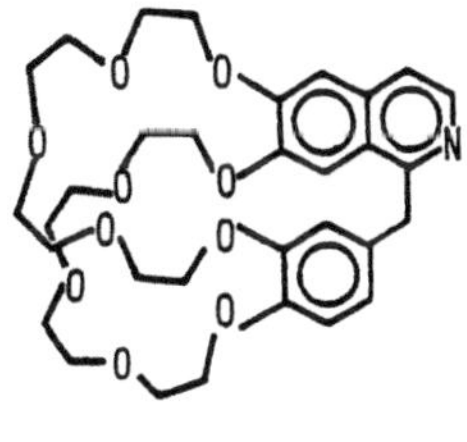

11

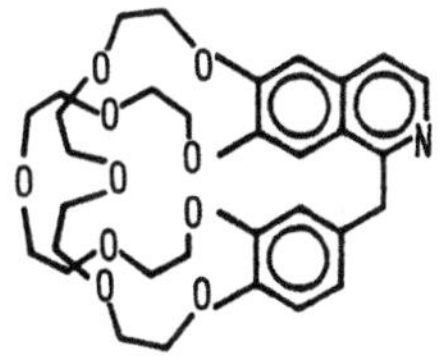

12

Hyde, Shaw and Shepherd[95], apparently unaware of the bis-crowns prepared by Smid *et al.*[22,228], reported a similar compound but by a remarkable route. In a survey of methods which might prove suitable for functionalizing the aromatic ring, these workers found that formylation could be effected by the Vilsmeier reaction ($POCl_3$ and N-methylformanilide). Reduction of 4'-formylmonobenzo-15-crown-5 using lithium aluminum hydride gave the corresponding alcohol in only 64% yield but the bis-crown resulting from ether formation in the side-chains was obtained in 25% yield. The authors state that they "cannot, however, suggest a mechanism to explain the formation of this dibenzyl ether"[95]. The synthesis is illustrated in Eq. (3.31).

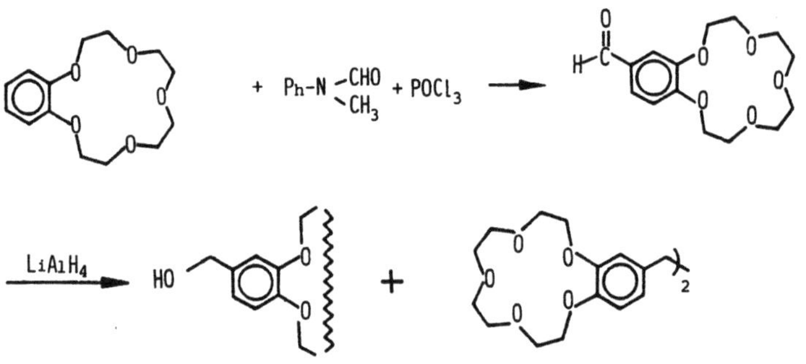

(3.31)

Recently, Wong and Ng[230] have prepared compounds similar to those shown above and to those previously prepared in collaboration with Smid[22,228]. These compounds were prepared by the method shown in Eq. (3.32).

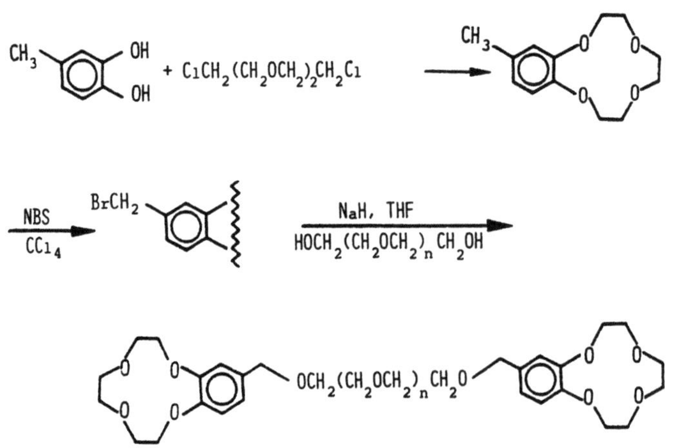

(3.32)

In the synthesis of the twelve-membered ring, $LiClO_4$ was used as a templating ion and the methylbenzo-12-crown-4 (mp 46–48°) was obtained in 15% yield. Alpha-bromination was effected in 70% yield and conversion to the bridged species occurred in 20% and 25% yield when n = 1or n = 2 respectively.

Wada and coworkers have recently prepared a number of very similar systems in which the bridging chain is bonded to the benzo group by oxygen[225]. The phenolic oxygen is incorporated by a Baeyer-Villager reaction on the acetyl derivative. These workers noted that attempted hydrolysis of the benzocrown diazonium salt and attempted oxidation of the benzocrown with $Tl(OCOCF_3)_3$ were both successful. The phenolic crowns were joined in a simple Williamson reaction leading to a number of bis-crown ethers. The synthesis of benzo-18-crown-6, bridged by an ethylene glycol unit is illustrated below in Eq. (3.33).

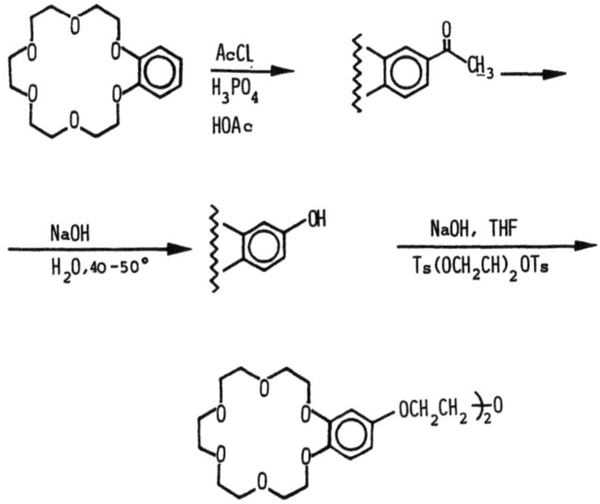

(3.33)

Frensch and Vögtle have recently appended three crown ether units to the cyclotriveratrylene unit[68]. Note that Hyatt had previously prepared the open-chained relatives of this structure (see Sect. 7.3 and Eq. 7.6). Whereas Hyatt prepared the cyclotriveratrylene skeleton and then appended polyethyleneoxy arms to it, Frensch and Vögtle conducted the condensation reaction (formaldehyde/HCl) on the preformed benzocrown. Thus benzo-15-crown-5 was converted into the corresponding tris-crown (*13*) (mp 203.5–205.5°) in 4% yield. The yield was somewhat higher for the condensation of benzo-18-crown-6, but in both cases, yield ranges were observed. These species formed 1:3 (ligand/salt) complexes with sodium and potassium ions.

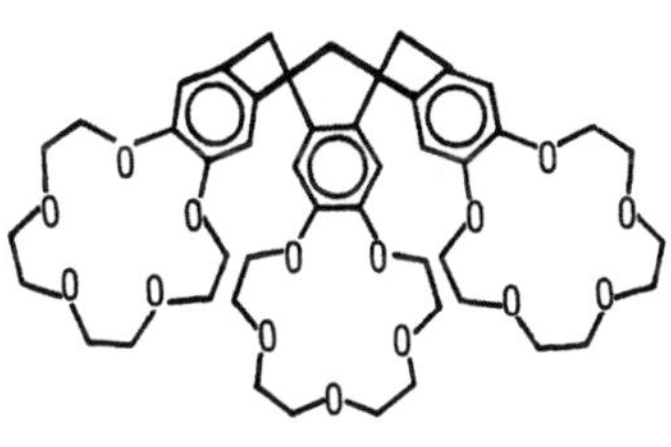

13

Another interesting variation on the linkage holding together two benzocrowns was reported by Shinkai and coworkers[194]. In their case, the benzo groups on the crowns were also part of a photolabile azobenzene system. The azo-linked bis-crown was prepared as follows (see Eq. 3.34). 4′-Nitrobenzo-15-crown-5 was reductively dimerized

(Zn powder, KOH) to give the trans-azolinked crown (mp 187–188°) in unspecified yield. The crown proved to be photolabile, undergoing facile trans-cis isomerism upon irradiation.

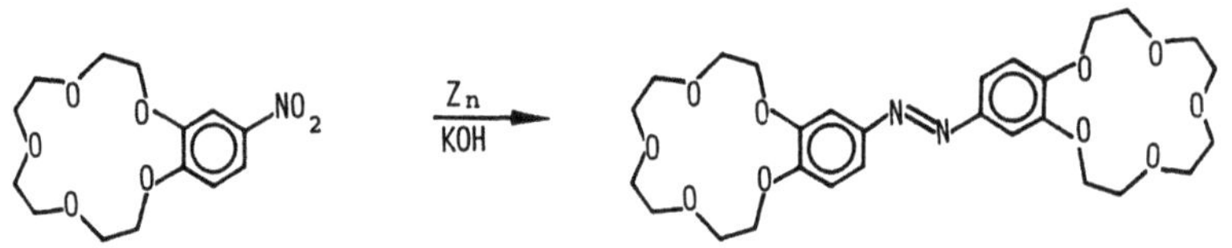

(3.34)

A final variation on the bis-crown theme is the group of compounds which Weber has prepared and dubbed "multi-loop crown ethers"[219]. The examples thus far prepared incorporate up to three macrorings joined by a spiro carbon rather than a single-strand link. The synthesis of these species is based on the pentaerythritol unit, which is partially protected as its benzaldehyde acetal. The remaining two hydroxyl groups are treated with NaH and a polyethylene glycol ditosylate. After having formed the first ring, the benzal protecting group is removed by hydrogenolysis (Pd/C, H_2). The newly exposed hydroxyl groups may either be converted into a second crown as described above, yielding a spiro-bis-crown, or treated with a polyethylene glycol monotosylate. In the latter case, one end of the glycol is protected as its tetrahydropyranyl ether. Upon hydrolysis, a two-armed crown is revealed which may then be treated with the bis-crown precursor identified above. The result will be a tris-crown system and the approach is illustrated below in Eq. (3.35).

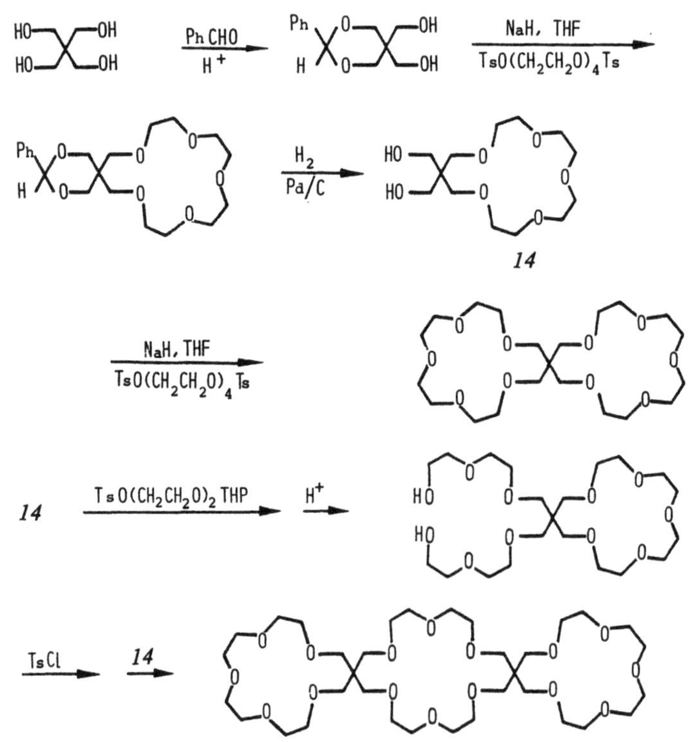

(3.35)

3.9 Lariat Ethers

The name "lariat ethers" has recently been suggested by Gokel and coworkers[58,71,74] to describe a class of compounds which are based on crown ethers and have secondary donor groups bound to the macroring by flexible arms (see Sect. 1.3.5). The name "lariat" is suggested by the analogy to the lassoes used to "rope and tie" animals since it is hoped that the more successful molecules of this class will be able to do likewise with sundry cationic species.

The notion of using additional donor groups to offer assistance in binding either cationic metals or organic species is not a new one and has been utilized before by a number of workers. In addition, arms have been appended to crowns for the purpose of providing additional lipophilicity or providing a binding link to a support or polymer backbone. Examples of some, but not all, of these applications may be found in Refs. 6, 14, 15, 31, 33, 36, 37, 88, 90, 105, 122, 128, 134, 167, 189, and 206. The concept of the lariat ether is intended to apply particularly to those neutral donor groups which can provide additional solvation but remain highly dynamic. This stands in contrast to use of arms intended to participate in reactions[31] or where apparently irreversible binding of a cation by the arm occurs[90].

In any event, the syntheses of the simple macrocycles so far consigned to this class are relatively straightforward. The glycerol unit has served several groups as an arm-bearing unit which may be incorporated into an alicyclic crown ether[47,58,71,74,134]. Montanari and Tundo[134] have utilized 3-*t*-butoxy-1,2-propanediol as the protected, armed glycerol unit. After incorporation into the macrocycle by typical Williamson reaction, the crown shown in Eq. (3.31) is formed (Z = *t*-Bu). Czech[47] and Gokel, Dishong and Diamond[74] have utilized benzyl rather than *t*-butyl as the protecting group. In the former case, the hydroxymethyl group is revealed by acid hydrolysis[134] and in the latter cases hydrogenolysis is used[47,74]. The synthesis of these molecules from alkyl or aryl glycidyl ethers is shown in Eq. (3.36). Synthesis of these ethers has been accomplished in all cases by alkylation of the appropriate alcohol or phenol with epichlorohydrin followed by hydrolysis of the epoxide. The diol may then be used directly as a nucleophile in the Williamson ether synthesis.

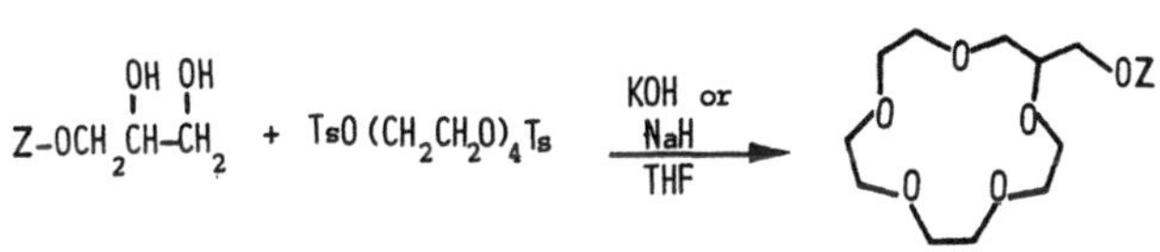

(3.36)

The nitrogen lariats[71] are readily prepared from the appropriately N-substituted diethanolamines (see Eq. 4.2 and Sect. 4.4).

3.10 Miscellaneous Crown Systems

So many different crown ether systems have been prepared over the recent decade that it sometimes seems that any of them could be placed in a miscellaneous category. On the other hand, each has its interesting features and probably merits a separate section for adequate discussion. Because both of these criteria cannot be met simultaneously, we have placed a number of compounds in this section which are fully deserving of detailed discussion, but not enough examples are yet available to group them separately.

The first of these compounds are the simple alicyclic crowns having either an olefin or carbonyl group as part of the ring. [See Sect. 3.4 for a description of compounds having alkenyl or carbonyl substituents appended to benzo units.] In 1970, Nerdel, Mamluk, and Weyerstahl[137] reported the synthesis of these molecules, accomplished using the base-initiated fragmentation of tribromopivalophenone. Treatment of the latter with 1,3-propanediol and sodium hydride in THF solution afforded the "2 + 2" product of macrocyclization and the benzoic acid derivative as a by-product. The olefins could be oxidatively cleaved (OsO_4, $NaIO_4$) to afford the bis-carbonyl derivative shown in Eq. (3.37). Reduction of the latter with lithium aluminum hydride, followed by tosylation of the alcohol and then another treatment with LAH afforded the non-substituted parent, 16-crown-4[137].

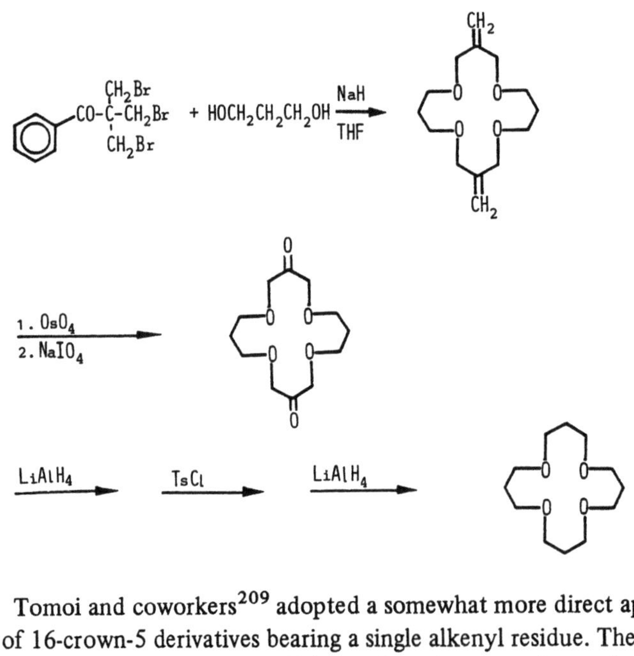

(3.37)

Tomoi and coworkers[209] adopted a somewhat more direct approach in their synthesis of 16-crown-5 derivatives bearing a single alkenyl residue. They hoped to obtain precursors to polymers which could be used as phase transfer catalysts. In this approach 1,1-bis-chloromethylethylene (α-chloromethallyl chloride) was allowed to react with the dianion of tetraethylene glycol (NaH/THF). By this method, methylene-16-crown-5 could be isolated in 66% yield after vacuum distillation. Ozonolysis led, in almost quantitative yield, to the formation of oxo-16-crown-5 as shown in Eq. (3.38). These authors prepared a number of other, closely related species by similar methods.

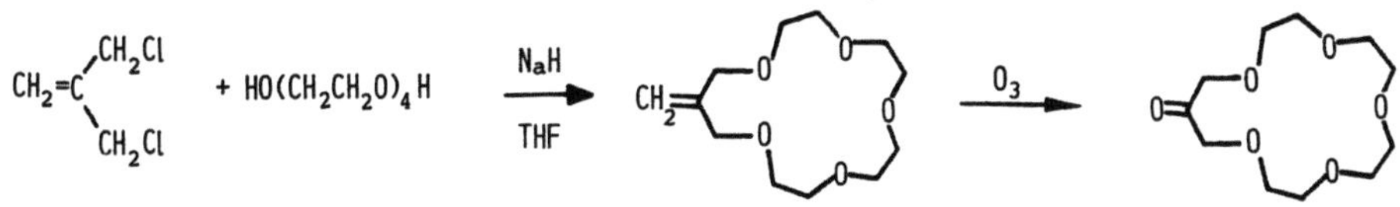

(3.38)

It was a "fascination with oxo-crowns . . . aroused by the speculation that the intramolecular interaction between a carbonyl group and a lone pair of electrons on heteroatoms in medium-sized ring compounds" might also be found in the crown relatives which prompted Beresford and Stoddart[17] to prepared a number of such species. The approach was somewhat different from those described above. Rather than preparing the alkene and then oxidatively cleaving to the ketone, the masked ketone was introduced in the first step. Thus, 2-(2′-hydroxyethyl)-1,3-dithiane was treated with *t*-butyldimethylsilyl chloride in DMF. The protected dithiane was then deprotonated and ethylene oxide added to the anion. After removal of the silyl protecting group (tetrabutylammonium fluoride), the diol was treated with tetraethylene glycol ditosylate and base to afford the dithiane derivative of oxo-18-crown-5. The latter was obtained by the usual oxidative hydrolysis of the dithiane group as illustrated in Eq. (3.39). It is interesting to note that oxo-18-crown-5 has a melting point (36–37°) virtually identical to that of 18-crown-6.

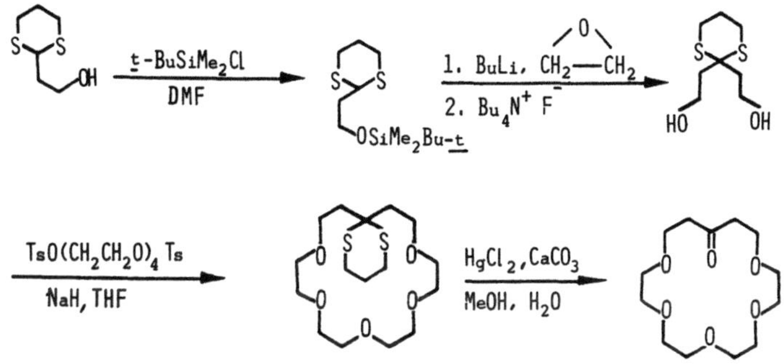

(3.39)

Unfortunately, the envisioned mode of binding proved unrealized, although the oxo-crowns did prove to be good complexing agents for ammonium salts[17].

Krespan[107–110] has prepared a number of macrocycles, having both aza- and oxa-linkages in them, based on the 3,3-dimethyleneoxetane unit (see also Sect. 8.4 and Eq. 8.12). Typically, 3,3-bis(chloromethyl)oxetane is treated with a diol as shown in Eq. (3.40), in the presence of base. Once the bicyclic system is formed, further treatment with other nucleophiles (e.g., ammonia) can lead to opening of the 4-membered ring.

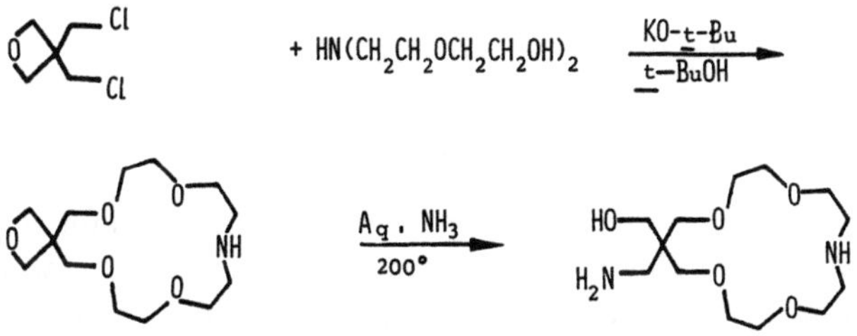

(3.40)

An enediol may be generated from molecules such as benzoin[132] or benzil[21] which can then be treated with a diol-ditosylate to afford the crown. Merz[132] demonstrated the utility of this approach in forming both the 9-crown-3 and 18-crown-6 derivatives (i.e., "1 + 1" and "2 + 2" products) of diphenylstilbenediol. The synthesis is remarkable since it utilizes phase transfer conditions to generate the crown ether. A benzene solution of benzoin is treated with 50% aqueous sodium hydroxide, diethylene glycol ditosylate and tetra-*n*-butylammonium bromide. The latter is intended to be the phase transfer catalyst for this reaction. The reaction is clearly successful: both 9- and 18-membered rings are isolated as shown in Eq. (3.41). One wonders, however, what roles the various reagents play in the synthesis. Presumably, the enediol of benzoin is generated at the phase boundary in accordance with the understanding of such phase transfer ether syntheses. The enediolate must then ion-pair with the quaternary ammonium salt and be transported into the benzene layer where reaction with diethylene glycol ditosylate occurs. Since hydroxide anion is virtually insoluble in benzene solution even when paired with tetrabutylammonium cation, and since extraction of NaOH from such a concentrated salt solution must be extremely difficult, it is not clear whether there is a templating ion present during the synthesis at all. One might argue that the incipient crown could complex sodium and draw it into benzene, but the incipient crown would be an alkoxide and the latter should be paired with tetrabutylammonium, not sodium. In any event, the success of the reaction is documented and shown schematically below.

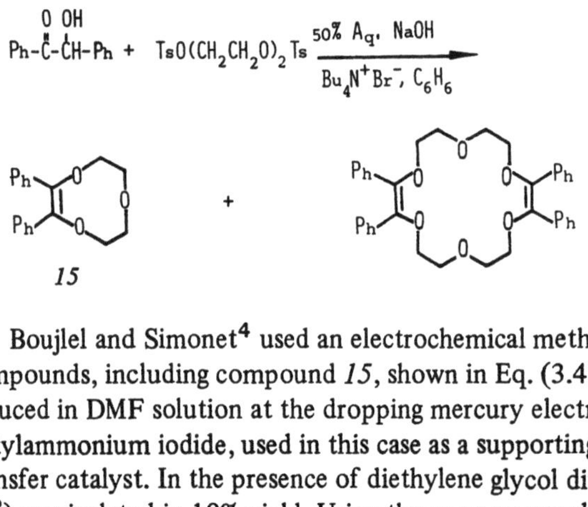

15

(3.41)

Boujlel and Simonet[4] used an electrochemical method to prepare a group of similar compounds, including compound *15*, shown in Eq. (3.41). In a typical case, benzil was reduced in DMF solution at the dropping mercury electrode in the presence of tetrabutylammonium iodide, used in this case as a supporting electrolyte rather than phase transfer catalyst. In the presence of diethylene glycol ditosylate, compound *15* (mp 77–78°) was isolated in 10% yield. Using the same approach, acenaphthenedione was reductively cyclized with triethylene glycol ditosylate to afford the product (mp 84–85°, 42% yield) shown in Eq. (3.42).

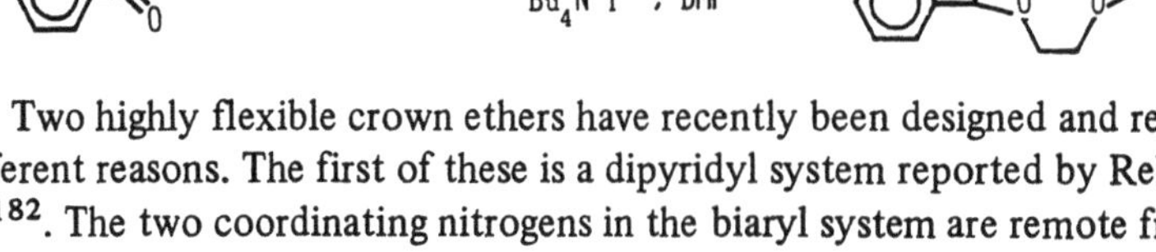

(3.42)

Two highly flexible crown ethers have recently been designed and reported for quite different reasons. The first of these is a dipyridyl system reported by Rebek and coworkers[182]. The two coordinating nitrogens in the biaryl system are remote from the all-oxygen macrocycle and respond to binding by different metal ions. Coordination of a transition metal in the bipyridyl subunit causes an alteration in the flexible crown ring. Note that Cram and coworkers have built in such flexibility in similar systems discussed primarily in Sect. 3.13, q.v. The syntheses of these crowns are fairly straightforward and is shown in Eq. (3.43).

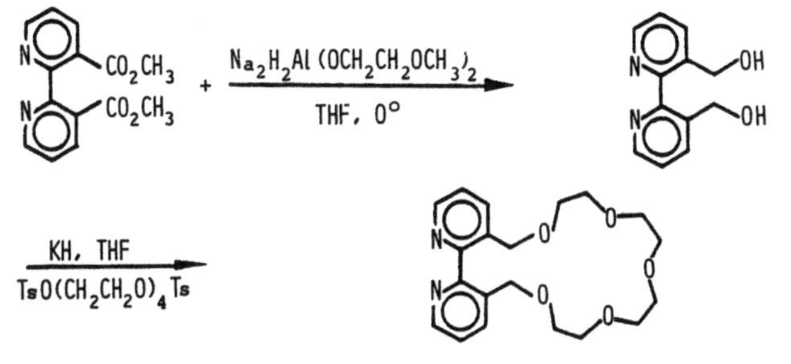

(3.43)

Schröder and Witt[190] have reported the synthesis of crown ethers having "fluctuating ring sizes" which they have termed "breathing crown ethers". The structures are based on the bullvalene subunit and, as the tetracyclic subunit undergoes Cope rearrangement, the size of the macroring likewise varies. The synthetic steps "follow the conventional routes used for the preparation of crown ethers"[190], and are illustrated in Eq. (3.44).

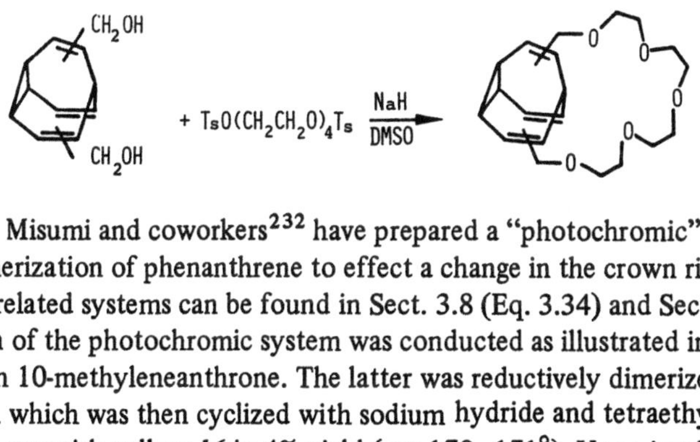

(3.44)

Misumi and coworkers[232] have prepared a "photochromic" crown using the photodimerization of phenanthrene to effect a change in the crown ring size. Examples of other, related systems can be found in Sect. 3.8 (Eq. 3.34) and Sect. 5.9 (Eq. 5.10). Construction of the photochromic system was conducted as illustrated in equation 3.45 starting with 10-methyleneanthrone. The latter was reductively dimerized to the ethylene-bridged diol which was then cyclized with sodium hydride and tetraethylene glycol ditosylate to give greenish-yellow *16* in 4% yield (mp 170–171°). Upon irradiation at an unspecified wavelength in benzene-d_6, the "photodimer" was obtained. The latter reverted to the former upon mild heating.

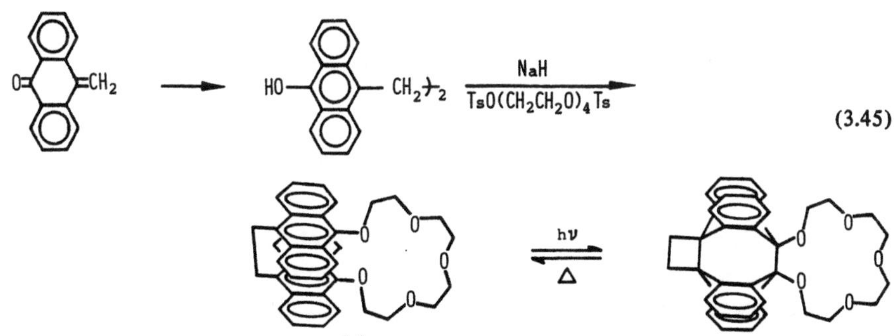

(3.45)

16

Two different all-aromatic crowns have been reported[55,104]. In 1975, de Jong, Siegel and Cram[55] reported the synthesis of a tri-binaphthyl system in which each oxygen was bound to a naphthalene ring, but aliphatic bridges were used to join the binaphthyl units. Relatives of this compound are discussed further in Sect. 3.13. The synthesis of this molecule (Structure *17*, below) was not simple, but was relatively straightforward. An interesting feature of it was the use of ethyl chloroacetate followed by LAH reduction to attach ethyleneoxy units to the naphthol unit.

The true, all-aromatic system (see *18*, below) described by Kime and Norymberski is unusual in the sense that all of the ether linkages bridge aromatic carbons[104]. Synthesis of *18*, therefore, required extensive use of copper mediated coupling reactions. As expected for such reactions, yields were generally low. The aromatics such as *18* were ineffective at binding either alkali metal or ammonium cations[104].

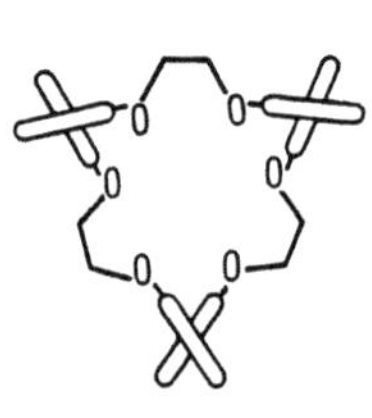

17

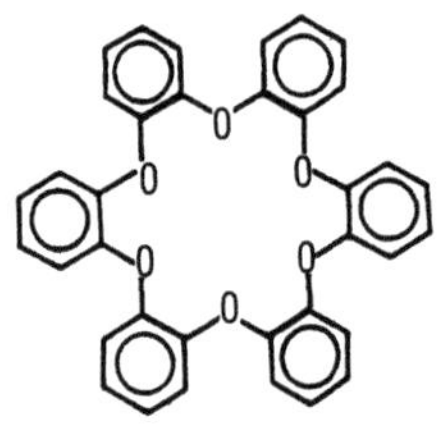

18

Finally, the 1,3-dione systems prepared by Cram and Alberts deserve special note[2,3]. These compounds, referred to as "hexahosts" are similar to the polymer-bound material illustrated as Compound *29* in Chap. 6. The synthesis is based on a methylene-bridged bis-dithiane unit. One of these may be cyclized with a polyethylene glycol, or more than one unit may be incorporated to give multiple 1,3-dione binding sites in the macrocycle. The former case is illustrated in Eq. (3.46).

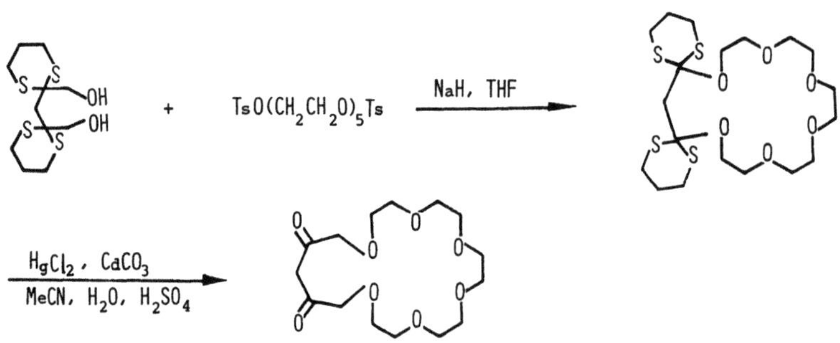

(3.46)

3.11 Newkome's Pyridine-containing Macrocycles

Newkome and his group have pioneered in the synthesis of aliphatic macrocycles containing the pyridine subcyclic unit[142–158]. Like the furan systems discussed in Sect. 3.6, these are macrocycles containing a potentially coordinating subcyclic unit whose donor ability will be determined by whether the heteroatom is turned inward or outward relative to the macroring. Newkome and coworkers have prepared numerous examples of both types and additional discussion is presented in Sects. 2.3, 4.11, 5.5 (esters containing the pyridyl subunit and 6.3), and a number of the compounds can be found in Tables 3.25, 3.26, 4.14, 4.16 and 4.18.

Although many of the crown-formation reactions are similar to those discussed in other sections of this chapter, we wish to direct special attention to Newkome's use of nucleophilic aromatic substitution reactions in the formation of C—O bonds. This approach is unusually accessible in pyridine chemistry since the heterocyclic unit is itself electron poor. An example is found in the double displacement on 2,6-dichloro-3-cyanopyridine, shown in Eq. (3.47). As in many of the cases discussed previously, the dianion of pentaethylene glycol is generated by reaction with sodium hydride in xylene. The choice of solvent was apparently dictated by the need for more thermal energy (as required by the aromatic electrophile) than could be provided by THF, the more common choice in reactions of this type. The crown (viscous oil) was isolated in 58% yield and less than 1% of the analog missing one ethyleneoxy unit was also isolated. The latter may arise from impure pentaethylene glycol, although the authors favor a fragmentation explanation[143].

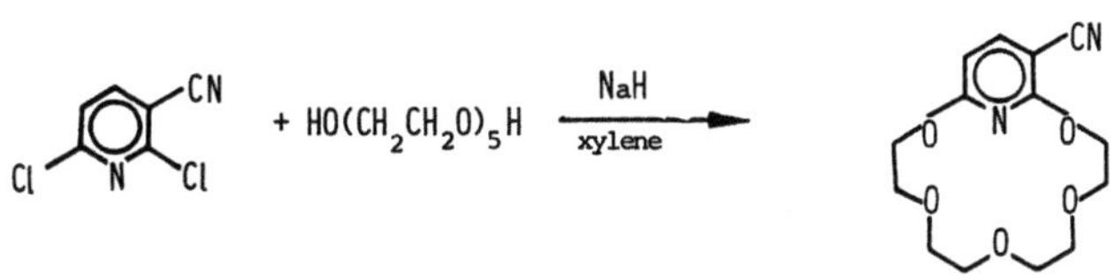

(3.47)

Other, closely related, nicotinic acid derivatives[143,144] and the unsubstituted system[153] itself have also been studied and undergo similar reactions. Moreover, the approach may be extended to 2,2′-bipyridyls. Newkome and his collaborators have used the 2,2′-bipyridyl unit (*19*) as an electrophile in which *ortho*-bromines served as leaving groups[149]. They have also used halomethyl systems and formed the macrocycles from these systems[148]. A compound derived from the latter starting material (*20*) is reported to form a cobalt complex, in which both nitrogens and only one of the oxygen atoms participate in the binding[148]. The two precursor units are shown below as *19* and *20*, respectively.

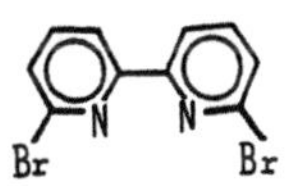

19

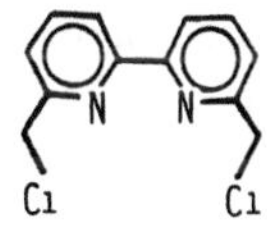

20

Pyrimidines have also served as electrophiles in crown synthesis from this group. 4,6-Dichloropyrimidine reacts with diethylene glycol and sodium hydride in anhydrous xylene solution to form the 20-crown-6 derivative as well as the other products shown in Eq. (3.48). Note that a closely related displacement on *sym*-trichlorotriazine has been reported by Montanari in the formation of "polypode" molecules (see Eq. 7.5).

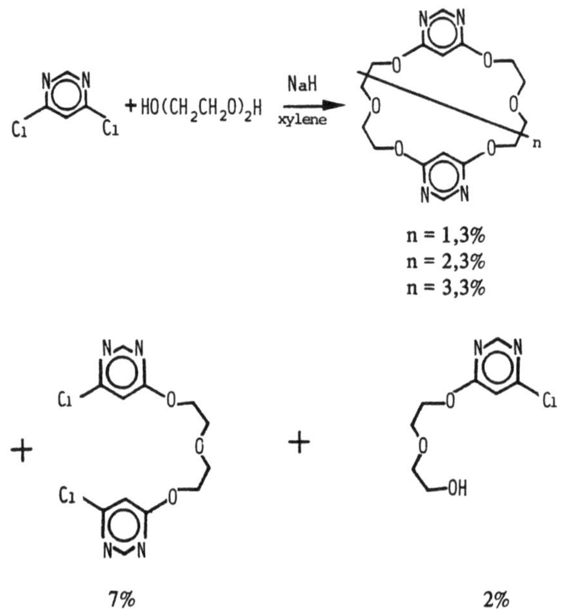

(3.48)

When 2-chloronicotinoyl chloride is allowed to react with pentaethylene glycol and sodium hydride in benzene solution, acylation followed by nucleophilic aromatic substitution occurs resulting in the formation of a macrocyclic lactone (48% yield) as shown in Eq. (2.8), in Chap. 2.

This group has also developed two ring-contraction systems of potential use in crown synthesis. In the first of these, extrusion of a phenylphosphine oxide unit results from treatment with alkoxide ion[142]. In the second, similar conditions initiated decarbonylation of a bis-pyridyl ketone[158]. Despite the apparent potential of these methods for crown synthesis, direct formation of crowns by processes which involve them do not appear to have enjoyed great success thus far.

3.12 Okahara's One-pot Approach

Okahara and his coworkers have pioneered a one-pot approach to crown synthesis which is also discussed in Sects. 2.3, 4.2, and 5.4. The reader is referred to those sections as well as to the information below. In the first report from this group[160], alkenes were treated with a halogenating agent such as N-bromosuccinimide in the presence of a polyethylene glycol. The bromohydrin is then purified and treated with base to effect ring closure. In a typical reaction, 1-hexene was treated with tetraethylene glycol and N-bromoacetamide. The bromohydrin, which was obtained in 62% yield, was then treated with sodium hydroxide in dioxane and the corresponding substituted 15-crown-5 was obtained in about 30% yield. Note that this reaction sequence[133,160], illustrated below as Eq. (3.49), is not strictly speaking, a one-pot synthesis. In principle, however, it might be and the related approaches described below do fall into this category.

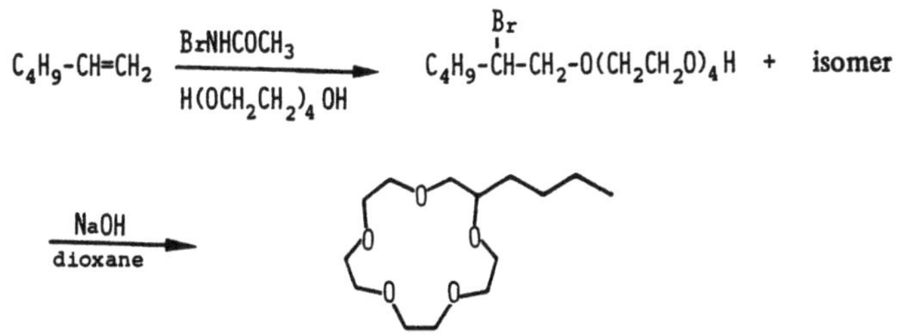

(3.49)

The first one-pot Williamson reaction which lead to a crown ether was reported in 1978 by Ping-Lin, Miki and Okahara[177]. In this approach, a diol is allowed to react simultaneously with a sulfonyl halide and base. The notion is that one end of the diol will be acylated and converted into a benzenesulfonate, tosylate or other leaving group. The remaining alcohol is then expected to deprotonate and attack the other end of the same molecule. In a typical reaction, pentaethylene glycol was treated with tosyl chloride and sodium hydroxide in dioxane at room temperature. The crown was obtained as the sodium tosylate complex and was freed by vacuum pyrolysis. 15-Crown-5 was isolated in 50% yield but detected (glpc) in 68% before isolation[177]. The corresponding preparation of 18-crown-6 is illustrated in Eq. (3.50). Note that the difficulty with this method is the requirement of penta- and hexaethylene glycols as precursors. Although a number of preparations have been published for these diols, their syntheses remain far from trivial.

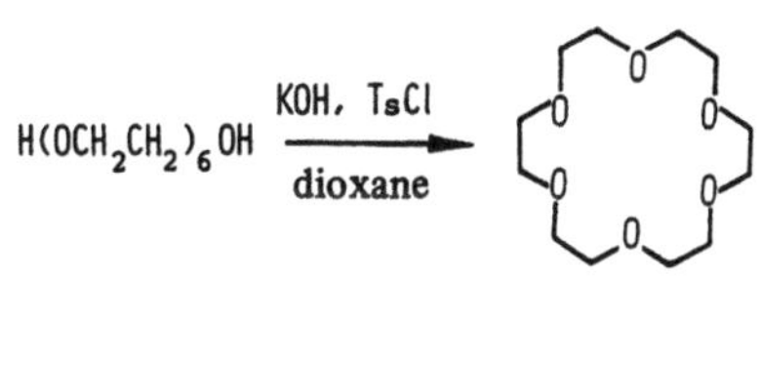

(3.50)

3.13 Chiral Crowns

A large number of chiral crowns have been prepared by numerous groups. The reader is directed to the tables at the end of this chapter to obtain an overview of these structures. It would not be useful to try to recount the synthetic approaches used in the preparation of all of these compounds; we have chosen rather to subdivide this mass of compounds into three principal groups. The groups are (1) Cram's chiral binaphthyl systems; (2) chiral crowns based on the tartaric acid unit; and (3) crowns incorporating sugar subunits. These are discussed in turn, below.

3.13.1 Cram's Chiral Binaphthyl Systems

Cram and his coworkers have pioneered the use of bis-binaphthyl crowns as chiral complexing agents for ammonium salts and amino acid salts. In these systems, the chiral binaphthyl unit provides a steric barrier within the macrocycle which allows discrimination of enantiomers. This work is remarkably extensive and can be found in Refs. 2, 3, 31, 36–38, 55, 59, 75, 76, 88–92, 105, 113–116, 135, 138–141, 165–167, 197, 198, 200, and 204–207. Our consideration of this work can really be broken into two parts: the formation of the macrorings and structural variation in the chiral barriers. These are considered in turn, below.

In Cram's first synthesis of a chiral bis-binaphthyl system[115], optically pure binaphthol and diethylene glycol ditosylate were heated at reflux in tetrahydrofuran solution for 15 h with potassium *t*-butoxide, two products were obtained. The "1 + 1" product (mp 230–231°) was isolated in 5% and the "2 + 2" product (mp 123–126°) was obtained in 31% yield. The reaction is shown in Eq. (3.51).

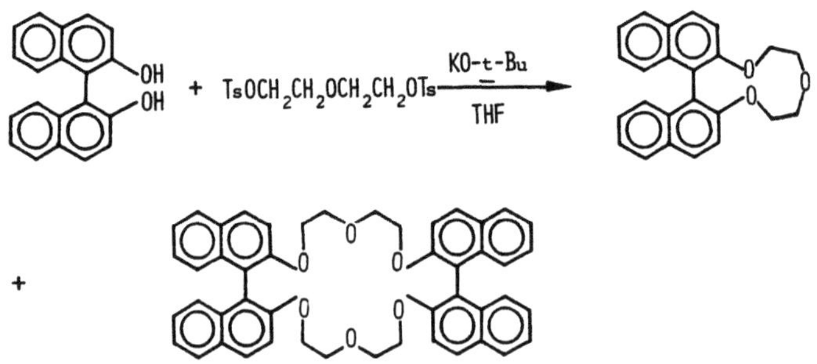

(3.51)

When phenolic nucleophiles were used, either potassium hydroxide or potassium *t*-butoxide was generally chosen as the base. When aliphatic hydroxyls constituted the nucleophiles, a stronger base was required and sodium hydride was generally chosen. An example of this is found in the synthesis of a crown containing a *meta*-xylylene unit (see also Sect. 3.5) and a "convergent" methoxyl group. 4-Methyl-2,6-bishydroxymethylanisole was treated with pentaethylene glycol and sodium hydride in THF solution to afford the crown (mp 71–73°) shown in Eq. (3.52) in 59% yield.

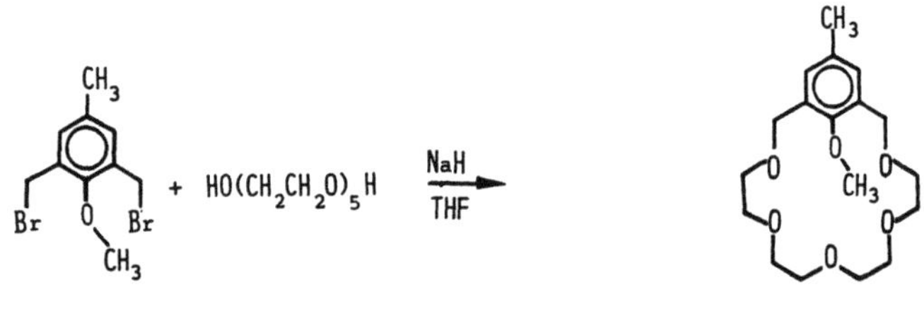

(3.52)

The structural variations reported by Cram and coworkers relate to an appreciable extent to the various ancillary functions which have been appended to the binaphthyl units or elsewhere in the macrocyclic system. Enhancements of the chiral barrier or functionalization through arms has generally been effected at the 3- or 6-positions. These positions are adjacent to the hydroxyl group or directly across the second ring from it, respectively.

Functionalization in the 3-position has been accomplished by a Mannich type condensation as shown in Eq. (3.53). The process is initiated by heating binaphthol with N-(*n*-butoxymethyl)morpholine for five days at 150°[90]. Under these conditions, a mixture of 3,3'-disubstituted and 3-monosubstituted product (61% and 15% respectively) is obtained. Further transformations are accomplished in the usual fashion. One particularly interesting interconversion involves heating the *bis*-morpholinomethyl compound with acetic anhydride at 140° for two weeks. Subsequent treatment of the acetate with lithium aluminum hydride affords the hydroxymethyl compound[90]. The latter may be converted into an "armed" crown by alkylation of the oxygen using ethyl bromoacetate.

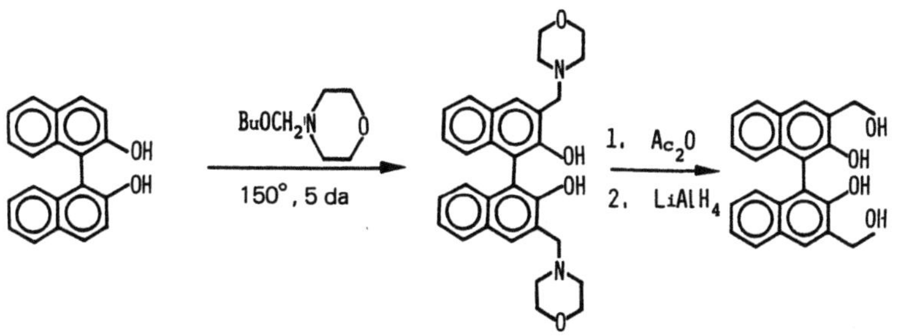

(3.53)

The six-position may be functionalized by electrophilic aromatic substitution. Either bromination (Br_2/CH_2Cl_2/-5°)[59] acetylation (acetyl chloride, aluminum chloride, nitrobenzene)[37] or chloromethylation (chloromethyl methyl ether, stannic chloride, -60°)[37] affords the 6,6'-disubstituted product. It should also be noted that treatment of the acetyl derivative with KOBr in THF affords the carboxylic acid in 84% yield. The brominated crown may then be metallated (n-BuLi) and treated with an electrophile to form a chain-extender. To this end, Cram has utilized both ethylene oxide[197] and dichlorodimethylsilane[59] in the conversion of bis-binaphthyl crowns into polymer-bound resolving agents. The acetylation/oxidation sequence is illustrated in Eq. (3.54).

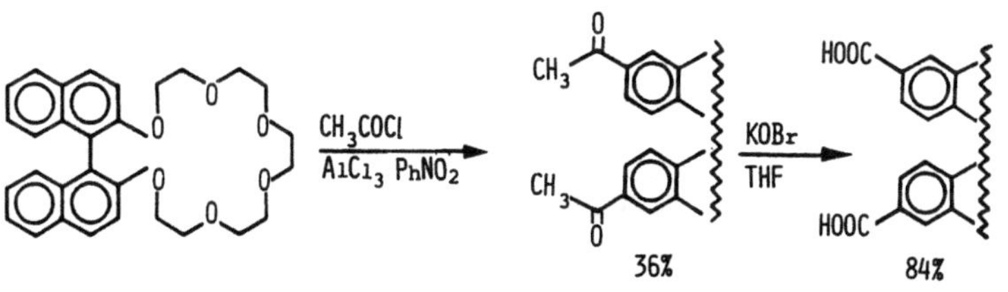

(3.54)

Further functionalization has been carried out in the mixed benzo-binaphtho crown shown in Eq. (3.55). Using 2-allylcatechol as starting material, the mixed crown was prepared in the usual fashion. The allyl group was isomerized to a propenyl substituent by treatment with potassium *t*-butoxide in a benzene/*t*-butanol mixture. Selective ozonolysis affords the aldehyde.

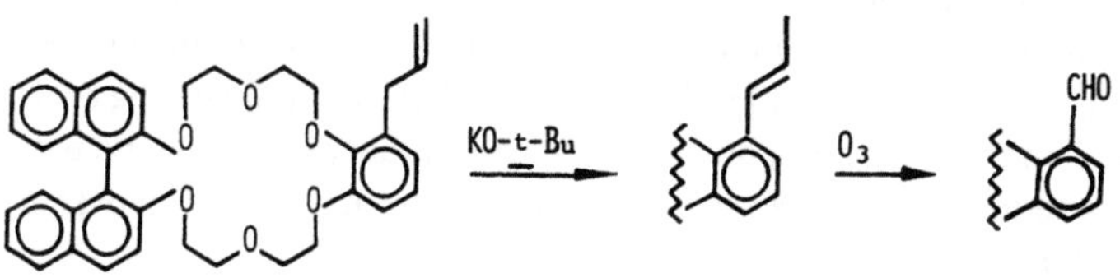

(3.55)

Just as selective oxidation can be carried out on these systems, reduction also occurs with considerable selectively. Hydrogenation of binaphthol (Pd catalyst) in glacial acetic acid at room temperature for seven days affords the octahydro (bis-tetrahydro) derivative in 92% yield with no apparent loss of optical activity when the reaction is conducted on optically pure material. The binaphthol may then be converted into the bis-binaphthyl crown in the usual fashion.

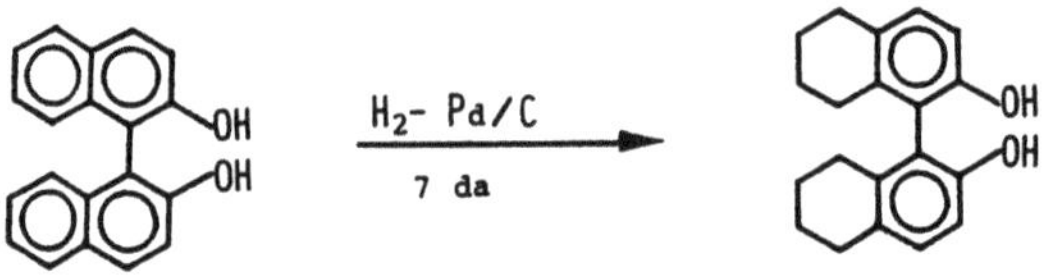

(3.56)

The principal variations on the normal crown synthesis methods were applied in preparing mixed crowns such as those shown in Eq. (3.55) and in forming isomers of the dibinaphthyl-22-crown-6 systems. The latter has been discussed in Sect. 3.5 (see Eq. 3.21)[75,76]. The binaphthyl unit was prepared to receive a non-naphthyl unit as shown in Eq. (3.57). Binaphthol was allowed to react with the tetrahydropyranyl ether or 2-chloroethoxyethanol. Cleavage of the THP protecting group followed by tosylation of the free hydroxyl afforded a two-armed binaphthyl unit which could serve as an electrophile in the cyclization with catechol. Obviously, the reaction could be accomplished in the opposite direction, beginning with catechol[116].

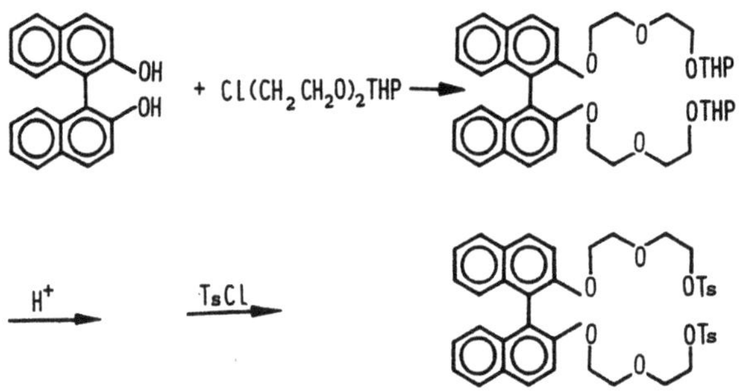

(3.57)

3.13.2 Crowns Incorporating the Tartaric or Lactic Acid Subunit

Although the crowns incorporating these subunits are quite numerous (see Refs. 6, 70, 122, 125, 126, 129, 165, 189, and 211) most rely on essentially the same methods for synthetic access. The nucleophiles are usually aliphatic alcohols and the base of choice has generally seemed to be sodium hydride. Electrophiles may be either halides or tosylates, but the latter have seemed to predominate. The difficulty, imagination and generality of these methods are demonstrated by the synthesis of chiral thiol-bearing crowns by Koga and his group[189] In this case, camphor provided the chirality for the steric barrier and catalytic thiols were introduced as a portion of the tartaric acid unit. The overall reaction sequence is shown in Eq. (3.58).

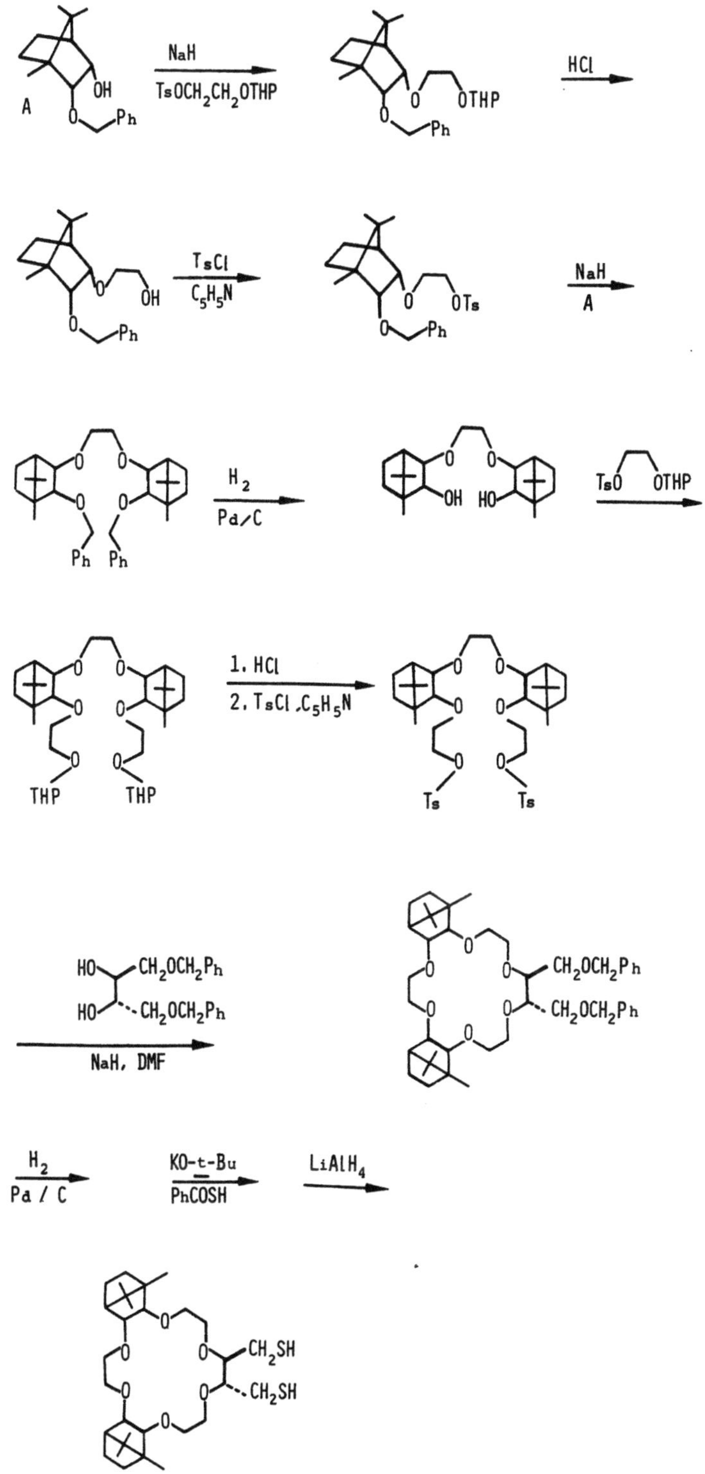
A
OH
O
Ph
NaH
TsOCH2CH2OTHP
OTHP
HCl
TsCl
C5H5N
OTs
NaH
A
H2
Pd/C
HO
TsO
THP
1. HCl
2. TsCl ,C5H5N
Ts
HO CH2OCH2Ph
HO CH2OCH2Ph
NaH, DMF
H2
Pd / C
KO-t-Bu
PhCOSH
LiAlH4
CH2SH

(3.58)

3.13.3 Crowns Incorporating Sugar Subunits

As in the previous categories in this section, there are numerous compounds which have been prepared based on a sugar subunit. Examples may be found in Refs. 7, 35, 42–45, 57, 82–85, 117–121, 175, 176, 193 and 208. Much of the work in these references has been reported by Stoddart and his coworkers, who have pioneered this field. As with the compounds prepared by Cram, the goal was to prepare a chiral receptor for ammonium ions which could be utilized in enzyme model studies.

Most of the cyclizations of aliphatic crowns reported in this volume have been conducted using sodium or potassium hydride as base, with the former predominating significantly. This is true for the sugar-based crowns, the principal variation being in choice of solvent from among THF, DMF and DMSO. In the first paper to report the incorporation of a sugar unit (mannitol)[42], the latter solvent was chosen. As usual, bridges were formed using diethylene glycol ditosylate as electrophile.

An interesting feature of the synthesis is the use of allyl as a two-carbon extension unit. This has been used in the stereospecific synthesis of dicyclohexano-18-crown-6 (see Eq. 3.13) and by Cram for formation of an aldehyde unit (see Eq. 3.55). In the present case, mannitol bis-acetonide was converted into its allyl ether which was ozonized (reductive workup) to afford the bis-ethyleneoxy derivative. The latter two groups were tosylated and the derivative was allowed to react with its precursor to afford the chiral crown. The entire process is shown below in Eq. (3.59).

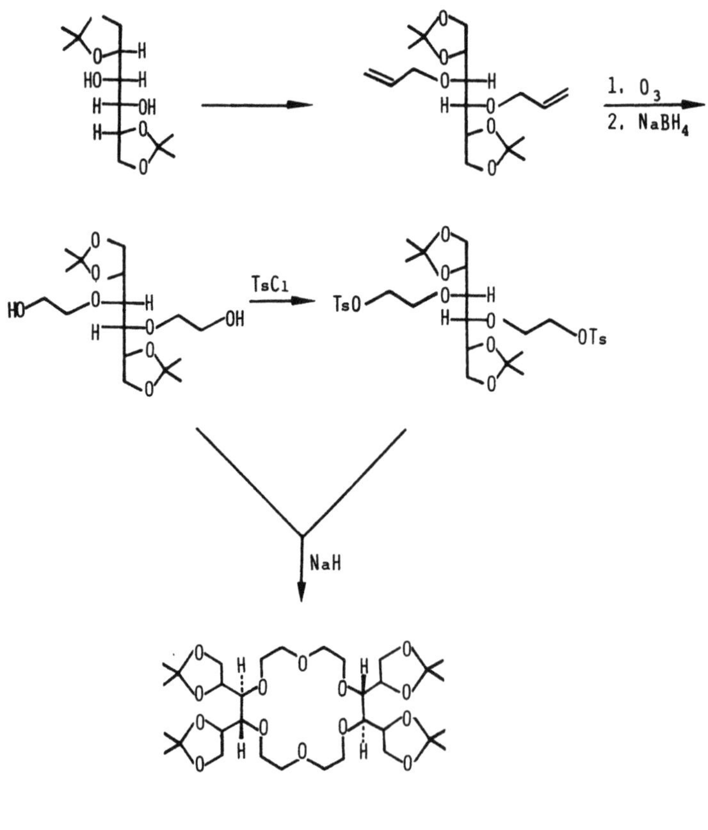

(3.59)

3.14 Organometallic Derivatives of Crowns

Crowns are well-known to complex alkali metal and alkaline earth metal cations and in this sense readily form organometallic complexes. Pannell, Hambrick and Lewandos[162] have prepared a true organometallic complex by irradiating (UV) dibenzo-18-crown-6 in the presence of chromium hexacarbonyl and isooctane. In this process, the chromium hexacarbonyl residue attaches to one or both aromatic rings yielding the π-arene complexes. As expected, the binding constants for both sodium and potassium cations are reduced by the substitution of the electron-withdrawing chromium tricarbonyl group. Binding constants are reduced more for the disubstituted crown than for the monosubstituted species. In the former case, cis-trans isomerism could have a significant effect on the binding constant but at the time of the report, the stereochemistry was unknown. The monosubstituted crown is illustrated below as *23*.

Bell and Hall[16] have incorporated an organometallic unit into a crown by using the ferrocenyl unit as part of the ring or as a third strand. The unit is incorporated either as the 1,1′-diformylferrocene or the corresponding acid. In the former case, the bis-imine is prepared and reduced to give the saturated crown (see structure *24*). In the latter case, the acid is converted into its corresponding chloride and thence into the diamide by reaction with a diamine. Diborane reduction affords the saturated amino-crown. Structure *24* could be prepared by either of these methods but the dialdehyde approach was reported to be poor compared to the amide approach which afforded the product in "ca. 60% yield"[16].

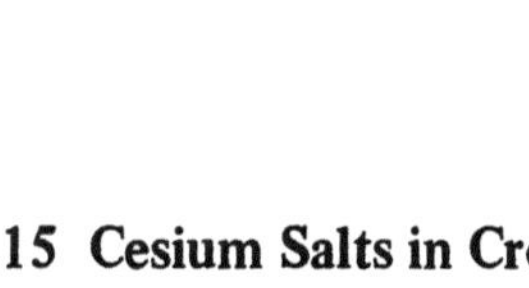

23

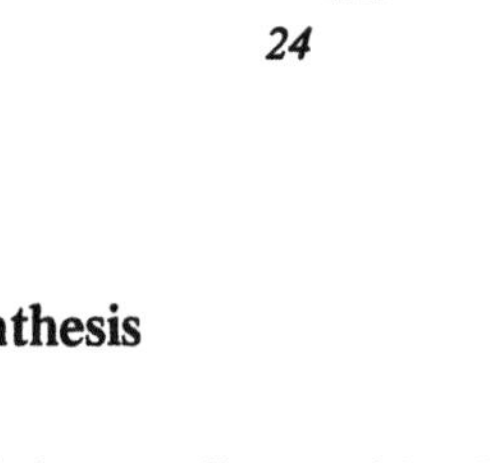

24

3.15 Cesium Salts in Crown Synthesis

In 1978, Piepers and Kellogg reported that excellent yields of macrocyclic lactones could be realized using cesium carboxylates (see Sect. 5.5 and Eq. 5.6). The following year, van Keulen, Kellogg and Piepers[217] as well as Reinhoudt, de Jong, and Tomassen[183] disclosed extensions of this method from macrocyclic lactones to macrocyclic ethers. Kellogg's group reported a variety of results with $CsCO_3$ as base[217] and the latter group surveyed a variety of metallic fluorides with special emphasis on the properties of CsF_2[183].

Using cesium carbonate in methanol, Kellogg and coworkers were able to convert catechol and pentaethylene glycol into benzo-18-crown-6 in 74% yield. They found that the cesium salts in these reactions were usually equal or superior to the potassium salts.

Further, they said: "In these reactions, the yields of the products increase with the increase in the ring size, although this has not been established as a general rule"[217].

The most demanding test of cesium carbonate as base was with 2,3-dihydroxypyridine (3-hydroxypyridone)[217]. The cesium salt was found to be fairly unstable, apparently oxidizing quite rapidly. Model reactions suggested that alkylation would occur 1,3 (N, O) to give the substituted pyridone. Nevertheless, on the basis of UV and ^{1}H-nmr analysis, the product of reaction between 2,3-dihydroxypyridine and tetraethylene glycol dibromide was assigned as the pyridocrown (23% yield, mp 77–78.5°) as shown in Eq. (3.60).

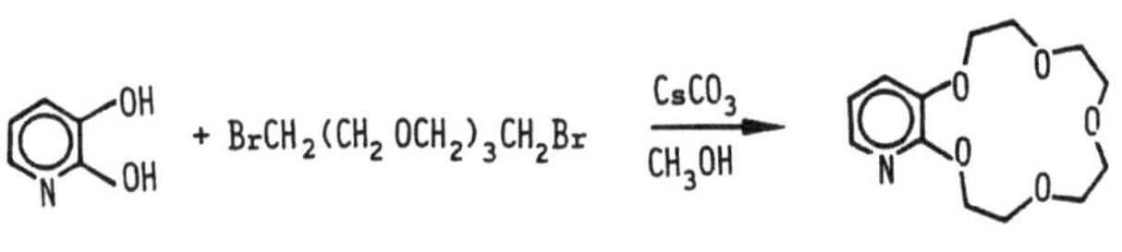

(3.60)

Reinhoudt, de Jong and Tomassen have explored the use of metallic fluorides as bases in the Williamson ether synthesis of crowns. They found that the efficacy order for the metal cations they examined was $Cs^+ > Rb^+ \geqslant K^+ \gg Na^+ \sim Li^+$. This order was determined for a model reaction involving catechol and pentaethylene glycol ditosylate in acetonitrile solution which affords benzo-18-crown-6 as product. The yields in this reaction ranged from 52%–67%. When tetrabutylammonium fluoride was substituted for a metallic fluoride salt, the yield was found to be only 5%. The authors noted that the yields of even the smaller crown ethers such as benzo-15-crown-5 and benzo-18-crown-6 were higher with cesium and rubidium fluoride than with potassium fluoride, and noted that "this seems to be in contrast with the accepted picture of the template cation ideally fitting into the cavity of the crown ether that is formed"[183]. This may not be so surprising considering the evidence presented in Sect. 2.3 (q.v.).

Although cesium salts appear to be excellent reagents for the formation of crown ethers, it should be noted that such salts are used at least in equivalent amounts[217] or in excess (4 eq.)[183]. These salts are relatively expensive and will probably be reserved for use in the synthesis of less readily accessible crowns.

3.16 Listing of Tables

3.17 Tables 3.1–3.29

Table 3.1. Derivatives of 12-crown-4

Cpd. No.	No. C's	Ring size	Structure		Mp or (bp) in °C	Yield (%)	Refs.
1	8	12	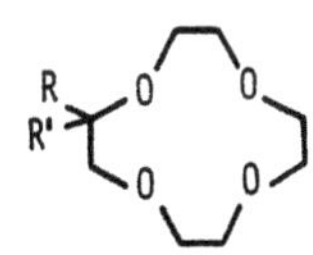	R = R′ = H	(67–70/0.05)	13	34
2	10	12		R = H, R′ = Et	nr	15	133
3	10	12		R = R′ = Me	(64/0.09)	12	133
4	10	12		R = Me, R′ = CH_2Cl	nr	14	133
5	11	12		R = Me, R′ = Et	(63/0.06)	23	133
6	12	12		R = Me, R′ = n-Pr	(71/0.09)	21	133
7	12	12	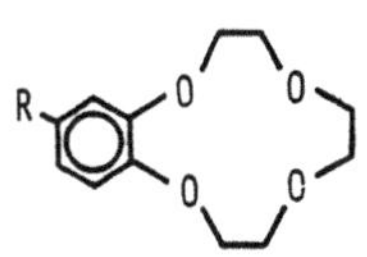	R = H	44–45.5	4	169, 173
8	13	12		R = Me	46 – 48	15	230
9	13	12		R = CH_2Br	83–85	70	230

Table 3.1 (continued)

Cpd. No.	No. C's	Ring size	Structure	Mp or (bp) in °C	Yield (%)	Refs.
10	20	12	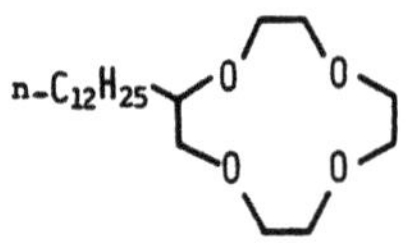	nr	nr	159
11	20		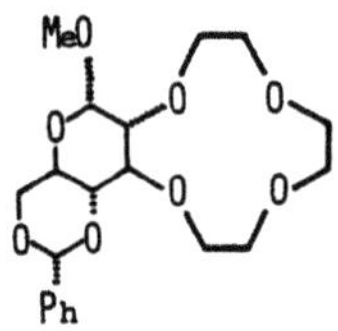	134	13	96
12	52	12	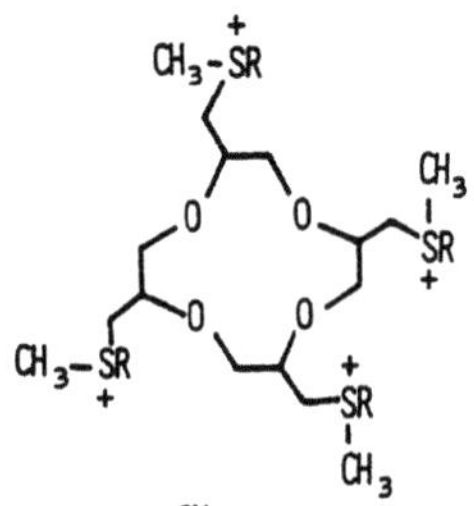 R = $(CH_2-CH(OH)-CH_2O)_3H$	nr	nr	124
13	64	12	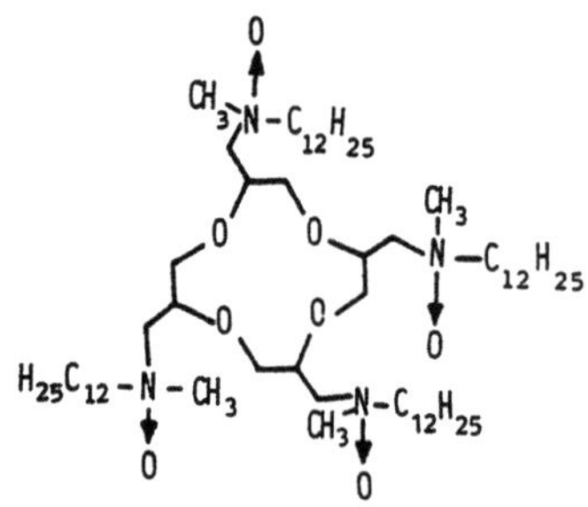	nr	nr	124
14	72	12	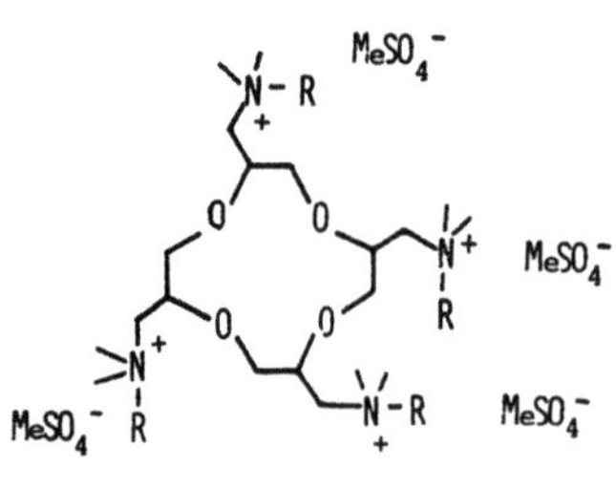 R = $C_{12}H_{25}$	nr	nr	124

Table 3.1 (continued)

Cpd. No.	No. C's	Ring size	Structure	Mp or (bp) in °C	Yield (%)	Refs.
15	84	12	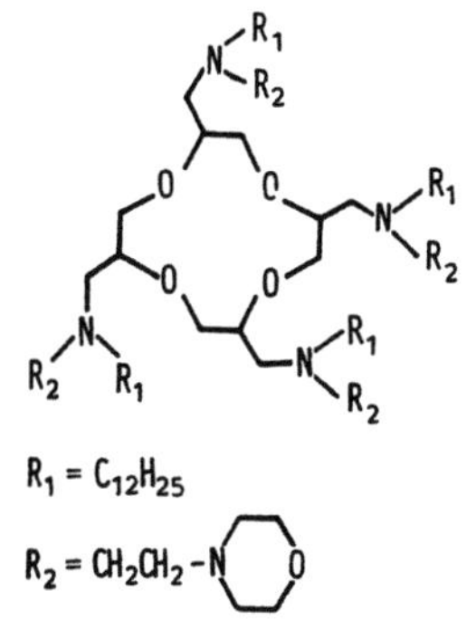	nr	nr	124
16	84	12	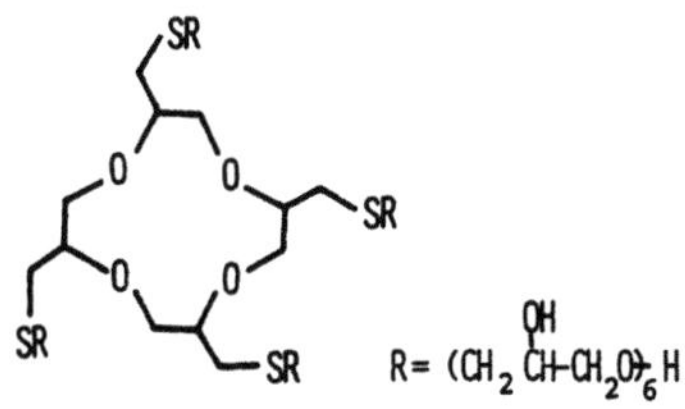	nr	nr	124
17	88	12	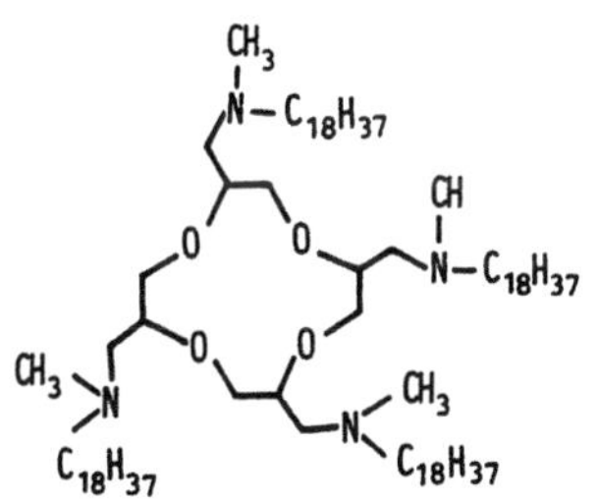	nr	nr	124
18	88	12	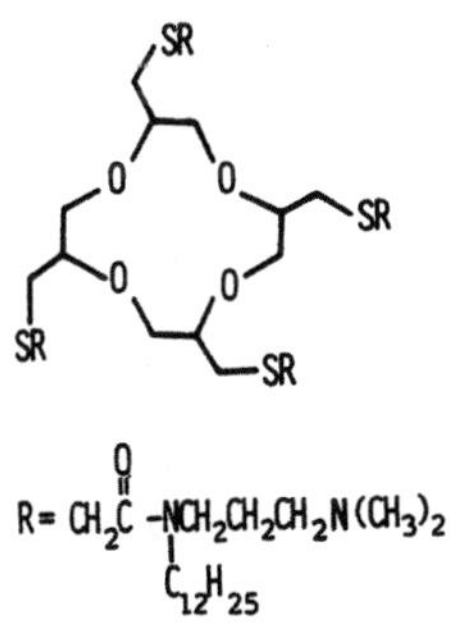	nr	nr	124

Table 3.1 (continued)

Cpd. No.	No. C's	Ring size	Structure	Mp or (bp) in °C	Yield (%)	Refs.
19	92	12	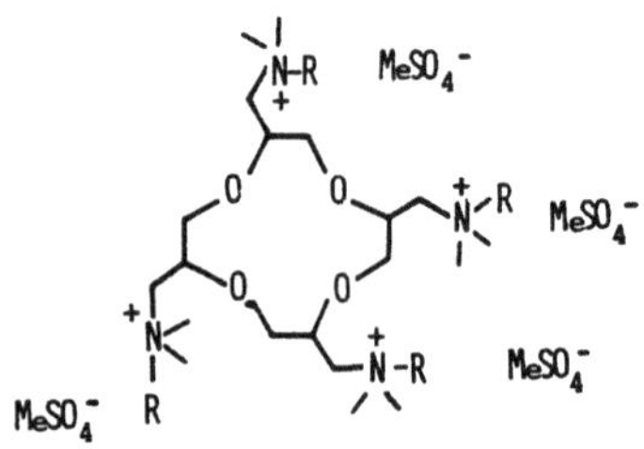 R= $C_{18}H_{37}$	nr	nr	124

Table 3.2. Derivatives of 15-crown-5

Cpd. No.	No. C's	Ring size	Structure	Mp or (bp) in °C	Yield (%)	Refs.
1	10	15	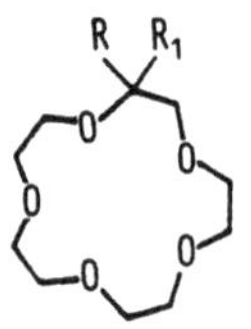 R = R' = H	(78/0.05)	14.2	34
				oil	20	52
				nr	38.7	98
				nr	nr	172
				nr	50	177
				nr	45	210
2	11	15	R = H, R' = CH_3	nr	41	177
3	11	15	R = H, R' = CH_2OH	oil	50	74
				oil	100	47
4	12	15	R = H, R' = CH_2OCH_3	oil	23	74
5	12	15	R = R' = CH_3	(89–91/0.05)	38	133
6	12	15	R = Me, R' = CH_2Cl	nr	25	133
7	13	15	R = H, R' = $CH_2OC(=O)CH_3$	oil	68	71
8	13	15	R = Me, R' = Et	(98–100/0.04)	39	133
9	14	15	R = H, R' = n-Bu	nr	12	133
				(120–125/0.20)	32	159, 160
10	14	15	R = H, R' = CH_2OSiMe_3	oil	68	71

Table 3.2 (continued)

Cpd. No.	No. C's	Ring size	Structure	Mp or (bp) in °C	Yield (%)	Refs.
11	14	15	R = H, R′ = $CH_2OCH_2CH_2CH_3$	nr	82	71
12	14	15	R = H, R′ = $CH_2OCH_2CH = CH_2$	nr (134–140/0.05)	41 34	177 74
13	14	15	R = H, R′ = $CH_2OCH_2CHOHCH_3$	oil	95	71
14	14	15	R = H, R′ = $CH_2OCH_2CH_2OCH_3$	oil	65	71
15	14	15	R = R′ = Et	(108–110/0.05)	22	133
16	14	15	R = Me, R′ = n-Pr	(103–105/0.10)	43	133
17	15	15	R = H, R′ = $CH_2O(CH_2)_3CH_3$	oil	64	71
18	15	15	R = H, R′ = $CH_2OC(CH_3)_3$	oil	46	74
19	15	15	R = H, R′ = $CH_2O\overset{O}{\overset{\Vert}{C}}(CH_2)_2CH_3$	oil	62	71
20	16	15	R = H, R′ = $(CH_2)_5CH_3$	nr (130–134/0.10)	35 29	133 159, 160
21	16	15	R = H, R′ = Ph	nr	nr	159
22	17	15	R = H, R′ = $CH_2O(CH_2)_2O(CH_2)_3CH_3$	oil	55	74
23	17	15	R = H, R′ = CH_2OPh	oil	32	74
24	18	15	R = H, R′ = $(CH_2)_7CH_3$	nr	6	133
25	18	15	R = H, R′ = CH_2OCH_2Ph	oil	62	74
26	18	15	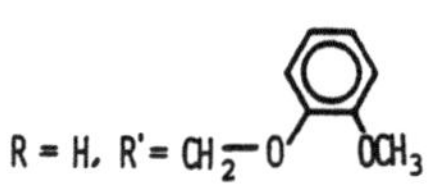	oil	70	74
27	18	15	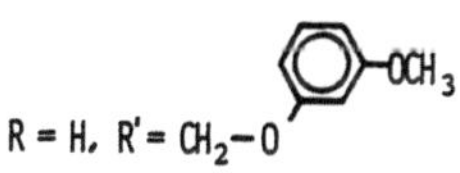	oil	57	71
28	18	15	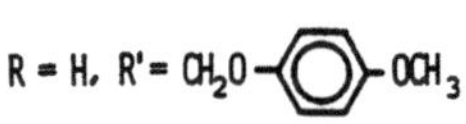	oil	29	74
29	18	15	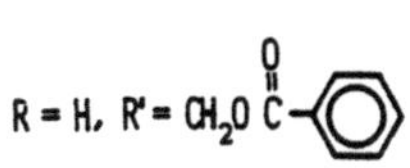	oil	80	74
30	18	15	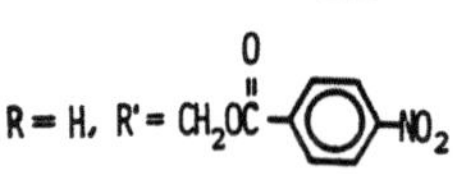	oil	78	71

Table 3.2 (continued)

Cpd. No.	No. C's	Ring size	Structure	Mp or (bp) in °C	Yield (%)	Refs.
31	19	15	R = H, R' = 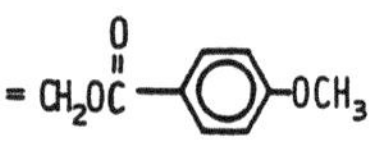	oil	74	71
32	20	15	R = H, R' = $(CH_2)_9CH_3$	nr	30	133
33	20	15	R = H, R' = 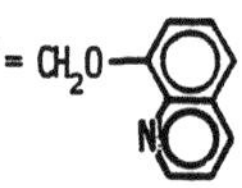	oil	57	71
34	21	15	R = H, R' = 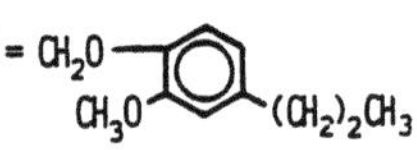	oil	99	71
35	21	15	R = H, R' = 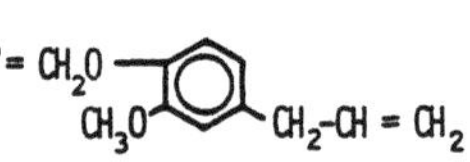	oil	63	71
36	21	15	R = H, R' = 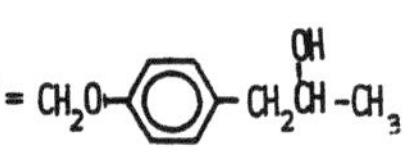	oil	nr	71
37	21	15	R = H, R' = 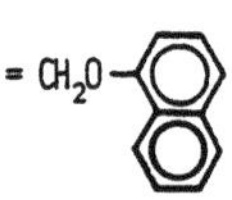	oil	nr	58
38	22	15	R = H, R' = $(CH_2)_{11}CH_3$	nr	36	133
39	27	15	R = H, R' = $CH_2O(CH_2)_{15}CH_3$	oil	84	71
40	27	15	R = H, R' = $CH_2OC(O)(CH_2)_{14}CH_3$	oil	nr	71
41	38	15	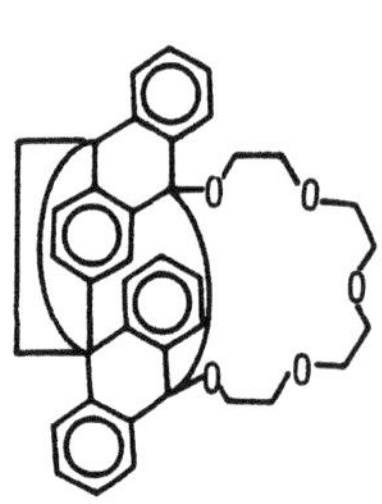	nr	nr	233

Table 3.3. Derivatives of 18-crown-6

Cpd. No.	No. C's	Ring size	Structure	Mp or (bp) in °C	Yield (%)	Refs.
1	12	18	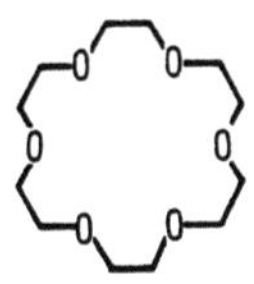	39	33	52
				36.5–38	~20	72
				36.5–38	~20	73
				39–40	30–60	81
				38–38.5	30	98
				(116/0.2)	30	98
				39–40	2	169, 172
				nr	75	177
				nr	32	210
2	13	18	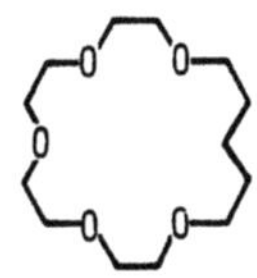	(138–140/0.15)	26	207
3	13	88	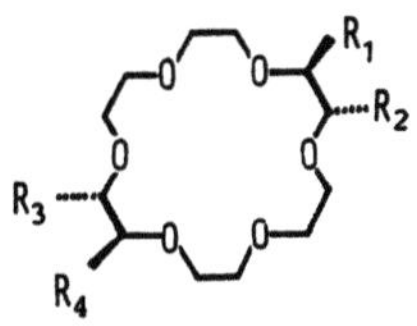 $R_1 = CH_3$ $R_2 = R_3 = R_4 = H$	nr	53	177
4	13	18	$R_1 = CH_2OH, R_2 = R_3 = R_4 = H$	oil	nr	47
				nr	nr	134
5	14	18	$R_1 = R_2 = CH_2SH, R_3 = R_4 = H$	nr	nr	128
				oil	30	129
6	16	18	$R_1 = R_2 = R_3 = R_4 = CH_2OH$	oil	65	43
				nr	4	17
7	16	18	$R_1 = R_2 = R_3 = R_4 = CO_2H$	211	nr	15
8	18	18	$R_1 = (CH_2)_5CH_3, R_2 = R_3 = R_4 = H$	nr	nr	159
9	18	18	$R_1 = R_2 = (CH_2)_3SH, R_3 = R_4 = H$	nr	nr	128
				oil	1.5	129
10	18	18	$R_1 = R_2 = CH_2O(CH_2)_2SH, R_3 = R_4 = H$	nr	nr	128
				oil	10	129
11	18	18	$R_1 = R_2 = CH_2O\overset{O}{\overset{\|}{C}}CH_2, R_3 = R_4 = H$	nr	16	6
12	20	18	$R_1 = R_2 = R_3 = R_4 = CHOHCH_2OH$	58–60	5	42
13	20	18	$R_1 = R_2 = R_3 = R_4 = CHOHCH_2OH$	69–71	90	43

Table 3.3 (continued)

Cpd. No.	No. C's	Ring size	Structure	Mp or (bp) in °C	Yield (%)	Refs.
14	20	18	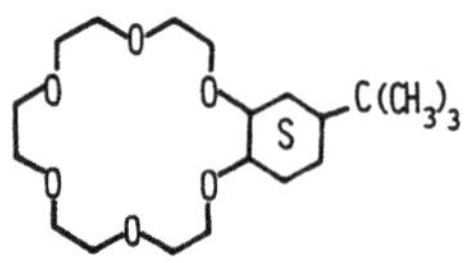	< 26	12	169, 173
15	20	18	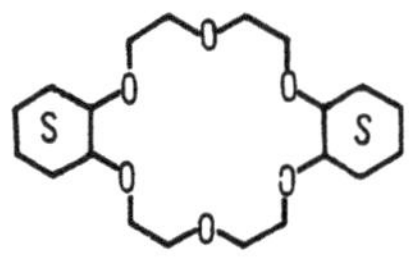 (cis-syn-cis)	68.5–69.5 61–62	45–67 nr	169, 171, 173 97
16	20	18	(cis-anti-cis)	83–84	nr	97
17	20	18	(trans-syn-trans)	120–121 120–121	nr 6	30 199
18	20	18	(trans-anti-trans)	77–80 77–80 77–80	31 14 nr	87 199 30
19	22	18	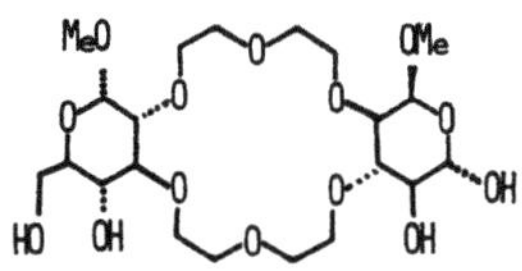	132–134	nr	208
20	22	18	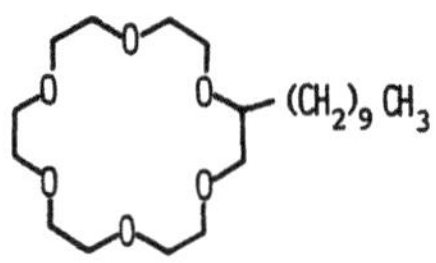	nr	28.6	33
21	24	18	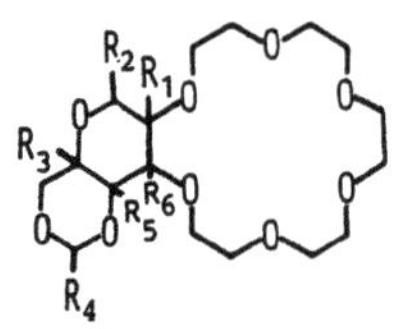 R_1 = ---H R_2 = ---OMe R_3 = ---H R_4 = ---Ph R_5 = –H R_6 = –H	nr	nr	176
22	24	18	R_1 = –H R_2 = ---OMe R_3 = ---H R_4 = –Ph R_5 = ---H R_6 = ---H	115	16	118
23	24	18	R_1 = –H R_2 = –OMe R_3 = ---H R_4 = –Ph R_5 = ---H R_6 = ---H	141–143	20	175

Table 3.3 (continued)

Cpd. No.	No. C's	Ring size	Structure	Mp or (bp) in °C	Yield (%)	Refs.
24	24	18	R_1 = –H R_2 = ---OMe R_3 = ---H R_4 = ---Ph R_5 = –H R_6 = ---H	52–56	45	11
25	24	18	R_1 = –H R_2 = –OMe R_3 = ---H R_4 = ---Ph R_5 = –H R_6 = ---H	78–80	46	175
26	24	18	R_1 = ---H R_2 = ---OMe R_3 = ---H R_4 = –Ph R_5 = –H R_6 = ---H	nr	nr	176
27	24	18	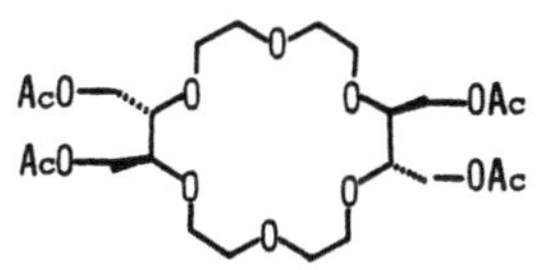	81 69–74	89 nr	43 17
28	24	18	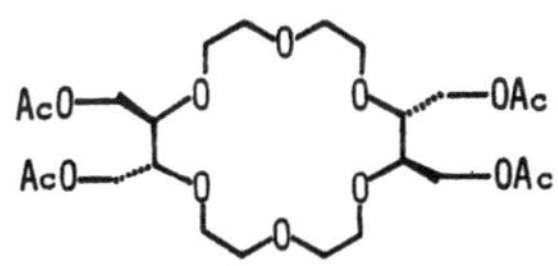	nr	nr	17
29	24	18	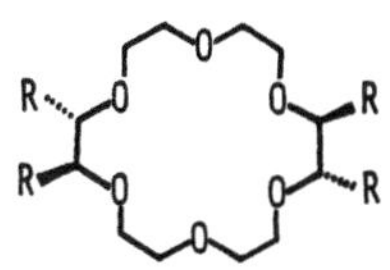 R = $\overset{O}{\overset{\parallel}{C}}NH(CH_2)_2NH_2$	174–177	66	14
30	24	18	R = $\overset{O}{\overset{\parallel}{C}}N(CH_3)_2$	186	20	70
31	26	18	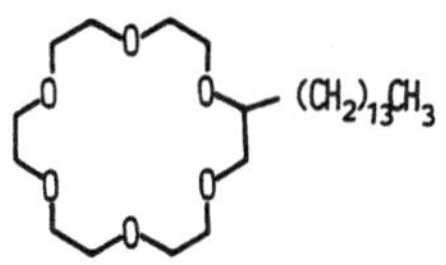	nr	18	33

Table 3.3 (continued)

Cpd. No.	No. C's	Ring size	Structure	Mp or (bp) in °C	Yield (%)	Refs.
32	28	18	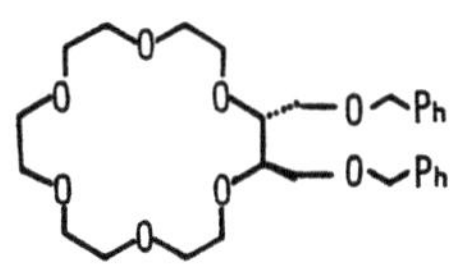	oil syrup	26 nr	6 193
33	28	18	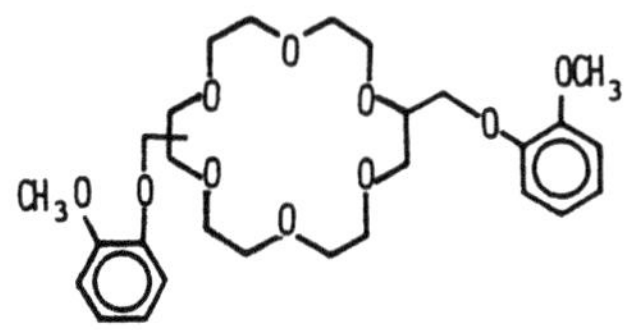	oil	33	71
34	28	18	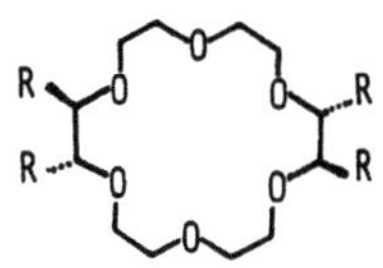 R = 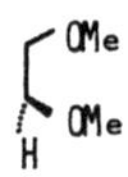	nr	56	43
35	28	18	R = 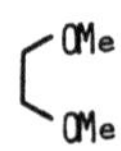	nr	nr	17
36	28	18	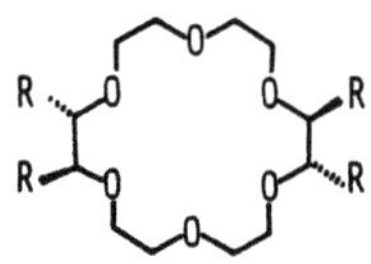 R = $\overset{O}{\|\|}{C}NHCH_2CO_2Me$	188	nr	15
37	28	18	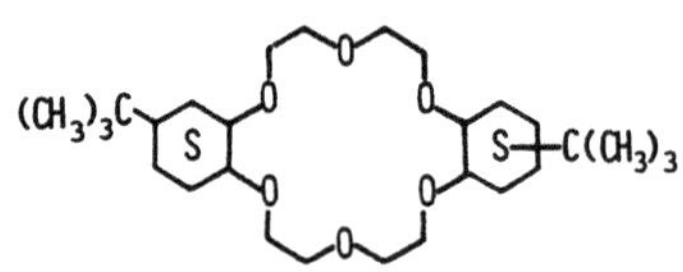	< 26	52	169
38	28	18	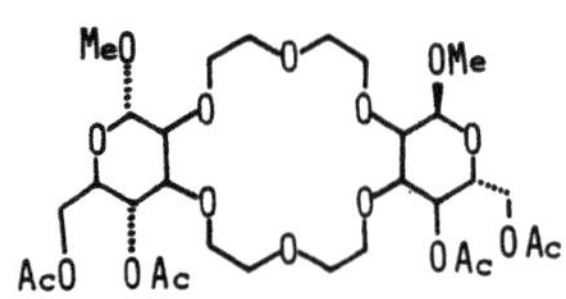	161–164	nr	208
39	29	18	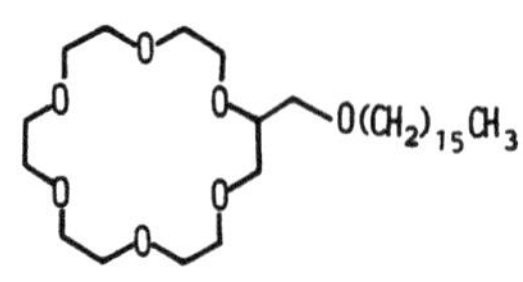	38–39	nr	134

Table 3.3 (continued)

Cpd. No.	No. C's	Ring size	Structure	Mp or (bp) in °C	Yield (%)	Refs.
40	30	18	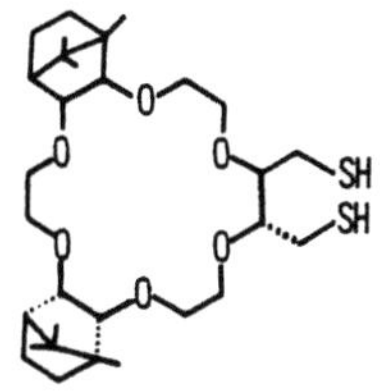	nr	nr	189
41	32	18	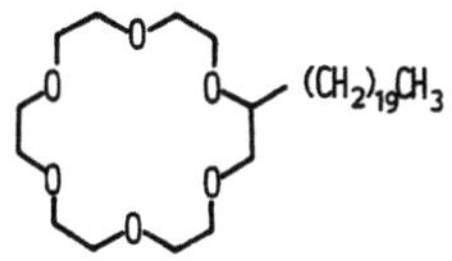	41–42	14.5	33
42	32	18	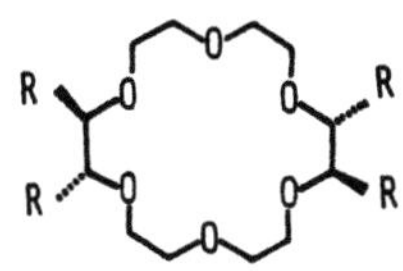 R = (LL) 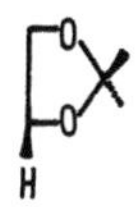	60–66	3.3	45
43	32	18	R = (DD) 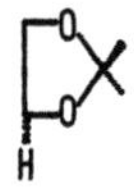	oil	24	43
44	32	18	R = (DD) 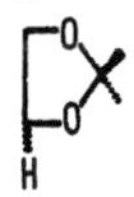	nr	nr	45
45	32	18	R = 	oil	5	17
46	32	18	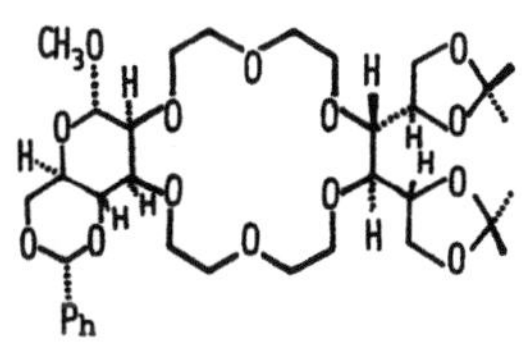	nr	34	176
47	32	18	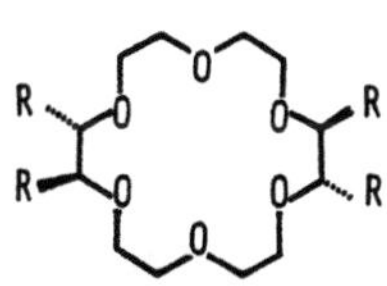 $R = \overset{O}{\overset{\|}{C}}NH-\underset{CH_2SH}{\underset{\mid}{CH}}-CO_2Me$	204–205	66	122

Table 3.3 (continued)

Cpd. No.	No. C's	Ring size	Structure	Mp or (bp) in °C	Yield (%)	Refs.
48	34	18	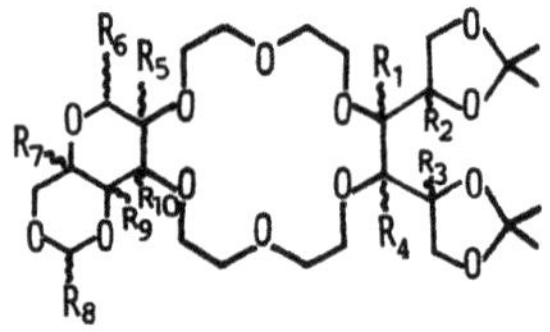 (DD) $R_1 = R_3 = R_5 = R_9 = -H$ $R_2 = R_4 = R_7 = R_{10} =$ ---H $R_6 = -OMe$, $R_8 =$ ---Ph	(240/0.05)	43	175
49	34	18	$R_1 = R_3 = R_5 = -H$ (DD) $R_2 = R_4 = R_7 = R_9 = R_{10} =$ ---H $R_6 = -OMe$, $R_8 = -Ph$	nr	15	175
50	34	18	$R_1 = R_3 = R_5 = -H$ (DD) $R_2 = R_4 = R_7 = R_9 = R_{10} =$ ---H $R_6 = H$, $R_8 = -Ph$	nr	29	118
51	34	18	$R_1 = R_3 = R_5 = R_9 = -H$ $R_2 = R_4 = R_7 = R_{10} =$ ---H $R_6 =$ ---OMe, $R_8 =$ ---Ph (DD)	44–46	40	118
52	36	18	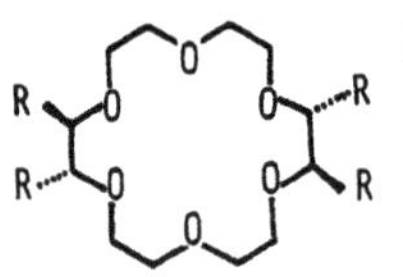 R = CH(OAc)CH$_2$OAc (DD)	oil	95	43
53	36	18	R = CH(OAc)CH$_2$OAc	nr	nr	17
54	36	18	R = 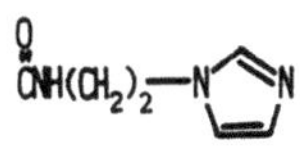	223	65	14
55	36	18	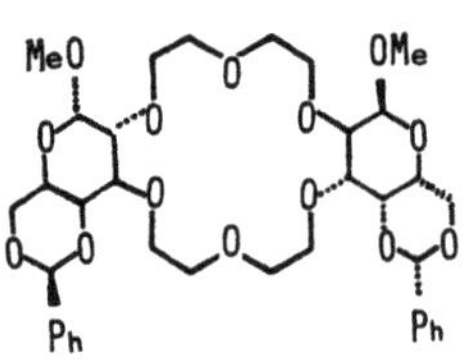	168–170 dec.	12	82
56	36	18	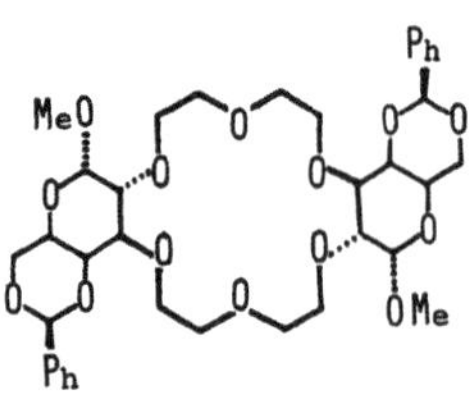	240 dec.	15	82

Table 3.3 (continued)

Cpd. No.	No. C's	Ring size	Structure	Mp or (bp) in °C	Yield (%)	Refs.
57	36	18	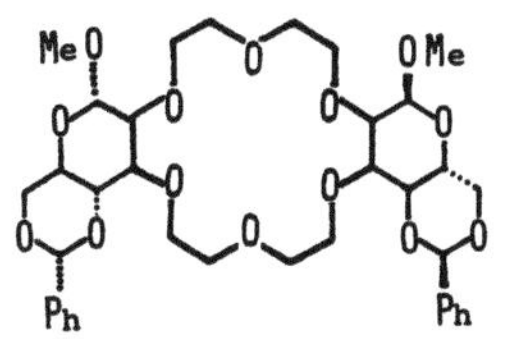	248–254 dec.	15	82
58	36	18	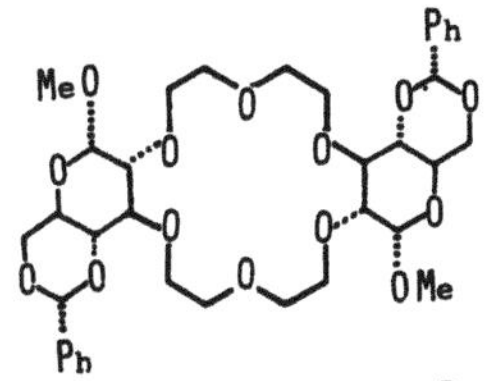	229–234 dec. nr	8 13	82 82
59	36	18	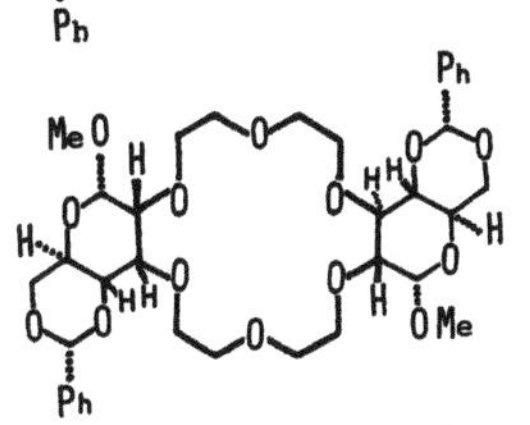	229–230	nr	121
60	36	18	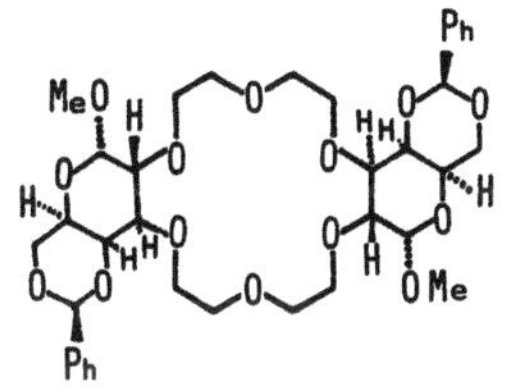	238–240	nr	121
61	36	18	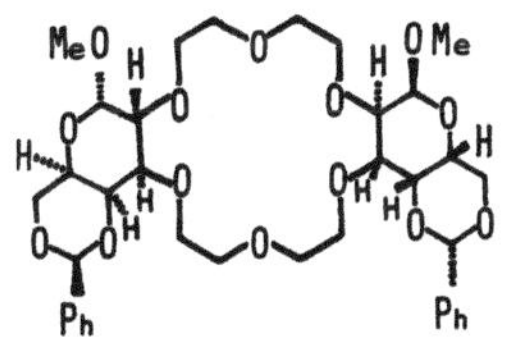	167–167.5	nr	121
62	36	18	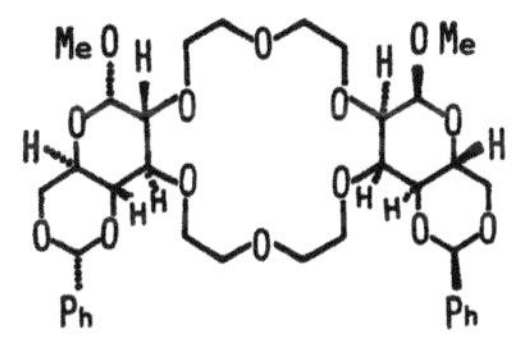 (DD)	258–261 233–234	15 nr	208 121
63	40	18	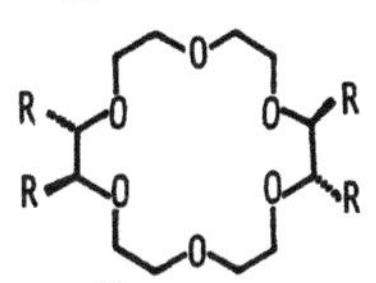 R = $\overset{O}{\overset{\|}{C}}NHCH_2CO_2^-NMe_4^+$	nr	nr	15

Table 3.3 (continued)

Cpd. No.	No. C's	Ring size	Structure		Mp or (bp) in °C	Yield (%)	Refs.
64	42	18	$R = CH_2OCH_2Ph$		oil	6	17
65	44	18	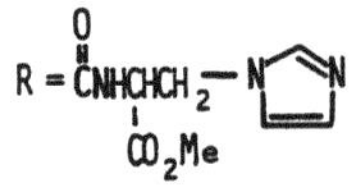	(L)	135–138 145–150	nr 70	15 211
66	44	18	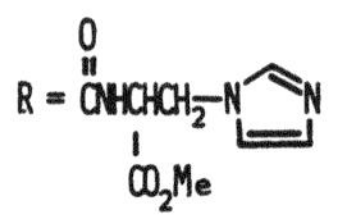	(LD) (DD)	145–150 135–138	70 77	205 211
67	44	18	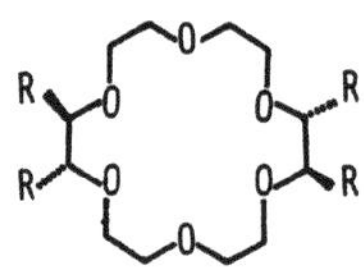	(LL)	135–138	65	211
			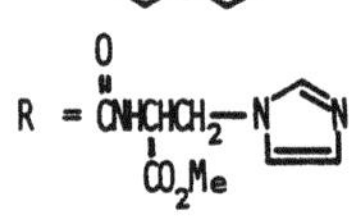	(L)			
68	48	18	$R = CH_2OCH_2Ph$		oil	11	43
69	56	18	$R = \overset{O}{\overset{\Vert}{C}}NH(CH_2)_2NHCO_2CH_2Ph$		214–215	70	14
70	56	18	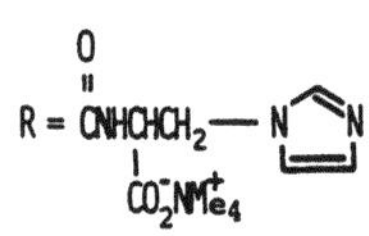	(L)	nr	nr	15
71	60	18	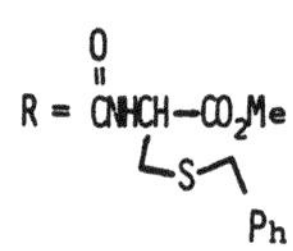		139–140	87	122
72	64	18	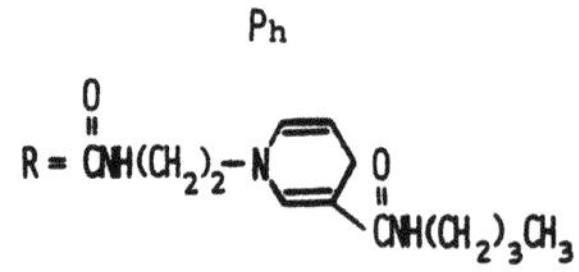		nr	nr	14
73	64	18	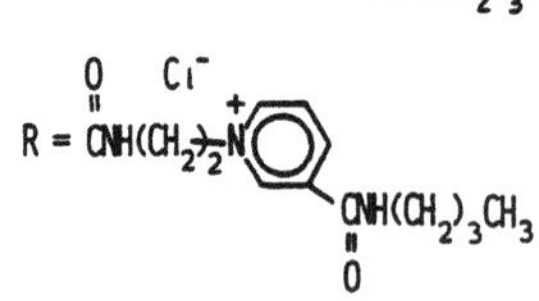		hygroscopic solid	33.3	14
74	80	18	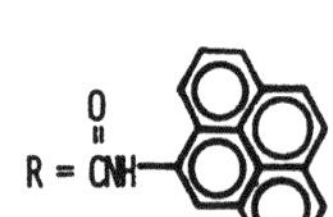	(L)	248 dec.	57	211

Table 3.3 (continued)

Cpd. No.	No. C's	Ring size	Structure		Mp or (bp) in °C	Yield (%)	Refs.
75	80	18	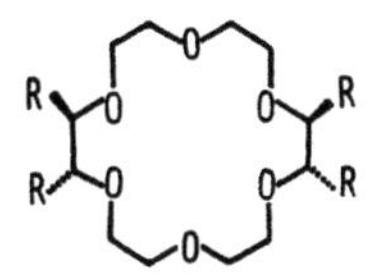 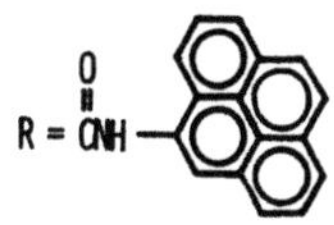	(D)	245 dec.	65	211
76	92	18	R = CH_2OCPh_3		nr	nr	17
				(LL)	nr	73	43

Table 3.4. Crowns containing fused cyclohexano subunits

Cpd. No.	No. C's	Ring size	Structure		Mp or (bp) in °C	Yield (%)	Refs.
1	12	12	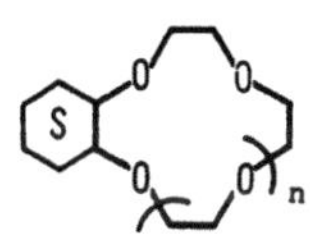	n = 1	< 26	48	169
2	14	15		n = 2	< 26	58	169, 173
				trans	oil	42	87
3	16	18		n = 3	< 26	46	169, 173
				trans	oil	37	87
4	16	14	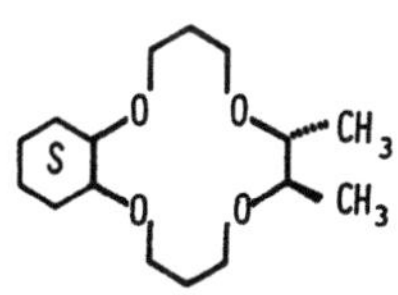	(RR)	nr	nr	126
5	18	15	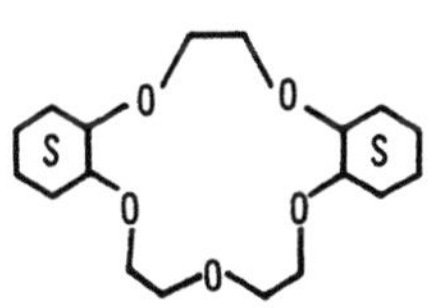		62–63	28	170
6	18	14	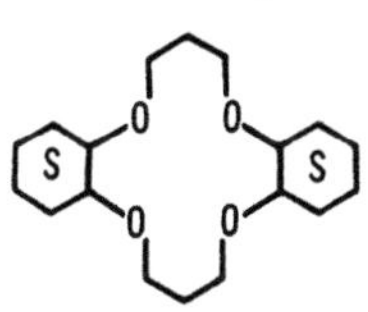		153.5–155	55	169, 173

Table 3.4 (continued)

Cpd. No.	No. C's	Ring size	Structure	Mp or (bp) in °C	Yield (%)	Refs.
7	18	15	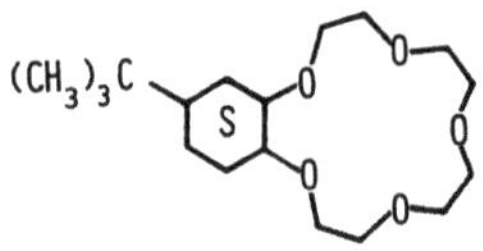	< 26	17	169, 173
8	18	15	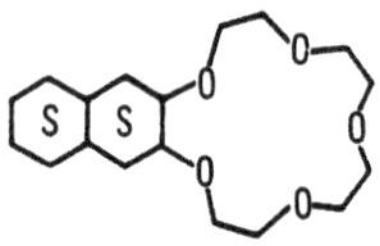	< 26	28	169
9	19	16	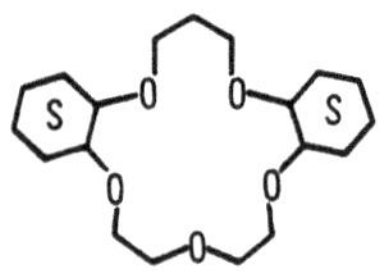	< 26	66	170
10	22	21	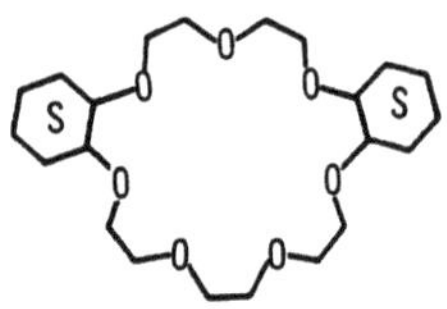	< 26	50	169
11	24	24	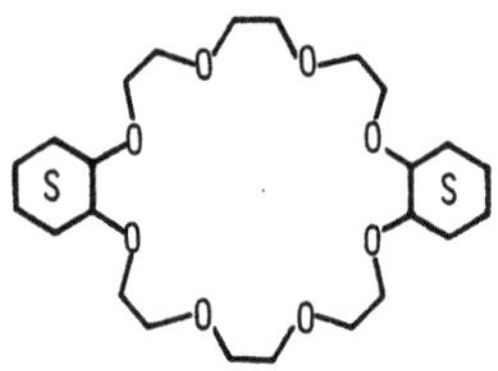	< 26	63	169
12	26	14	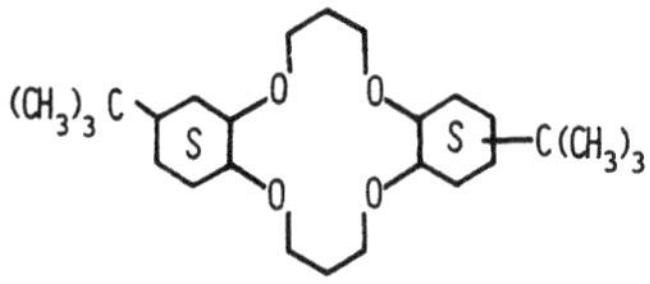	glass	41	169, 173
13	28	30	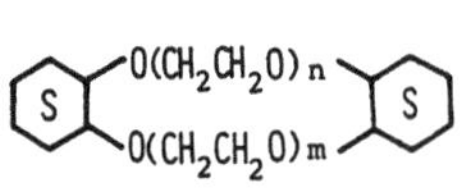 n = m = 4	< 26	61	169, 173
14	40	48	n = 5, m = 9	< 26	58	169, 173
15	48	60	n = m = 9	< 26	45	169, 173

Table 3.5. Miscellaneous alicyclic crowns

Cpd. No.	No. C's	Ring size	Structure		Mp or (bp) in °C	Yield (%)	Refs.
1	6	9	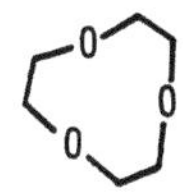		(168–172)	nr	49
2	10	14	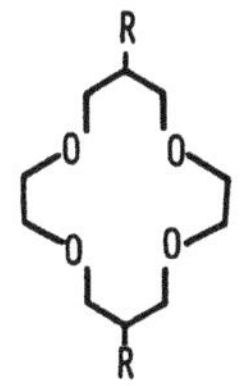	R = H	23–24	23	137
3	10	14		R = OH	75.5 (80–90/0.02)	89	137
4	11	16	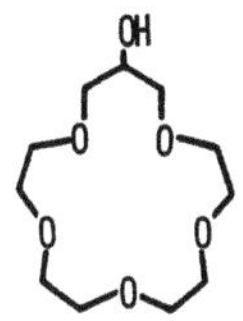		oil	63	209
5	12	14	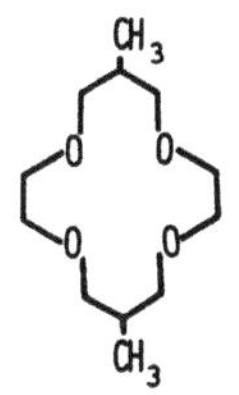		(70–75/0.01)	78	137
6	12	16	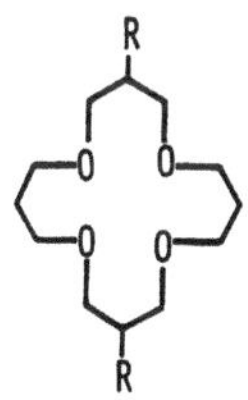	R = H	67.5	39	137
7	12	16		R = OH	94.7	91	137
8	12	16	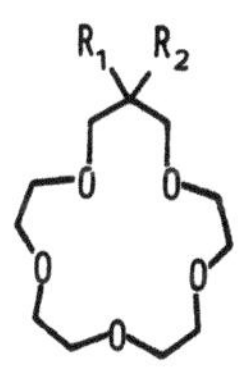	$R_1, R_2 = CH_2$	(115–120/0.005)	45	209
9	12	16		$R_1 = H, R_2 = CH_2OH$	oil	88	209
10	12	16		$R_1 = CH_3, R_2 = OH$	oil	78	209
11	14	16	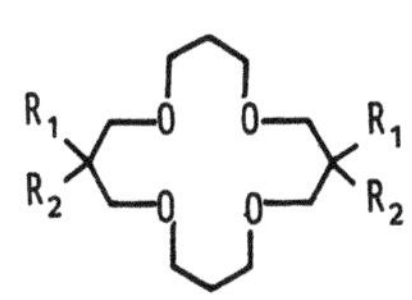	$R_1, R_2 = CH_2$	32–33	14	137
12	14	16	$R_1 = CH_3, R_2 = H$		(80–90/0.1)	75	137

Table 3.5 (continued)

Cpd. No.	No. C's	Ring size	Structure	Mp or (bp) in °C	Yield (%)	Refs.
13	14	19	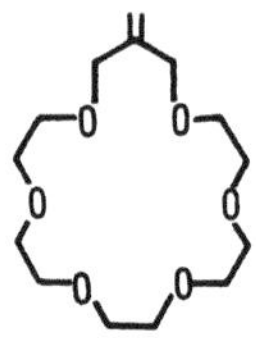	(120–125/0.004)	66	209
14	14	21	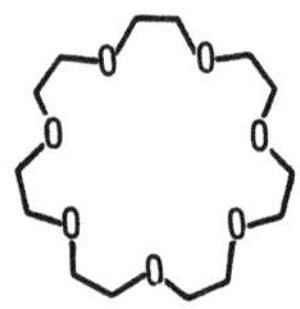	nr	18	81
				nr	nr	172
				oil	26	52
15	15	19	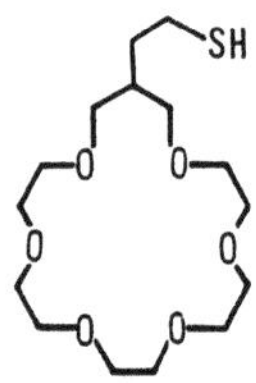	nr	nr	181
16	16	9		108–108.5	64	21
17	16	9	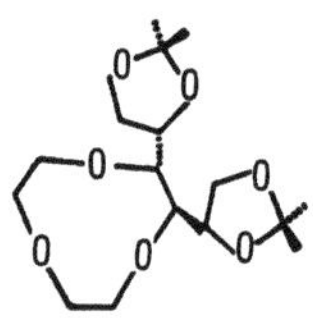 (D)	oil	13	43
18	16		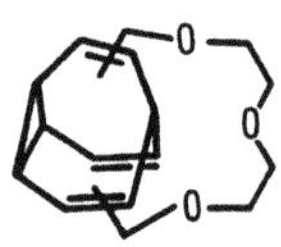	84	9	190
19	16	16	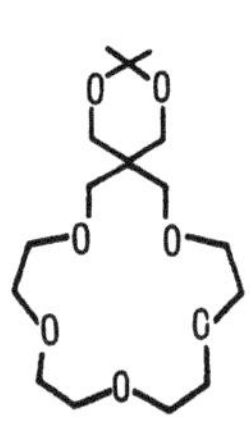	nr	64	210

Table 3.5 (continued)

Cpd. No.	No. C's	Ring size	Structure		Mp or (bp) in °C	Yield (%)	Refs.
20	16	24	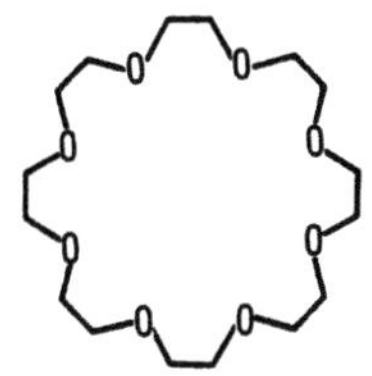		nr 19 nr	nr 15 15	172 52 81
21	18	9	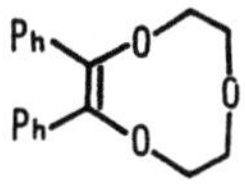		77–78	10	21
22	18	12	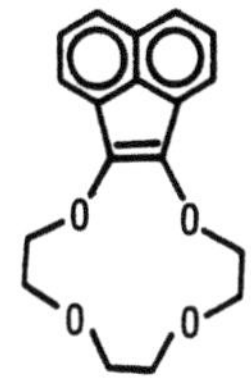		84–85	42	21
23	18	20	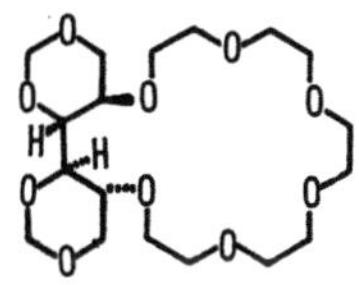		(207/0.01)	34	120
24	20	17	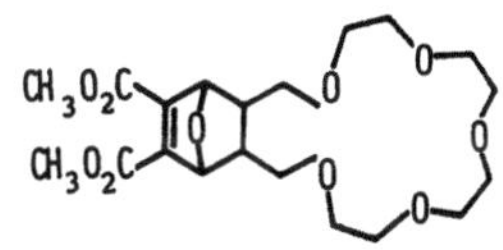		nr	nr	185
25	20	30	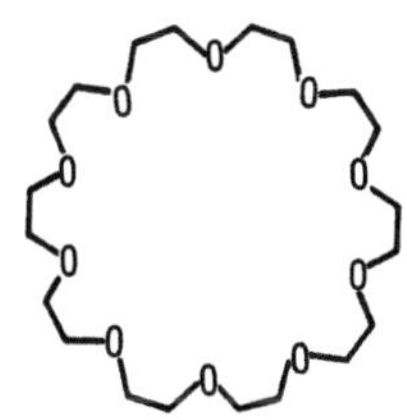		nr	nr	172
26	22		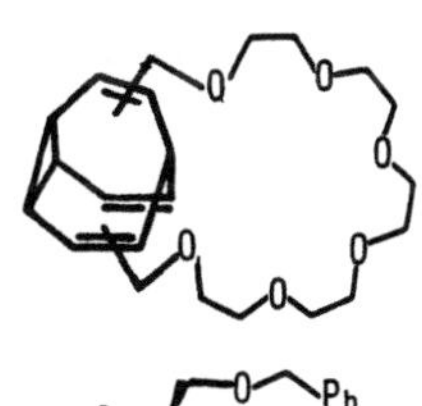		(110/0.001)	20	190
27	22	9	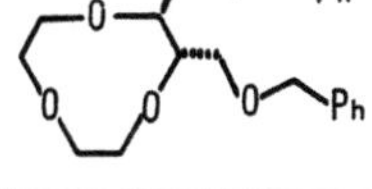	(L)	~40	3	43

Table 3.5 (continued)

Cpd. No.	No. C's	Ring size	Structure	Mp or (bp) in °C	Yield (%)	Refs.
28	24	17	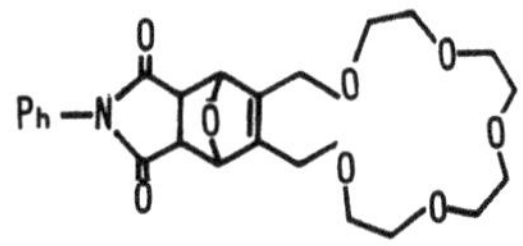	nr	nr	185
29	24	22	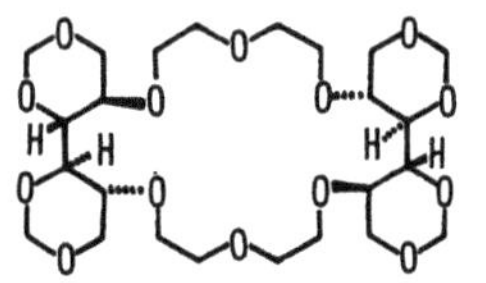 (DD)	192–193	15	120
30	26	19	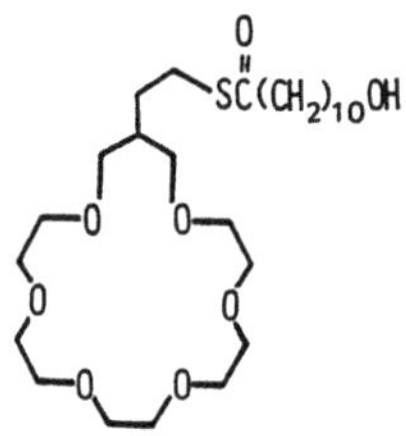	nr	nr	181
31	26	32	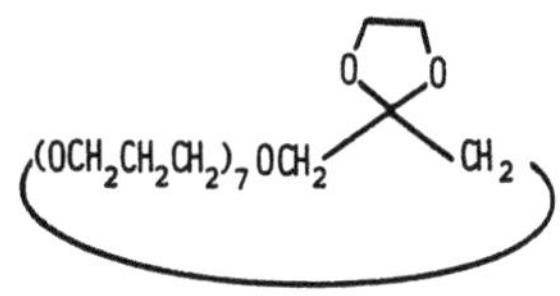	nr	nr	210
32	32	24	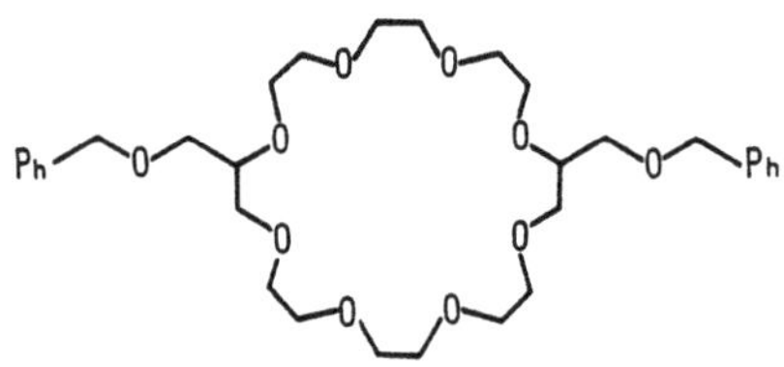	nr	nr	84
33	32	24	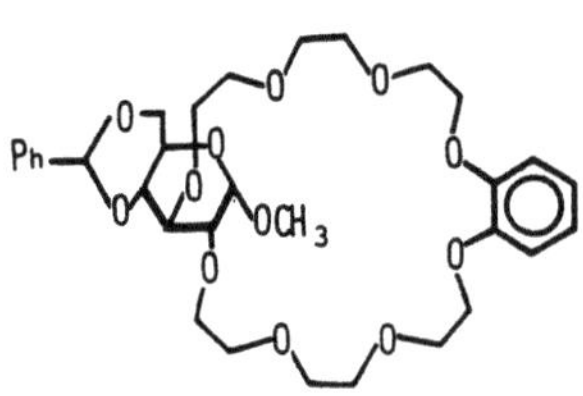	105–107	70	57
34	36	18	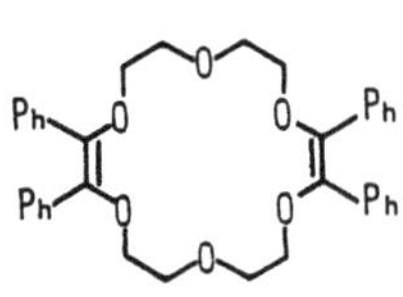	173–175	19.5	132

Table 3.5 (continued)

Cpd. No.	No. C's	Ring size	Structure	Mp or (bp) in °C	Yield (%)	Refs.
35	36	27	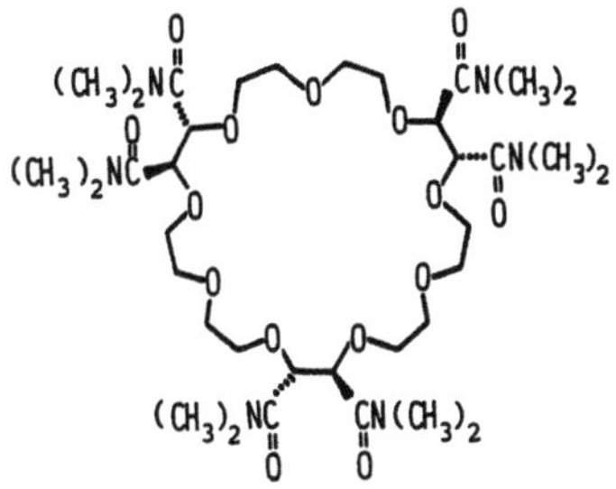	glass	10	123
36	38	17	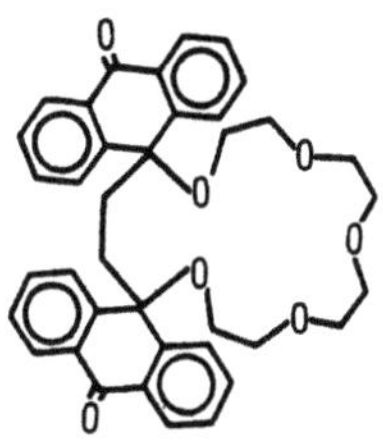	217–218	18	232
37	38	20	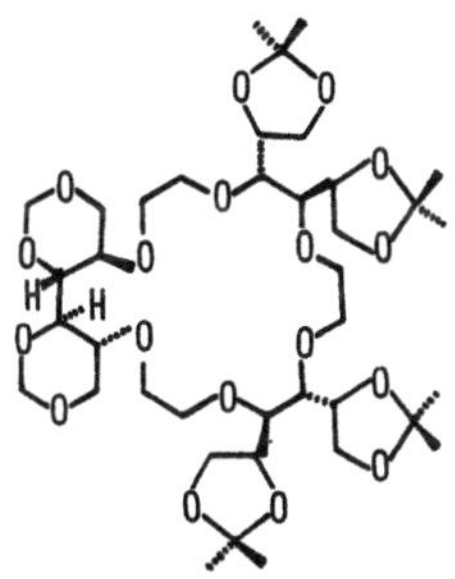 (DDD)	50–54	4	7
38	48	27	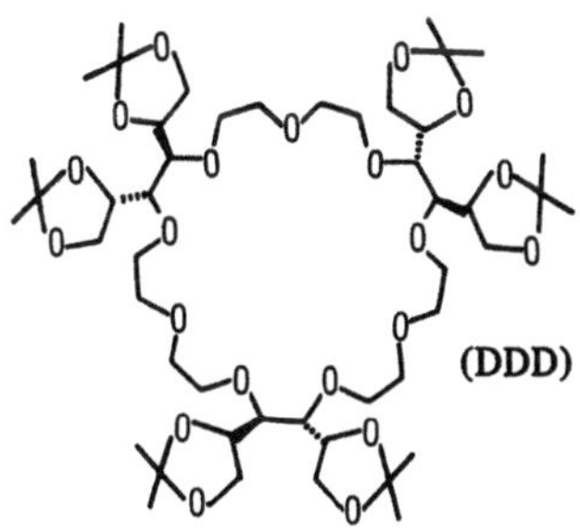 (DDD)	oil	10	43

Table 3.6. Crowns containing a ketonic carbonyl as part of the macroring

Cpd. No.	No. C's	Ring size	Structure	Mp or (bp) in °C	Yield (%)	Refs.
1	9	12	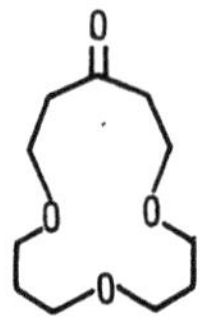	(83–85/0.02)	21	17
2	10	14	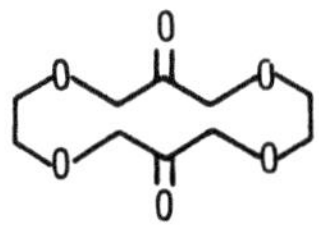	120	73	137
3	11	16	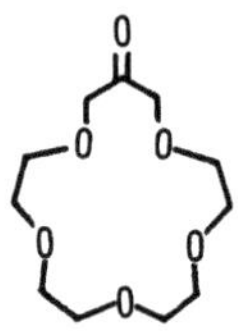	oil	98	209
4	12	16	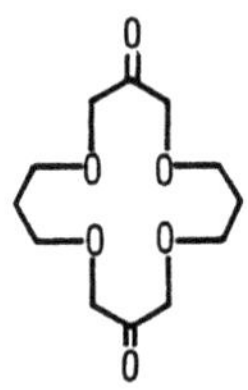	87.7	75	137
5	13	18	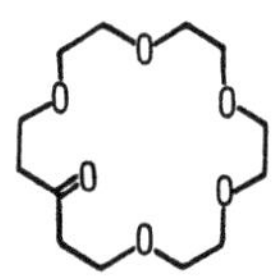	36.7	nr	17
6	15	21	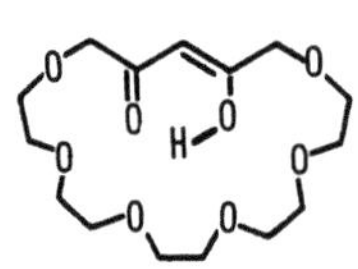	nr	10	2
7	18	24	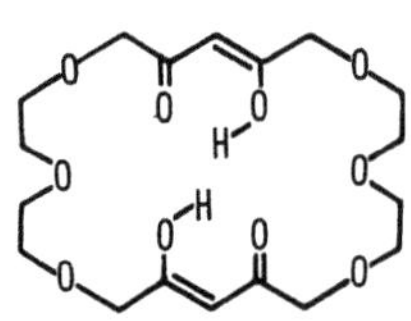	nr	10	2
8	18	20	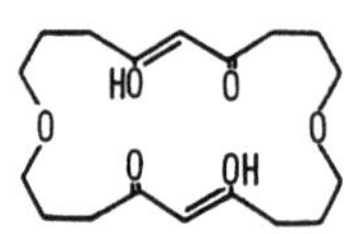	nr	11	3

Table 3.6 (continued)

Cpd. No.	No. C's	Ring size	Structure	Mp or (bp) in °C	Yield (%)	Refs.
9	26	20	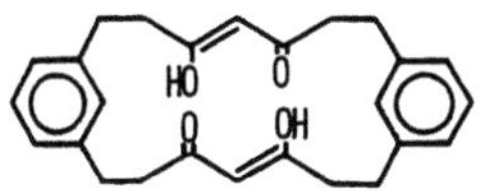	nr	12	3
10	27	36	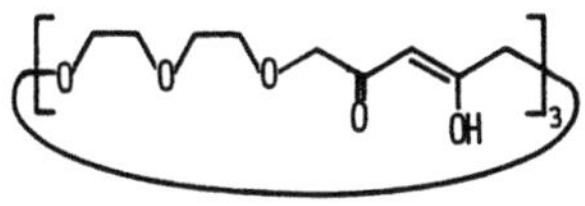	nr	7	2

Table 3.7. Benzo-15-crown-5 derivatives

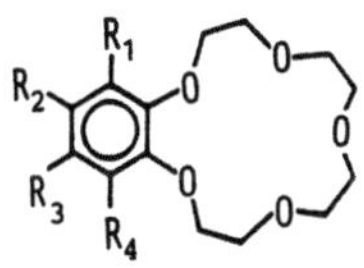

Cpd. No.	No. C's	Ring size	Structure	Mp or (bp) in °C	Yield (%)	Refs.
1	14	15	$R_1 = R_2 = R_3 = R_4 = H$	79–79.5	62	169,173
				nr	61	183
				75.5–76.5	50	217
2	14	15	$R_2 = NH_2, R_1 = R_3 = R_4 = H$	73–74	93	213
				nr	nr	61
3	14	15	$R_2 = NO_2, R_1 = R_3 = R_4 = H$	84–85	77	213
4	14	15	$R_2 = OH, R_1 = R_3 = R_4 = H$	101–103	93	225
5	14	15	$R_2 = Br, R_1 = R_3 = R_4 = H$	66–68	57	213
6	14	15	$R_2 = I, R_1 = R_3 = R_4 = H$	87–88	nr	95
7	14	15	$R_1 = R_2 = NO_2, R_3 = R_4 = H$	168	20	213
8	14	15	$R_1 = R_2 = R_3 = R_4 = F$	oil	38	83
9	15	15	$R_2 = CH_2OH, R_1 = R_3 = R_4 = H$	52–54	nr	95
10	15	15	$R_2 = CH_2Cl, R_1 = R_3 = R_4 = H$	65–66	nr	95

Table 3.7 (continued)

Cpd. No.	No. C's	Ring size	Structure	Mp or (bp) in °C	Yield (%)	Refs.
11	15	15	$R_2 = CHO, R_1 = R_3 = R_4 = H$	78–81 78–79	nr 40	95 213
12	15	15	$R_2 = CH_3, R_3 = I, R_1 = R_4 = H$	69–71	nr	95
13	16	15	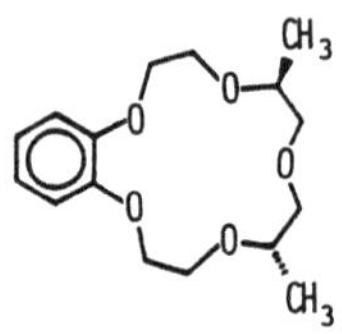 (S,S)	43	nr	125
14	16	15	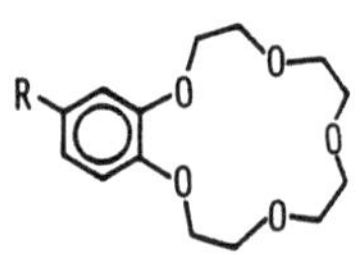 $R = O\overset{O}{\overset{\parallel}{C}}CH_3$	57–58	82	225
15	16	15	$R = CH_2CH_2NH_2$	90–92	nr	220
16	16	15	$R = CH{=}CH_2$	43.5–44.2 nr	21–24 22	103 196
17	16	15	$R = \overset{O}{\overset{\parallel}{C}}-CH_3$	95–96.5	22	164
18	16	15	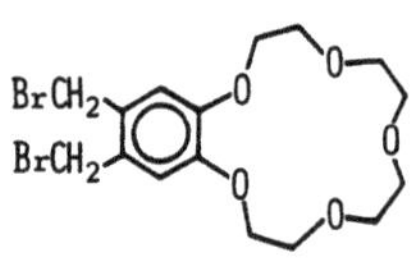	173–176	85	46
19	17	15	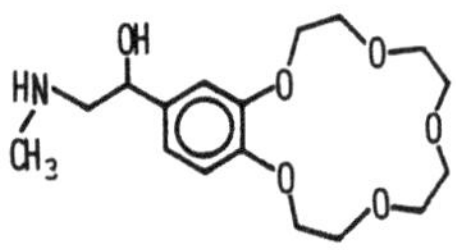	61	nr	220
20	17	15	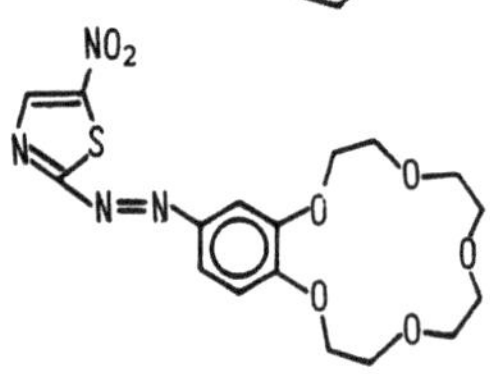	169–171	12	61
21	18	15	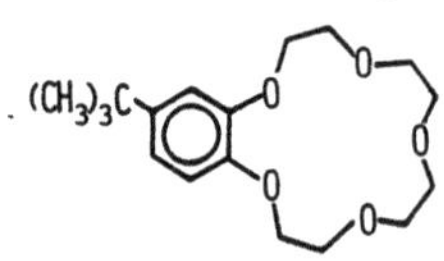	43.5–44.5	61	169,173
22	18	15	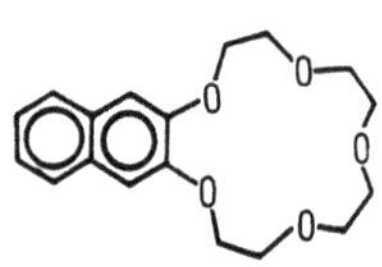	117–119	20	169

Table 3.7 (continued)

Cpd. No.	No. C's	Ring size	Structure		Mp or (bp) in °C	Yield (%)	Refs.
23	19	15	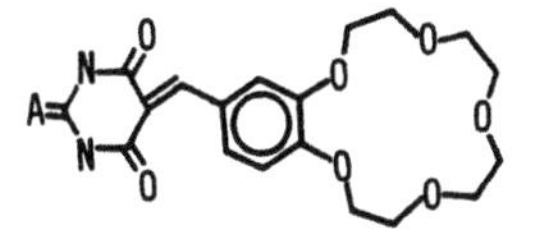	A = O	248–250	86	62
24	19	15		A = S	296–298	82	62
25	20	15	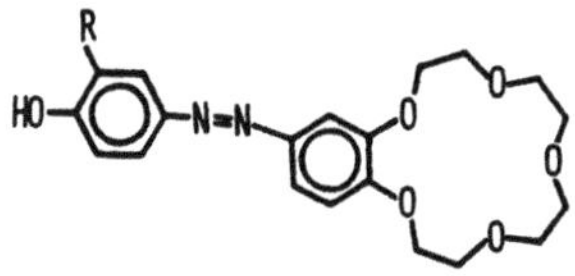	R = H	72–74	10	234
26	20	15		R = Cl	62–64	10	234
27	20	15	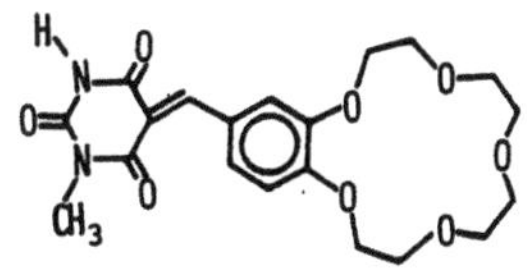		234	78	62
28	20	15	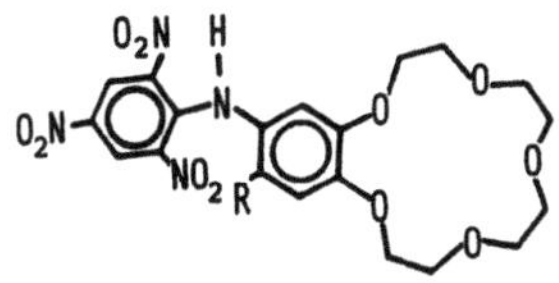	R = H	nr	72	136
29	20	15		R = NO_2	208–209	82	136
30	20	15		R = Br	216–217	60	136
31	21	15	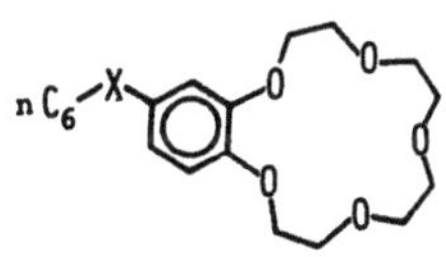	X = CHOH	oil	92	164
32	21	15		X = C=O	49–51	55	164
33	31	15	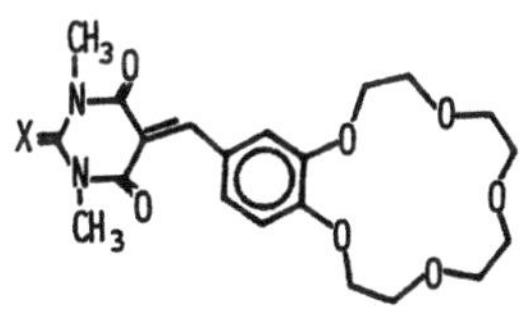	X = O	172–173	76	62
34	21	15		X = S	204	81	62
35	22	15	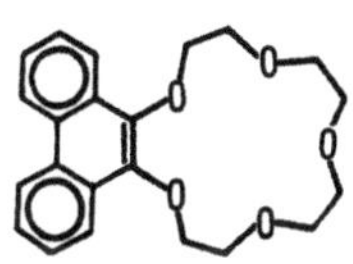		104–105	55	21
36	24	15	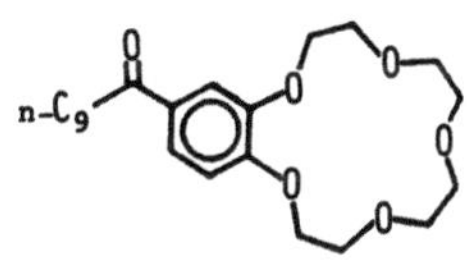		nr	50	164
37	24	15	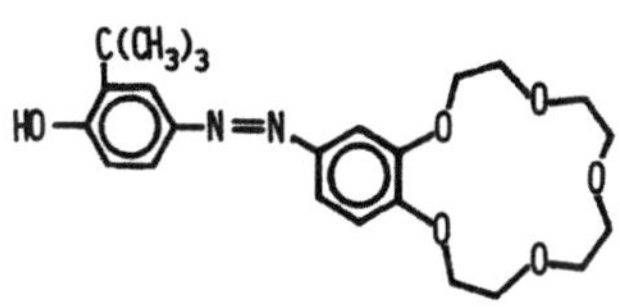		107–108	31	234

Table 3.7 (continued)

Cpd. No.	No. C's	Ring size	Structure	Mp or (bp) in °C	Yield (%)	Refs.
38	24	15	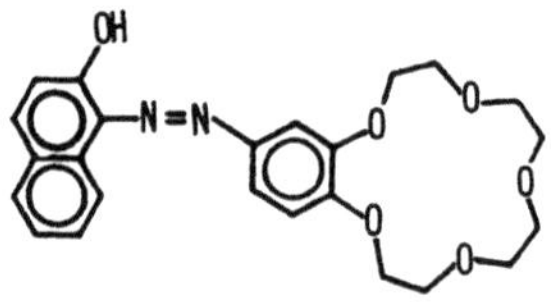	162–163	59	234
39	24	15	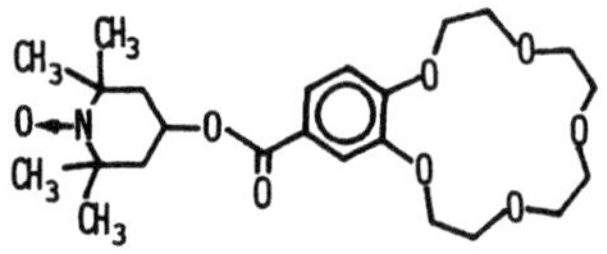	113–114	nr	96
40	25	15	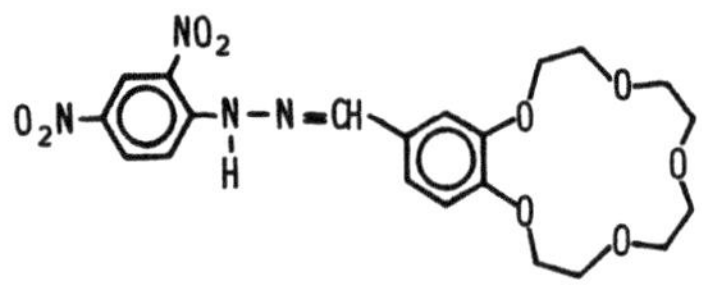	222–225	nr	95
41	25	15	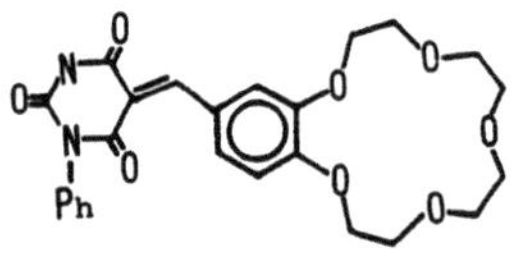	258	79	62
42	25	15	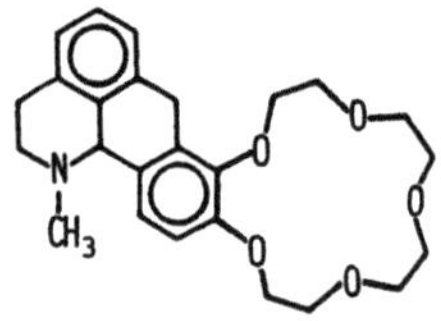	nr	nr	220
43	26	15	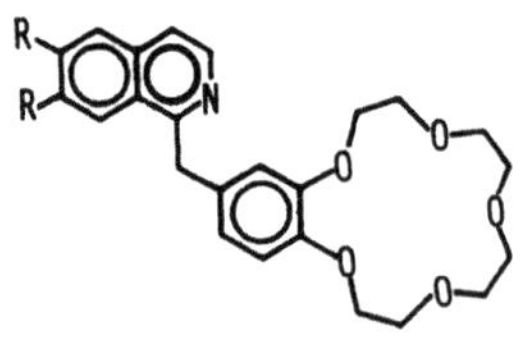 R = OCH_3	104	25	219
44	26	15	R = $\overset{O}{\overset{\|}{C}}CH_3$	104	25	62
45	26	15	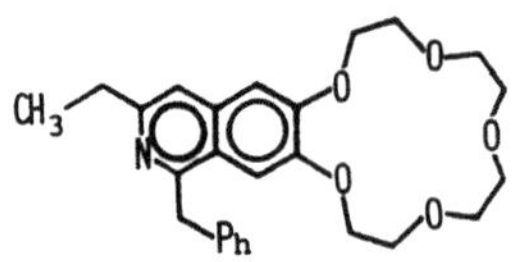	108–110	40	62
46	27	15	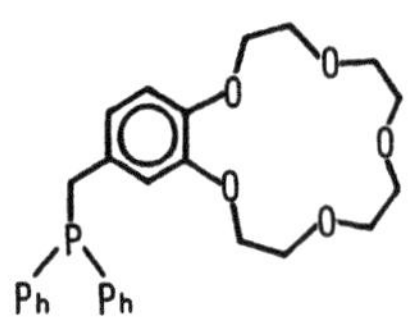	108–109	nr	95

Table 3.7 (continued)

Cpd. No.	No. C's	Ring size	Structure	Mp or (bp) in °C	Yield (%)	Refs.
47	31	15	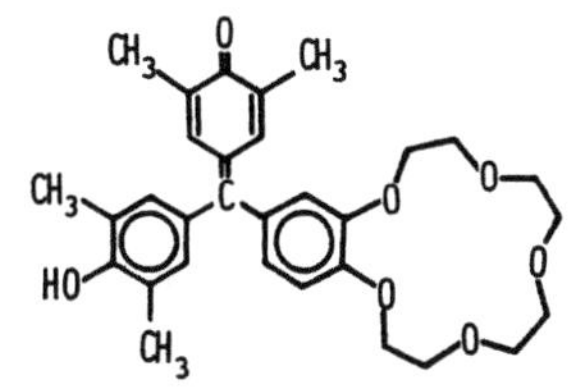	244–245	38	234
48	33	15	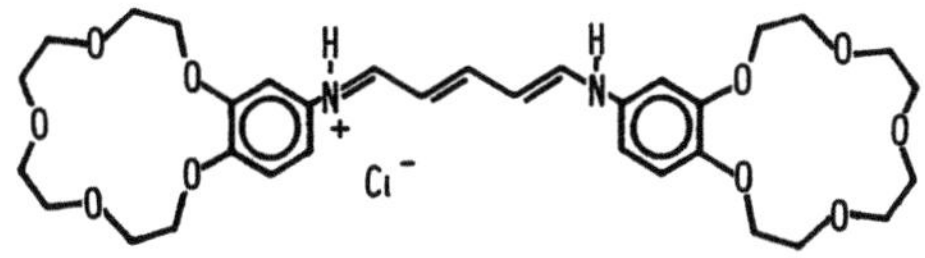	112	50	61

Table 3.8. Derivatives of monobenzo-18-crown-6

Cpd. No.	No. C's	Ring size	Structure	Mp or (bp) in °C	Yield (%)	Refs.
1	16	18	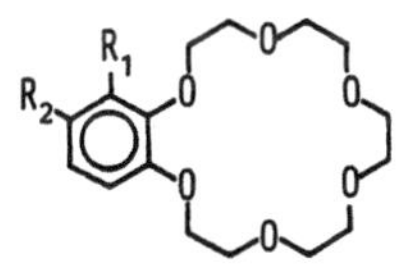 $R_1 = R_2 = H$	43–44 nr 43–43.5	60 67 74	169, 173 183 217
2	16	18	$R_1 = H, R_2 = NO_2$	70–72	56	213
3	16	18	$R_1 = H, R_2 = OH$	52.3–52.6	82	225
4	16	18	$R_1 = H, R_2 = I$	92–94	nr	95
5	16	18	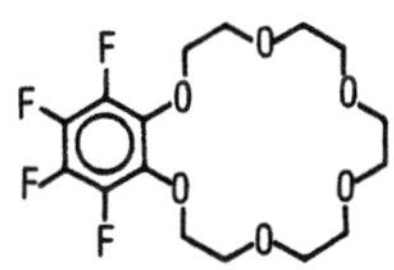	oil	45	83
6	17	18	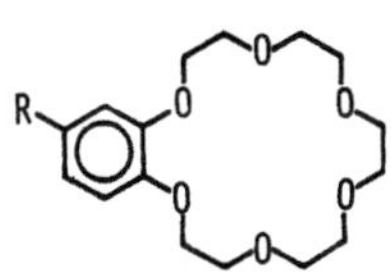 $R = CH_2OH$	nr	47	183
7	17	18	$R = CHO$	nr 60–62	57 25	18 213

Table 3.8 (continued)

Cpd. No.	No. C's	Ring size	Structure	Mp or (bp) in °C	Yield (%)	Refs.
8	17	18	$R = C(=N{-}OH){-}H$	138.5–139	nr	99
9	17	18	$R = CO_2H$	118–120	80	213
10	18	18	$R = CH{=}CH_2$	59–61	20	196
11	18	18	$R = OC(=O)CH_3$	48.5–50	65	225
				nr	52	182
12	18	18	$R = C(=N{-}OH){-}CH_3$	110.5–111.5	nr	99
				115.4–116.0	nr	99
13	18	18	$R = CH{=}N{-}OMe$	81.5–82.0	nr	99
			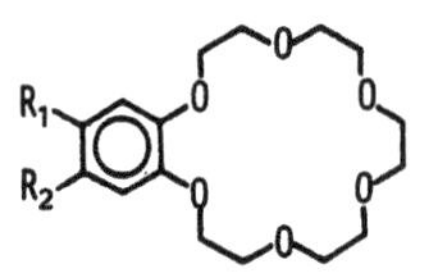			
14	18	18	$R_1 = C(=N{-}OH)CH_3, R_2 = NH_2$	nr	nr	99
15	18	18	$R_1 = C(=N{-}OH)CH_3, R_2 = NO_2$	nr	nr	99
16	18	18	$R_1 = C(=N{-}OH)CH_3, R_2 = Cl$	nr	nr	99
			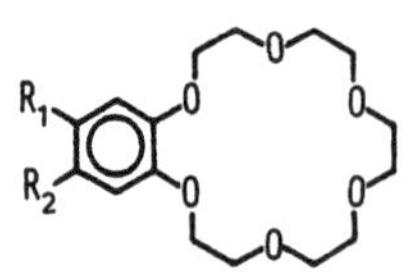			
17	19	18	$R_1 = C(=N{-}OH)CH_2CH_3, R_2 = H$	77.1–81.8	nr	99

Table 3.8 (continued)

Cpd. No.	No. C's	Ring size	Structure	Mp or (bp) in °C	Yield (%)	Refs.
18	19	18	$R_1 = C(=NOCH_3)CH_3, R_2 = H$	79.0–79.8	nr	99
19	19	18	$R_1 = C(=NOH)CH_2CH_2F, R_2 = H$	nr	nr	99
20	19	18	$R_1 = C(=O)OCH_2CH_3, R_2 = H$	nr	68	183
21	19	18	$R_1 = C(=NOH)CH_3, R_2 = CH_3$	86–87.5	7	99
22	19	18	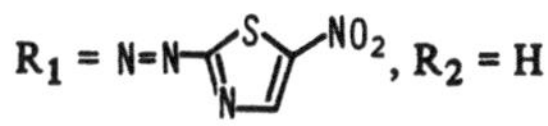	162–164	7	61
23	20	18	$R_1 = C(CH_3)_3, R_2 = H$	35–37	62	169, 173
24	20	18	$R_1 = C(=NOC(=O)CH_3)CH_3, R_2 = H$	65.6–66.7	nr	99
25	20	18	$R_1 = C(=NOCH_2CH_3)CH_3, R_2 = H$	71.2–73.4	nr	99
26	20	18	$R_1 = C(=NOCH_3)CH_2CH_3, R_2 = H$	nr	nr	99
27	20	18	$R_1 = C(=NOH)(CH_2)_2CH_3, R_2 = H$	104.0–105.1	nr	99
28	20	18	$R_1 = C(=NOH)(CH_2)_3Cl, R_2 = H$	nr	nr	99
29	20	18	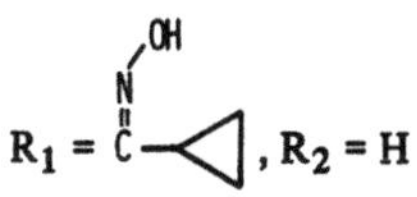	64.4–66.8	nr	99
30	20	18	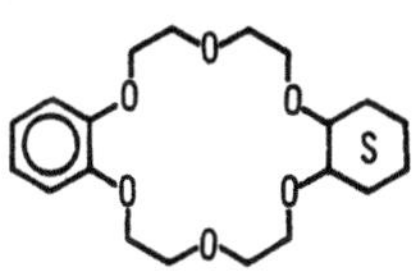	< 26	8	170

Table 3.8 (continued)

Cpd. No.	No. C's	Ring size	Structure	Mp or (bp) in °C	Yield (%)	Refs.
31	20	18	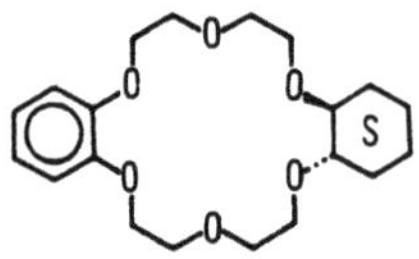	73–74	70	57
32	20	18	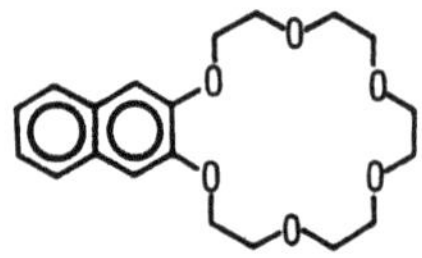	110–111.5	25	169
33	20	18	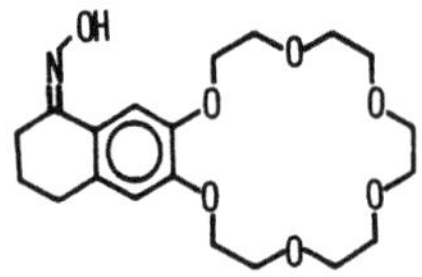	124–126	nr	99
			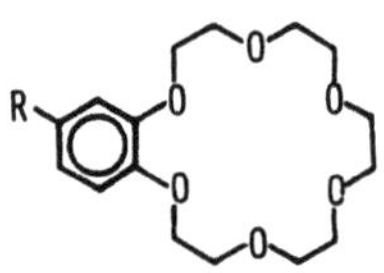			
34	21	18	$R = CH(OH)CH_2NHCH(CH_3)_2$	82.5–84.5	10	62
35	21	18	R = 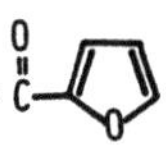	66.1–68.4	nr	100
36	21	18	R = 	81.0–81.8	nr	100
37	21	18	R = 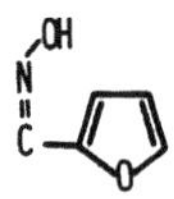	nr	nr	99
38	21	18	$R = \overset{O}{\overset{\|}{C}}NH(CH_2)_3CH_3$	97–98	83	213
39	21	18	R = 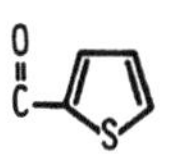	105.1–108.4	nr	100
40	21	18	R = 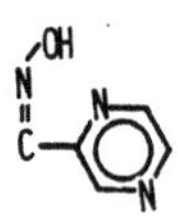	nr	nr	99
41	21	18	R = 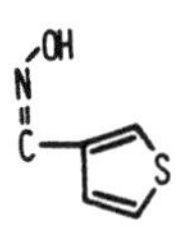	nr	nr	99

Table 3.8 (continued)

Cpd. No.	No. C's	Ring size	Structure	Mp or (bp) in °C	Yield (%)	Refs.
42	22	18	R = C(=N–OH)$(CH_2)_4CH_3$	nr	nr	99
43	22	18	R = C(=N–$OC(CH_3)_3$)CH_3	62.7–64.9	nr	99
44	22	18	R = 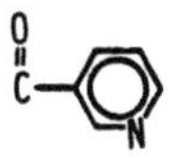	98–100	nr	100
45	22	18	R = 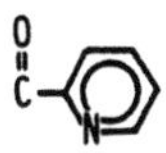	78–80	nr	100
46	22	18	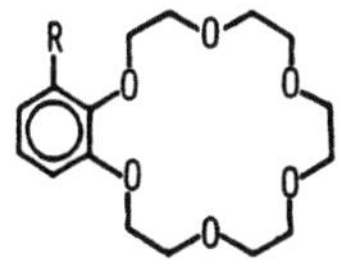 R= 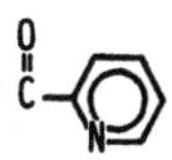	75–76	nr	100
47	22	18	R = 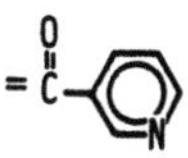	(208–216/0.15)	nr	100
48	22	18	R = 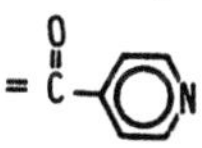	(205–215/0.1)	nr	100
			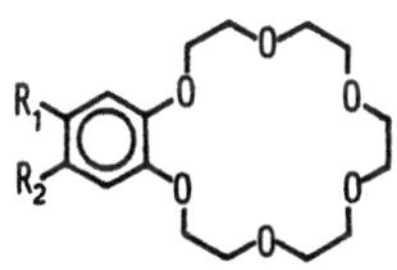			
49	22	18	R_1 = 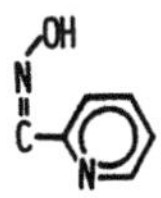 , R_2 = H	nr	nr	99
50	22	18	R_1 = 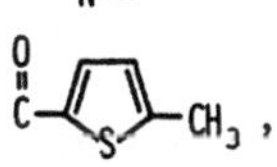 , R_2 = H	78.4–79.3	nr	100
51	22	18	R_1 = 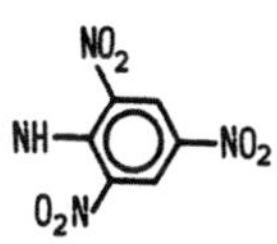 , R_2 = H	nr	nr	214
52	22	18	R_1 = 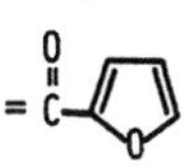 , R_2 = CH_3	72–74	nr	100
53	22	18	R_1 = 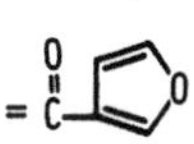 , R_2 = CH_3	99–102 (· H_2O)	nr	100

Table 3.8 (continued)

Cpd. No.	No. C's	Ring size	Structure	Mp or (bp) in °C	Yield (%)	Refs.
54	22	18	R_1 = 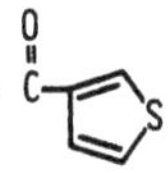, $R_2 = CH_3$	98–100	nr	100
55	22	18	R_1 = 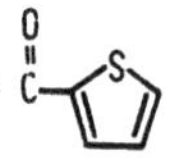, $R_2 = CH_3$	98–108	nr	100
56	22	18	R_1 = , $R_2 = CH_3$	nr	nr	99
57	22	18	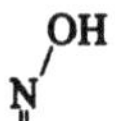 $R_1 = CCH_3$, R_2 = n-Bu	nr	nr	99
58	22	18	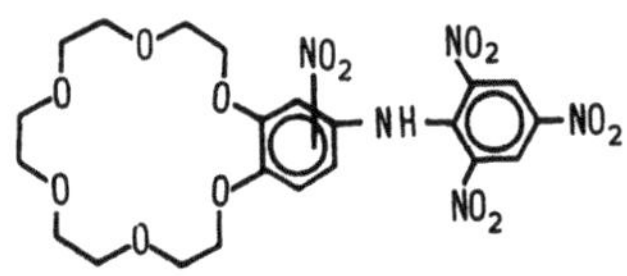	nr	nr	214
59	22	18	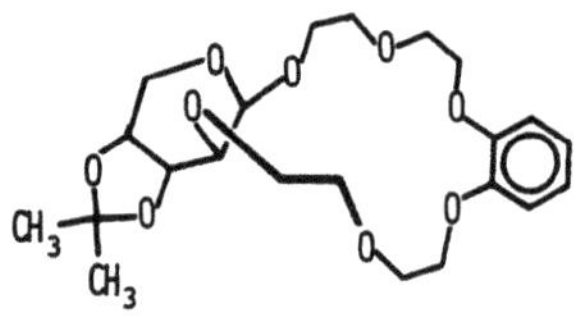	gum	72	57
60	22	18	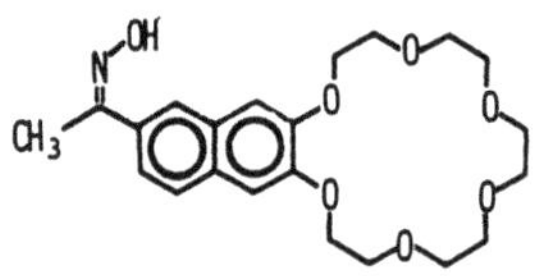	138.1–140.2	nr	99
61	22	18	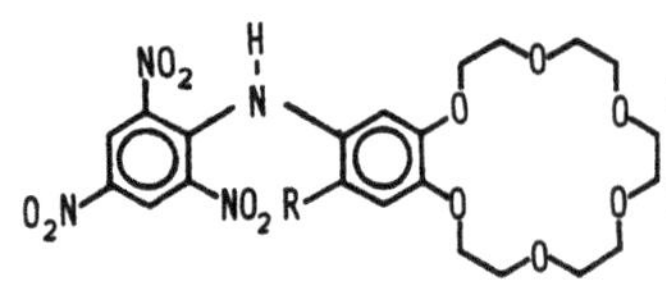 R = H	55–56	45	234
62	22	18	R = NO_2	160–161	32	234
			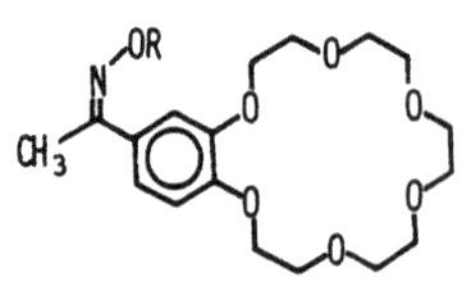			
63	23	18	R = $(CH_2)_4CH_3$	46.6–48.5	nr	99
64	23	18	R = $(CH_2)_3N(CH_3)_2$	nr	nr	99

Table 3.8 (continued)

Cpd. No.	No. C's	Ring size	Structure	Mp or (bp) in °C	Yield (%)	Refs.
65	23	18	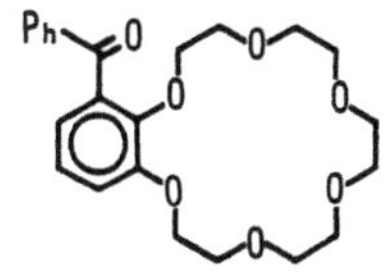	(215–220/0.15)	nr	100
66	23	18	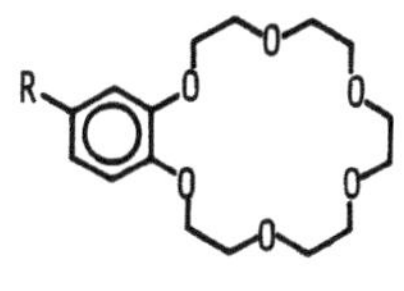 R = C(=O)Ph	112.2–113.8	nr	100
67	23	18	R = 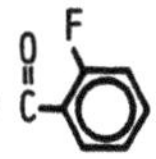	117.8–119.3	nr	100
68	23	18	R = 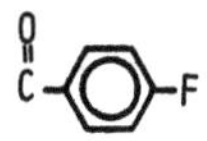	126.8–128.5	nr	100
69	23	18	R = 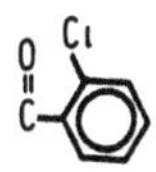	134–136.5	nr	100
70	23	18	R = 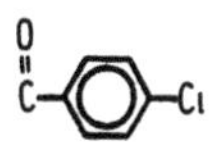	142.5–144.1	nr	100
71	23	18	R = 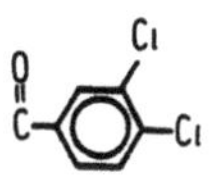	152–153.4	nr	100
72	23	18	R = 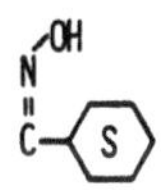	128–130.1	nr	99
73	23	18	R = 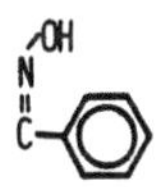	209–210 as $NH_2OH \cdot 2H_2O$ complex	nr	99
74	23	18	R = 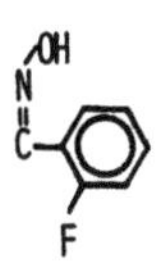	nr	nr	99
			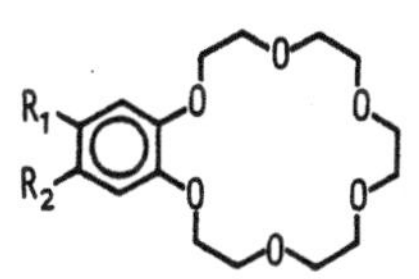			
75	23	18	R_1 = C(=O)Ph, R_2 = NH_2	58 dec.	nr	100
76	23	18	R_1 = C(=O)Ph, R_2 = NO_2	151.4–153.8	nr	100

Table 3.8 (continued)

Cpd. No.	No. C's	Ring size	Structure	Mp or (bp) in °C	Yield (%)	Refs.
77	23	18	$R_1 = \overset{O}{\overset{\|}{C}}Ph, R_2 = Cl$	104–106	nr	100
78	23	18	$R_1 = \overset{O}{\overset{\|}{C}}Ph, R_2 = Br$	114–116	nr	100
79	23	18	R_1 = 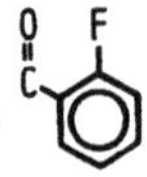, $R_2 = NH_2$	oil	nr	100
80	23	18	R_1 = 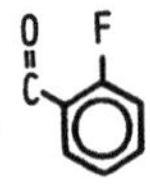, $R_2 = NO_2$	124.0–125.2	nr	100
81	23	18	R_1 = 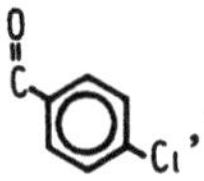, $R_2 = NH_2$	oil	nr	100
82	23	18	R_1 = 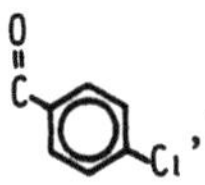, $R_2 = NO_2$	132.5–134.5	nr	100
83	23	18	R_1 = 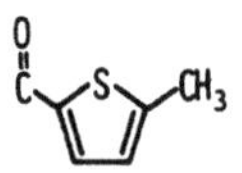, $R_2 = CH_3$	104.5–106	nr	100
84	23	18	R_1 = 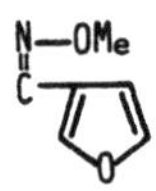, $R_2 = CH_3$	nr	nr	99
85	23	18	R_1 = 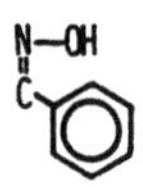, $R_2 = Br$	nr	nr	99
86	23	18	R_1 = 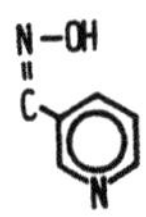, $R_2 = CH_3$	nr	nr	99
87	23	18	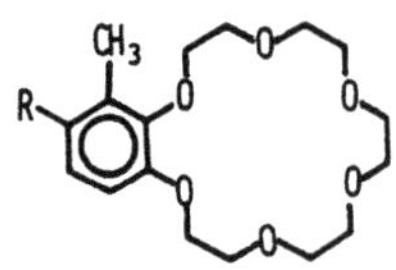 R = 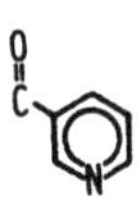	nr	nr	100
88	23	18	R = 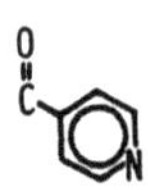	nr	nr	100

Table 3.8 (continued)

Cpd. No.	No. C's	Ring size	Structure	Mp or (bp) in °C	Yield (%)	Refs.
89	23	18	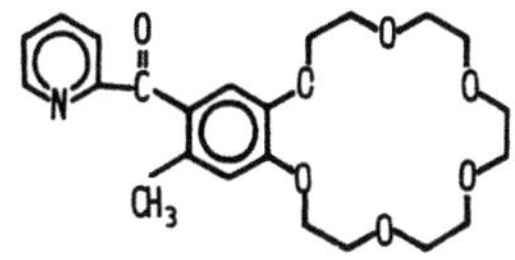	97–98	nr	100
90	24	18	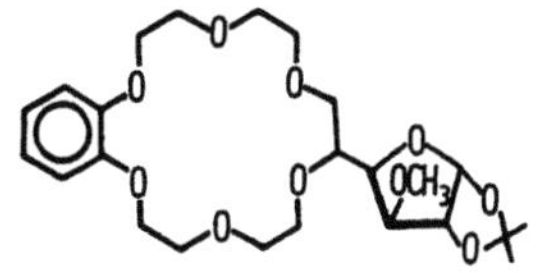	gum	65	57
91	24	18	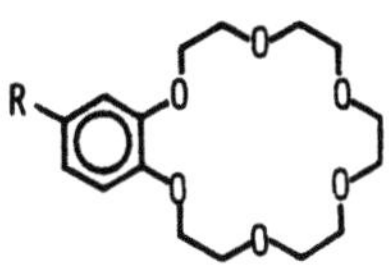 R = 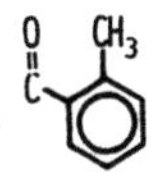	107.1–109.5	nr	100
92	24	18	R = 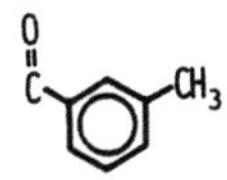	75.1–77.5	nr	100
93	24	18	R = 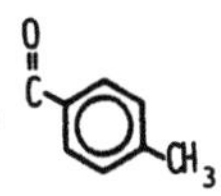	100.1–103.2	nr	100
94	24	18	R = 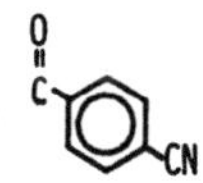	168–171	nr	100
95	24	18	R = 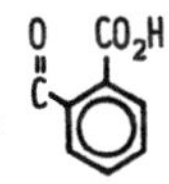	70–78 · ½ H_2O	nr	100
96	24	18	R = 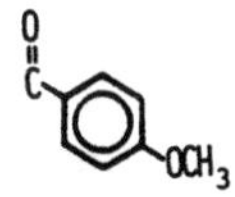	117.4–117.9	nr	100
97	24	18	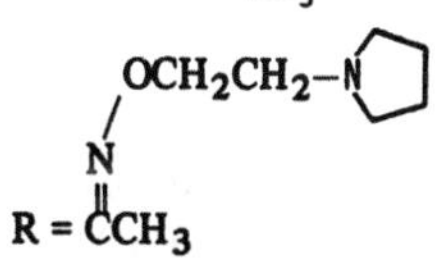	nr	nr	99
98	24	18	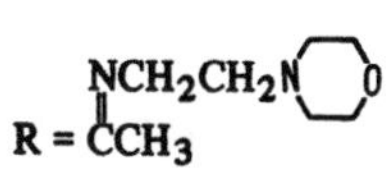	nr	nr	99
99	24	18	R = C(=N–OMe)Ph	69.8–70.9	3	99
100	24	18	R = C(=N–OH)CH_2Ph	nr	nr	99

Table 3.8 (continued)

Cpd. No.	No. C's	Ring size	Structure	Mp or (bp) in °C	Yield (%)	Refs.
101	24	18	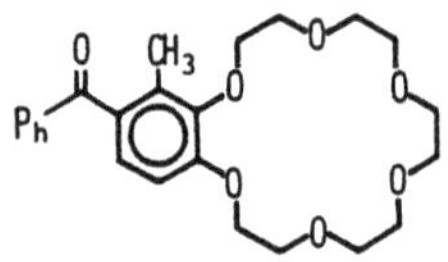	oil	nr	100
102	24	18	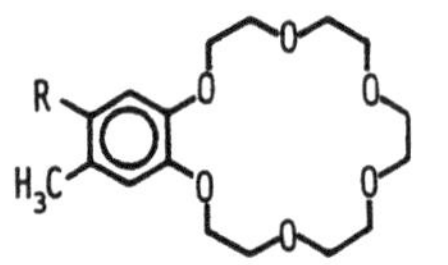 R = 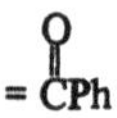	88.5–89	nr	100
103	24	18	R = 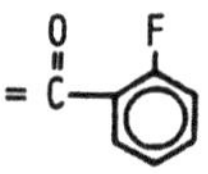	107–109	nr	100
104	24	18	R = 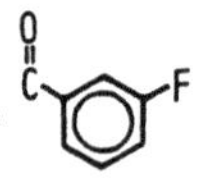	89.5–91	nr	100
105	24	18	R = 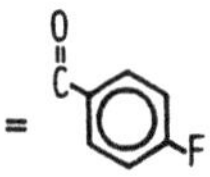	82.5–84.5	nr	100
106	24	18	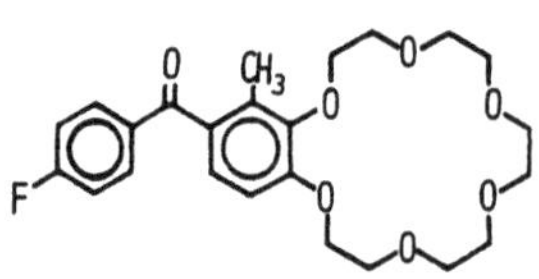	(226–232/0.1)	nr	100
107	24	18	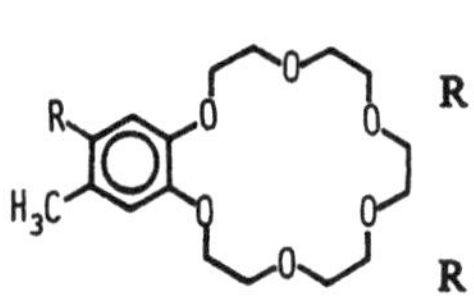 R = 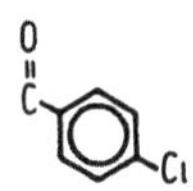	96–97	nr	100
108	24	18	R = 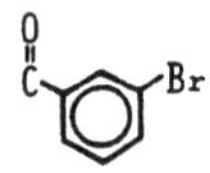	96–98	nr	100
109	24	18	R = 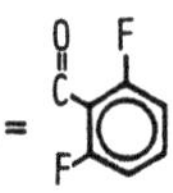	105.5–107.5	nr	100
110	24	18	R = 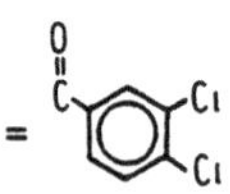	85–95 $\cdot H_2O$	nr	100
111	24	18	R = 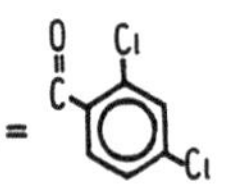	103–104.5	nr	100
112	24	18	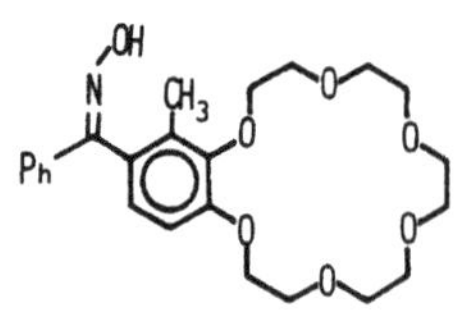	123–125	nr	99

Table 3.8 (continued)

Cpd. No.	No. C's	Ring size	Structure	Mp or (bp) in °C	Yield (%)	Refs.
113	25	18	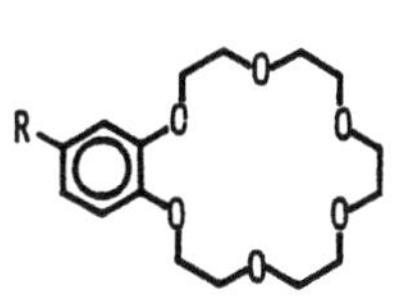 R = 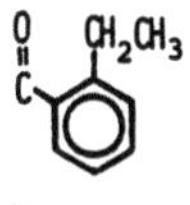	115–116	nr	100
114	25	18	R = 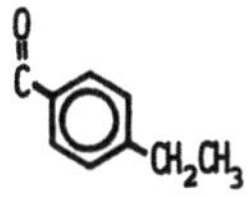	104.5–106.5	nr	100
115	25	18	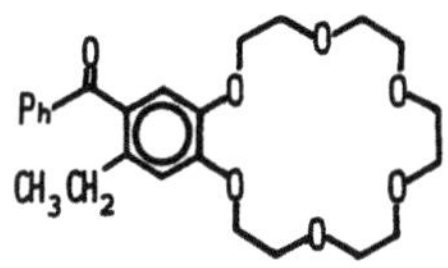	96–97	nr	100
116	25	18	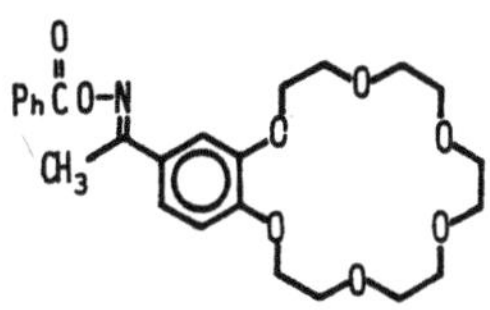	65.5–80.3 · H_2O	nr	99
117	25	18	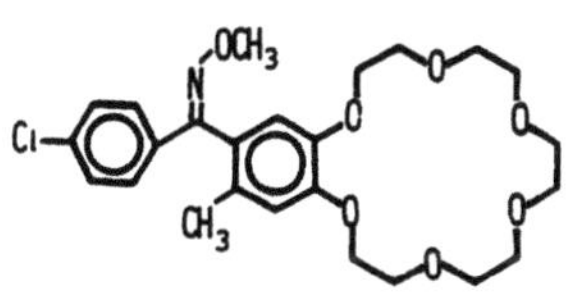	nr	nr	99
118	25	18	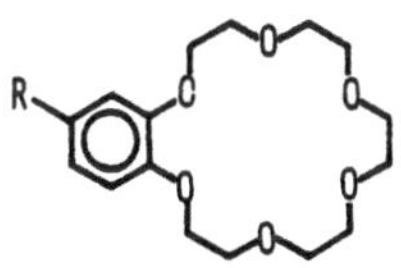 $R = \overset{NOCH_2Ph}{\overset{\Vert}{C}}CH_3$	65.5–68.2	nr	99
119	25	18	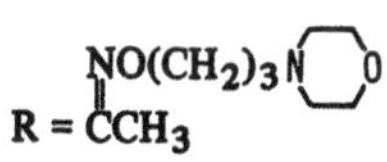	112.8–114	nr	99
120	25	18	R = 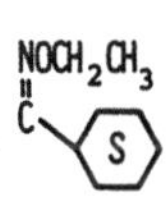	nr	nr	99
121	25	18	R = 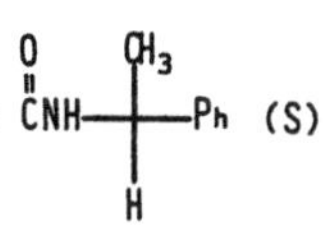	116	nr	193
122	26	18	R = 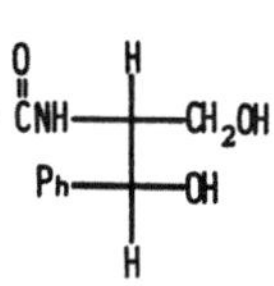 (S, S)	134–135	nr	193

Table 3.8 (continued)

Cpd. No.	No. C's	Ring size	Structure	Mp or (bp) in °C	Yield (%)	Refs.
123	26	18	$R = \overset{O}{\overset{\Vert}{C}}NH\overset{CH_2OH}{\overset{\vert}{C}}HCH(OH)Ph$	101–103	nr	193
124	27	18	R = 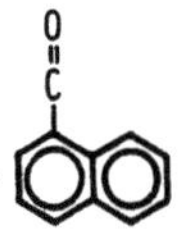	108.2–111.4	nr	100
125	27	18	R = 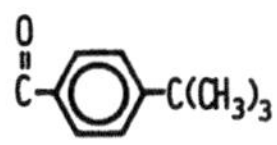	61.4–68.2	nr	100
126	27	18	R = 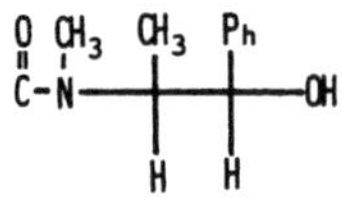 (1R, 2S)	syrup	nr	193
127	27	18	R = 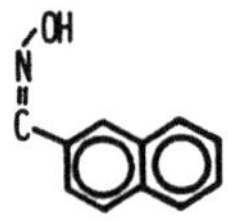	nr	nr	99
128	27	18	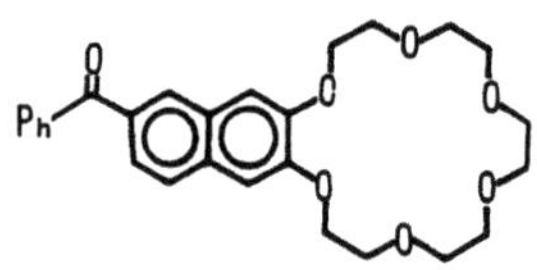	145.2–147.4	nr	100
129	27	18	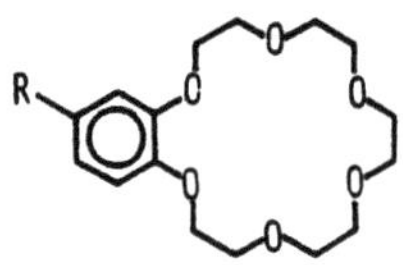 $R = \overset{NOH}{\overset{\Vert}{C}}(CH_2)_9CH_3$	nr	nr	99
130	28	18	R = 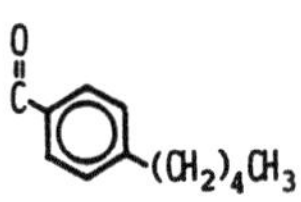	62.5–64	nr	100
131	28	18	R = 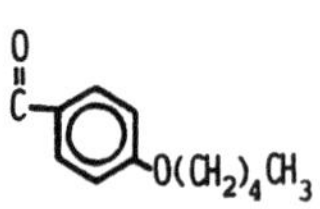	83.7–84.9	nr	100
132	28	18	$R = \overset{N-O(CH_2)_9CH_3}{\overset{\Vert}{C}}CH_3$	41.7–42.6 $\cdot H_2O$	nr	99
133	28	18	R = 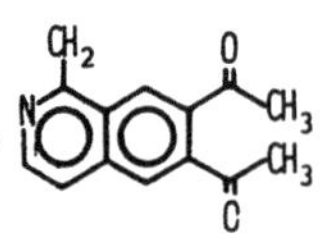	99–102	24	62

Table 3.8 (continued)

Cpd. No.	No. C's	Ring size	Structure	Mp or (bp) in °C	Yield (%)	Refs.
134	28	18	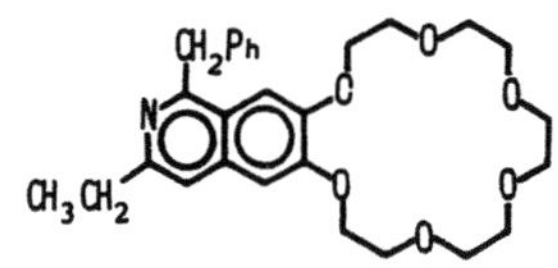	84–85	40	62
135	28	18	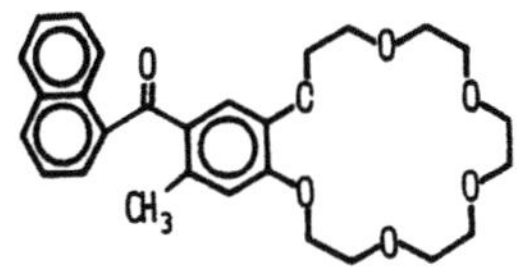	80–93	nr	100
136	28	18	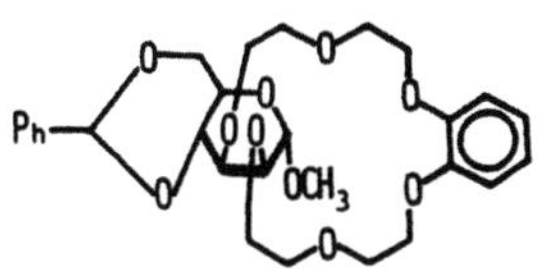	gum	70	57
137	28	18	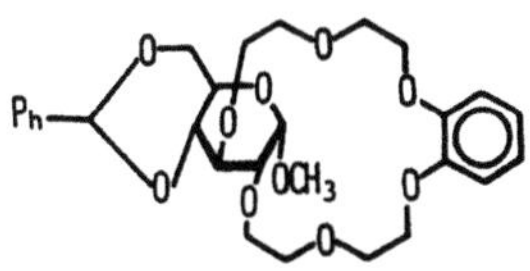	129–130	68	57
			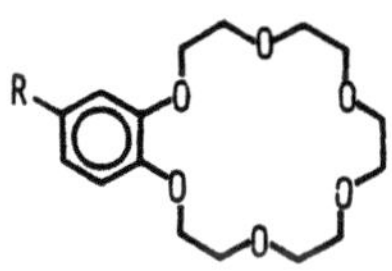			
138	29	18	R = 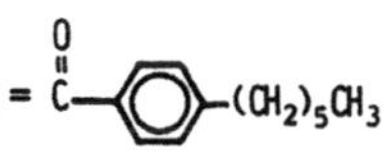	70.0–70.8	nr	100
139	30	18	R = C(=NOCH$_2$Ph)Ph	nr	nr	99
140	29	18	R = 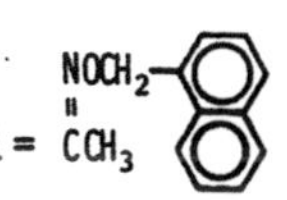	nr	nr	99
141	29	18	R = 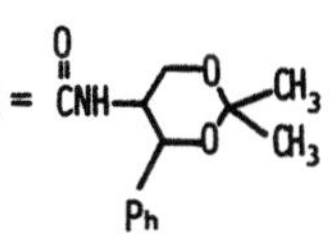	syrup	nr	193
			(S, S)	syrup	nr	193
142	34	18	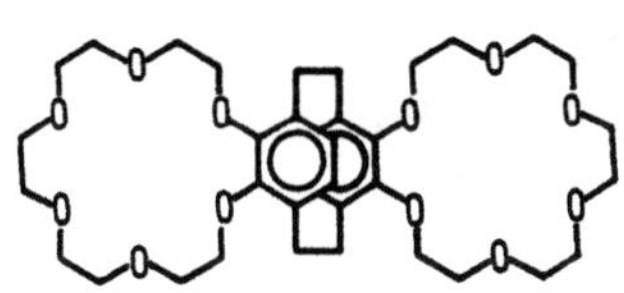	157–158	14	89, 91

Table 3.9. Derivatives of dibenzo-18-crown-6

Cpd. No.	No. C's	Ring size	Structure	Mp or (bp) in °C	Yield (%)	Refs.
			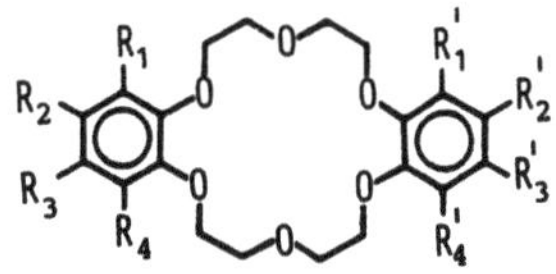			
1	20	18	$R_1 = R_2 = R_3 = R_4 = R'_1 = R'_2 = R'_3 = R'_4 = H$	164	80	169, 171, 173
2	20	18	$R_1 = R_3 = R_4 = R'_1 = R'_3 = R'_4 = H$, $R_2 = R'_2 = NH_2$	180–184	nr	65
3	20	18	$R_1 = R_3 = R_4 = R'_1 = R'_2 = R'_4 = H$, $R_2 = R'_3 = NH_2$	199–203	83	4, 65
4	20	18	$R_1 = R_3 = R_4 = R'_1 = R'_2 = R'_4 = H$, $R_2 = R'_3 = NO_2$	247–252	34.7	4, 65
5	20	18	$R_1 = R_3 = R_4 = R'_1 = R'_3 = R'_4 = H$, $R_2 = R'_2 = NO_2$	206–232	nr	65
6	20	18	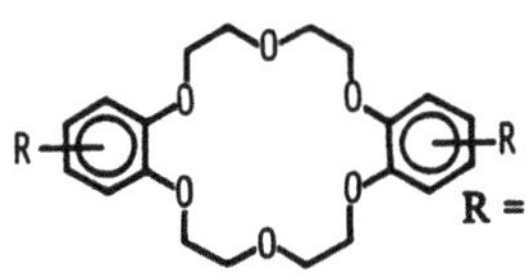 $R = NO_2$	177–236	nr	173
7	20	18	$R = SO_3H \cdot H_2O$	139	nr	173
8	20	18	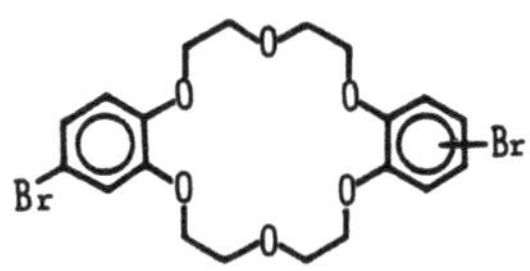	nr	83	191
9	20	18	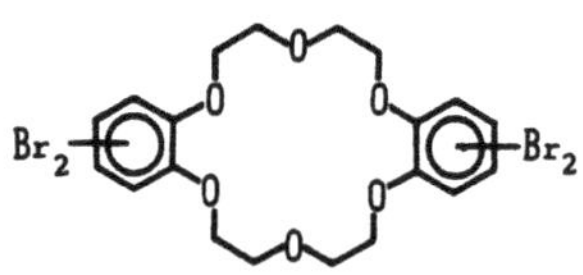	200–215	50	163
10	20	18	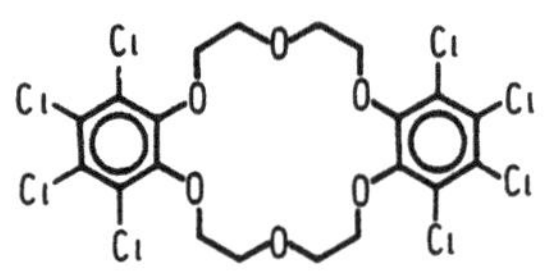	283	25	163
11	22	18	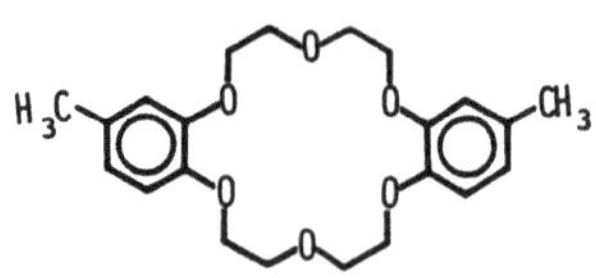	132–135	40	229

Table 3.9 (continued)

Cpd. No.	No. C's	Ring size	Structure	Mp or (bp) in °C	Yield (%)	Refs.
12	23	18	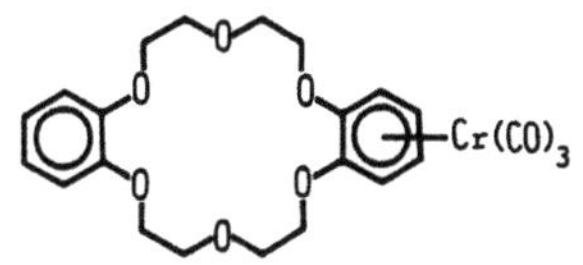	nr 141–143	~70 2	162 163
13	24	18	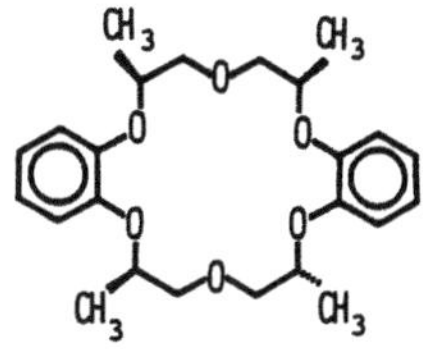 (R, R, R, S)	84–86	nr	165
14	24	18	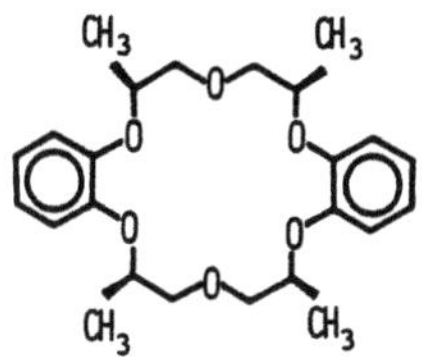 (R, S, R, S)	109–110	6.5	165
15	24	18	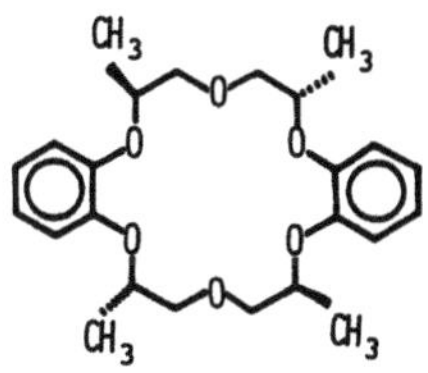 (R, R, S, S)	121–122	nr	165
16	24	18	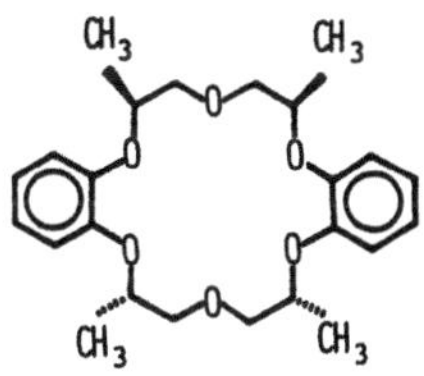 (R, S, S, R)	137	5.6	165
17	24	18	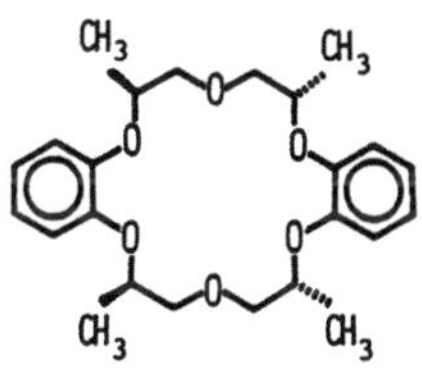 (R, R, R, R)	136–137	3.0	165
18	24	18	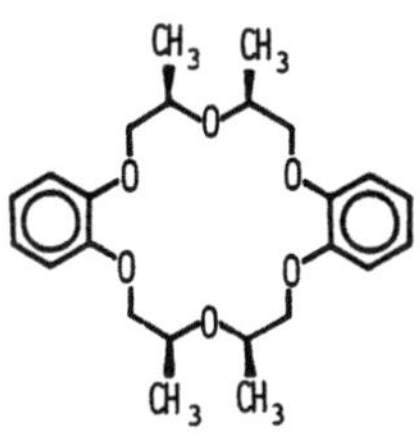 (R, S, R, S)	92	1.9	165

Table 3.9 (continued)

Cpd. No.	No. C's	Ring size	Structure	Mp or (bp) in °C	Yield (%)	Refs.
19	24	18	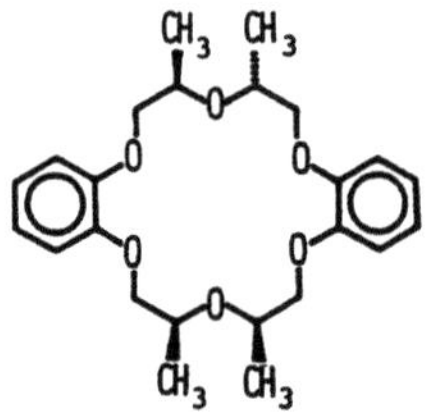 (R, R, R, S)	120–122	31.1	165
20	24	18	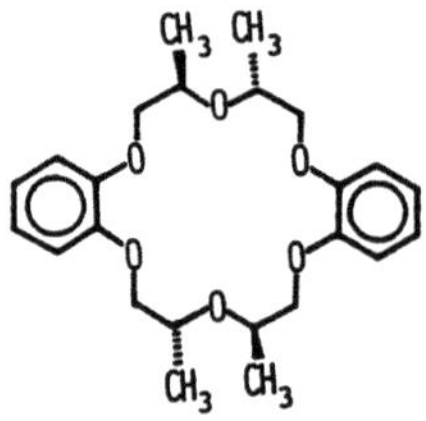 (R, R, R, R)	134	3.8	165
21	24	18	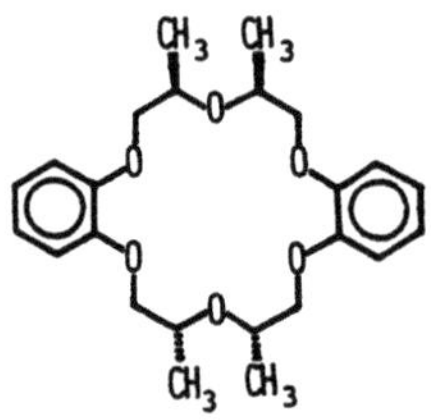 (R, S, S, R)	160	8.4	165
22	24	18	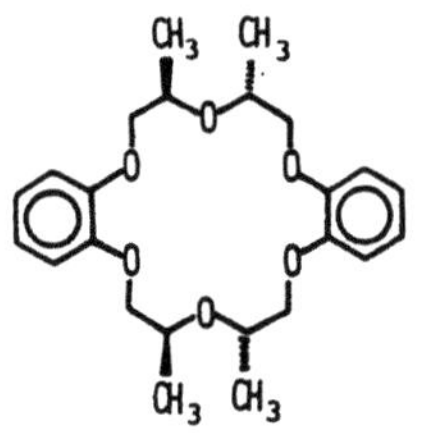 (R, R, S, S)	199–200	6.7	165
23	24	18	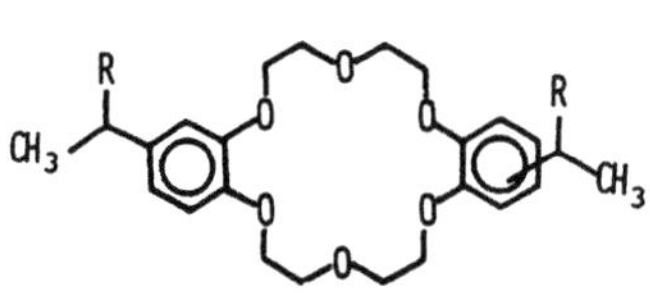 R = H	nr	nr	202
24	24	18	R = OH	164–167	84	164
			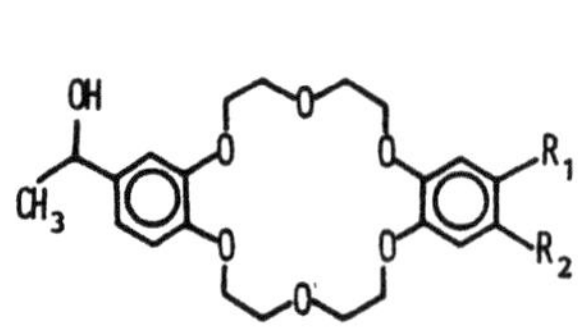			
25	24	18	R_1 = H, R_2 = CH(OH)CH_3	nr	nr	203
26	24	18	R_1 = CH(OH)CH_3, R_2 = H	nr	nr	203

Table 3.9 (continued)

Cpd. No.	No. C's	Ring size	Structure	Mp or (bp) in °C	Yield (%)	Refs.
27	24	18	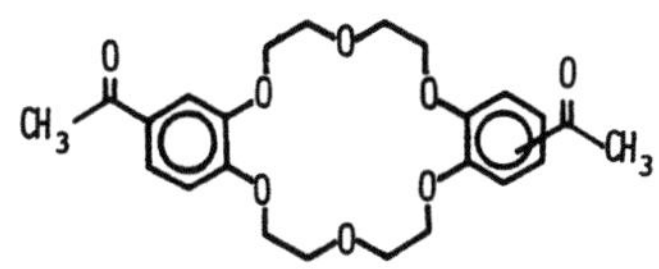	200–208 nr	86 85	164 201, 202
			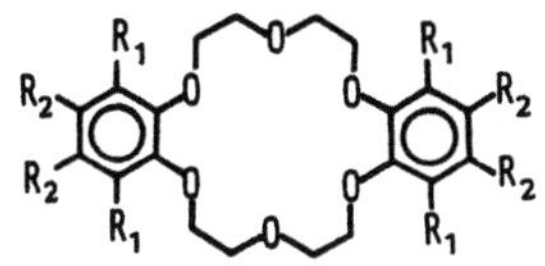			
28	24	18	R_1 = H, R_2 = CH_2Br	212–215	75	46
29	24	18	R_1 = CN, R_2 = H	220	5	130
30	24	18	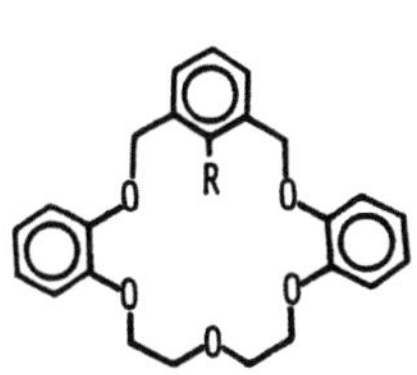 R = H	158–160	5	80
31	24	18	R = NO_2	179–180	36	80
32	24	18	R = F	178–179	3	80
33	25	18	R = OMe	215–216	9	80
34	25	18	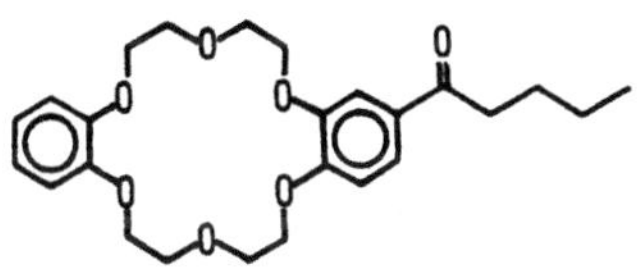	123–124	nr	86
35	26	18	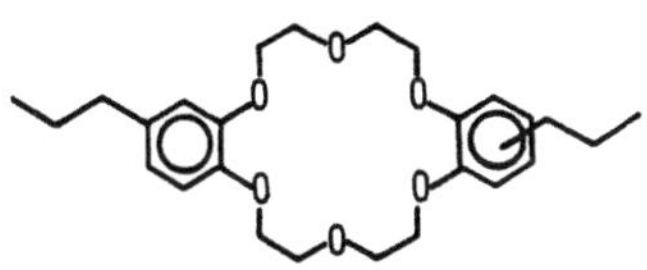	nr	nr	202
			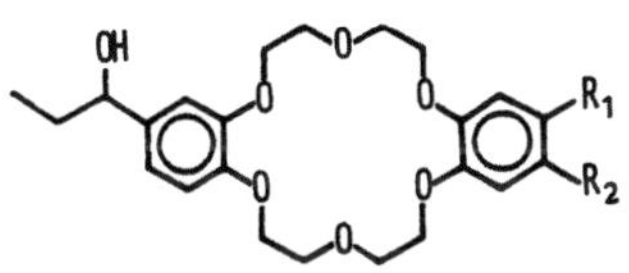			
36	26	18	R_1 = H, R_2 = $CH(OH)CH_2CH_3$	nr	nr	203
37	26	18	R_1 = $CH(OH)CH_2CH_3$, R_2 = H	nr	nr	203
38	26	18	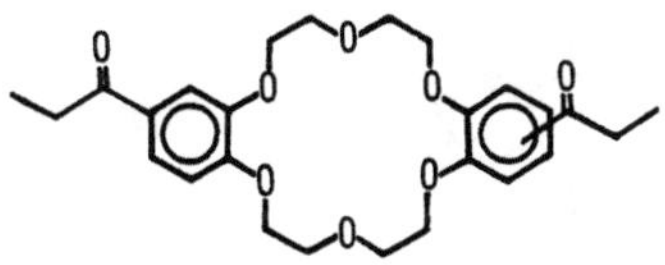	nr	70	201, 202

Table 3.9 (continued)

Cpd. No.	No. C's	Ring size	Structure	Mp or (bp) in °C	Yield (%)	Refs.
39	26	18	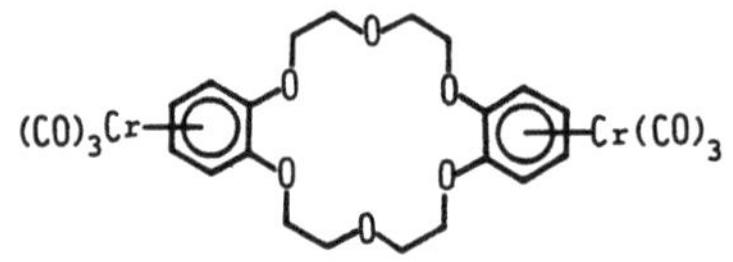	155–158	~70	162, 163
			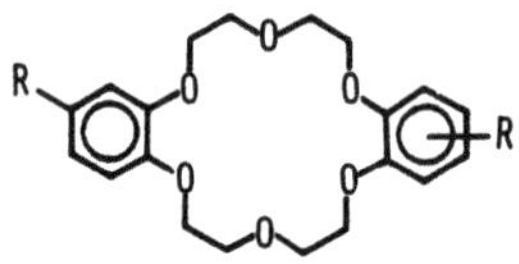			
40	28	18	$R = (CH_2)_3CH_3$	nr	nr	202
41	28	18	$R = C(CH_3)_3$	135–137	17	169, 173
42	28	18	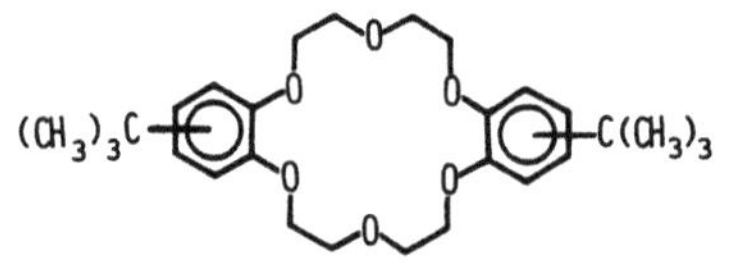	60–90	91	46
			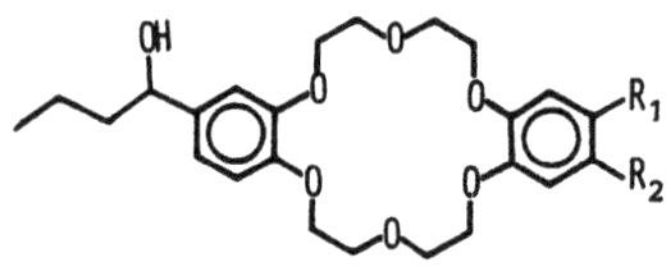			
43	28	18	$R_1 = H, R_2 = CH(OH)CH_2CH_2CH_3$	nr	nr	203
44	28	18	$R = CH(OH)CH_2CH_2CH_3, R_2 = H$	nr	nr	203
			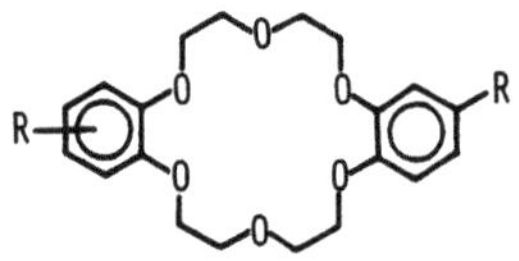			
45	28	18	$R = \overset{O}{\overset{\parallel}{C}}CH_2CH_2CH_3$	149–159 nr	72 70	164 201, 202
46	28	18	$R = \overset{O}{\overset{\parallel}{C}}CH(CH_3)_2$	154–160	88	164
47	28	18	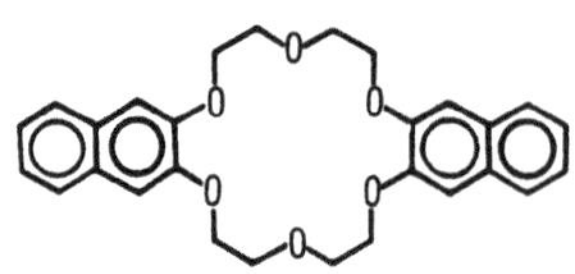	244–246	25	169, 173

Table 3.9 (continued)

Cpd. No.	No. C's	Ring size	Structure	Mp or (bp) in °C	Yield (%)	Refs.
			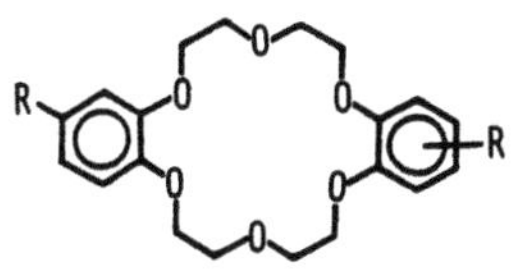			
48	30	18	$R = (CH_2)_4CH_3$	nr	nr	202
49	30	18	$R = CH(OH)CH_2CH_2CH_2CH_3$	nr	nr	203
50	30	18	$R = \overset{O}{\overset{\|}{C}}(CH_2)_3CH_3$	nr	72	201, 202
51	30	18	$R = \overset{O}{\overset{\|}{C}}C(CH_3)_3$	146–159	85	164
52	30	18	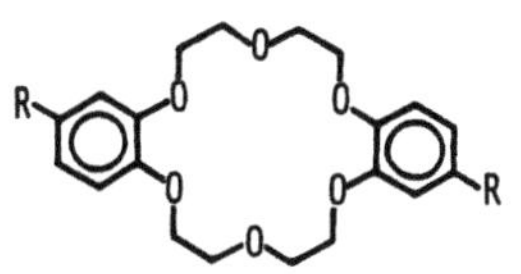 R = 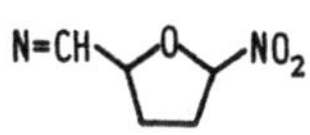	nr	nr	178
			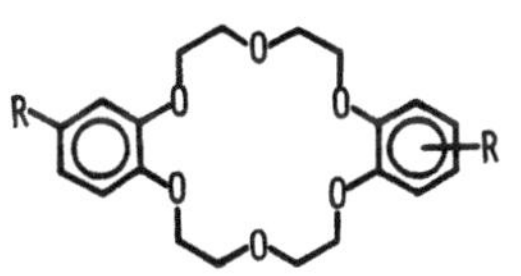			
53	32	18	$R = (CH_2)_5CH_3$	nr	nr	202
54	32	18	$R = CH(OH)(CH_2)_4CH_3$	nr	nr	203
55	32	18	$R = \overset{O}{\overset{\|}{C}}(CH_2)_4CH_3$	nr	nr	202
56	32	18	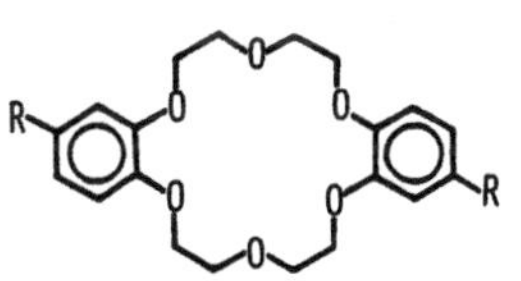 R = 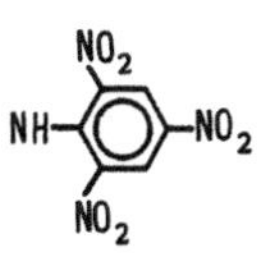	222–226	67	83

Table 3.9 (continued)

Cpd. No.	No. C's	Ring size	Structure	Mp or (bp) in °C	Yield (%)	Refs.
			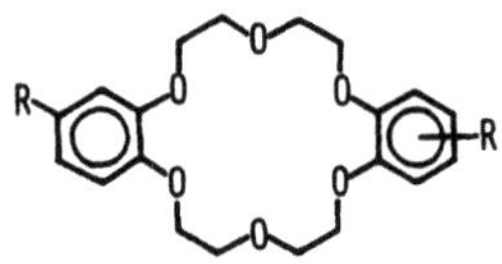			
57	34	18	$R = (CH_2)_6CH_3$	nr	nr	202
58	34	18	$R = CH(OH)(CH_2)_5CH_3$	nr	nr	203
59	34	18	$R = \overset{O}{\overset{\|}{C}}(CH_2)_5CH_3$	nr	nr	202
60	34	18	$R = CH_2Ph$	nr	nr	202
61	34	18	$R = \overset{O}{\overset{\|}{C}}Ph$	nr 166–169	68 91	201, 202 164
			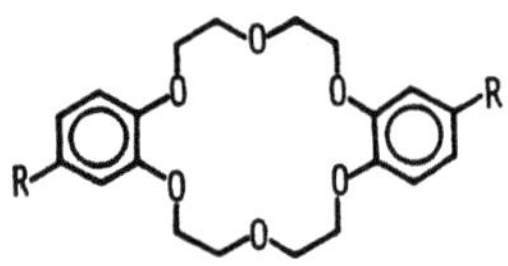			
62	34	18	R = 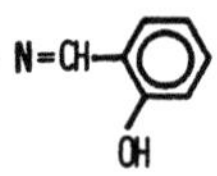	nr	nr	178
63	34	18	$R = NH\overset{S}{\overset{\|}{C}}NHPh$	nr	nr	178
			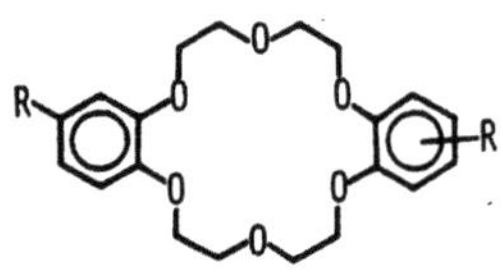			
64	36	18	$R = (CH_2)_7CH_3$	nr	nr	202
65	36	18	$R = CH(OH)(CH_2)_6CH_3$	nr	nr	203
66	36	18	$R = \overset{O}{\overset{\|}{C}}(CH_2)_6CH_3$	nr	nr	202
67	36	18	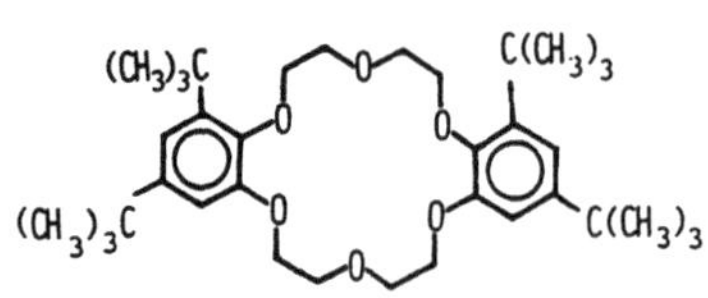	nr	12	212

Table 3.9 (continued)

Cpd. No.	No. C's	Ring size	Structure	Mp or (bp) in °C	Yield (%)	Refs.
			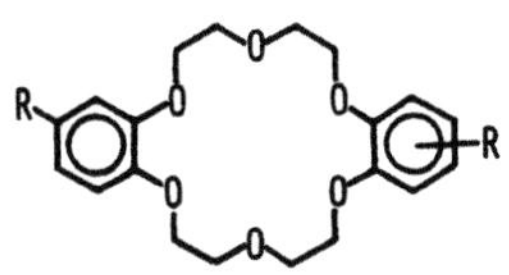			
68	36	18	R = 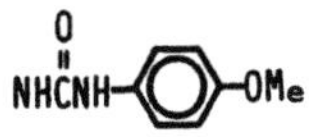	nr	nr	178
69	40	18	R = $CH(OH)(CH_2)_8CH_3$	120–122	100	164
70	40	18	R = $C(=O)(CH_2)_8CH_3$	117–129	75	164
71	48	18	R = $CH(OH)(CH_2)_{12}CH_3$	111–117	100	164
72	48	18	R = $C(=O)(CH_2)_{12}CH_3$	123–127	71	164
73	52	18	R = $(CH_2)_{15}CH_3$	nr	28	212
74	56	18	R = $C(=O)(CH_2)_{16}CH_3$	112–122	75	164

Table 3.10. Miscellaneous dibenzo crowns

Cpd. No.	No. C's	Ring size	Structure		Mp or (bp) in °C	Yield (%)	Refs.
1	16	12	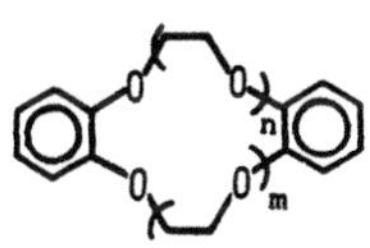	n = m = 1	208–209	15	169
2	18	15		n = 1, m = 2	113.5–115	43	169, 173
3	18	14	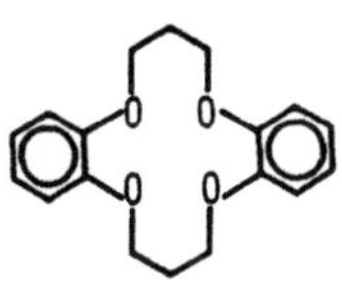		150–152	27	169, 173

Table 3.10 (continued)

Cpd. No.	No. C's	Ring size	Structure	Mp or (bp) in °C	Yield (%)	Refs.
4	19	16	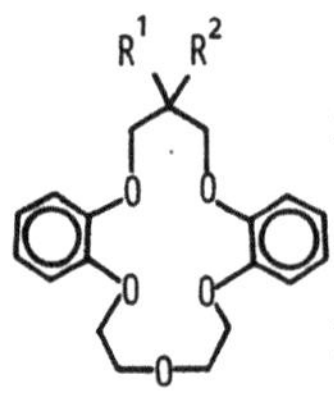	117–118	25	229
			$R^1 = R^2 = H$	117–118	18	169, 173
5	20	16	$R^1, R^2 = CH_2$	94–95	40	209
6	20	16	$R^1 = CH_2OH, R^2 = H$	110–111	75	209
7	20	19	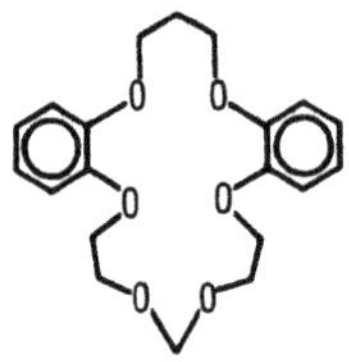	118	31	170, 174
8	20	18	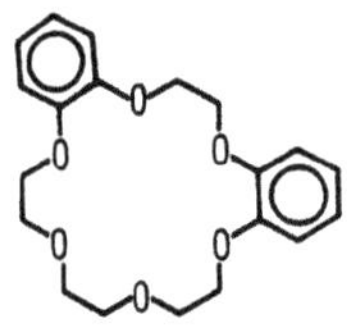	117–118	25	169, 173
9	21	18	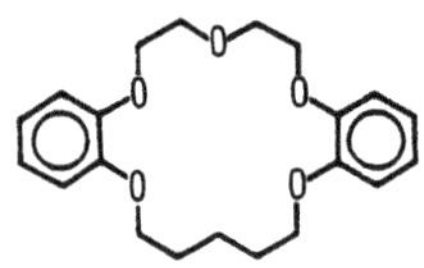	157–158	28	169
10	21	19	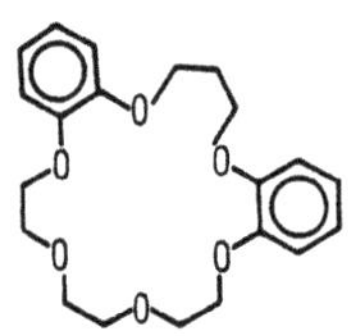	84.5–86	16	169, 173
11	21	21	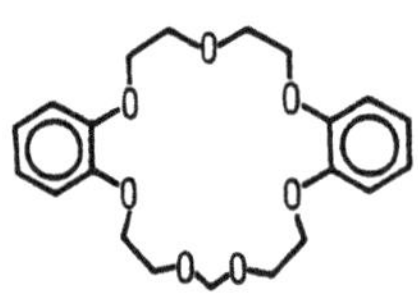	151–151	62	170, 174
12	22	24	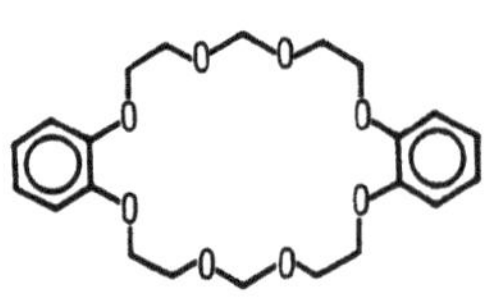	166–167	8	170, 174

Table 3.10 (continued)

Cpd. No.	No. C's	Ring size	Structure	Mp or (bp) in ′C	Yield (%)	Refs.
13	22	21	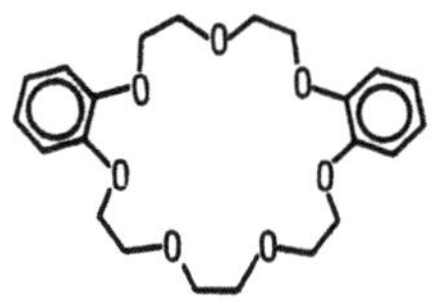	106.5–107.5	36	169
14	24	20	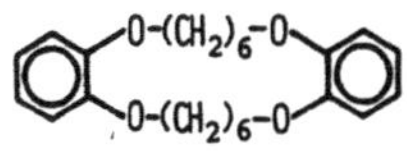	139–141	4	169
15	24	22	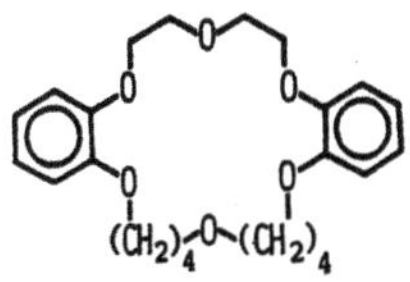	82–83	9	169
16	24	24	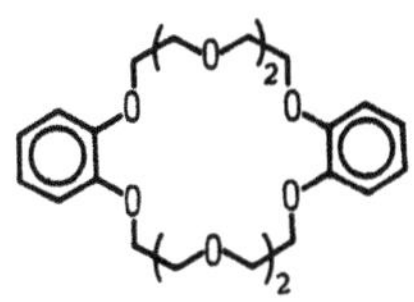	113–114	38	169,173
17	26	14	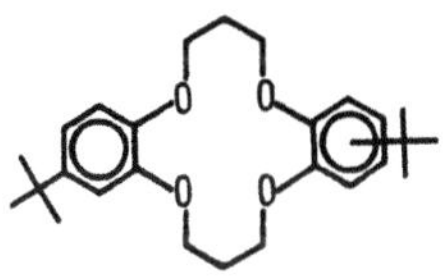	149–152	11	169, 173
18	26	18	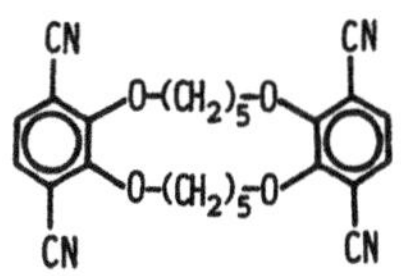	187–188	nr	131
19	28	26	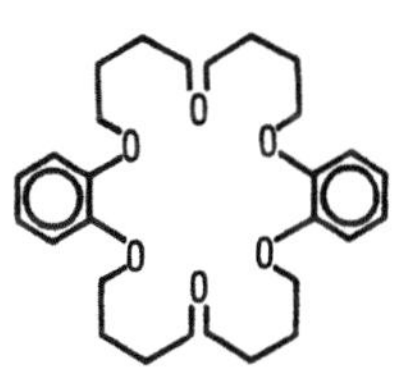	125–127	17	169
20	28	24	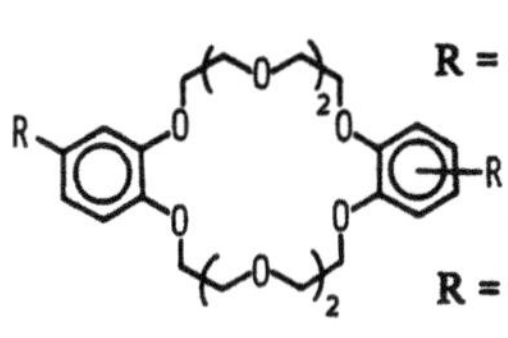 R = $CHOHCH_3$	oil	89	164
21	28	24	R = $\overset{O}{\parallel}{C}CH_3$	150–160	85	164

Table 3.10 (continued)

Cpd. No.	No. C's	Ring size	Structure	Mp or (bp) in °C	Yield (%)	Refs.
22	28	30	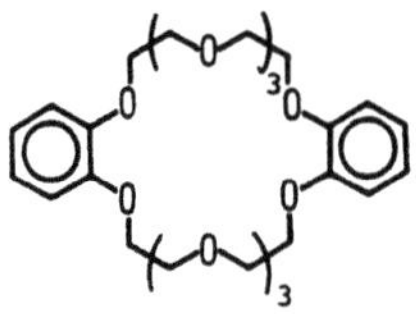	106–107.5	nr	169,173
23	32	24	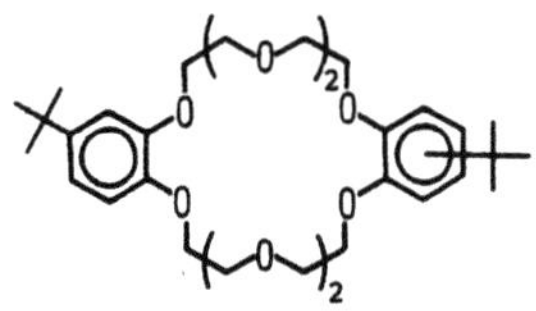	52–74	93	46
24	32	24	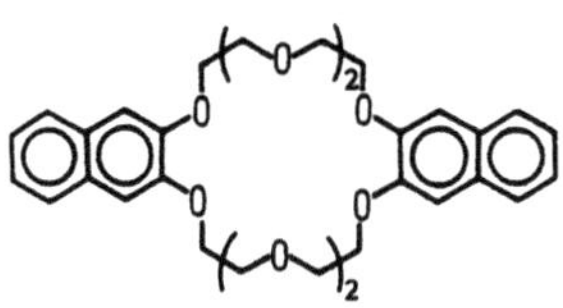	190–191.5	63	169
25	32	28	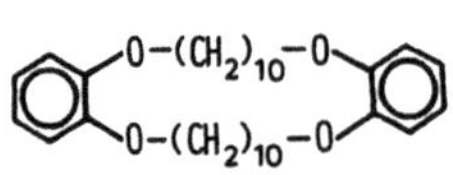	137–138.5	3	169
26	32	34	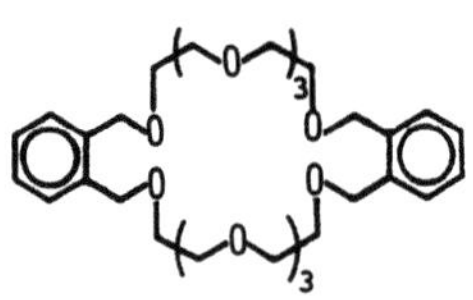	nr	8	185,186
27	34	20	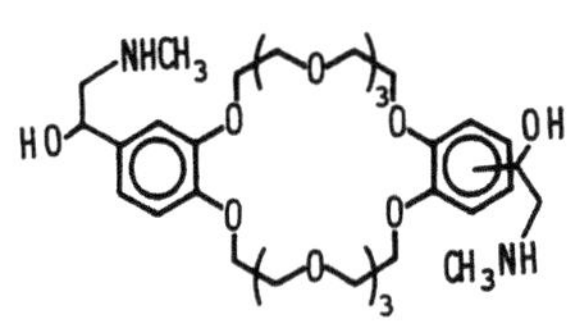	190–192	nr	220
28	40	48	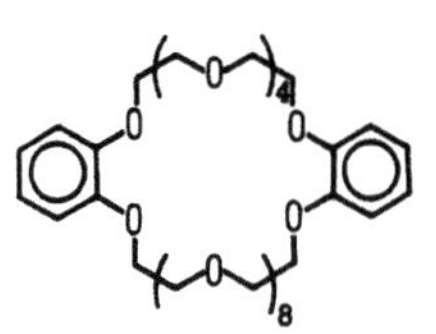	<26	32	169,173
29	44	24	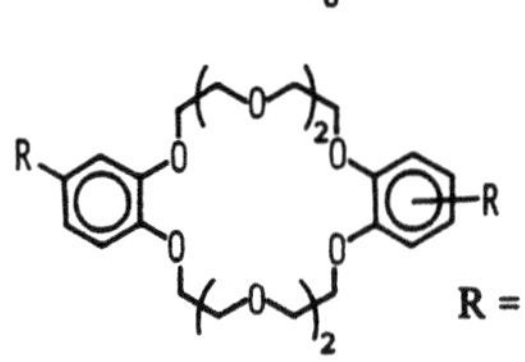 R = C(=O)$(CH_2)_8CH_3$	78–91	84	164
30	44	24	R = CHOH$(CH_2)_8CH_3$	88–92	75	164

Table 3.10 (continued)

Cpd. No.	No. C's	Ring size	Structure	Mp or (bp) in °C	Yield (%)	Refs.
31	48	60	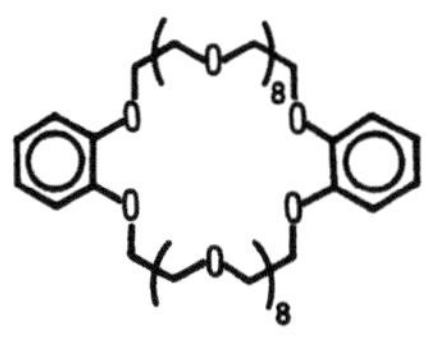	<26	41	169, 173

Table 3.11. Crowns containing two or more fused benzo units

Cpd. No.	No. C's	Ring size	Structure	Mp or (bp) in °C	Yield (%)	Refs.
1	24	17	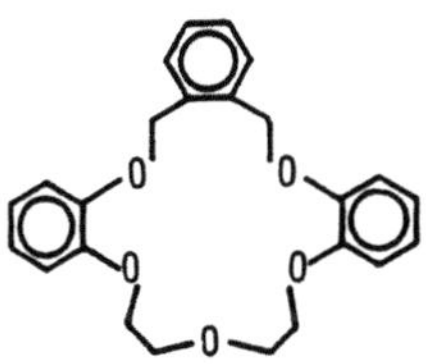	131–132	8	80
2	24	18	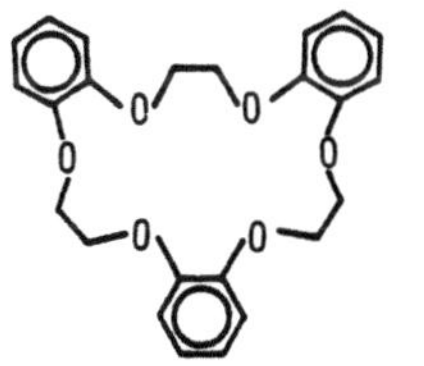	190–192	28	169, 173
3	24	19	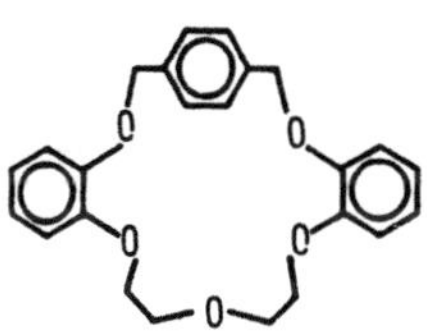	121–122	2	80
4	25	19	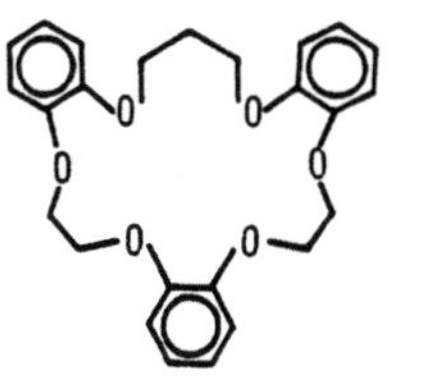	147–149	16	169, 173

Table 3.11 (continued)

Cpd. No.	No. C's	Ring size	Structure	Mp or (bp) in °C	Yield (%)	Refs.
5	26	21	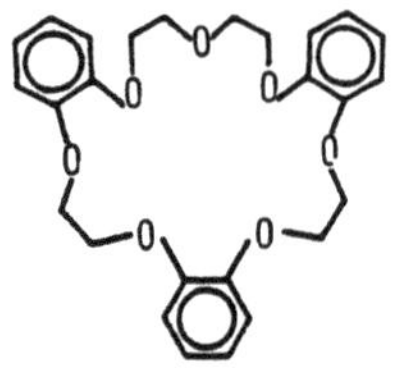	98.5–100	19	169
6	28	18	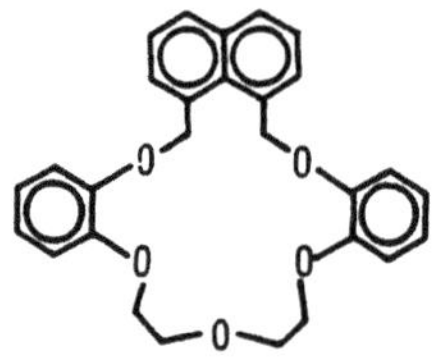	152–153	10	80
7	28	24		oil	nr	173
8	32	16	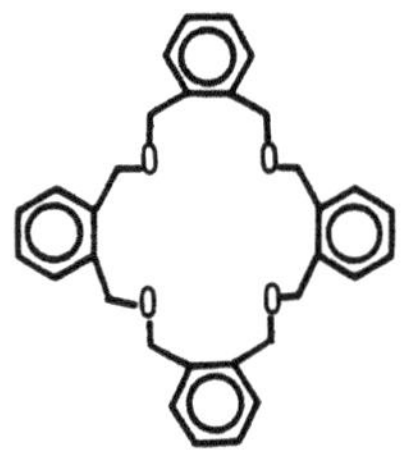	194	40	224
9	32	24	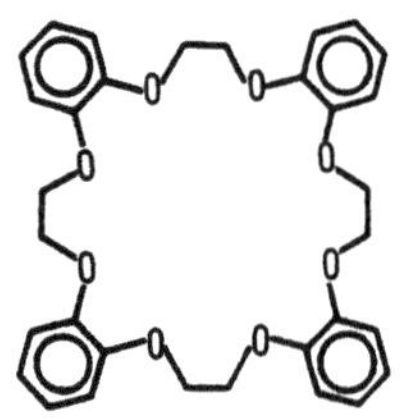	150–152	19	169,173
10	36	18	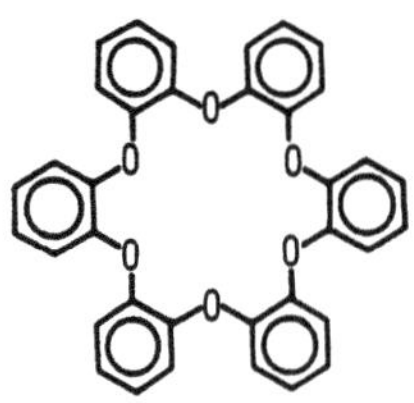	282–283	nr	104

Table 3.11 (continued)

Cpd. No.	No. C's	Ring size	Structure	Mp or (bp) in °C	Yield (%)	Refs.
11	42	21	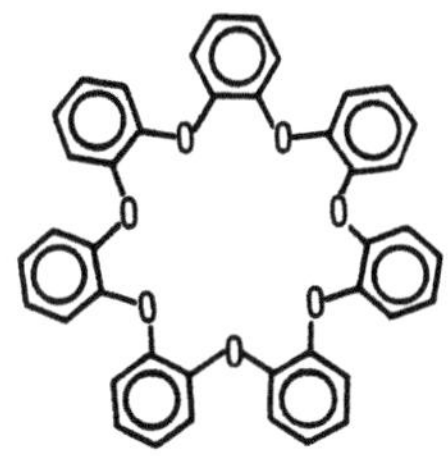	198–200	nr	104

Table 3.12. Crowns having an *ortho*-xylyl subunit as part of the macroring

Cpd. No.	No. C's	Ring size	Structure		Mp or (bp) in °C	Yield (%)	Refs.
1	12	11	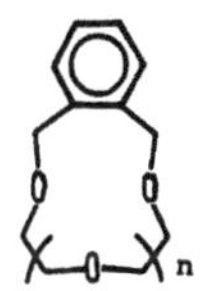	n = 1	nr	1	185, 186
2	14	14		n = 2	oil	8–10	185
					nr	25	186
3	16	17		n = 3	oil	34	185, 186
4	17	17	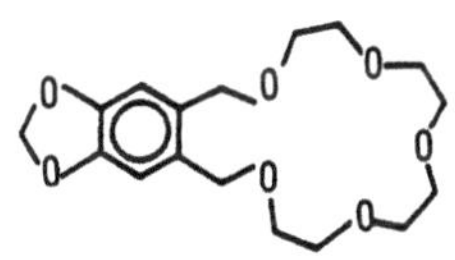		nr	29	185
5	18	20	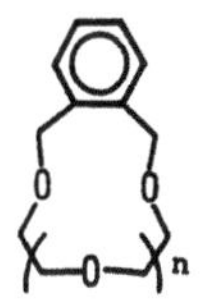	n = 4	nr	53	185, 186
6	20	23		n = 5	(200/0.01)	50	185, 186
7	21	23	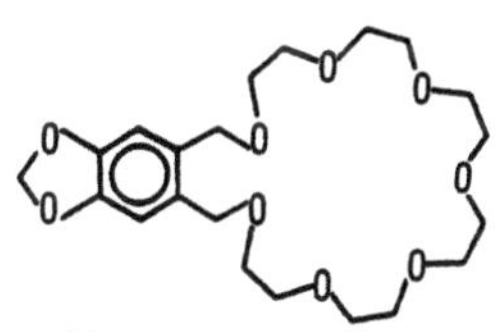		nr	25	185, 186
8	22	26		n = 6	nr	43	185, 186
9	24	29		n = 7	nr	28	185

Table 3.13. Miscellaneous fused *ortho*-benzo crowns

Cpd. No.	No. C's	Ring size	Structure	Mp or (bp) in °C	Yield (%)	Refs.
1	10	9	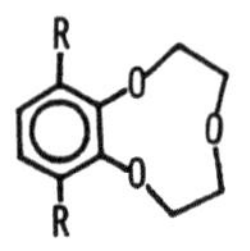 R = H	67–69	5	169
2	12	9	R = CN	193	24	131
3	16	14	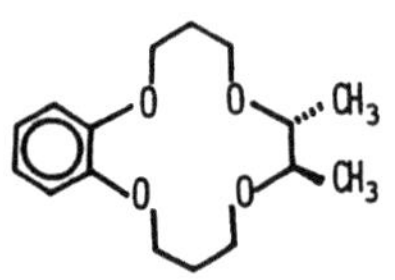 (RR)	nr	nr	126
4	18	9	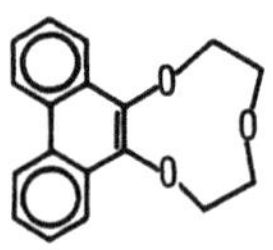	71–72	81	21
5	18	19	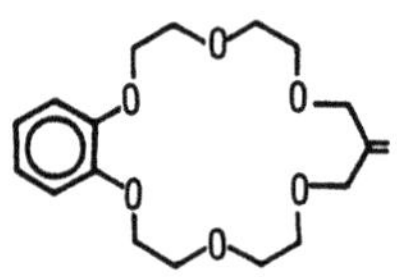	oil	39	209
6	18	21	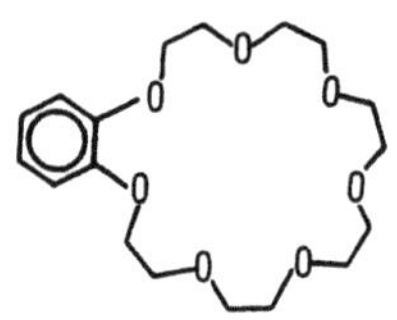	nr	65	182
7	20	12	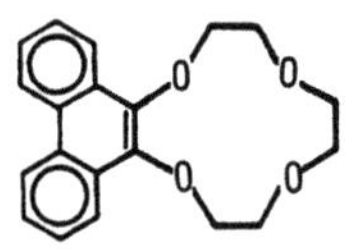	143–144	68	21
8	20	24	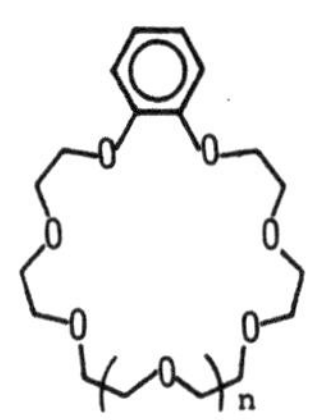 n = 1	nr	59	183
9	22	27	n = 2	nr	44	183

Table 3.14. Crowns having a *meta*-xylyl subunit as part of the macroring

Cpd. No.	No. C's	Ring size	Structure	Mp or (bp) in °C	Yield (%)	Refs.
1	12	12	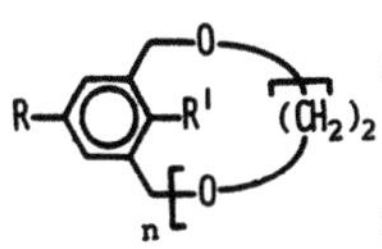 n = 2, R = R' = H	nr	2	80
				nr	2	184
2	14	15	n = 3, R = R' = H	nr	16	184
				nr	16	80
3	14	15	n = 3, R = H, R' = OH	nr	40	127
4	14	15	n = 3, R = NO_2, R' = OH	105–106	nr	127
5	15	15	n = 3, R = H, R' = CO_2H	106–112	98	138,140
6	16	15	n = 3, R = H, R' = CO_2Me	oil	34	138, 140
7	16	15	n = 3, R = Me, R' = OMe	oil	58	105
8	16	18	n = 4, R = H, R' = H	nr	67	80
				nr	60	135
				nr	60	138
				nr	67	184
				43–45	30	207
9	16	18	n = 4, R = Br, R' = H	45–60	62	135
10	16	18	n = 4, R = H, R' = OH	48–49	52	127
11	16	18	n = 4, R = H, R' = Cl	oil	53	138,140
12	16	18	n = 4, R = H, R' = Br	oil	7	138,140
13	16	18	n = 4, R = NO_2, R' = OH	91–91.5	nr	127
14	17	18	n = 4, R = CN, R' = H	oil	62	135
15	17	18	n = 4, R = OMe, R' = H	wax	86	135
16	17	18	n = 4, R = SMe, R' = H	oil	30	135
17	17	18	n = 4, R = H, R' = CN	oil	10	138,140
18	17	18	n = 4, R = H, R' = CH_2OH	nr	80	138,140
19	17	18	n = 4, R = H, R' = CO_2H	100–101	100	138,140
20	18	18	n = 4, R = H, R' = CH_2OMe	nr	50	138
21	18	18	n = 4, R = H, R' = CO_2Me	oil	82	138,140

Table 3.14 (continued)

Cpd. No.	No. C's	Ring size	Structure	Mp or (bp) in °C	Yield (%)	Refs.
22	18	18	n = 4, R = CH_3, R' = OMe	70–72	49	105
23	18	21	n = 5, R = H, R' = H	nr nr	49 49	80 184
24	19	18	n = 4, R = CO_2Et, R' = H	glass	50	135
25	19	21	n = 5, R = H, R' = CO_2H	86–95	98	138,140
26	20	18	n = 4, R = *t*-Bu, R' = H	nr	61	135
27	20	21	n = 5, R = H, R' = CO_2Me	oil	68	138,140
28	20	21	n = 5, R = CH_3, R' = OMe	71–73	59	105
29	20	24	n = 6, R = H, R' = H	nr	18	80
30	22	27	n = 7, R = H, R' = H	nr nr	21 21	80 184
31	24	18	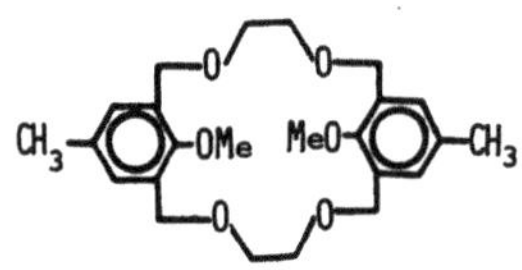	148–151	14	105
32	24	30	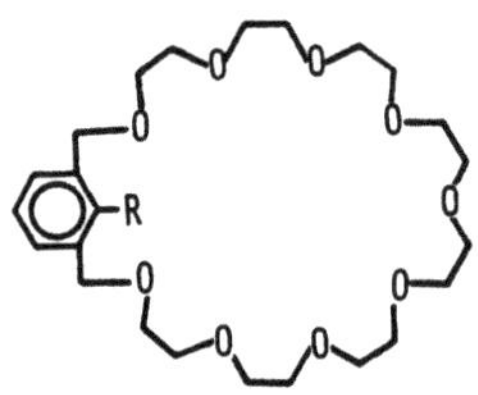 R = H	nr nr	21 21	80 184
33	25	30	R = CO_2H	oil	98	138,140
34	26	30	R = CO_2Me	oil	34	138,140
35	28	20	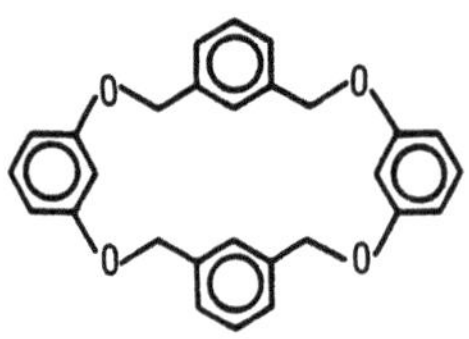	243–244	3	222
36	28	24		88–89	26	105

Table 3.14 (continued)

Cpd. No.	No. C's	Ring size	Structure	Mp or (bp) in °C	Yield (%)	Refs.
37	28	30	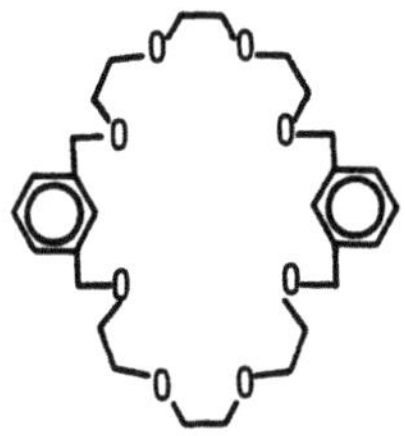	nr	16	80
38	32	18	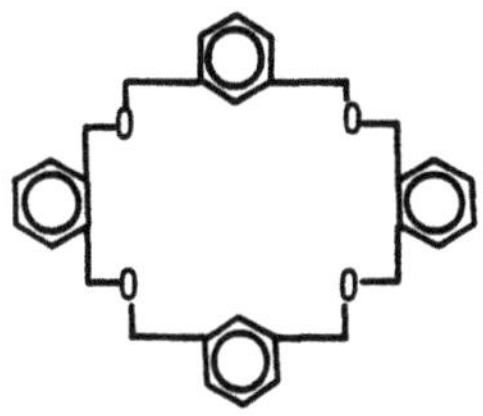	142–143	nr	224
39	52	22	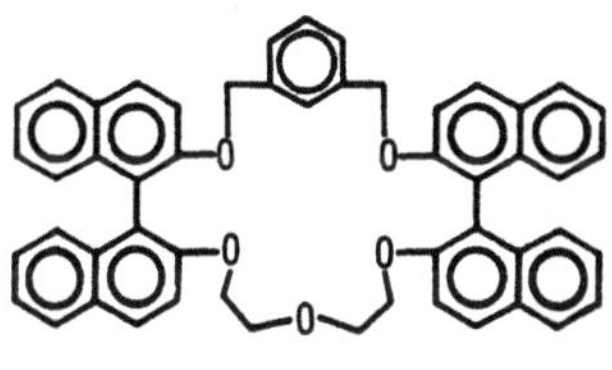	foam	13	114
				nr	13	76

Table 3.15. Crowns having a *meta*-benzo unit as part of the macroring

Cpd. No.	No. C's	Ring size	Structure		Mp or (bp) in °C	Yield (%)	Refs.
1	14	16	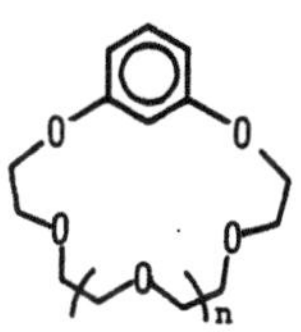	n = 1	oil	20	217
2	16	19		n = 2	oil	69	217
3	18	22		n = 3	oil	70	217

Table 3.15 (continued)

Cpd. No.	No. C's	Ring size	Structure		Mp or (bp) in °C	Yield (%)	Refs.
4	19	16	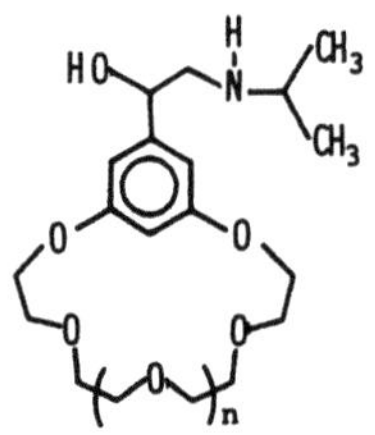	n = 1	114	1	62
5	21	19		n = 2	107.5–109.5	5	62

Table 3.16. Crowns having a *para*-benzo subunit as part of the macroring

Cpd. No.	No. C's	Ring size	Structure		Mp or (bp) in °C	Yield (%)	Refs.
1	12	13	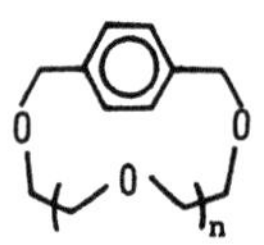	n = 1	nr	4	186
2	14	16		n = 2	nr	35	186
					56–56.5	2.6	101
					nr	4	80
3	16	19		n = 3	nr	16	186
					59–60	10	101
					nr	35	80
4	16	18	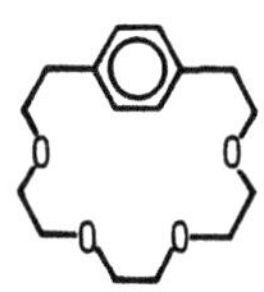		oil	17	101
5	16	20	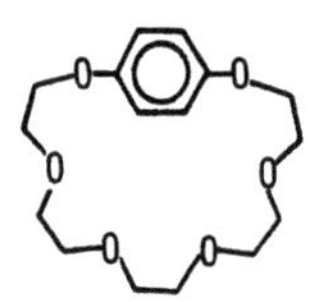		oil	2	172
					oil	nr	89
					oil	2.2	101
6	18	21	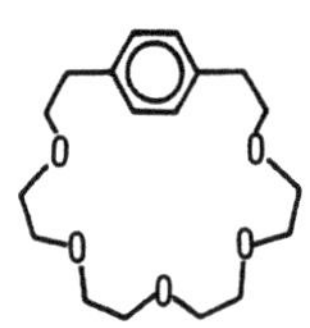		oil	17	101

Table 3.16 (continued)

Cpd. No.	No. C's	Ring size	Structure		Mp or (bp) in °C	Yield (%)	Refs.
7	18	20	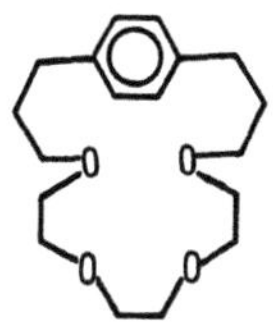		36.0–37.0	2.9	101
8	18	22		n = 1	nr	16	80
					nr	6	186
9	20	25		n = 2	nr	6	80
					nr	7	186
10	20	24	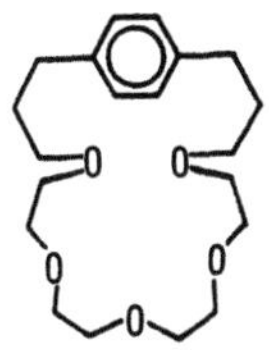		oil	3	101
11	20	28	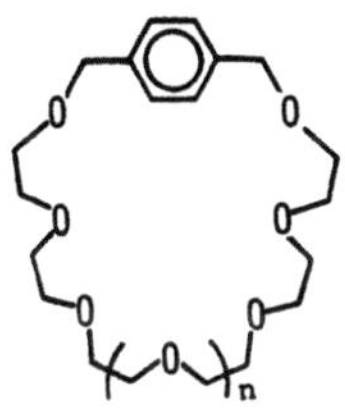	n = 2	nr	13	186
					nr	7	80
12	24	31		n = 3	nr	13	80

Table 3.17. Miscellaneous crowns containing the biphenyl subunit

Cpd. No.	No. C's	Ring size	Structure		Mp or (bp) in °C	Yield (%)	Refs.
1	16	11	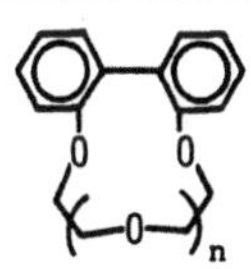	n = 1	190–192	27	5
2	22	20		n = 4	nr	23	183
					64–65	12	116
3	22	19	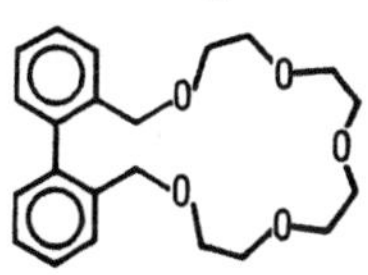		oil	30	182

Table 3.17 (continued)

Cpd. No.	No. C's	Ring size	Structure	Mp or (bp) in °C	Yield (%)	Refs.
4	28	20	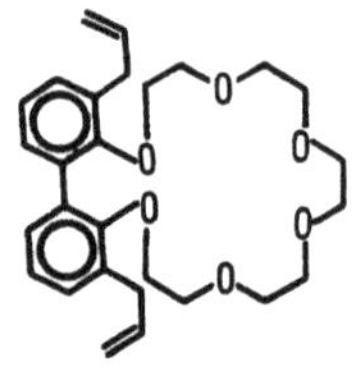	nr	47	183
5	30	19	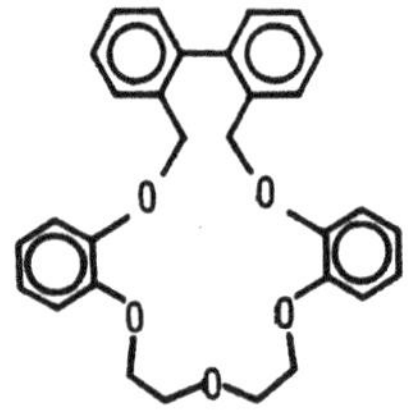	112–115	3	80

Table 3.18. Crowns containing the naphthalene subunit as part of the macroring

Cpd. No.	No. C's	Ring size	Structure	Mp or (bp) in °C	Yield (%)	Refs.
1	18	14	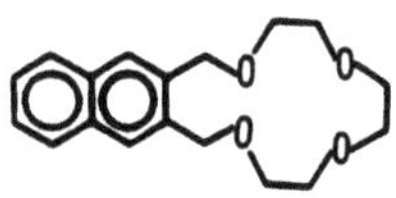	86–88	15.2	24
2	18	16	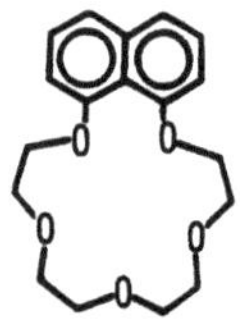	112–115	28	170
3	18	15		77–77.5	20.5	24
4	20	17	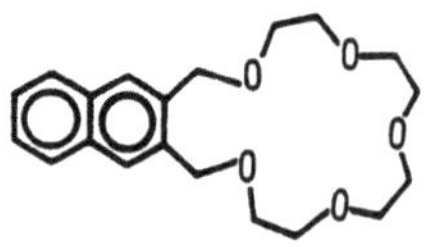	51–53	18.8	24

Table 3.18 (continued)

Cpd. No.	No. C's	Ring size	Structure	Mp or (bp) in °C	Yield (%)	Refs.
5	20	18		69.5–70	6.9	24
6	20	20	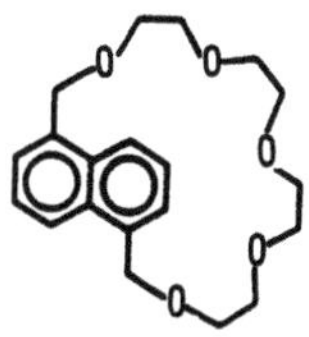	(110/0.05)	11.3	24
7	22	20	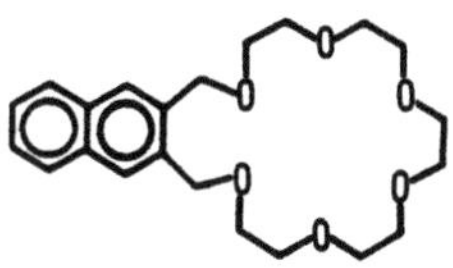	58–58.5	35.6	24
8	22	20	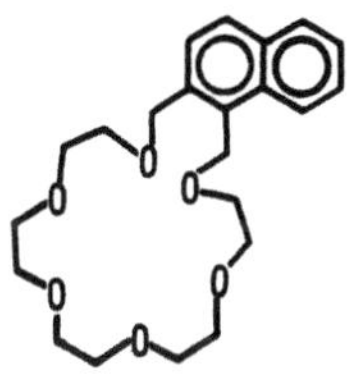	58–58.5	4.4	24
9	22	21		53.5–55.5	38.4	24
10	22	21	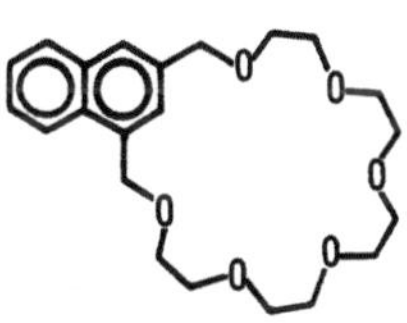	(120/0.05)	16.2	24
11	22	22		71.5–72.5	9.2	24

Table 3.18 (continued)

Cpd. No.	No. C's	Ring size	Structure	Mp or (bp) in °C	Yield (%)	Refs.
12	22	24	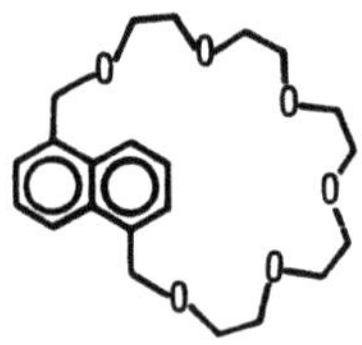	55–56	14.2	24

Table 3.19. Crown compounds containing a single 2,2′-binaphthyl subunit

Cpd. No.	No. C's	Ring size	Structure	Mp or (bp) in °C	Yield (%)	Refs.
1	24	11	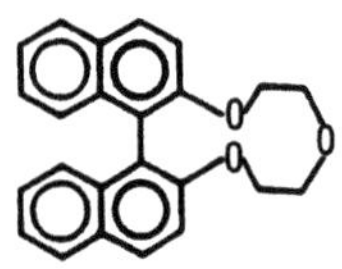	nr	nr	36
2	24	11	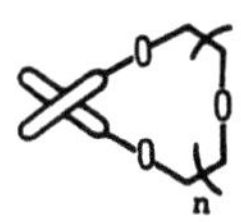 m = 1, (+) –(S)	230–231	5	114, 115
3	26	14	m = 2, (–) –(S)	gum	65	36
4	28	17	m = 3, (–) –(S)	gum	52	36
			racemic mixture	107–108.5	45	36, 116
5	28	18	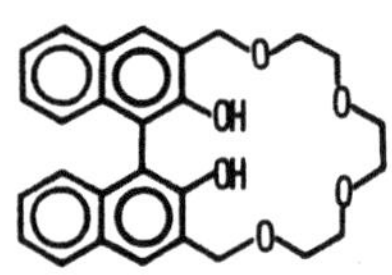	214–215	40	105
			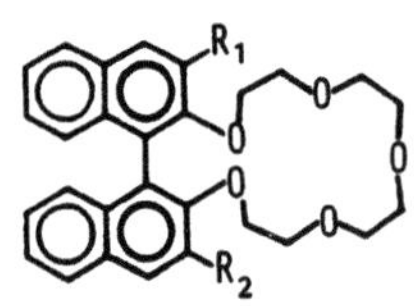			
6	29	17	$R_1 = CH_2OH$, $R_2 = H$	151–152	31	36
7	30	17	$R_1 = R_2 = CH_2OH$	oil	10	36, 90
8	30	17	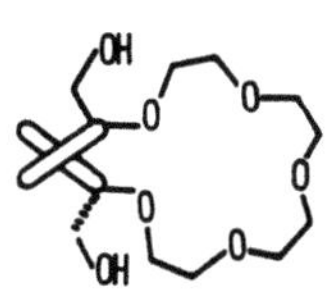	glass	6	36

Table 3.19 (continued)

Cpd. No.	No. C's	Ring size	Structure		Mp or (bp) in °C	Yield (%)	Refs.
9	30	18	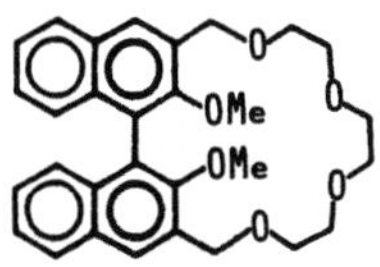		glass	40	105
10	30	20	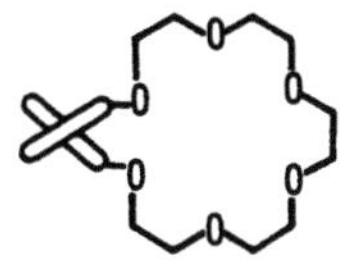	(–) –(S) (–) –(S) racemic	70–71 oil 130–130.5	29 64 60	182 36 115, 116
11	30	20	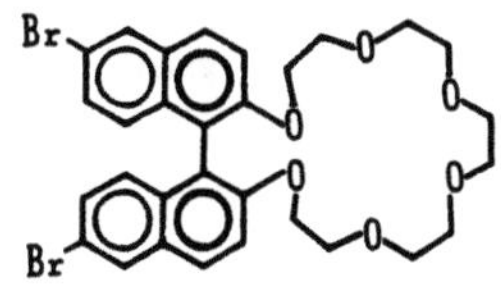		138–139.5	67	36
12	30	21	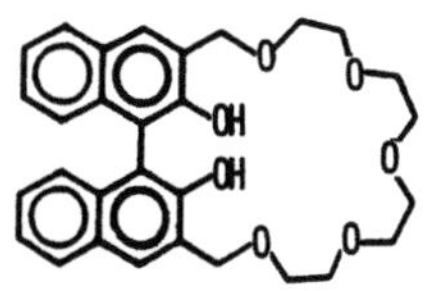		188–189	30	105
13	31	17	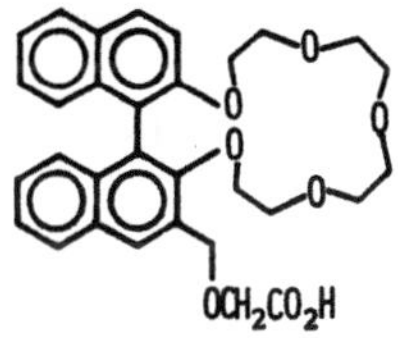		128–130	35	36
			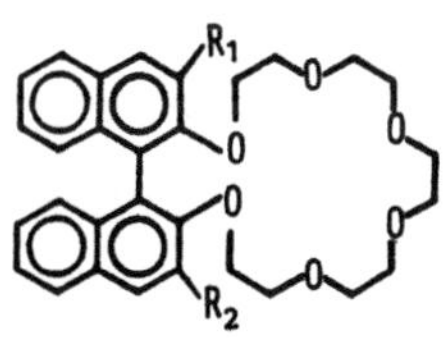				
14	31	20	$R_1 = CH_2OH, R_2 = H$		136–137	50	36, 90
15	32	20	$R_1 = CH_2OH, R_2 = CH_3$		159	59	36
16	32	20	$R_1 = R_2 = CH_2OH$		132	60	31, 36 90
17	32	20	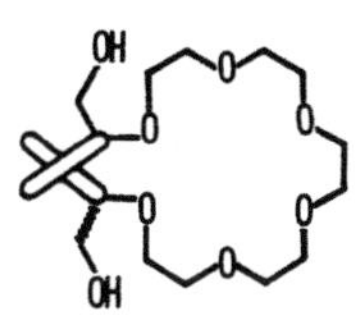	(S)	glass	55	36

Table 3.19 (continued)

Cpd. No.	No. C's	Ring size	Structure		Mp or (bp) in °C	Yield (%)	Refs.
18	32	20	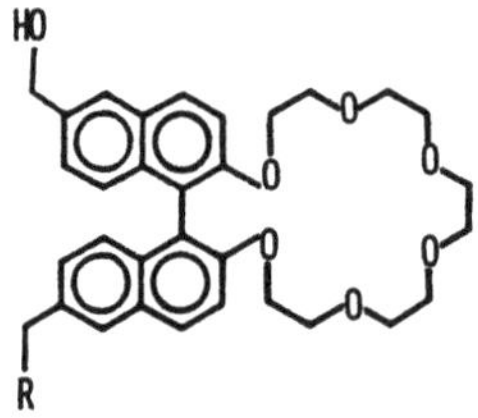	R = OH	glass	74	37, 38
19	32	20		R = Cl	glass	74	36
20	32	20	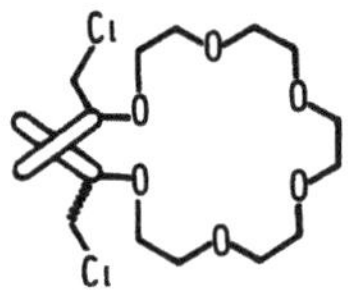	(S)	glass	81	36
				racemic	glass	91	36
			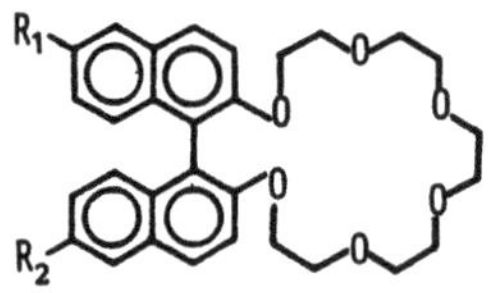				
21	32	20	$R_1 = R_2 = CH_2Cl$		glass	50	36
22	32	20	$R_1 = CO_2H, R_2 = CH_2Cl$		291–292	84	36
23	32	20	$R_1 = R_2 = CO_2H$		291–292	84	37, 38
24	32	20	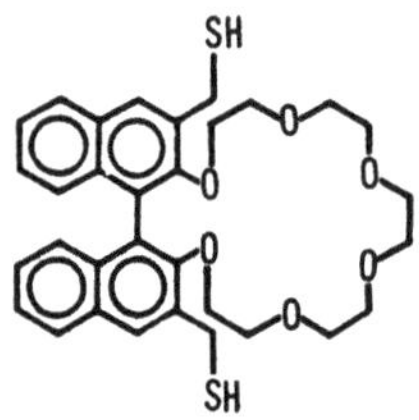	(S)	nr	nr	31
25	32	21	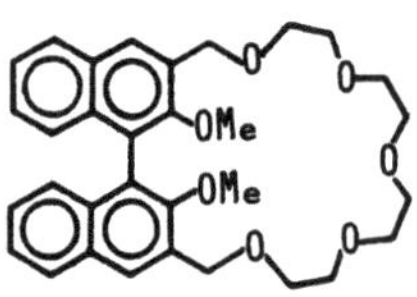		glass	40	105
26	32	23	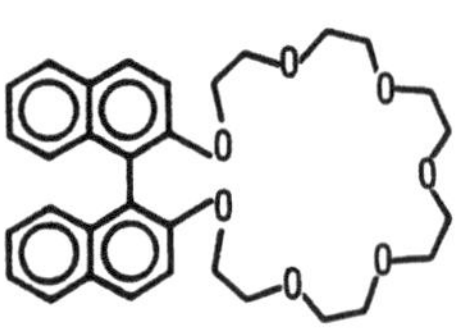		oil	48	36, 116

Table 3.19 (continued)

Cpd. No.	No. C's	Ring size	Structure	Mp or (bp) in °C	Yield (%)	Refs.
27	32	24	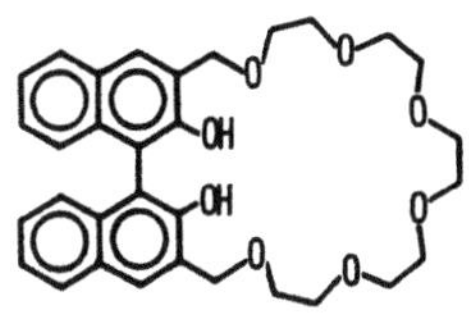	162–163	15	105
28	33	20	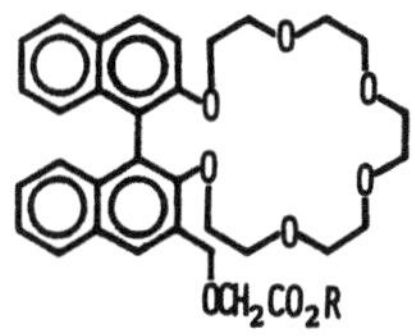 R = H	glass	70	36, 90
29	34	20	R = CH_3	glass	70	36, 90
30	34	17	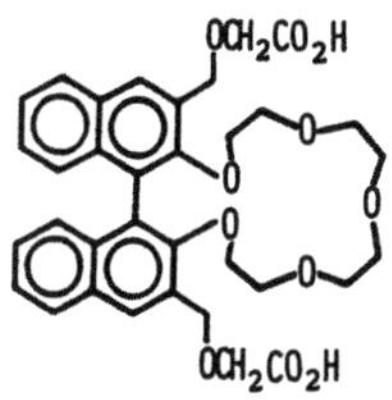	glass	85	36, 90
31	34	17	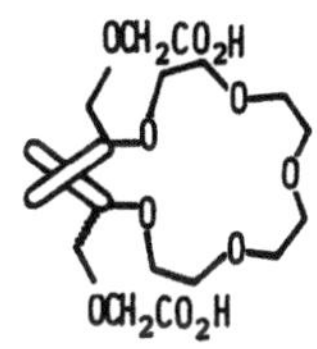	glass	65	36, 88
32	34	20	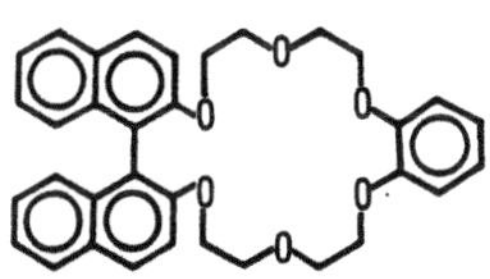	147–148	41	116
33	34	20	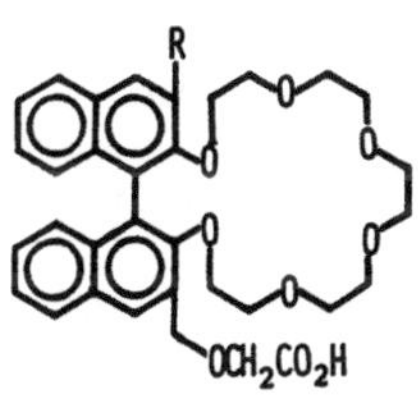 R = CH_3	glass	85	36
34	34	20	R = CH_2OH	glass	77	36
35	34	20	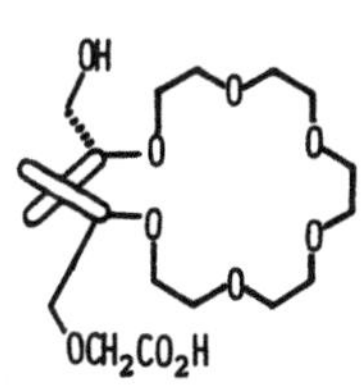 (R)	glass	82	96

Table 3.19 (continued)

Cpd. No.	No. C's	Ring size	Structure	Mp or (bp) in °C	Yield (%)	Refs.
			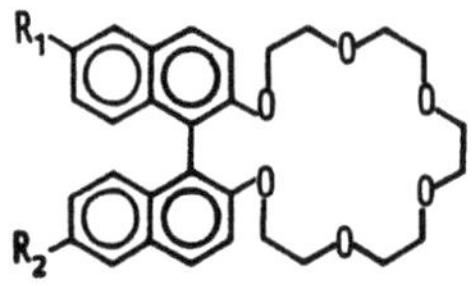			
36	34	20	$R_1 = CH_2OCH_2CO_2H$, $R_2 = CH_2Cl$	glass	75	36
37	34	20	$R_1 = CH_2SCH_2CO_2H$, $R_2 = CH_2Cl$	oil	75	36
38	34	20	$R_1 = R_2 = \overset{O}{\overset{\|}{C}}CH_3$	103–104	36	36
39	34	20	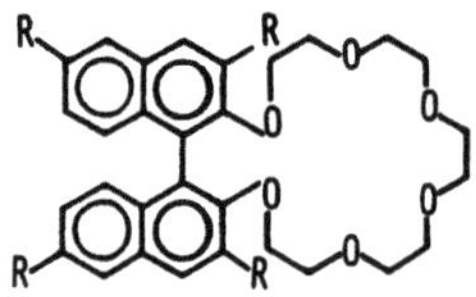 R = CH_2OH	glass	68	36
40	34	20	R = CH_2Cl	glass	82	36
41	34	23	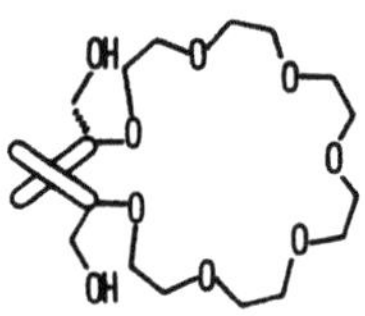 (R)	glass	57	36
42	34	23	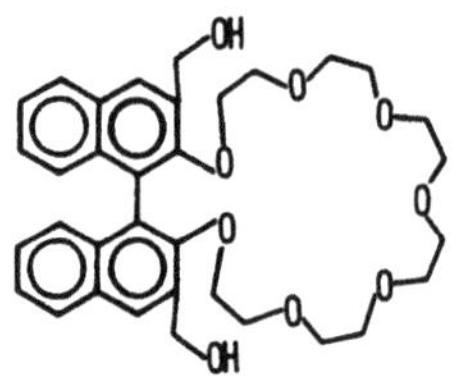	glass	50	36, 90
43	34	20	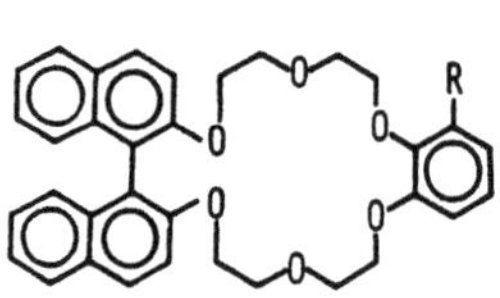 R = H	147–148	41	36
44	35	20	R = CH_2N_3	glass	80	36
45	35	20	R = CH_2OH	90–100	80	36
46	35	20	R = CH_2Cl	90–100	100	36
47	35	20	R = CHO	nr	nr	36
48	34	24	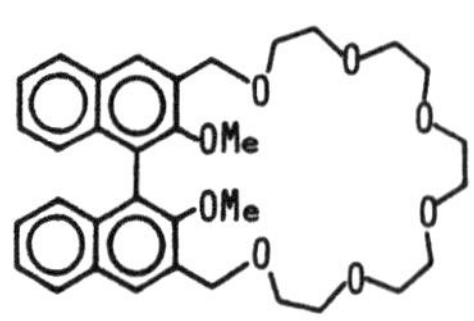	glass	20	105

Table 3.19 (continued)

Cpd. No.	No. C's	Ring size	Structure	Mp or (bp) in °C	Yield (%)	Refs.
49	36	17	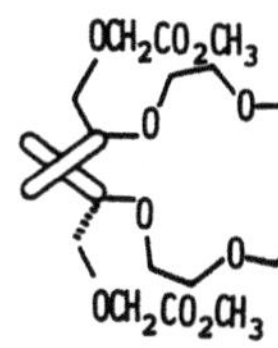 (S)	glass	51	36
50	36	17	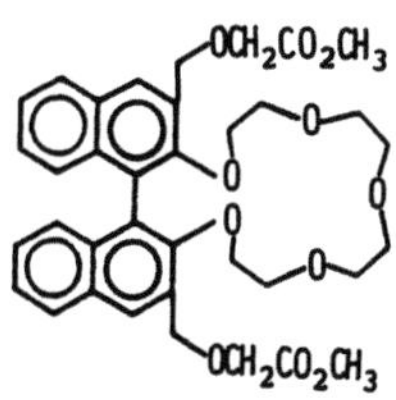	glass	55	36, 90
51	36	20	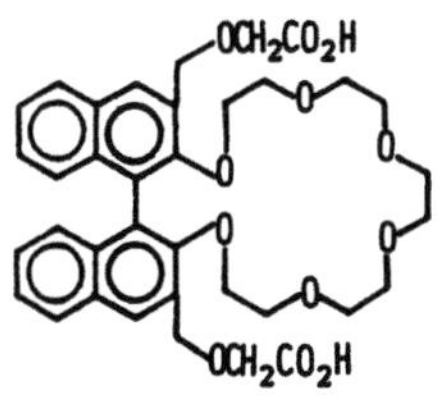	glass	80	36,90
52	36	20	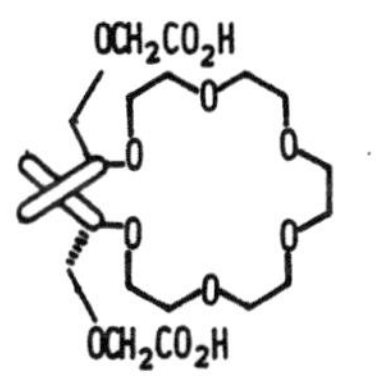 (S)	glass	90	36, 88
53	36	20	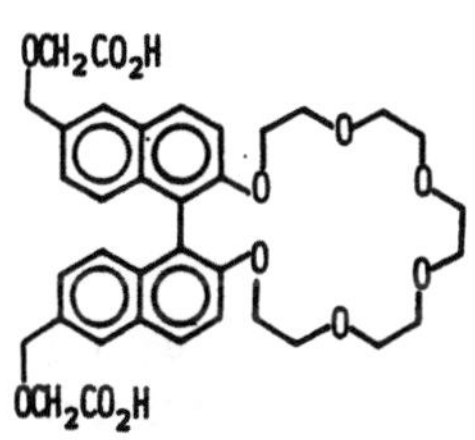	glass	55	37
54	36	20	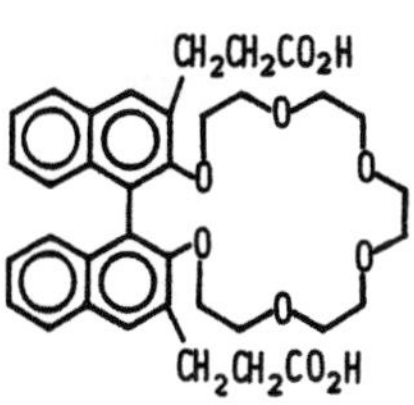	glass	92	36

Table 3.19 (continued)

Cpd. No.	No. C's	Ring size	Structure	Mp or (bp) in °C	Yield (%)	Refs.
55	36	20	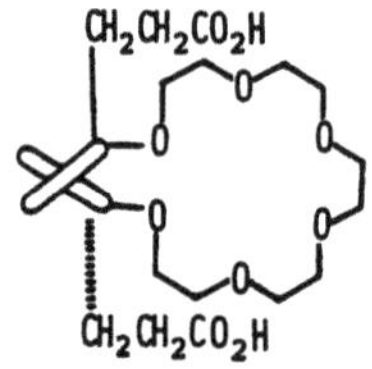 (S)	glass	85	36
56	36	20	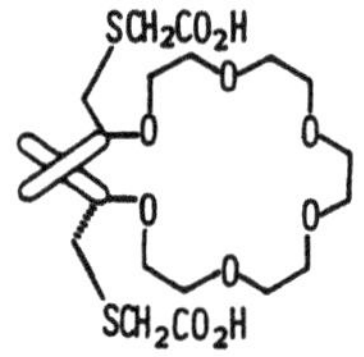 (S)	glass	72	36
57	36	20	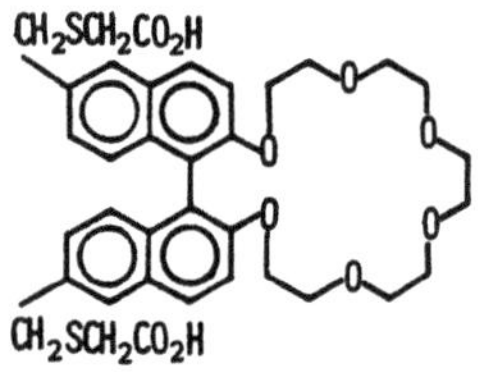	oil	75	37, 38
			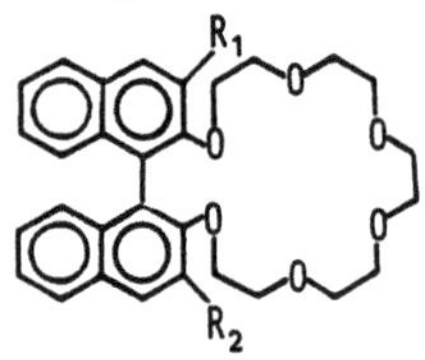			
58	36	20	$R_1 = R_2 = CH_2N(CH_3)_2$	glass	64	36
59	36	20	$R_1 = CH_2OH$, $R_2 = CH_2$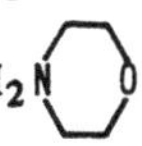	glass	55	36, 90
			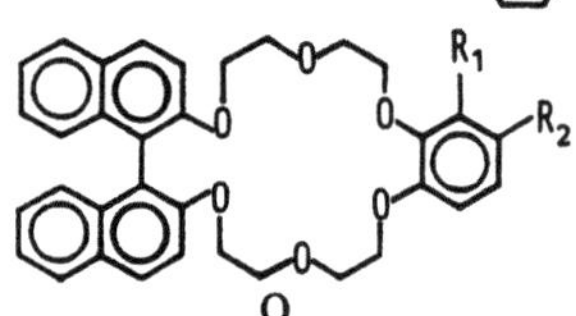			
60	37	20	$R_1 = CH_2NHC(=O)CH_3$, $R_2 = H$	193–194 193–194	36 57	116 36
61	37	20	$R_1 = H$, $R_2 = (CH_2)_3OH$	oil glass	16 46	116 36
62	37	20	$R_1 = H$, $R_2 = (CH_2)_3Cl$	nr	95	36
63	37	20	R_1 = cis-CH=CHCH$_3$, $R_2 = H$	oil	41	116
64	37	20	R_1 = trans-CH=CHCH$_3$, $R_2 = H$	glass	95	36

Table 3.19 (continued)

Cpd. No.	No. C's	Ring size	Structure	Mp or (bp) in °C	Yield (%)	Refs.
65	38	20	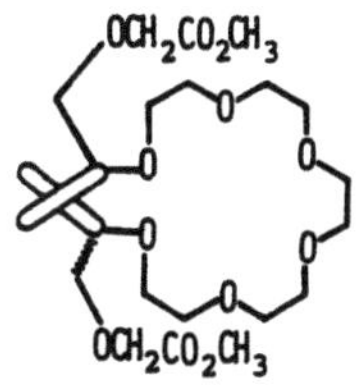 (S)	glass	44	36
66	38	20	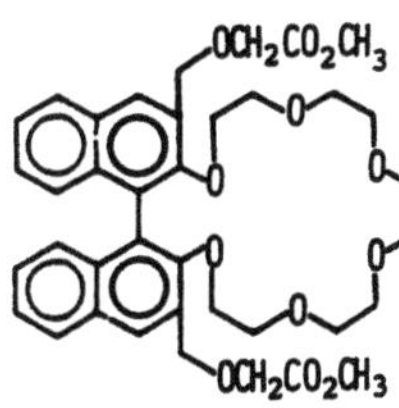	glass	60	36, 90
67	38	20	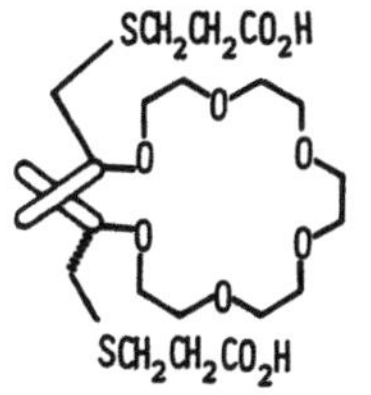 (S)	glass	58	36
			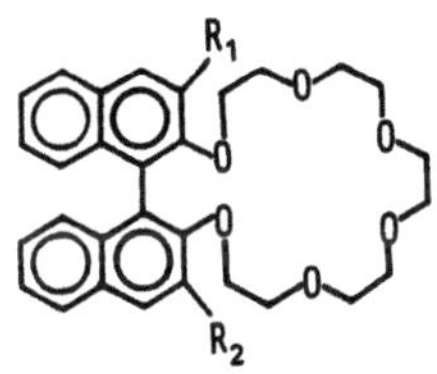			
68	38	20	$R_1 = R_2 = CH_2SCH_2CH_2CO_2H$	glass	97	36
69	38	20	$R_1 = R_2 = CH_2CH(CO_2H)_2$	140	91	36
70	38	20	$R_1 = CH_2OCH_2CO_2H$, $R_2 = CH_2$	glass	65	36, 90
71	38	20	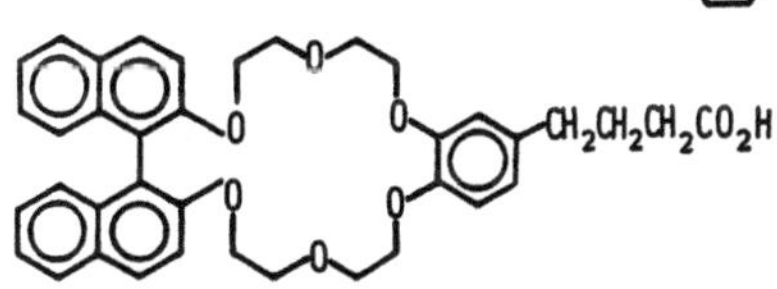	glass	74	36
72	38	23	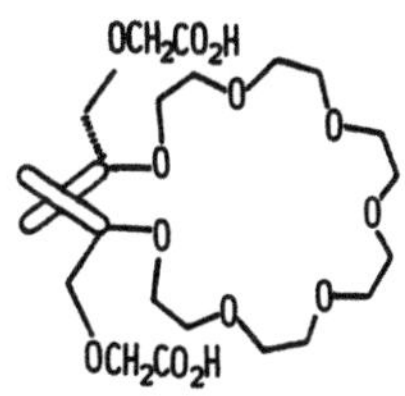 (R)	glass	50	36

Table 3.19 (continued)

Cpd. No.	No. C's	Ring size	Structure	Mp or (bp) in °C	Yield (%)	Refs.
73	38	23	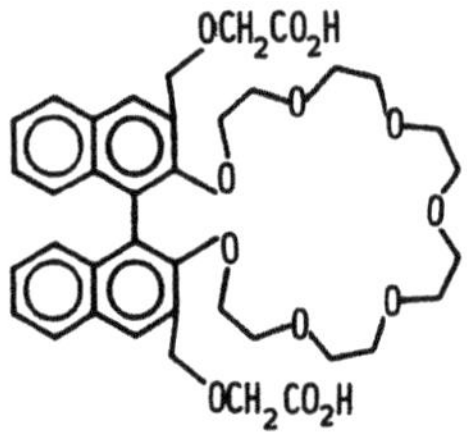	glass	75	36, 90
74	39	20	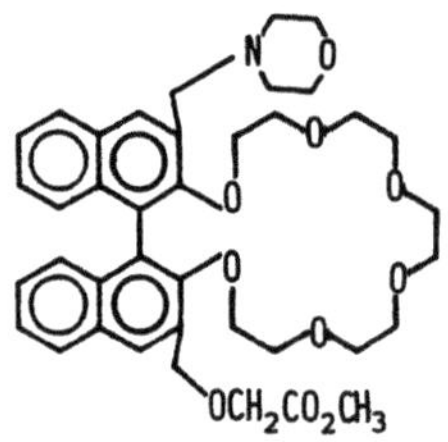	glass	35	36, 90
75	40	23	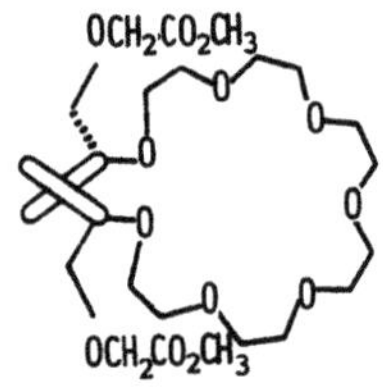 (R)	glass	54	36
76	40	23	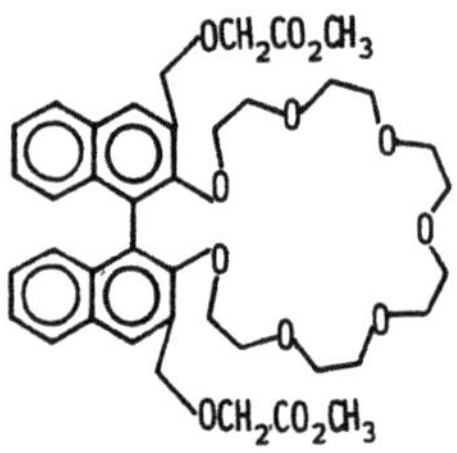	glass	50	36, 90
77	40	20	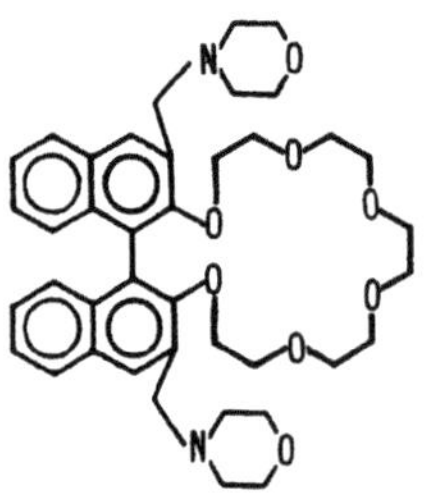	glass	65	36, 90
78	40	20	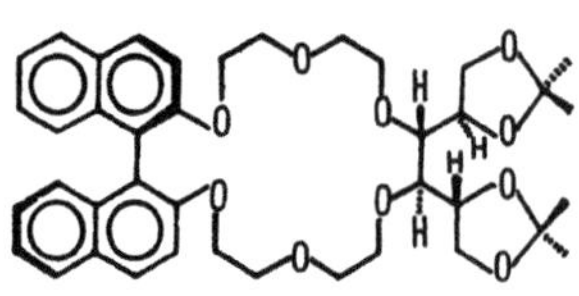	nr	7	44

Table 3.19 (continued)

Cpd. No.	No. C's	Ring size	Structure	Mp or (bp) in °C	Yield (%)	Refs.
79	40	20	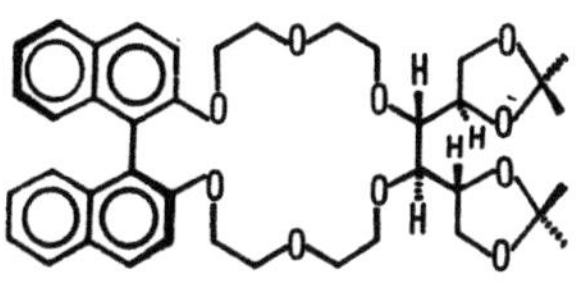 [(S)-D-S]	nr	6	44
80	42	20	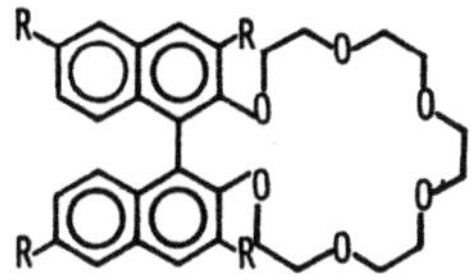 R = $CH_2SCH_2CO_2H$	glass	96	36

Table 3.20. Crown compounds containing two 2,2′-binaphthyl subunits

Cpd. No.	No. C's	Ring size	Structure		Mp or (bp) in °C	Yield (%)	Refs.
1	44	16	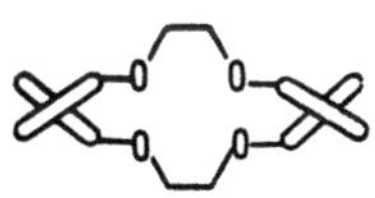	(SR)	355 dec.	~1	36, 114
2	48	22	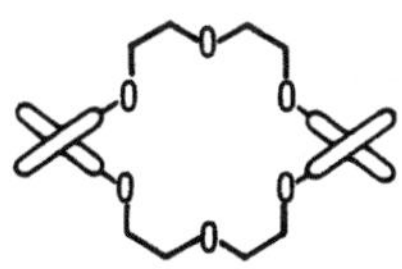	(SS)	glass	31	36, 115
3	48	22	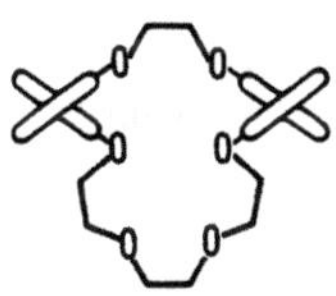	(SS) racemic	foam nr	19 nr	36, 114 75
4	48	22	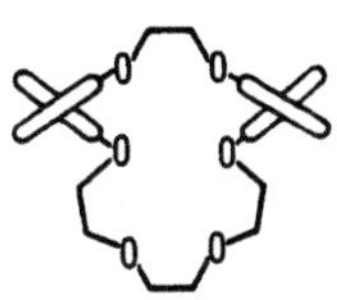	(SR)	nr	15	36, 114

Table 3.20 (continued)

Cpd. No.	No. C's	Ring size	Structure		Mp or (bp) in °C	Yield (%)	Refs.
5	48	22	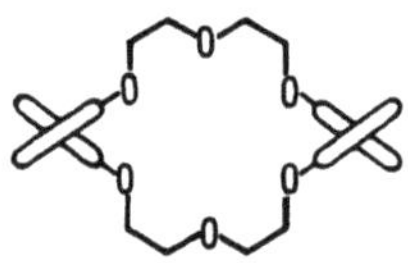	(SS)	123–126 nr nr nr	31 nr 47 nr	92 38 76 38
6	48	22	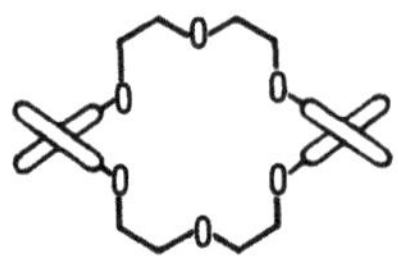	(RR)	nr 123–126	nr 22	38 92
7	48	22	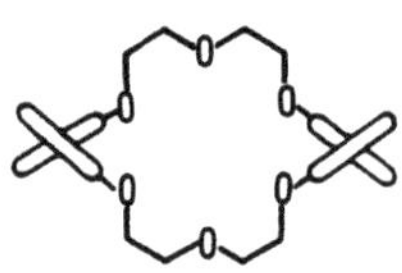	(RS) (meso)	nr 283–284	nr 2	38 115
8	48	22	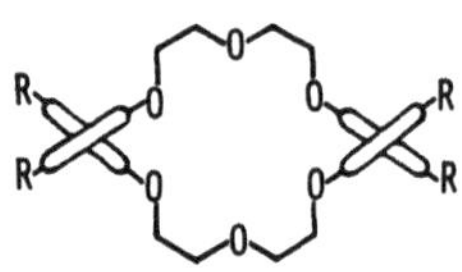 R = $SO_3(Ba)_{1/2}$	(SS)	nr	50	38
9	48	22	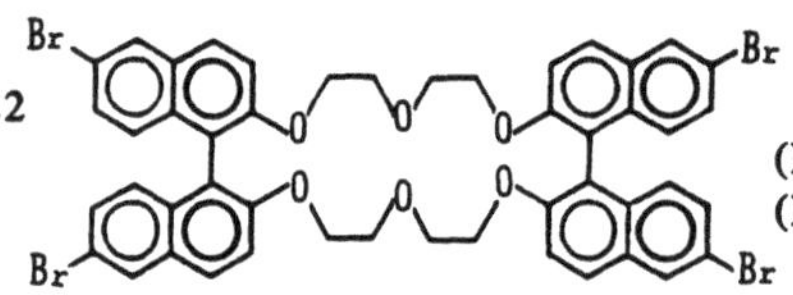	 (RR) (RR) (SR)	299–300 179–180 189–191 334–335	52 91 91 90	36, 59 36 59 36, 59
10	49	22	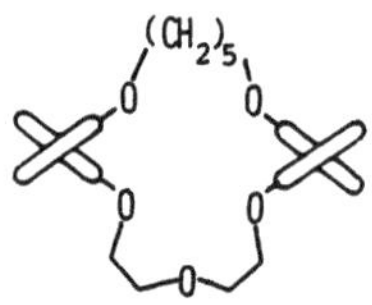	(SS)	nr foam	41 41	76 114
11	50	22	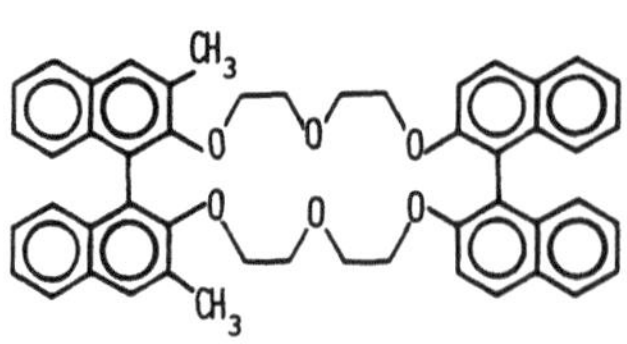	(SR) (SS)	foam foam	16 64	38 38

Table 3.20 (continued)

Cpd. No.	No. C's	Ring size	Structure		Mp or (bp) in °C	Yield (%)	Refs.
			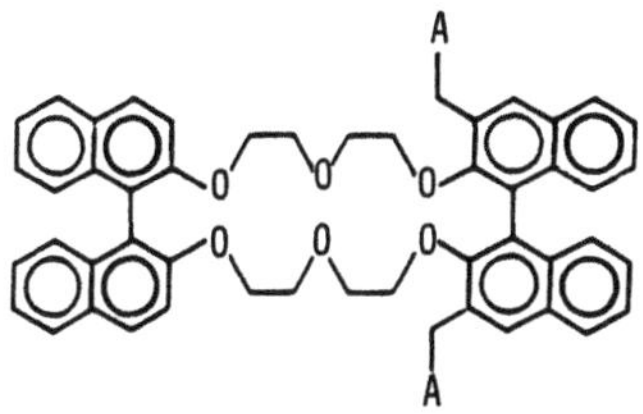				
12	50	22	A = OH		230–231	16	36
			A = OH	(+) –(RR)	glass	28	36, 92
			A = OH	(SR) (RS)	197–198	8	36
13	50	22	A = Cl	(+) –(RR)	glass	76	36, 92
14	50	22	A = H	(–) –(SR)	195–196	10	36
				(SR) (RS)	249	12	36
				(SS) (RR)	222–223	10	36
				(+) –(RR)	glass	80	36, 92
15	50	22	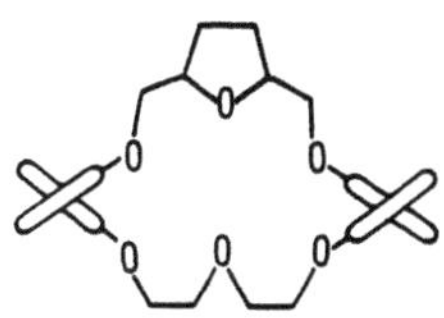	(–) –(SS)	glass	26	114
			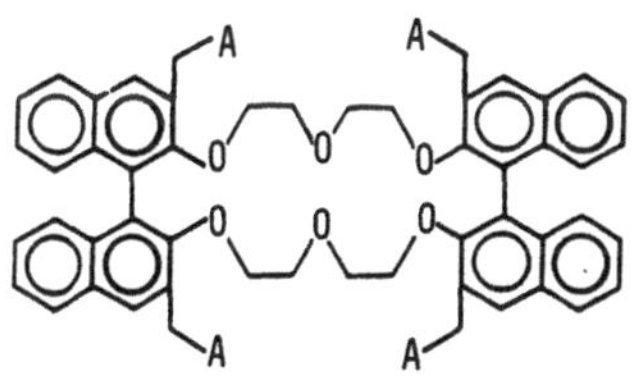				
16	52	22	A = OH	(RR) (SS)	168–170	10	38
				(RS)	168–170	10	38
17	52	22	A = H	(SS) (RR)	158–160	10	36
				(+) –(RR)	glass	28	38
				(–) –(SS)	glass	28	38
				(SR)	314–315	7	36
18	52	22	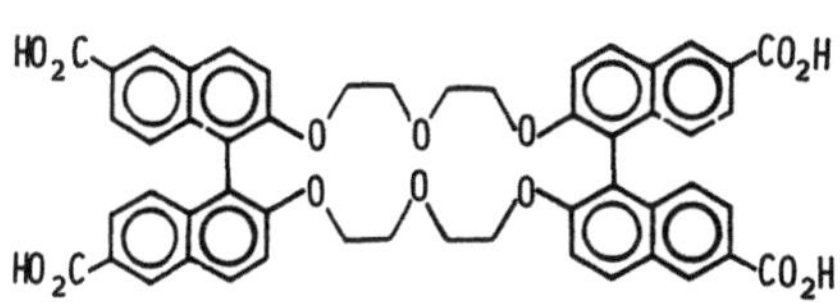	(SS) (RR)	380 dec.	94	36

Table 3.20 (continued)

Cpd. No.	No. C's	Ring size	Structure		Mp or (bp) in °C	Yield (%)	Refs.
19	52	22	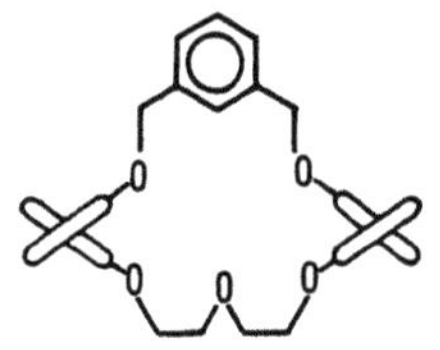	(–) –(SS)	foam	13	114
20	52	22		(SS)	nr	13	76
			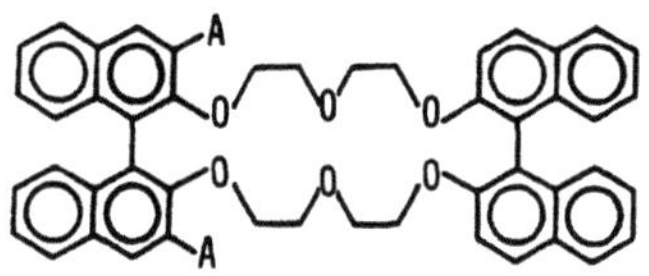				
21	54	22	$A = CH_2OCH_2CO_2H$	(RR) (SS)	glass	83	38
				(+) –(RR)	glass	94	36
				(SR) (RS)	glass	94	36
22	54	22	$A = CH(CH_3)_2$	(+) –(RR)	foam	75	38
23	54	22	$A = C(OH)(CH_3)_2$	(+) –(RR)	foam	28	38
			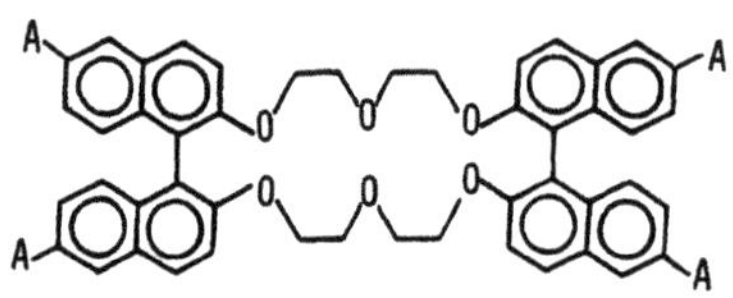				
24	56	22	$A = \overset{O}{\overset{\|}{C}}CH_3$	(RR) (SS)	300–301	77	38
				(RR) (SS)	340–341	74	36
				(–) –(SS)	264–265	nr	38
25	56	22	$A = CH_2CH_2OH$	(SS)	foam	90	38
26	56	22	$A = CH{=}CH_2$	(–) –(SS)	glass	47	38
27	56	22	$A = Si(CH_3)_2OCH_3$	(RS)	95–96	84	36, 59

Table 3.20 (continued)

Cpd. No.	No. C's	Ring size	Structure		Mp or (bp) in °C	Yield (%)	Refs.
28	56	22	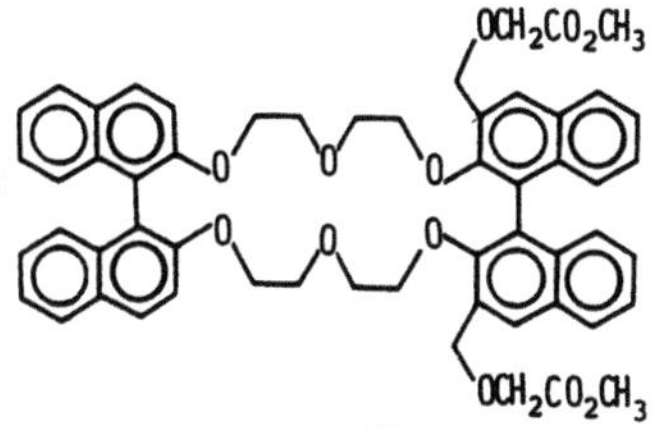	(SS) (RR) (+) –(RR) (SR) (RS)	glass glass glass	71 97 50	36 36 36
29	56	18	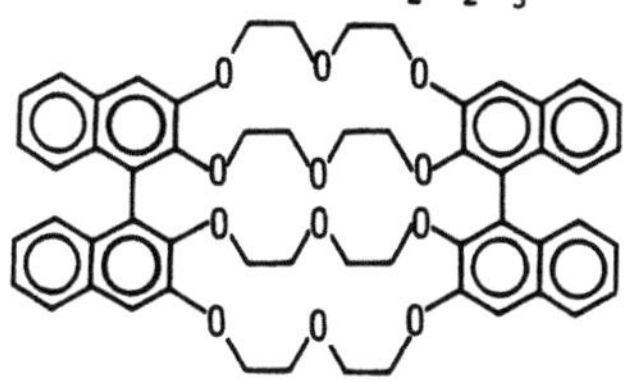	(RS)	159.5–161 164–165	30 2–3	200
			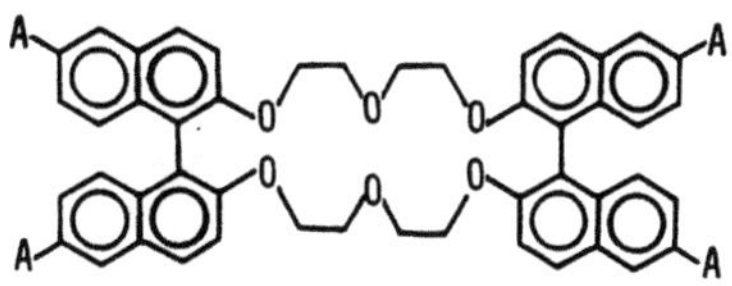				
30	60	22	A = $CH_2CO_2CH_3$	(–) –(SS)	foam	98	38
31	64	22	A = $C(CH_3)_3$	(–) –(SS)	289–291	45	38
32	66	24	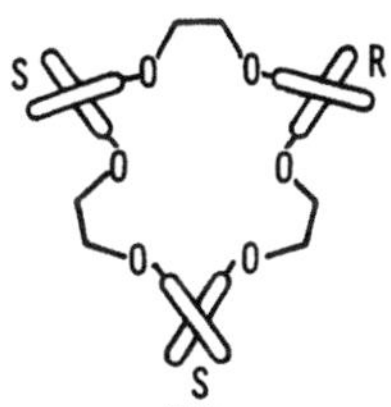	(SSR)	247–249	nr	36
			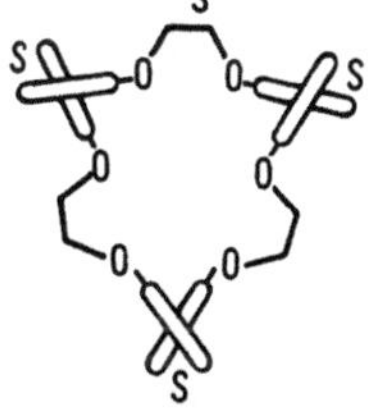	(SSS)	170–190	nr	55, 114
33	68	22	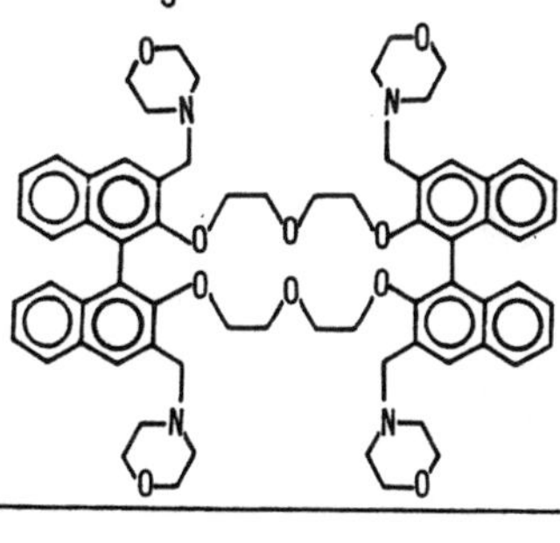		130–132	51	38

Table 3.20 (continued)

Cpd. No.	No. C's	Ring size	Structure	Mp or (bp) in °C	Yield (%)	Refs.
34	–	22	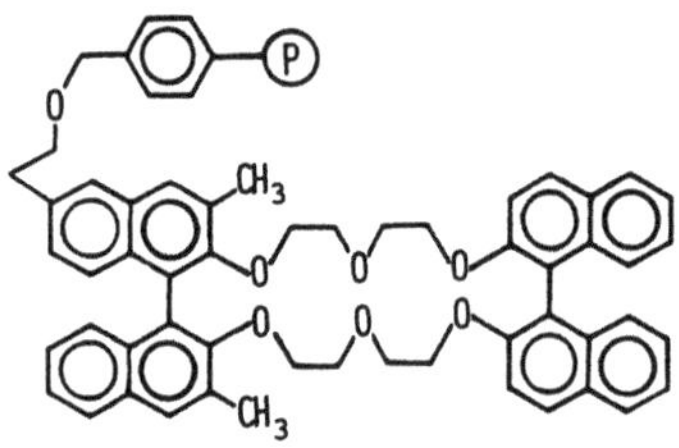 (RR)	nr	nr	197

Table 3.21. Crowns containing miscellaneous aromatics as part of the macroring

Cpd. No.	No. C's	Ring size	Structure	Mp or (bp) in °C	Yield (%)	Refs.
1	24	28	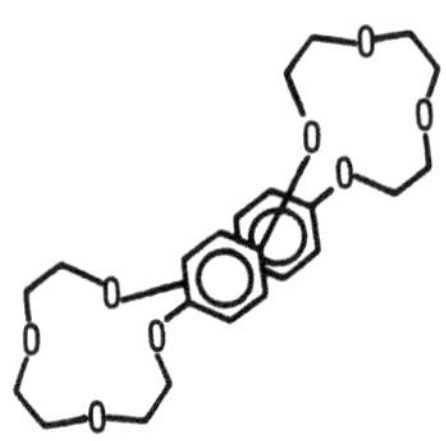	96–97	7	89, 91
2	26	18	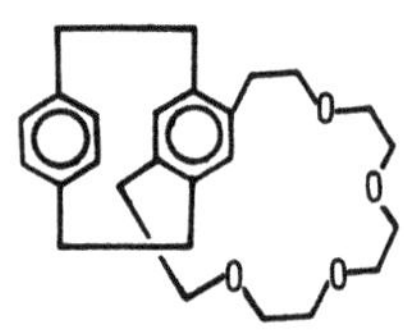	oil	46	101
3	26	23	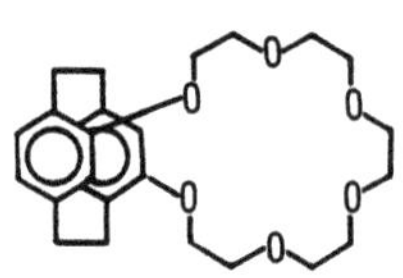	nr	nr	89
4	26	15	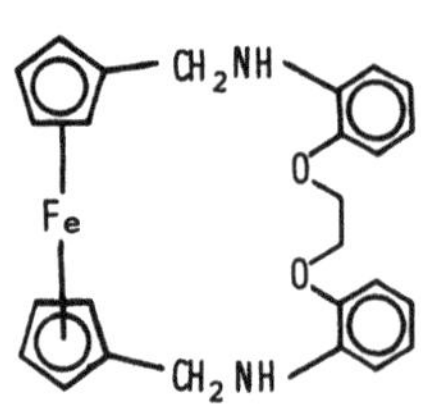	nr	~60	16

Table 3.21 (continued)

Cpd. No.	No. C's	Ring size	Structure	Mp or (bp) in °C	Yield (%)	Refs.
5	28	34	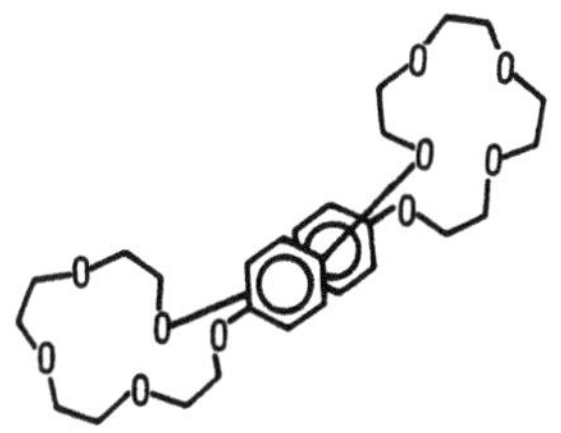	93–94	8	89, 91
6	28	21	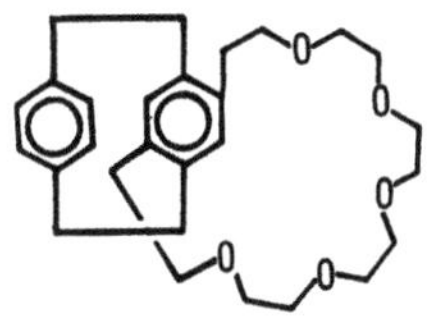	94.0–94.5	25	101
7	32	40	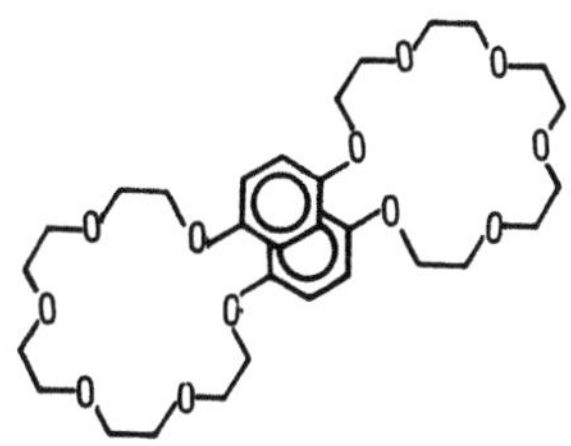	67–69	7	89, 91
8	36	23	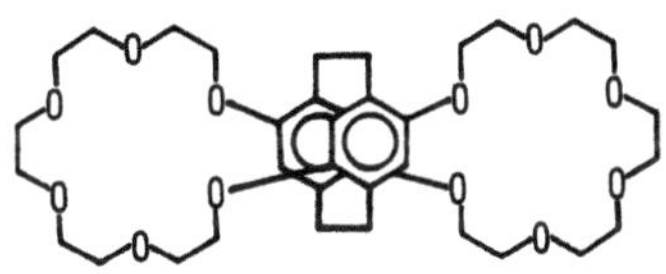	121–122	25	89, 91
9	36	24	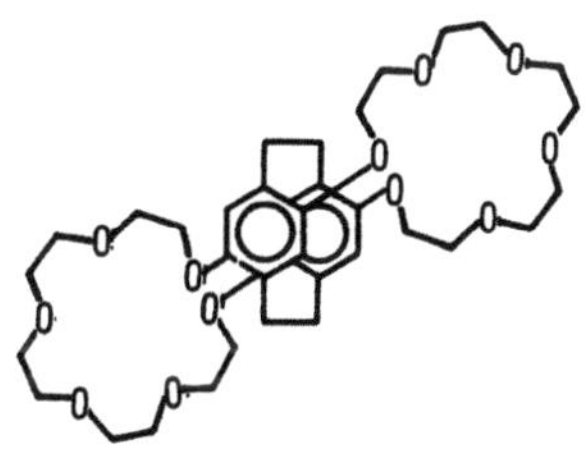	133–134	23	89, 91
10	38	31		170–171	4	232, 233

Table 3.22. Crown compounds having a spiro-3,3-oxetane subunit in the macroring

Cpd. No.	No. C's	Ring size	Structure		Mp or (bp) in °C	Yield (%)	Refs.
1	9	10	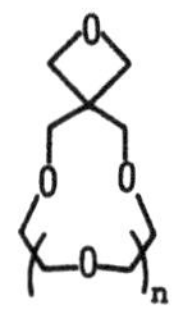	n = 1	nr	0.2	107
2	11	13		n = 2	nr 97–98	33 30–35	210 107
3	13	16		n = 3	(105–108/ 2×10^{-4})	53	107
4	14	14	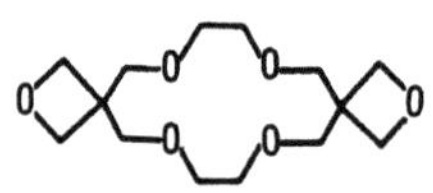		163–165	0.7	107
5	15	19	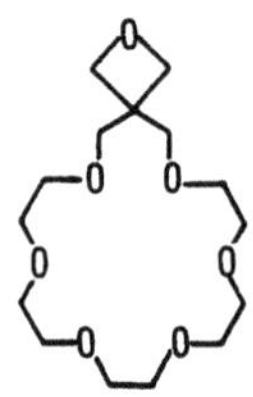		nr (136–137/ 1×10^{-3})	65 60	210 107
6	18	20	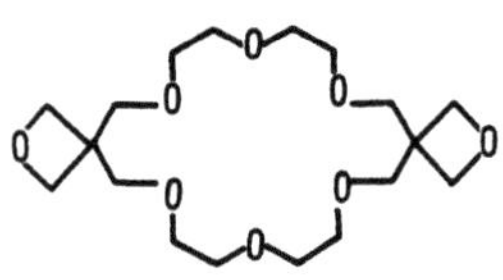		86–87	47	107
7	19	25	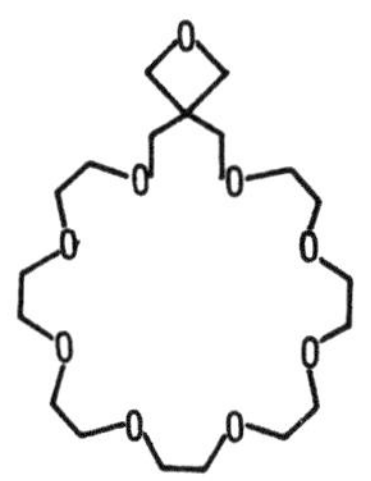		(182–183/ 3×10^{-4})	35	107
8	22	26	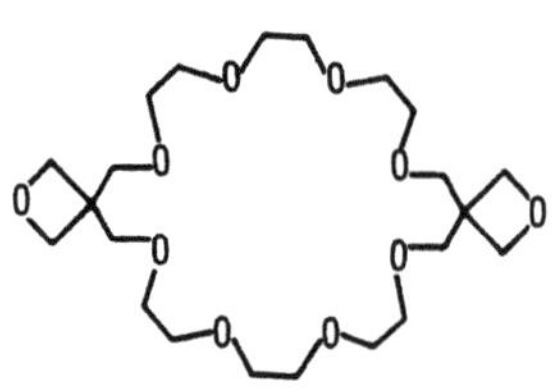		52–53	20–25	107

Table 3.22 (continued)

Cpd. No.	No. C's	Ring size	Structure		Mp or (bp) in °C	Yield (%)	Refs.
9	23	31	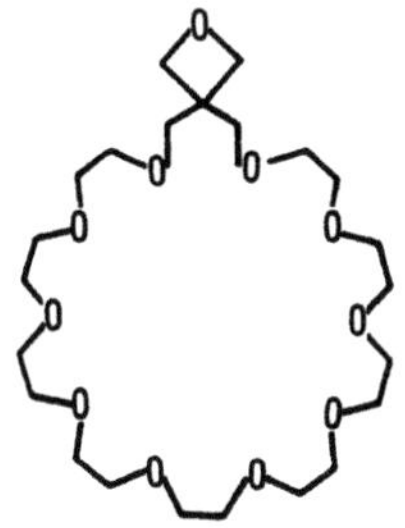		(250–260/ 1×10^{-3})	11	107
10	38	50	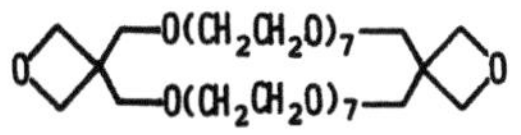		42–43	3	107

Table 3.23. Crown compounds containing the furan subunit

Cpd. No.	No. C's	Ring size	Structure		Mp or (bp) in °C	Yield (%)	Refs.
1	12	14	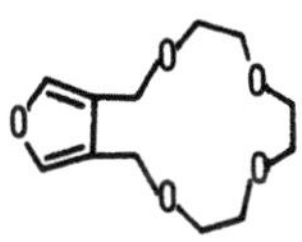		nr	7–8	185
				+ dimer	nr	18	185
2	12	15	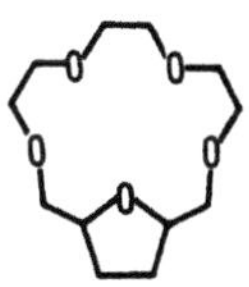	(Z)	oil	23	207
3	12	15			nr	10–30	184
4	14	17	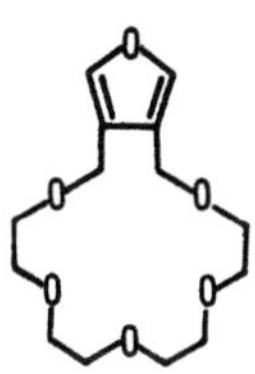		nr	19–31	185
				+ dimer	nr	1–8	185

Table 3.23 (continued)

Cpd. No.	No. C's	Ring size	Structure		Mp or (bp) in °C	Yield (%)	Refs.
5	14	18	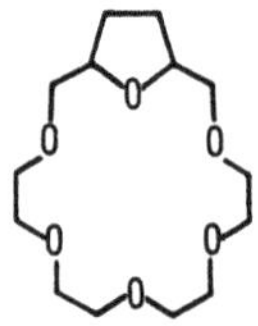	(Z)	oil	97	207
6	14	18	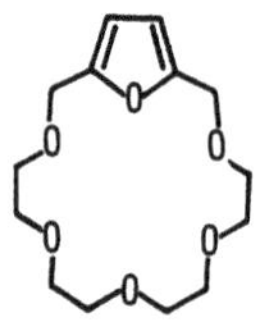		nr ~0	10–30 36	184 204
7	14	18	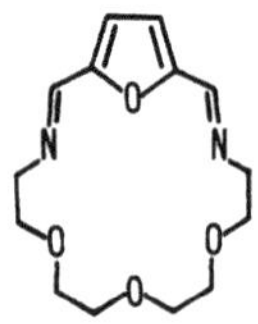		nr	nr	66
8	16	18	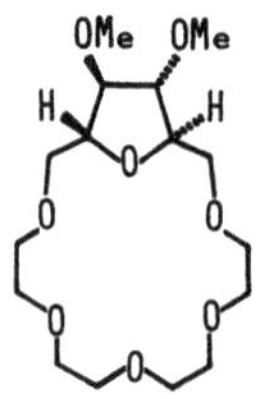		(120–122/0.03)	19	85
9	16	20			nr	30	185
10	16	18	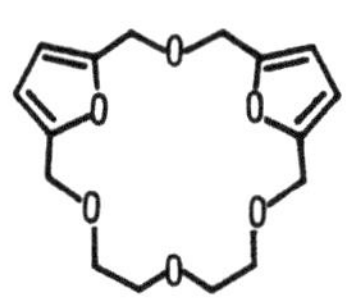		69–70	35	204, 207
11	16	18	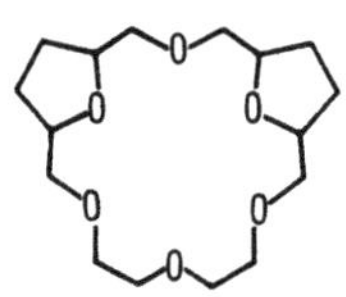	(syn-cis-cis) (anti-cis-cis)	oil	95	207

Table 3.23 (continued)

Cpd. No.	No. C's	Ring size	Structure		Mp or (bp) in °C	Yield (%)	Refs.
12	16	18	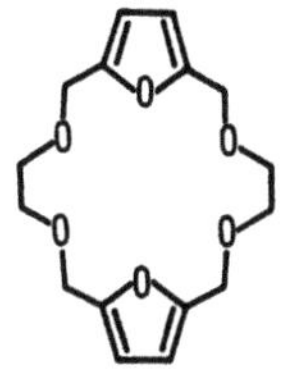		109–111	11	204
13	17	14	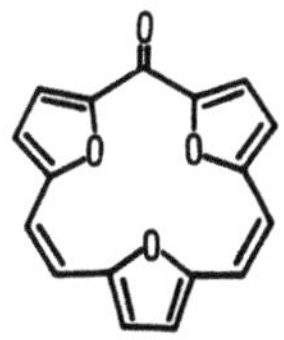		236–237	9.6	41
14	18	15	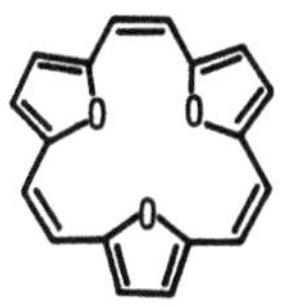		>200 dec.	<1	9
15	18	18	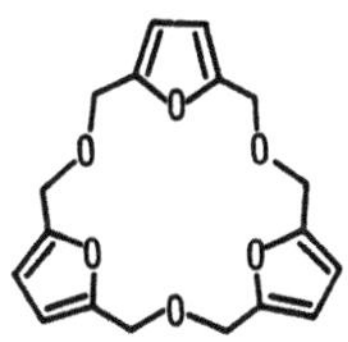		124–126	10	204
16	18	18	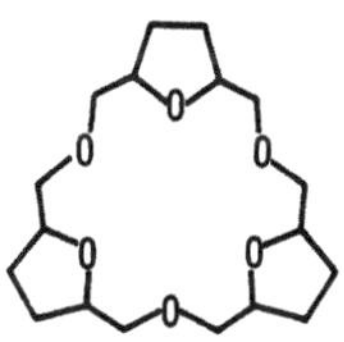	(syn-cis-cis-cis)	oil	97	207
17	18	23	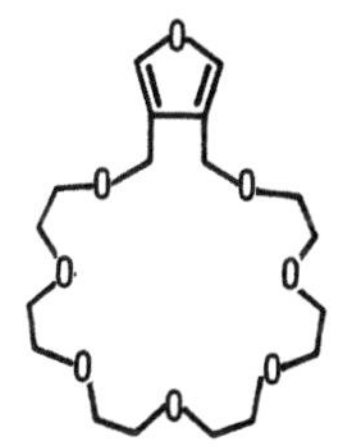		nr	44	185
18	18	24	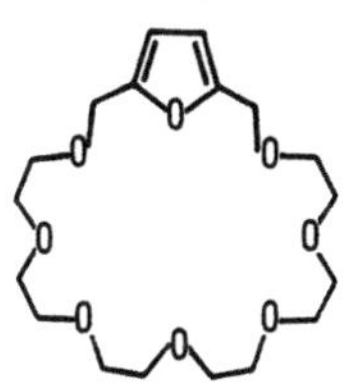		nr	10–30	184

Table 3.23 (continued)

Cpd. No.	No. C's	Ring size	Structure	Mp or (bp) in °C	Yield (%)	Refs.
19	20	22	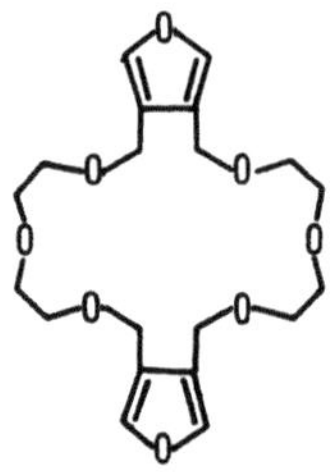	nr	5	186
20	20	26	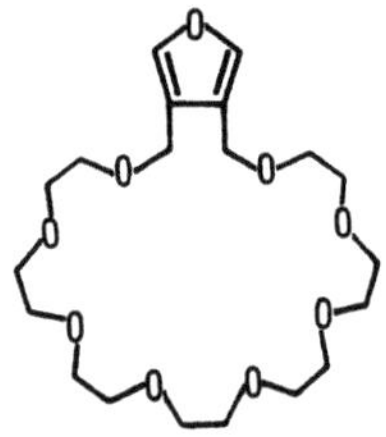	nr	43	185
21	24	20	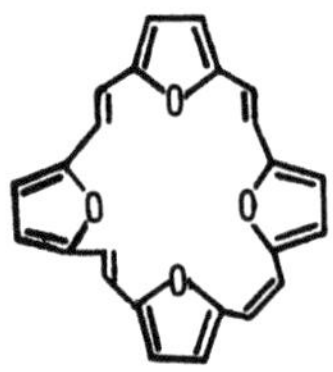	216–217	0.7	63
22	24	20	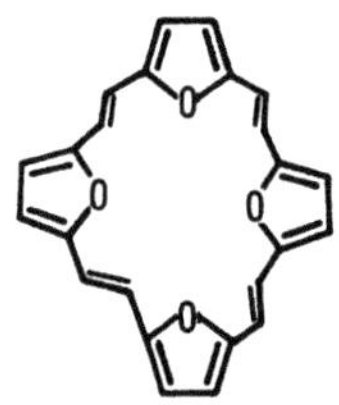	269–270	0.8	63
23	24	30	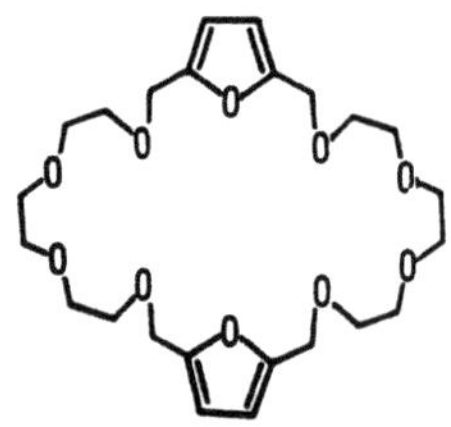	nr	~10	184
24	28	16	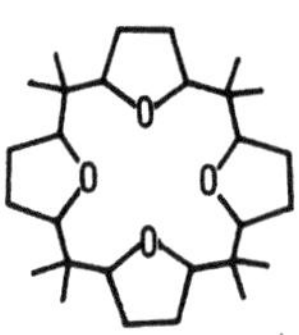	nr	nr	32

Table 3.23 (continued)

Cpd. No.	No. C's	Ring size	Structure	Mp or (bp) in °C	Yield (%)	Refs.
25	28	16	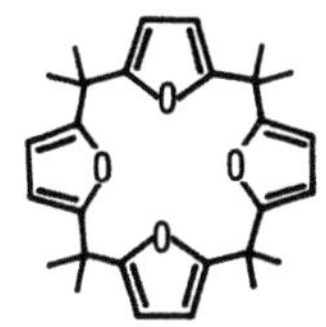	243	18.5	1
				223	13.5	25
				nr	40–45	32
				243–244	1–48	187
26	28	16	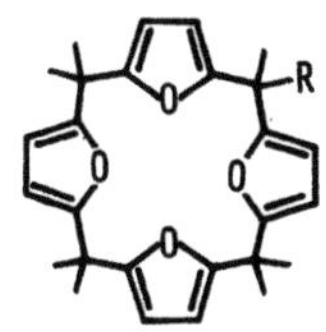 R = CH_2Cl	219.5–220	56	26
27	28	16	R = COOH	250 dec.	nr	26
28	28	16	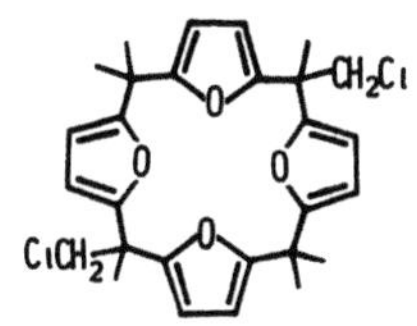	211–211.5	6.0	26
29	28	36	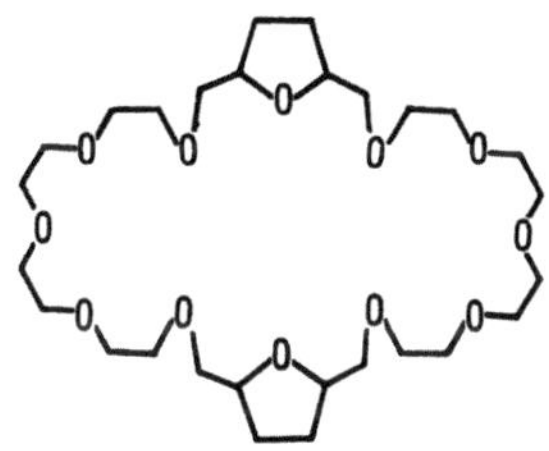	oil	8	207
30	29	16	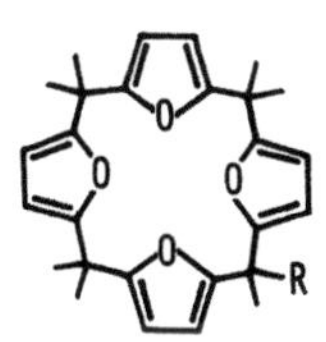 R = CH_2CO_2H	248.5–249.5	nr	26
31	29	16	R = CO_2CH_3	172.5	15	26
32	30	16	R = $CH_2CO_2CH_3$	179	17	26
33	30	16	R = $CO_2CH_2CH_3$	169.5	39	26
34	30	16	R = $CH_2CH_2CO_2H$	225.5–226	45	26
35	30	20	 isomer a	218–220	0.4	63
			isomer b	192–194	0.15	63

Table 3.23 (continued)

Cpd. No.	No. C's	Ring size	Structure	Mp or (bp) in °C	Yield (%)	Refs.
36	31	16	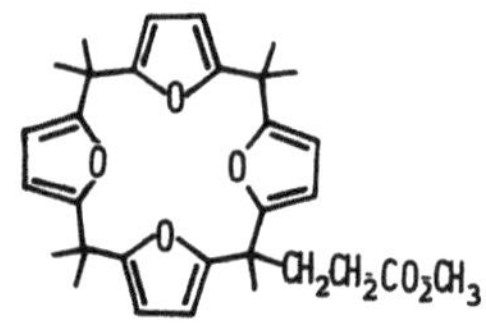	157.5	34	26
37	31	16	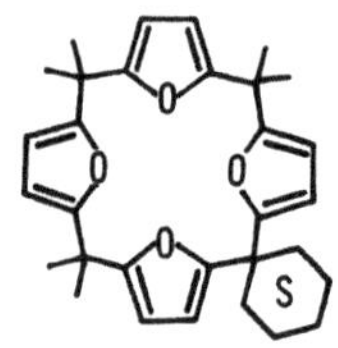	182.3–183.3	21	27
38	32	16	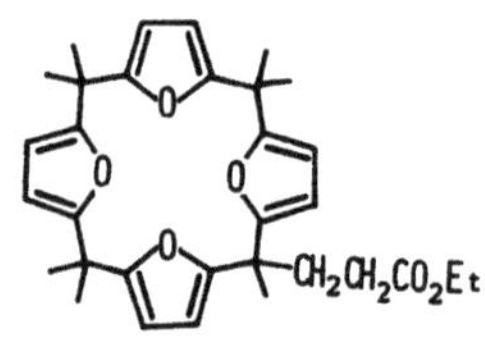	153	63	26
39	32	16	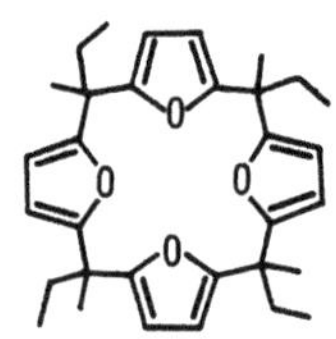	174 174	11.5 27	1 26
40	34	16	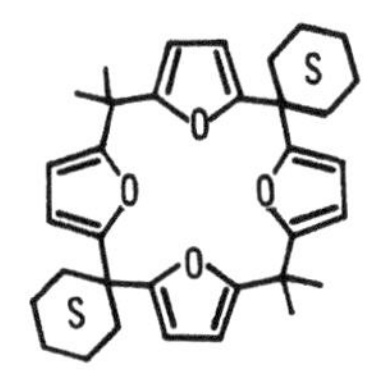	208.2–209.2	16	27
			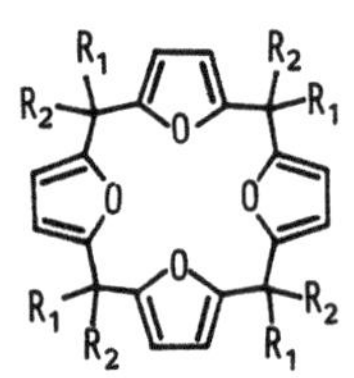			
41	36	16	$R_1 = CH_3$, $R_2 = CH_2CH_2CH_3$	148	5.5	1
42	36	16	$R_1 = R_2 = CH_2CH_3$	249	20	12

Table 3.23 (continued)

Cpd. No.	No. C's	Ring size	Structure	Mp or (bp) in °C	Yield (%)	Refs.
43	36	30	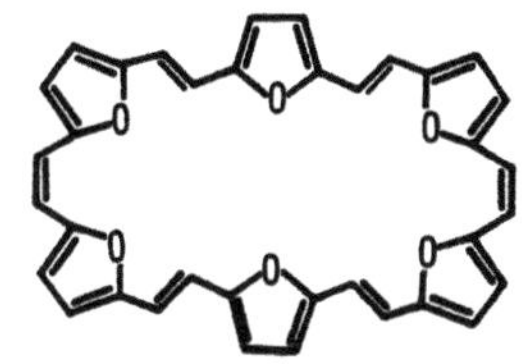	250–252	nr	63
44	40	16		263–264	30	27
			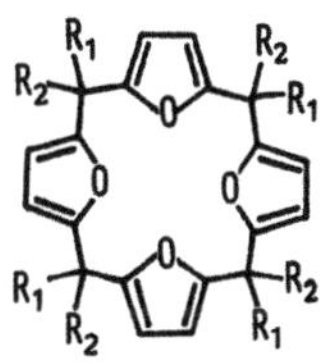			
45	40	16	$R_1 = CH_3$, $R_2 = CH_2CH(CH_3)_2$	241	6.3	1
46	44	16	$R_1 = CH_3$, $R_2 = (CH_2)_4CH_3$	198	6.1	1
47	50	22	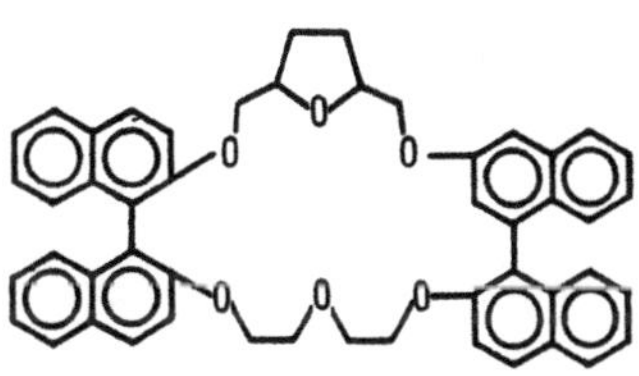	glass	26	114

Table 3.24. Crowns containing a 3,4-thienyl subunit

Cpd. No.	No. C's	Ring size	Structure		Mp or (bp) in °C	Yield (%)	Refs.
1	14	14	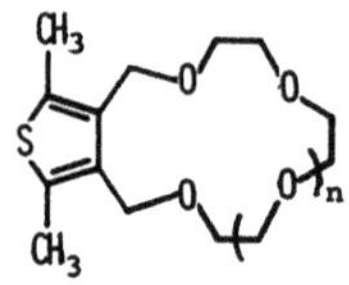	n = 1	nr	7	185
2	16	17		n = 2	nr	35	185
3	18	20		n = 3	nr	38	185
					nr	48	186
4	20	23		n = 4	nr	43	185
5	22	26		n = 5	nr	42	185
6	24	29		n = 6	nr	22	185

Table 3.25. Crowns containing the 4,6-pyrimidine subunit

Cpd. No.	No. C's	Ring size	Structure		Mp or (bp) in °C	Yield (%)	Refs.
1	10	13	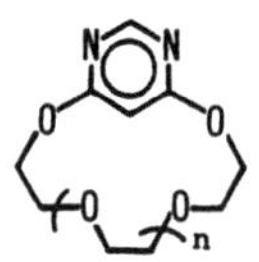	n = 1	114–115	4	151
2	12	16		n = 2	(189–193/0.1)	6	151
3	14	19		n = 3	(210–215/0.1)	7	151
4	16	22		n = 4	(216–220/0.1)	9	151
5	16	20	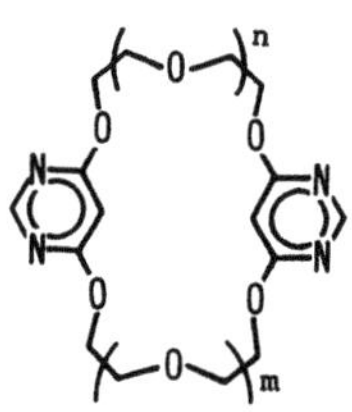	n = m = 1	149–150	3	151
6	20	26		n = 2, m = 2	71–72	4	151
7	22	29		n = 2, m = 3	65–67	3	151
8	24	32		n = m = 3	94–95	6	151
9	24	30	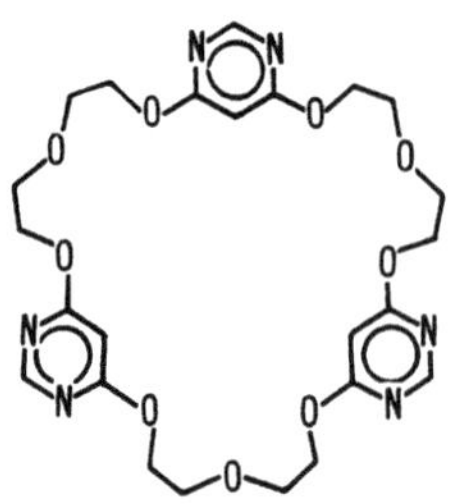		156–158	3	151

Table 3.25 (continued)

Cpd. No.	No. C's	Ring size	Structure		Mp or (bp) in °C	Yield (%)	Refs.
10	32	40	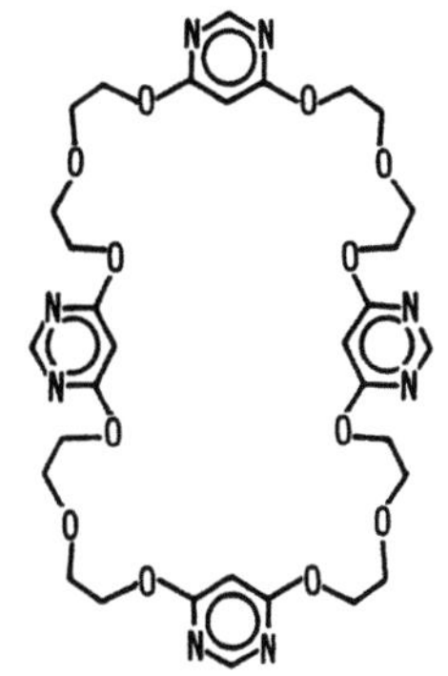		120–121	3	151

Table 3.26. Crown compounds containing pyridyl or bipyridyl subunits

Cpd. No.	No. C's	Ring size	Structure		Mp or (bp) in °C	Yield (%)	Refs.
1	13	15	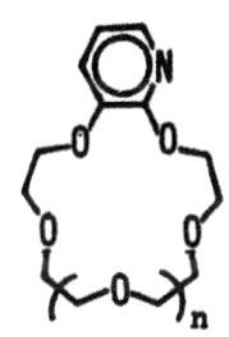	n = 1	64.5–65.5	14	217
2	15	18		n = 2	77–78.5	23	217
3	17	21		n = 3	oil	31	217
4	18	16	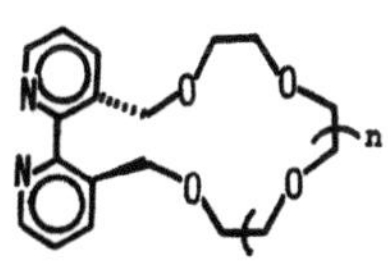	n = 1	103	50	182
5	20	19		n = 2	wax	25	182
6	22	22		n = 3	nr	15	182

Table 3.27. Crown compounds containing two macrocyclic rings

Cpd. No.	No. C's	Ring size	Structure		Mp or (bp) in °C	Yield (%)	Refs.
1	17	13	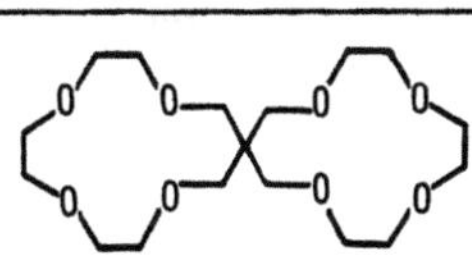		78–79	nr	226

Table 3.27 (continued)

Cpd. No.	No. C's	Ring size	Structure	Mp or (bp) in °C	Yield (%)	Refs.
2	21	16	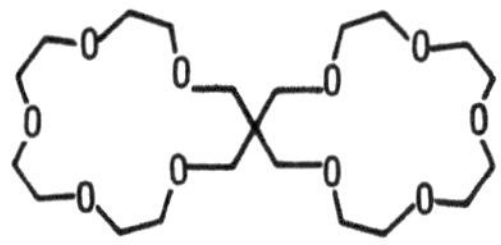	oil	nr	226
3	21	19	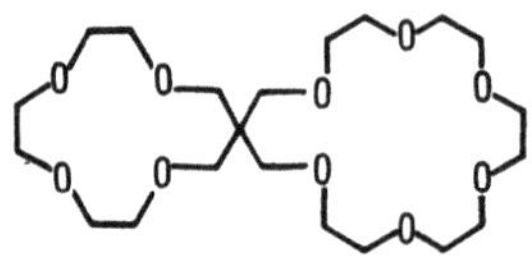	oil	nr	226
4	23	16	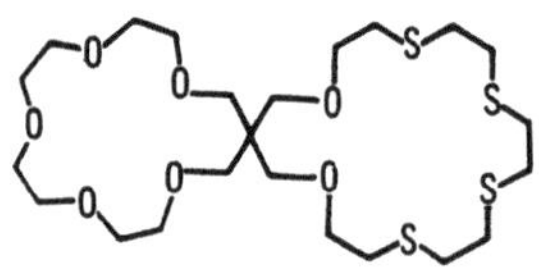	oil	nr	226
5	25	19	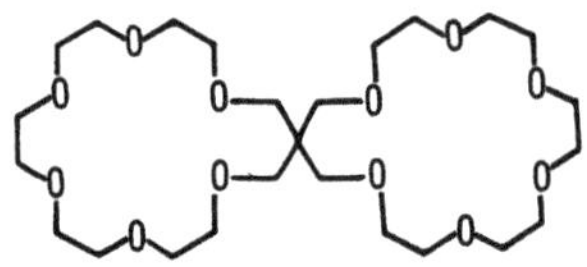	oil	nr	226
6	25	25	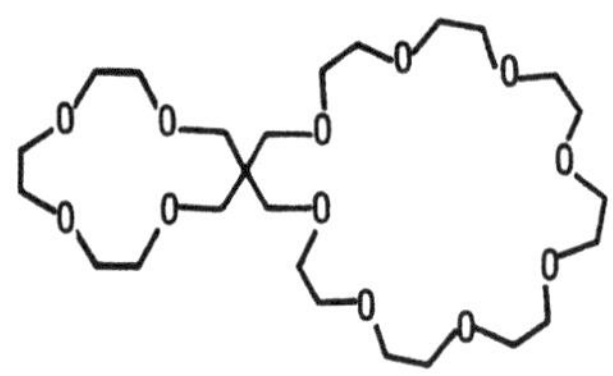	nr	nr	226
7	28	15	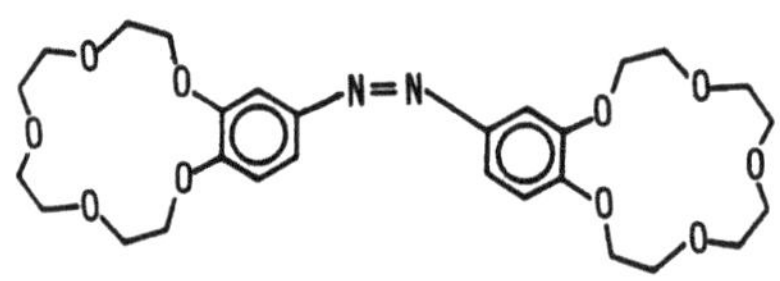 (E)	187–188	nr	226
8	29	19	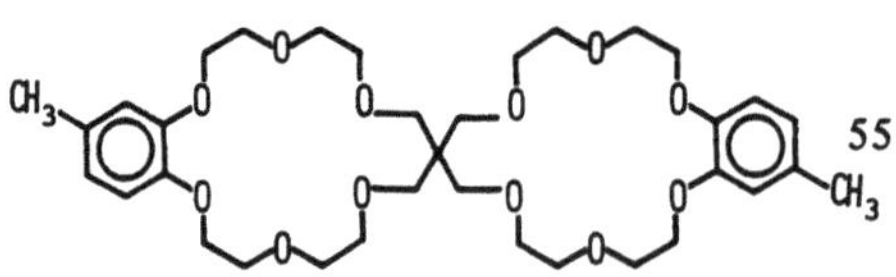	55–57	nr	226
9	30	12	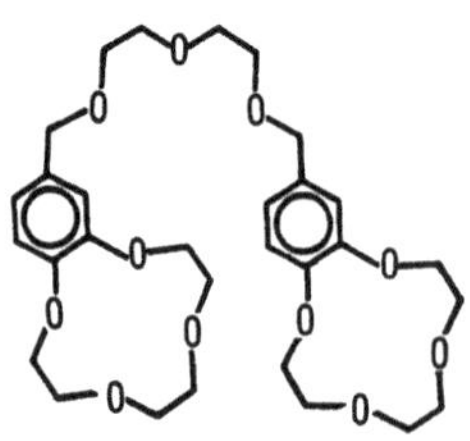	oil	20	230

Table 3.27 (continued)

Cpd. No.	No. C's	Ring size	Structure	Mp or (bp) in °C	Yield (%)	Refs.
10	30	15	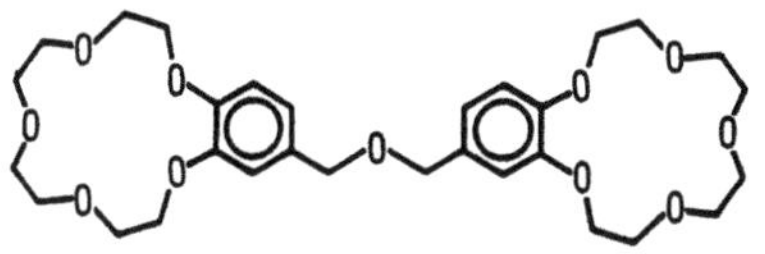	107–109	nr	95
11	31	15	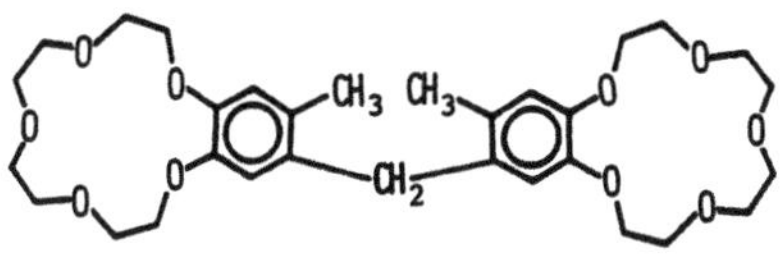	111–112	nr	95
12	32	12	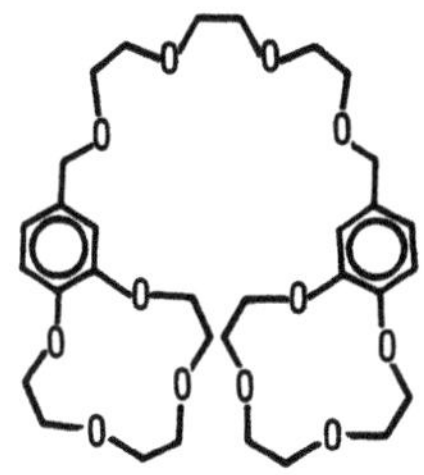	oil	25	230
13	32	15	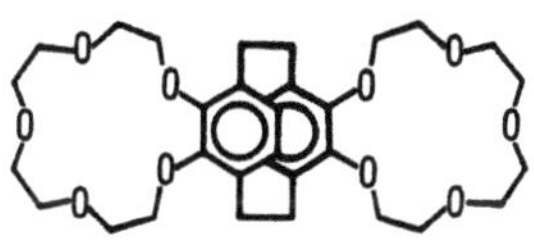	192–193	6	89, 91
14	32	15	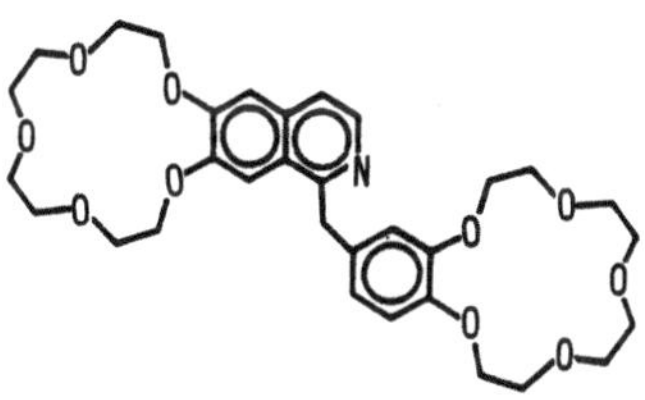	109–110	20	219
15	32	15	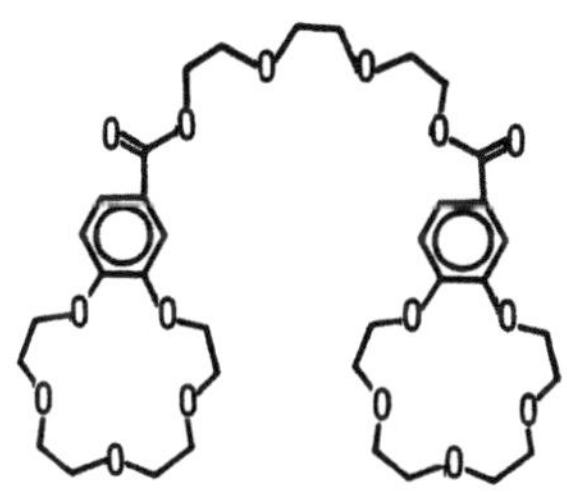	124–125	87	22
16	32	15	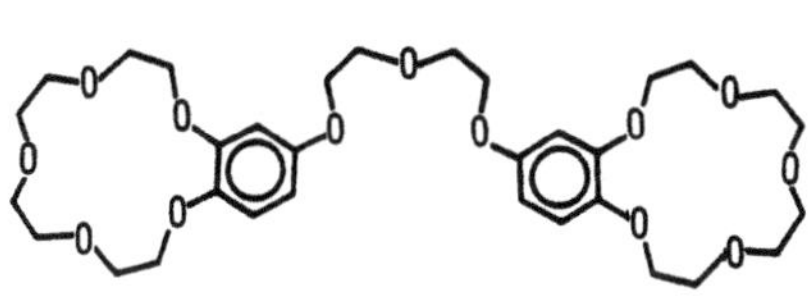	73–74.5	67	225

Table 3.27 (continued)

Cpd. No.	No. C's	Ring size	Structure	Mp or (bp) in °C	Yield (%)	Refs.
17	32	20	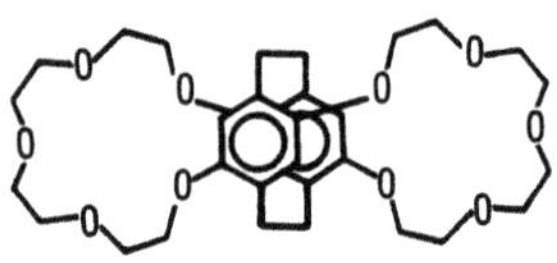	118–119	3	89, 91
18	33	19	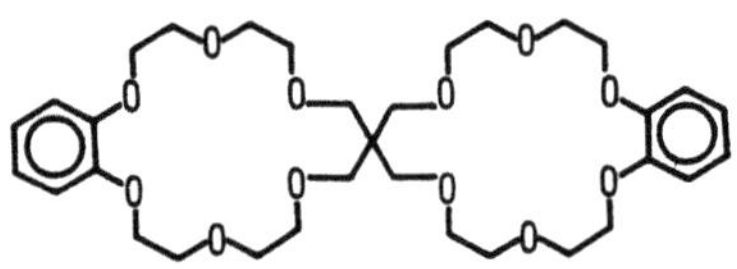	77–78	nr	226
19	34	15	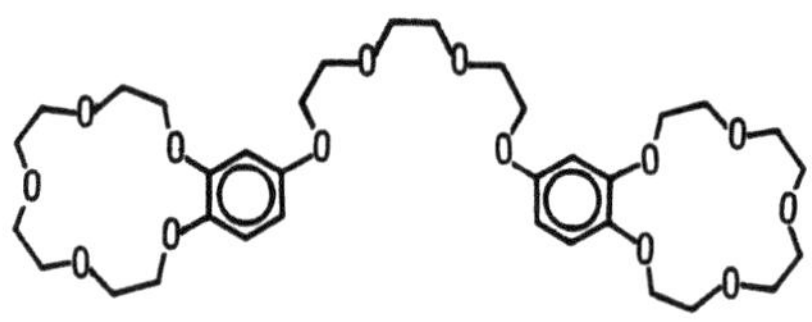	88.9–92.2	83	225
20	34	15	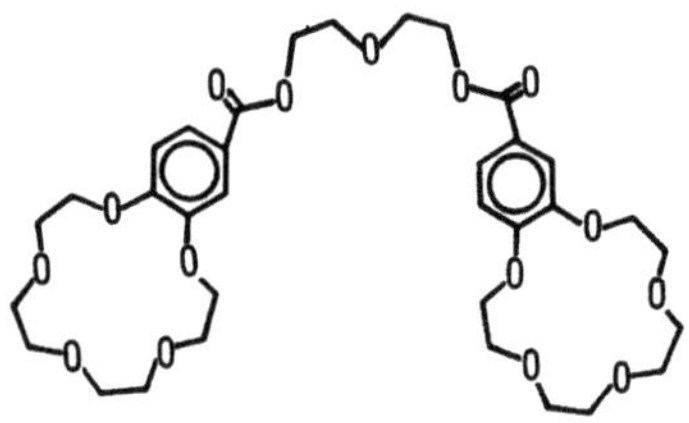	89–91	64	22
21	34	20	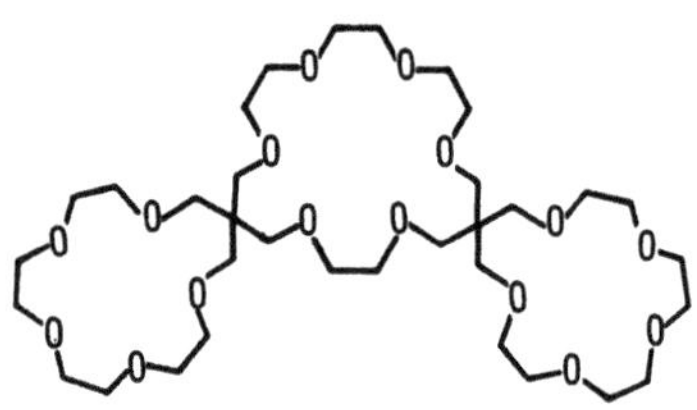	oil	nr	226
22	34	36	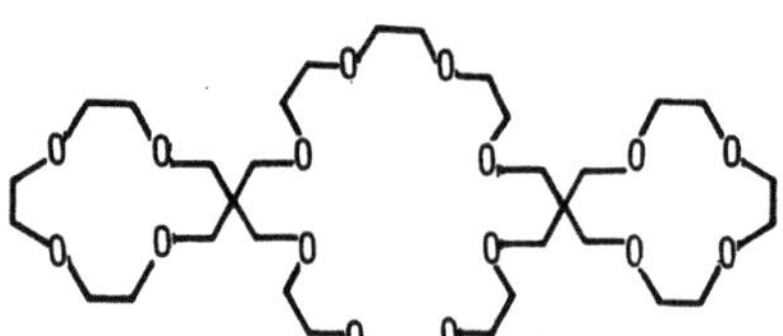	nr	nr	226
23	35	15	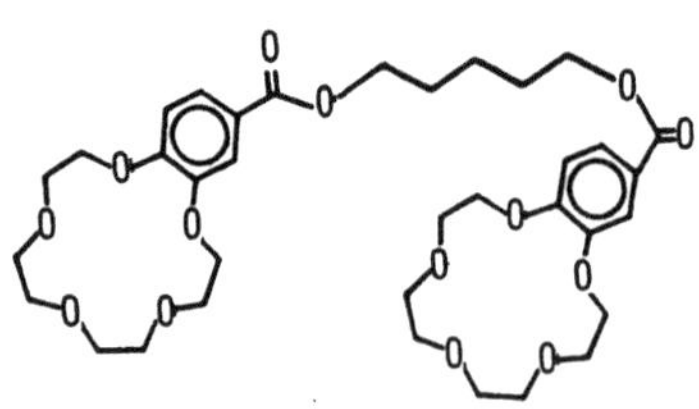	88–89	60	22

Table 3.27 (continued)

Cpd. No.	No. C's	Ring size	Structure	Mp or (bp) in °C	Yield (%)	Refs.
24	36	15	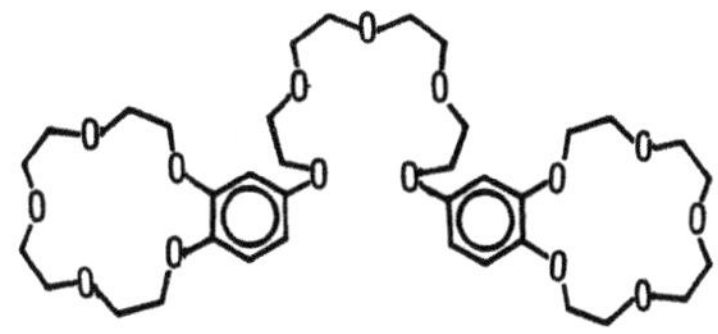	73–74	21	225
25	36	15	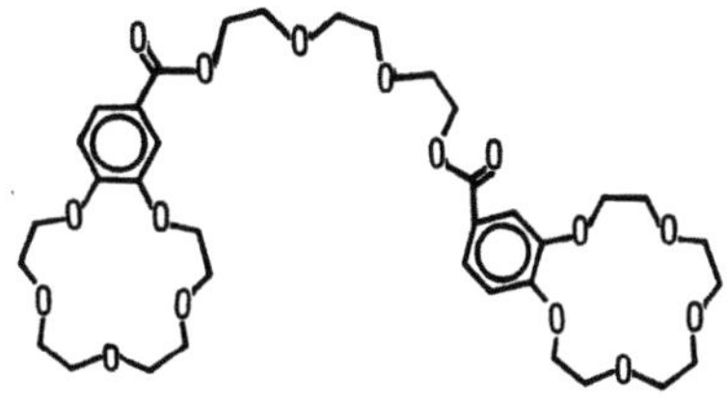	83–84	48	22
26	36	18	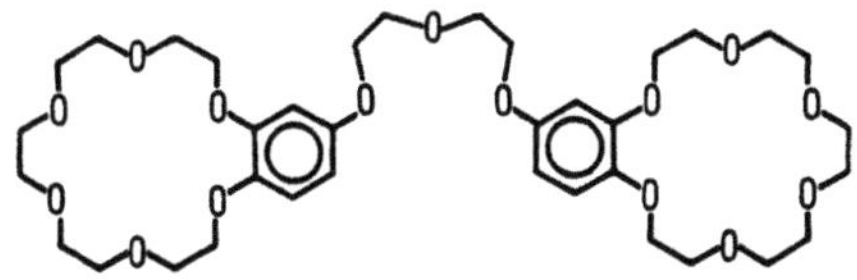	74.6–77.5	52	225
27	38	15	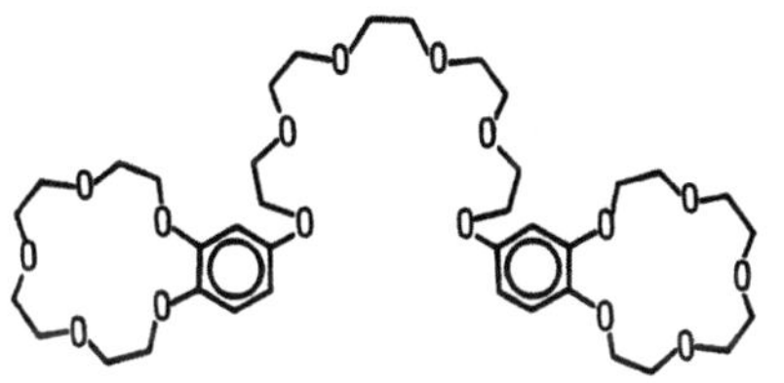	oil	63	225
28	38	15	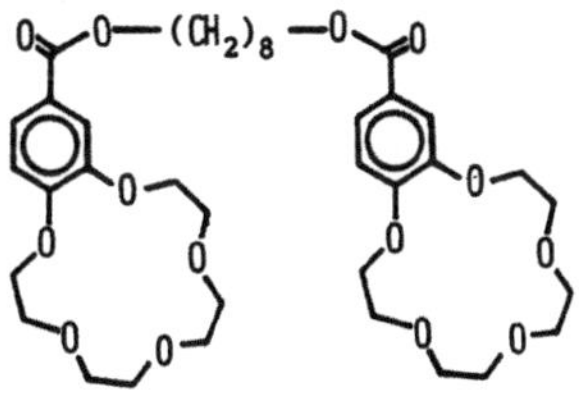	82	56	22
29	38	18	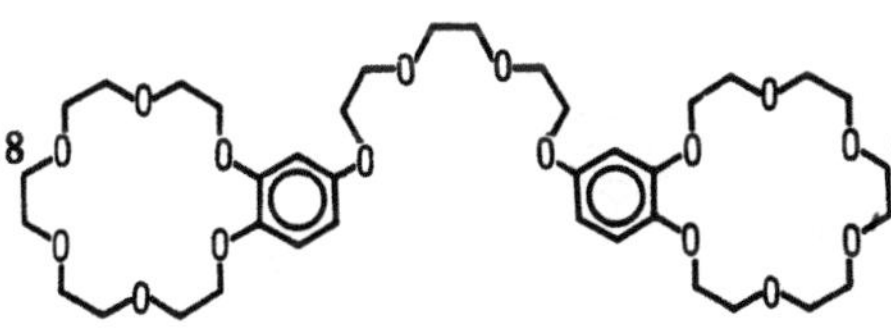	69.5–72.0	78	225
30	38	20	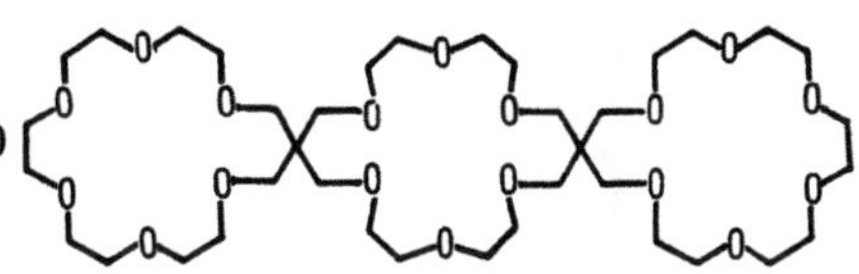	oil	nr	226

Table 3.27 (continued)

Cpd. No.	No. C's	Ring size	Structure	Mp or (bp) in °C	Yield (%)	Refs.
31	38	22	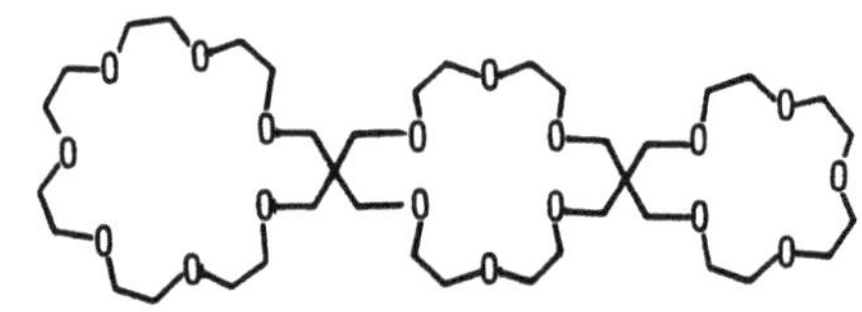	nr	nr	226
32	38	26	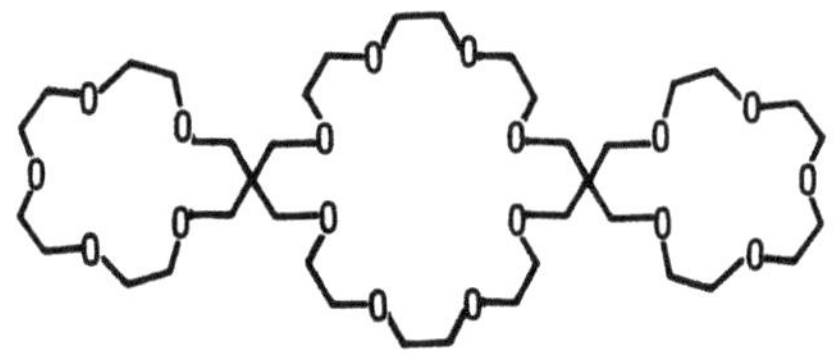	nr	nr	226
33	40	18	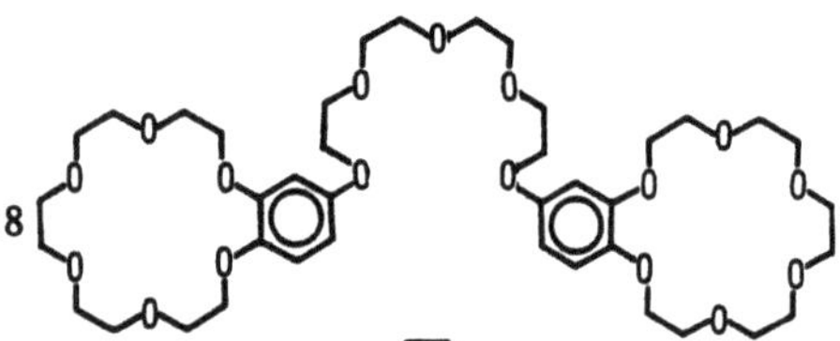	52–54.5	50	225
34	42	18	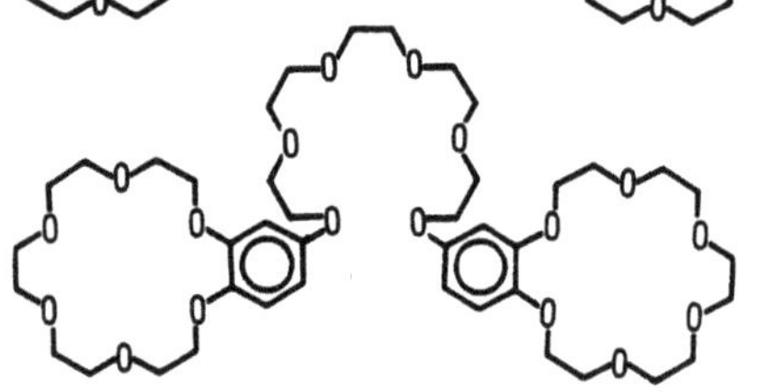	oil	58	225
35	44	20	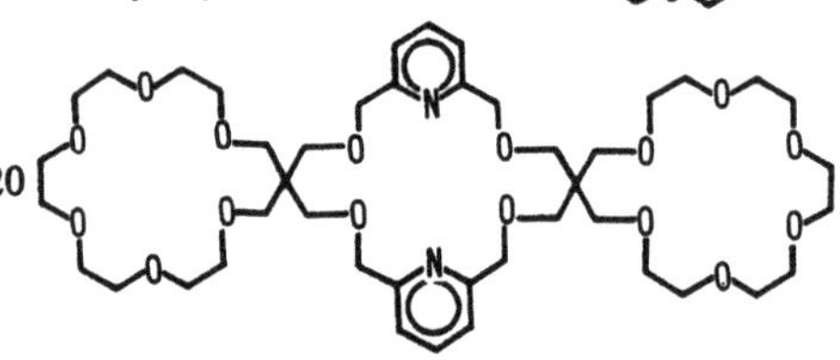	123–125	nr	226
36	45	15	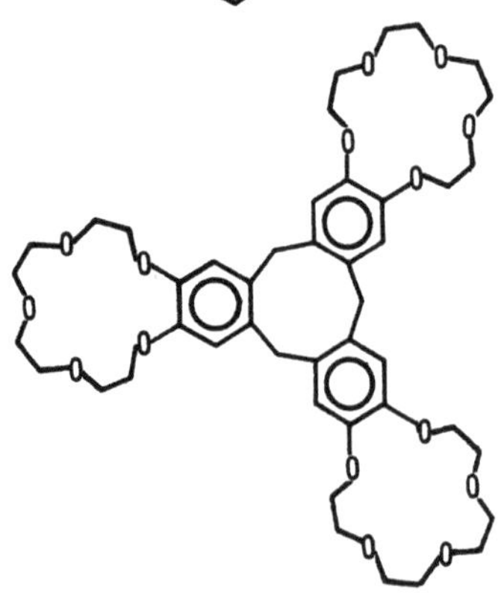	203.5–205.5	4	68

Table 3.27 (continued)

Cpd. No.	No. C's	Ring size	Structure	Mp or (bp) in °C	Yield (%)	Refs.
37	51	18	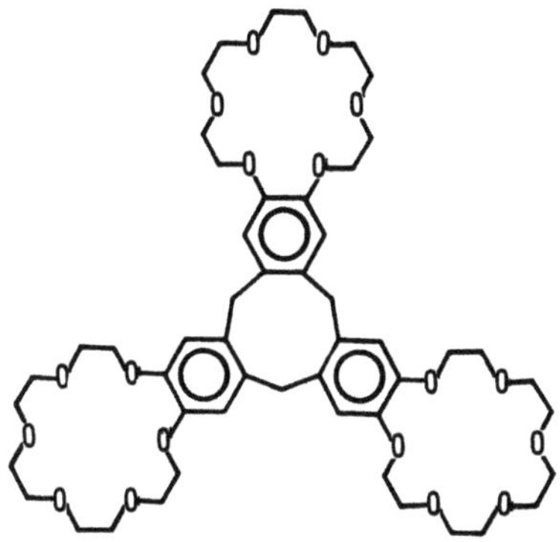	169–172	32	68

Table 3.28. Macrocyclic oligomers

Cpd. No.	No. C's	Ring size	Structure	Mp or (bp) in °C	Yield (%)	Refs.
1	12	12	$[CH_2CH(CH_2Cl)O]_4$	(185/0.2)	nr	80
2	28	32	$(C_7H_{14}O_4)_4$	nr	nr	65
3	35	40	$(C_7H_{14}O_4)_5$	19	nr	65
4	42	48	$(C_7H_{14}O_4)_6$	38	nr	65
5	49	56	$(C_7H_{14}O_4)_7$	nr	nr	65
6	56	64	$(C_7H_{14}O_4)_8$	nr	nr	65
7	56	12	$(CH_2-CH(CH_2S(CH_2)_{10}-COOH)-O)_4$	60	nr	80
8	56	12	$(CH_2-CH(CH_2S(CH_2)_{11}-OSO_3H)-O)_4$	nr	nr	80
9	56	12	$(CH_2-CH(CH_2S(CH_2)_{11}-OSO_3NH_4)-O)_4$	nr	nr	80
10	56	12	$(CH_2-CH(CH_2S(CH_2)_{11}OH)-O)_4$	nr	96	80

Table 3.28 (continued)

Cpd. No.	No. C's	Ring size	Structure	Mp or (bp) in °C	Yield (%)	Refs.
11	60	12	$(CH_2-CH-O)_4$ $CH_2S(CH_2)_{11}OSO_2CH_3$	nr	nr	80
12	64	12	$(CH_2-CH-O)_4$ $CH_2S(CH_2)_{11}NMe_2$	nr	nr	80
13	64	12	$(CH_2-CH-O)_4$ $CH_2S(CH_2)_{11}SCH_2COOH$	nr	nr	80
14	68	12	$(CH_2-CH-O)_4$ $CH_2S(CH_2)_{11}NMeCH_2CH_2OH$	nr	nr	80
15	72	12	$(CH_2-CH-O)_4$ $CH_2S(CH_2)_{11}SCH_2CO_2Et$	nr	nr	80
16	72	12	$(CH_2-CH-O)_4$ $CH_2S(CH_2)_{11}\overset{+}{N}Me_2CH_2COO^-$	nr	nr	80
17	76	12	$(CH_2-CH-O)_4$ $CH_2S(CH_2)_{11}\overset{+}{N}Me_2CH_2CH_2OH\,^-O_3SCH_3$	nr	nr	80
18	80	12	$(CH_2-CH-O)_4$ $CH_2S(CH_2)_{11}OSO_3^-\overset{+}{N}H(CH_2CH_2OH)_3$	nr	nr	80
19	84	12	$(CH_2-CH-O)_4$ $CH_2S(CH_2)_{11}SCH_2\overset{O}{\overset{\|}{C}}NH(CH_2)_3NMe_2$	nr	nr	80
20	124	12	$(CH_2-CH-O)_4$ $CH_2S(CH_2)_{11}S\overset{O}{\overset{\|}{C}}(OCH_2CH_2)_6NEt_2$	nr	nr	80
21	128	12	$(CH_2-CH-O)_4$ $CH_2S(CH_2)_{11}-(CH_2\overset{OH}{\overset{\|}{C}}H-CH_2O)_6H$	nr	nr	80
22	128	12	$(CH_2-CH-O)_4$ $CH_2S(CH_2)_{11}-O(-CH_2-\underset{CH_2OH}{\underset{\|}{C}H}-O)_2-(CH_2-\underset{CH_2SCH_2CHOHCH_2OH}{\underset{\|}{C}H}-O-)_2H$	nr	nr	80

Table 3.29. Polymeric crowns

Cpd. No.	No. C's	Ring size	Structure	Mp or (bp) in °C	Yield (%)	Refs.
1		18	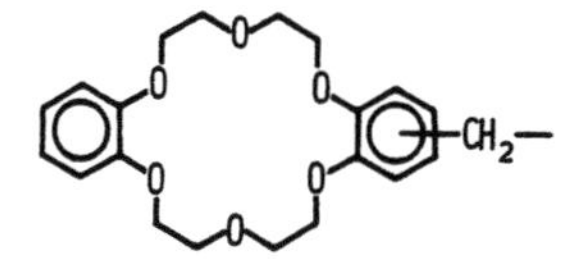	nr	nr	173
2		18	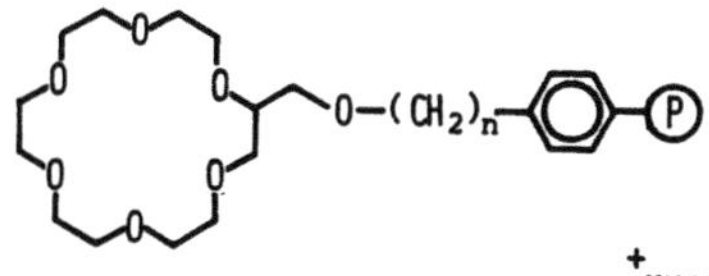 n = 1	nr	nr	134
3		18	n = 6	nr	nr	134
4		18	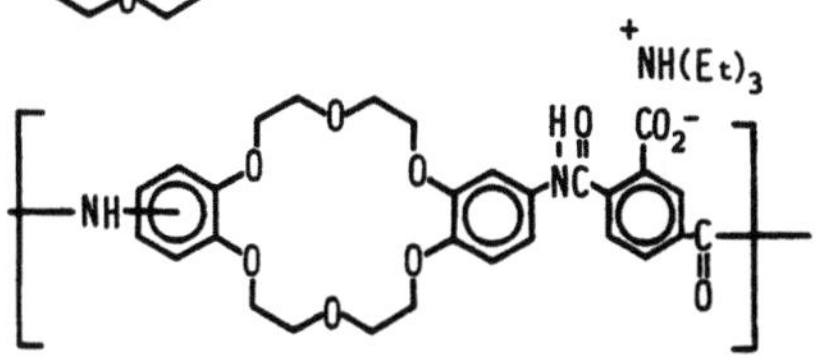	nr	nr	69
5		18	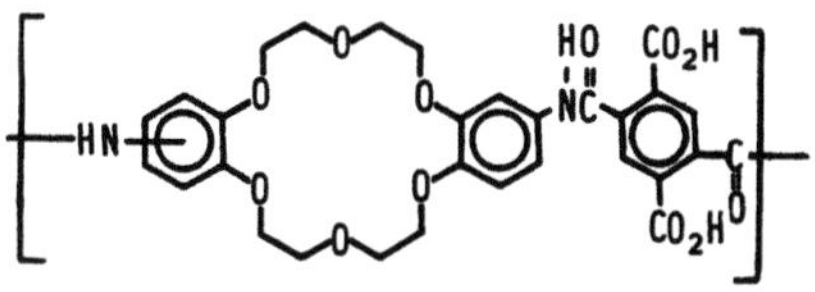	nr	nr	69
			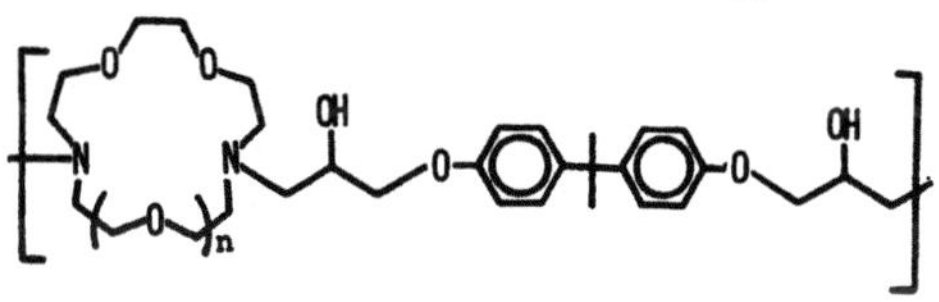			
6		15	n = 1	nr	83	78
7		18	n = 2	nr	83	78
8		18	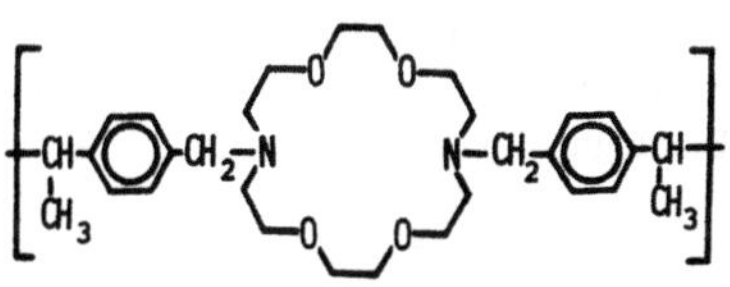	nr	nr	18
9		18	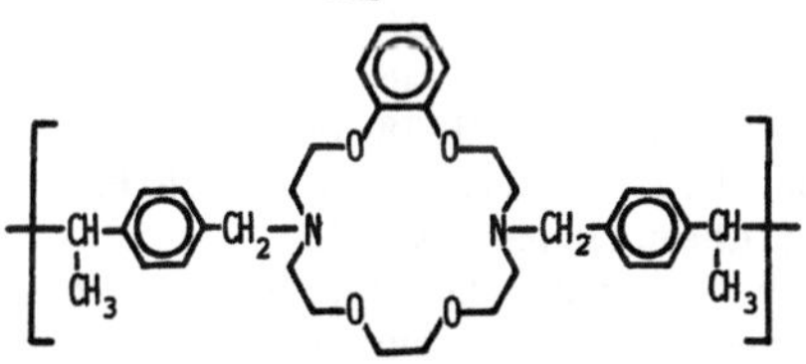	nr	nr	18
10		18	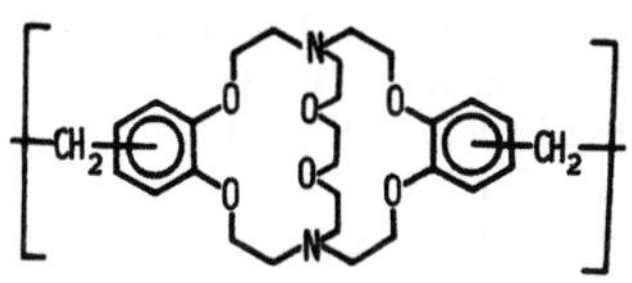	nr	nr	18

Table 3.29 (continued)

Cpd. No.	No. C's	Ring size	Structure	Mp or (bp) in °C	Yield (%)	Refs.
11		18	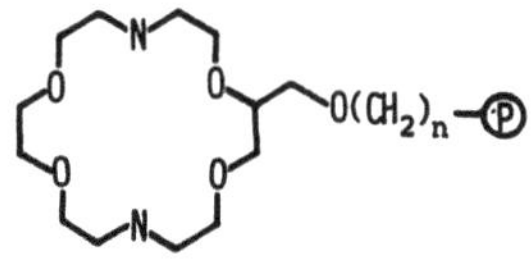 n = 1	nr	nr	134
12		18	n = 6	nr	nr	134
13		18	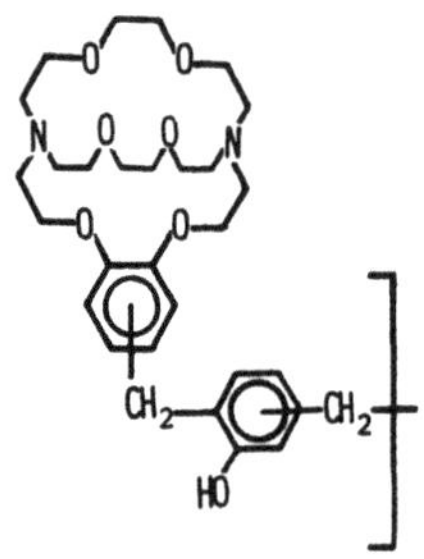	nr	nr	18

3.18 References

1. Ackman, R. G., Brown, W H., Wright, G. F.: J. Org. Chem. (1955) *20,* 1147
2. Alberts, A. H. and Cram, D. J.: J. C. S. Chem. Comm. (1976) 958
3. Alberts, A. H. and Cram, D. J.: J. Amer. Chem. Soc. (1977) *99,* 3880
4. Alekseeva, I. A., Popova, V. A. and Podgornaya, I. V.: Zh. Obshch. Khim. (1978) *48,* 1630
5. Allen, D. W., Braunton, P. N., Millar, I. T. and Tebby, J. C.: J. Chem. Soc. (C) (1971) 3454
6. Ando, N., Yamamoto, Y., Oda, J. and Inouye, Y.: Synthesis (1978) 688
7. Andrews, D. G., Ashton, P. R., Laidler, D. A., Stoddart, J. F. and Wolstenholme, J. B.: Tetrahedron Letters (1979) 2629
8. Ashby, J., Hull, R., Cooper, M. J. and Ramage, E. M.: Syn. Commun. (1974) *4,* 113
9. Badger, G. M., Elix, J. A. and Lewis, G. E.: Aust. J. Chem. (1966) *19,* 1221
10. Badger, G. M., Elix, J. A., Lewis, G. E., Singh, V. P. and Spotswood, T. M.: J. C. S. Chem. Comm. (1965) 269
11. Bartsch, R. A. and Juri P. N.: J. Org. Chem. (1980) *45,* 1011
12. Beals, R. E. and Brown, W. H : J. Org. Chem. (1956) *21,* 447
13. Beckford, H. F., King, R. M. and Stoddart, J. F.: Tetrahedron Letters (1978) 171
14. Behr, J-P. and Lehn, J. M.: J. C. S. Chem. Comm. (1978) 143
15. Behr, J-P., Lehn, J. M. and Vierling, P.: J. C. S. Chem. Comm. (1976) 621
16. Bell, A. P. and Hall, C. D.: J. C. S. Chem. Comm. (1980) 163
17. Beresford, G. D. and Stoddart, J. F.: Tetrahedron Letters (1980) *21,* 867
18. Blasius, E., Janzen, K. P. and Neumann, W.: Mikrochimica Acta (1977) *II*, 279
19. Blasius, E., Adrian, W., Janzen, K. P. and Klautke, G.: J. Chromatogr. (1974) *96,* 89
20. Bonthrone, W. and Cornforth, J. W.: J. Chem. Soc. (C) (1969) 1202
21. Boujlel, K. and Simonet, J.: Tetrahedron Letters, (1979) 1497
22. Bourgoin, M., Wong, K. H., Hui, J. Y. and Smid, J.: J. Amer. Chem. Soc. (1975) *97,* 3462
23. Bradshaw, J. S , Asay, R. E., Maas, G. E., Izatt, R. M. and Christensen, J. J.: J. Heterocyclic Chem. (1978) *15,* 825

24. Brown, H. S., Muenchausen, C. P. and Sousa, L. R.: J. Org. Chem. (1980) *45,* 1682
25. Brown, W. H. and French, W. N.: Can. J. Chem. (1958) *36,* 371
26. Brown, W. H. and French, W. N.: Can. J. Chem. (1958) *36,* 537
27. Brown, W. H., Hutchinson, B. J. and MacKinnon, M. H.: Can. J. Chem. (1971) *49,* 4017
28. Buhleier, E., Wehner, W. and Vögtle, F.: Chem. Ber. (1978) *111,* 200
29. Buhleier, E., Wehner, W. and Vögtle, F.: Liebigs. Ann. Chem. (1978) 537
30. Burden, I. J., Coxon, A. C., Stoddart, J. F. and Wheatley, C. M.: J. C. S. Perkin Trans. I (1977) 220
31. Chao, Y. and Cram, D. J.: J. Amer. Chem. Soc. (1976) *98,* 1015
32. Chastrette, M. and Chastrette, F.: J. C. S. Chem. Comm. (1973) 534
33. Cinquini, M. and Tundo, P.: Synthesis (1976) 516
34. Cook, F. L., Caruso, T. C., Byrne, M. P., Bowers, C. W., Speck, D. H. and Liotta, C. L.: Tetrahedron Letters (1974) 4029
35. Coxon, A. C., Laidler, D. A , Pettman, R. B. and Stoddart, J. F.: J. Am. Chem. Soc. (1978) *100,* 8260
36. Cram, D. J : Ger. Offen. 2,414,188, 21 Nov. 1974; U.S.P. 4,001,279 4 Jan. 1977
37. Cram, D. J., Helgeson, R. C., Koga, K., Kyba, E. P., Madan, K., Sousa, L. R., Siegel, M. G., Moreau, P., Gokel, G. W., Timko, J. M. and Sogah, G. D. Y.: J. Org. Chem. (1978) *43,* 2758
38. Cram, D. J., Helgeson, R. C., Peacock, S. C., Kaplan, L. J., Domeier, L. A., Moreau, P., Koga, K., Mayer, J. M , Chao, Y., Siegel, M. G., Hoffman, D. H. and Sogah, G. D. Y.: J. Org. Chem (1978) *43,* 1930
39. Cresp, T. M. and Sargent, M. V.: J. C. S. Chem. Comm. (1971) 1458
40. Cresp, T. M. and Sargent, M. V.: J. C. S. Perkin Trans. I (1973) 1786
41. Cresp, T. M. and Sargent, M. V.: J. C. S. Perkin Trans. I (1973) 2961
42. Curtis, W. D., Laidler, D A., Stoddart, J. F. and Jones, G. H.: J. C. S. Chem. Comm. (1975) 833
43. Curtis, W. D., Laidler, D. A., Stoddart, J. F. and Jones, G. H.: J. C. S. Perkin Trans. I (1977) 1756
44. Curtis, W. D., King, R. M., Stoddart, J. F. and Jones, G. H.: J. C. S. Chem. Comm. (1976) 285
45. Curtis, W. D., Laidler, D. A., Stoddart, J. F., Wolstenholme, J. B. and Jones, G. J.: Carbohydr. Res. (1977), *57,* C17
46. Cygan, A., Biernat, J. F. and Chadzynski, H.: Pol. J. Chem. (1979) *53,* 929
47. Czech, B : Tetrahedron Letters (1980) 4197
48. Dale, J.: Ger. Off. 2,523,542, Dec. 4, 1975
49. Dale, J , Borgen, G. and Daasvatn, K.: Acta Chem. Scand. (1974) *B28,* 378
50. Dale, J. and Daasvatn, K.: J. C. S. Chem. Comm. (1976) 295
51. Dale, J. and Daasvatn, K.: U.S.P. 3,997,563, 14 Dec. 1976
52. Dale, J. and Kristiansen, P. O.: Acta Chem. Scand. (1972) *26,* 1471
53. de Jong, F., Reinhoudt, D N. and Smit, C. J.: Tetrahedron Letters (1976) 1371
54. de Jong, F., Reinhoudt, D. N., Smit, D. J. and Huis, R.: Tetrahedron Letters (1976) 4783
55. de Jong, F., Siegel, M. G. and Cram, D. J.: J. C. S. Chem. Comm. (1975) 551
56. DeVries, J. G. and Kellogg, R. M.: J. Am. Chem. Soc. (1979) *101,* 2759
57. DiCesare, P. and Gross, B.: Synthesis (1979) 458
58. Dishong, D. M. and Gokel, G. W.: unpublished results (1980)
59. Dotsevi, G., Sogah, Y. and Cram, D. J.: J. Amer. Chem. Soc. (1975) *97,* 1259
60. Down, J. L., Lewis, J., Moore, B. and Wilkinson, G.: J. Chem. Soc. (1959) 3767
61. Dix, J. P. and Vögtle, F.: Chem. Ber. (1980) *113,* 457
62. Elben, U., Fuchs, H. B., Frensch, K. and Vögtle, F.: Liebigs Ann. Chem. (1979) *8,* 1102–11
63. Elix, J. A.: Aust. J. Chem. (1969) *22,* 1951
64. Elix, J. A.: J. C. S. Chem. Comm. (1968) 343
65. Feigenbaum, W. M. and Michel, R. H.: J. Polym. Sci. Part A (1971) 817
66. Fenton, D. E , Cook, D. H., Nowell, I. W. and Walker, P. E.: J. C. S. Chem. Comm. (1977) 623
67. Frensch, K. and Vögtle, F.: Liebigs Ann. Chem. (1979) 2118
68. Frensch, K. and Vögtle, F.: Liebigs Ann. Chem. (1979) 2121
69. Frost, L. W.: U.S.P. 3,956,136, 11 May 1976
70. Girodeau, J. M., Lehn, J. M. and Sauvage, J. P.: Angew. Chem. Int. Ed. Engl. (1975) *14,* 764

71. Gokel, G. W.: Unpublished results (1980)
72. Gokel, G. W., Cram, D. J., Liotta, C. L., Harris, H. P. and Cook, F. L.: J. Org. Chem. (1974) *39*, 2445
73. Gokel, G. W., Cram, D. J., Liotta, C. L., Harris, H. P. and Cook, F. L.: Org. Synth. (1977) *57*, 30
74. Gokel, G. W., Dishong, D. M. and Diamond, C. J.: J. C. S. Chem. Commun. (1980) 1053
75. Gokel, G. W., Timko, J. M. and Cram, D. J.: J. C. S. Chem. Commun. (1975) 394
76. Gokel, G. W., Timko, J. M. and Cram, D. J.: J. C. S. Chem. Commun. (1975) 444
77. Gomes, A. D. S. and Oliveira, C. M. F.: Org. Mass Spectrom. (1977) *12*, 407
78. Gramain, P. and Frere, Y.: Macromolecules (1979) *6*, 1038
79. Gray, R. T. and Reinhoudt, D. N.: Tetrahedron Letters (1975) 2109
80. Gray, R. T., Reinhoudt, D. N., Smit, C. J. and Veenstra, I.: Recl. Trav. Chim. Pays-Bas (1976) *95*, 258
81. Greene, R. N.. Tetrahedron Letters (1972) 1793
82. Hain, W., Lehnert, R., Röttele, H. and Schröder, G.: Tetrahedron Letters (1978) 625
83. Haines, A. H.: Tetrahedron Letters (1980) 285
84. Haines, A. H. and Karntiang, P.: Carbohydr. Res. (1980) *78*, 205
85. Haslegrave, J. A., Stoddart, J. F. and Thompson, D. J.: Tetrahedron Letters (1979) 2279
86. Hautala, R. R. and Hastings, R. H.: J. Amer. Chem. Soc. (1978) *100*, 648
87. Hayward, R. C., Overton, C. H. and Whitham, G. H.: J. C. S. Perkin Trans. I (1976) 2413
88. Helgeson, R. C., Koga, K., Timko, J. M. and Cram, D. J.: J. Amer. Chem. Soc. (1973) *95*, 3021
89. Helgeson, R. C., Tarnowski, T. L., Timko, J. M. and Cram, D. J.: J. Amer. Chem. Soc. (1977) *99*, 6411
90. Helgeson, R. C., Timko, J. M. and Cram, D. J.: J. Amer. Chem. Soc. (1973) *95*, 3023
91. Helgeson, R. C., Timko, J. M. and Cram, D. J.: J. Amer. Chem. Soc. (1974) *96*, 7380
92. Helgeson, R. C., Timko, J. M., Moreau, P., Peacock, S. C., Mayer, J. M. and Cram, D. J.: J. Amer. Chem. Soc. (1974) *96*, 6762
93. Htay, M. M. and Meth-Cohn, O.: Tetrahedron Letters (1976) 79
94. Htay, M. M. and Meth-Cohn, O.: Tetrahedron Letters (1976) 469
95. Hyde, E. M., Shaw, B. L. and Shepherd, I.: J. C. S., Dalton Trans. (1978) 1696
96. Ishizu, K., Kohama, H., and Mukai, K.: Chem. Lett. (1978) 227
97. Izatt, R. M., Haymore, B. L., Bradshaw, J. S. and Christensen, J. J.: Inorg. Chem. (1975) *14*, 3132
98. Johns, G., Ransom, C. J. and Reese, C. B.: Synthesis (1976) 515
99. Kauer, J. C.: U.S.P. 3,997,565, 14 Dec. 1976
100. Kauer, J. C.: U.S.P. 4,024,158, 17 May 1977
101. Kawashima, N., Kawashima, T. Otsubo, T. and Misumi, S.: Tetrahedron Letters (1978) 5025
102. Kellogg, R. M., van Bergen, T. J., van Doren, H., Hedstrand, D., Kool, J., Kruizinga, W. H. and Troostwijk, C. B.: J. Org. Chem. (1980) *45*, 2854
103. Kikukawa, K., Nagira, N. and Matsuda, T.: Bull. Chem. Soc. Jpn. (1977) *50*, 2207
104. Kime, D. E. and Norymberski, J. K.: J. C. S. Perkin Trans. I (1977) 1048
105. Koenig, K. E , Helgeson, R. C. and Cram, D. J.: J. Amer. Chem. Soc. (1976) *98*, 4018
106. Kossai, R., Simonet, J. and Jeminet, G.: Tetrahedron Letters (1979) 1059
107. Krespan, C. G.: J. Org. Chem. (1974) *39*, 2351
108. Krespan, C. G.: J. Org. Chem. (1975) *40*, 1205
109. Krespan, C. G.: U.S.P. 3,952,015, 20 April 1976
110. Krespan, C. G.: J. Org. Chem. (1980) *45*, 1177
111. Kudrya, T. N., Talanova, G. G., Bonadyk, S. V. and Sinitsa, A. D.: Dokl. Adad. Nauk Ukr. RSR, Ser. B: Geol., Khim. Biol. Nauki (1979) 434
112. Kuwamura, T.: Nippon Kagaku Kaishi (1980) *3*, 427–34
113. Kyba, E. P., Helgeson, R. C., Madan, K., Gokel, G. W., Tarnowski, T. L., Moore, S. S. and Cram, D. J.: J. Amer. Chem. Soc. (1977) *99*, 2564
114. Kyba, E. P., Gokel, G. W., deJong, F., Koga, K., Sousa, L. R., Siegel, M. G., Kaplan, L., Sogah, G. D. Y. and Cram, D. J.: J. Org. Chem. (1977) *42*, 4173
115. Kyba, E. P., Koga, K., Sousa, L. R., Siegel, M. G. and Cram, D. J.: J. Amer. Chem. Soc. (1973) *95*, 2692

116. Kyba, E. P., Siegel, M. G., Sousa, L. R., Sogah, D. G. Y. and Cram, D. J.: J. Amer. Chem. Soc. (1973) *95*, 2691
117. Laidler, D. A. and Stoddart, J. F.: J. C. S., Chem. Commun. (1976) 979
118. Laidler, D. A. and Stoddart, J. F.: Carbohydr. Res. (1977) *55*, Cl
119. Laidler, D. A. and Stoddart, J. F.: J. C. S., Chem. Commun. (1977) 481
120. Laidler, D. A. and Stoddart, J. F.: Tetrahedron Letters (1979) 453
121. Laidler, D. A., Stoddart, J. F. and Wolstenholme, J. B.: Tetrahedron Letters (1979) 465
122. Lehn J. M. and Sirlin, C.: J. C. S., Chem. Commun. (1978) 949
123. Lehn, J. M., Vierling, P. and Hayward, R. C.: J. C. S., Chem. Commun. (1979) 296
124. L'Oreal SA: Fr. Demande, 2359165, 17 Feb. 1978
125. Mack, M. P., Hendrickson, R. R., Palmer, R. A. and Ghirardelli, R. G.: J. Amer. Chem. Soc. (1976) *98*, 7830
126. Malpass, G. D., Palmer, R. A. and Ghirardelli, R. G.: Tetrahedron Letters (1980) *21*, 1489
127. McKervey, M. A. and Mulholland, D. L.: J. C. S., Chem. Commun. (1977) 438
128. Matsui, T. and Koga, K.: Tetrahedron Letters (1978) 1115
129. Matsui, T. and Koga, K.: Chem. Pharm. Bull. (1979) *27*, 2295
130. Matsumoto, S. and Nakagawa, D.: Jpn. Kokai 79:90,189, 17 July 1979
131. Mazur, S., Dixit, V. M. and Gerson, F.: J. Amer. Chem. Soc. (1980) *102*, 5343
132. Merz A.: Angew. Chem. Int. Ed. Engl. (1977) *16*, 467
133. Mizuno, T., Nakatsuji, Y., Yanagida, S. and Okahara, M.: Bull. Chem. Soc. Jpn. (1980) *53*, 481
134. Montanari, F. and Tundo, P.: Tetrahedron Letters (1979) 5055
135. Moore, S. S., Tarnowski, T. L., Newcomb, M. and Cram, D. J.: J. Amer. Chem. Soc. (1977) *99*, 6398
136. Nakamura, H., Takagi, M. and Ueno, K.: Talanta, (1979) *26*, 921
137. Nerdel, F., Mamluk, M. and Weyerstahl, P.: Liebigs Ann. Chem. (1970) *736*, 75
138. Newcomb, M. and Cram, D. J.: J. Amer. Chem. Soc. (1975) *97*, 1257
139. Newcomb, M., Gokel, G. W. and Cram, D. J.: J. Amer. Chem. Soc. (1974) *96*, 6810
140. Newcomb, M., Moore, S. S. and Cram, D. J.: J. Amer. Chem. Soc. (1977) *99*, 6405
141. Newcomb, M., Timko, J. M., Walba, D. M. and Cram, D. J.: J. Amer. Chem. Soc. (1977) *99*, 6392
142. Newcome, G. R. and Hager, D. C.: J. Amer. Chem. Soc. (1978) *100*, 5567
143. Newkome, G. R. and Kawato, T.: J. Org. Chem. (1979) *44*, 2693
144. Newkome, G. R. and Kawato, T.: J. Amer. Chem. Soc. (1979) *101*, 7088
145. Newkome, G. R. and Kawato, T.: Tetrahedron Letters (1978) 4643
146. Newkome, G. R., Kawato, T. and Benton, W. H.: J. Org. Chem. (1980) *45*, 626
147. Newkome, G. R., Kawato, T. and Nayak, A.: J. Org. Chem. (1979) *44*, 2697
148. Newkome, G. R., Kohli, D. K. and Fronczek, F.: J. C. S., Chem. Commun. (1980) 9
149. Newkome, G. R., Nayak, A., Fronczek, F., Kawato, T., Taylor, H. C. R., Meade, L. and Mattice, W.: J. Amer. Chem. Soc. (1979) *101*, 4472
150. Newkome, G. R., Nayak, A., Sauer, J. D., Mattschei, P. K., Watkins, S. F., Fronczek, F. and Benton, W. H.: J. Org. Chem. (1979) *44*, 3816
151. Newkome, G. R., Nayak, A., Sorci, M. G. and Benton, W. H.: J. Org. Chem. (1979) *44*, 3812
152. Newkome, G. R., McClure, G. L., Simpson, J. B. and Danesh-Khoshboo, F.: J. Am. Chem. Soc. (1975) *97*, 3232
153. Newkome, G. R., Nayak, A., McClure, G. L., Danesh-Khoshboo, F. and Broussard-Simpson, J.: J. Org. Chem. (1976) *42*, 1500
154. Newkome, G. R., Nayak, A., Otemaa, J., Van, A. and Benton, W. H.: J. Org. Chem. (1978) *43*, 3362
155. Newkome, G. R. and Robinson, J. M.: J. C. S., Chem. Commun. (1973) 831
156. Newkome, G. R. and Robinson, J. M.: J. Org. Chem. (1976) *41*, 2536
157. Newkome, G. R., Sauer J. D., Mattschei, P. K. and Nayak, A.: Heterocycles (1978) *9*, 1555
158. Newkome G. R. and Taylor, H. C. R.: J. Org. Chem. (1979) *44*, 1362
159. Okahara, M., Ikeda, I. and Yanagida, S.: Jpn. Kokai, 78:98,985, 29 Aug. 1978
160. Okahara, M., Miki, M., Yanagida, S. and Ikeda, I.: Synthesis (1977) 854
161. Orvik, J. A.: J. Amer. Chem. Soc. (1976) *98*, 3322
162. Pannell, K. H., Hambrick, D. C. and Lewandos, G. S.: J. Organomet. Chem. (1975) *99*, C21

163. Pannell, K. H., Yee, W., Lewandos, G. S., Hambrick, D. C.: J. Amer. Chem. Soc. (1977) *99*, 1457
164. Parish, W. W., Stott, P. E., McCausland, C. W. and Bradshaw, J. S.: J. Org. Chem. (1978) *43*, 4577
165. Parsons, D. G.: J. C. S., Perkin Trans. I (1975) 245
166. Peacock, S. C. and Cram, D. J.: J. C. S., Chem. Commun. (1976) 282
167. Peacock, S. C., Domeier, L. A., Gaeta, F. C. A., Helgeson, R. C., Timko, J. M. and Cram, D. J.: J. Amer. Chem. Soc. (1978) *100*, 8190
168. Pedersen, C. J.: J. Amer. Chem. Soc. (1967) *89*, 2495
169. Pedersen, C. J.: J. Amer. Chem. Soc. (1967) *89*, 7017
170. Pedersen, C. J.: J. Amer. Chem. Soc. (1970) *92*, 391
171. Pedersen, C. J.: Org. Synth. (1972) *52*, 66
172. Pedersen, C. J.: U.S.P. 3,562,295 2/9/71
173. Pedersen, C. J.: U.S.P. 3,687,978 8/29/72
174. Pedersen, C. J.: U.S.P. 3,987,061 10/19/76
175. Pettman R. B. and Stoddart, J. F.: Tetrahedron Letters (1979) 457
176. Pettman, R. B. and Stoddart, J. F.: Tetrahedron Letters (1979) 461
177. Ping-Lin, K., Miki, M. and Okahara, M.: J. C. S., Chem. Commun. (1978) 504
178. Popova, V. A., Podgornaya, I. R., Postovskii, I. Y. and Frolova, N. N.: Khim-Farm. Zh. (1976) *10*, 66
179. Prochaska, R. J.: U.S.P. 3,274,214 9/20/66
180. Rasshofer, W., Muller, W. M., Oepen, G. and Vögtle, F.: J. Chem. Res.(S) (1978) 72
181. Rastetter, W. H. and Phillion, D. P.: Tetrahedron Letters (1979) 1469
182. Rebek, J., Trend, J. E., Wattley, R. V. and Chakravorti, S.: J. Amer. Chem. Soc. (1979) *101*, 4333
183. Reinhoudt, D. N., deJong, F. and Tomassen, H. P. M.: Tetrahedron Letters (1979) 2067
184. Reinhoudt, D. N. and Gray, R. T.: Tetrahedron Letters (1975) 2105
185. Reinhoudt, D. N., Gray, R. T., Smit, C. J. and Veenstra, I.: Tetrahedron (1976) *32*, 1161
186. Reinhoudt, D. N. and Gray, R. T.: Ger. Offen. 2,610,498, 30 Sep. 1976
187. Rest, A. J., Smith, S. A. and Tyler, I. D.: Inorg. Chim. Acta (1976) *16*, L1
188. Robinson, I. M. and Pruckmayr, G.: Macromolecules (1980) *12*, 1043
189. Sasaki, S. and Koga, K.: Heterocycles (1979) 12(10), 1305
190. Schröder, G. and Witt, W.: Angew. Chem. Int. Ed. Engl. (1979) *18*, 311
191. Shchori, E. and Jagur-Crodzinski, J.: J. Amer. Chem. Soc. (1972) *94*, 7957
192. Shchori, E., Jagur-Grodzinski, J. and Shporer, M.: J. Amer. Chem. Soc. (1973) *95*, 3842
193. Shida, Y, Ando, N., Yamamoto, Y, Oda, J. and Inouye, Y.: Agric. Biol. Chem. (1979) *95*, 1797
194. Shinkai, S., Ogawa, T., Nakaji, T. and Manabe, O.: J. C. S., Chem. Commun. (1980) 375
195. Sieger, H. and Vögtle, F.: Liebigs Ann. Chem. (1980) 425
196. Smid, J., El Haj, B., Majewica, T., Nonni, A. and Sinta, R.: Org. Prep. Proc. Int. (1976) *8*, 193
197. Sogah, G. D. Y. and Cram, D. J.: J. Amer. Chem. Soc. (1976) *98*, 3038
198. Sousa, L. R., Hoffman, D. H., Kaplan, L. and Cram, D. J.: J. Amer. Chem. Soc. (1974) *96*, 7100
199. Stoddart, J. F. and Wheatley, C. M.: J. C. S., Chem. Commun. (1974) 390
200. Tarnowski, T. L. and Cram, D. J., J. C. S., Chem. Commun. (1976) 661
201. Tashmukhamedova, A. K., Abdullaeva, R. A., Stempnevskaya, I. A., Saifullina, N. Zh. and Adilbekov, M. T.: Bioorg. Khim. (1978) *4*, 806
202. Tashmukhamedova, A. K., Saifullina, N. Zh., Stempnevskaya, I. A. and Abdullaeva, R. A.: Bioorg. Khim. (1978) 1232
203. Tashmukhamedova, A. K., Stempnevskaya, I. A., Saifullina, N. Zh. and Abdullaeva, R. A.: Bioorg, Khim (1978) *4*, 1414
204. Timko, J. M. and Cram, D. J.: J. Amer. Chem. Soc. (1974) *96*, 7159
205. Timko, J. M., Helgeson, R. C., Newcomb, M., Gokel, G. W. and Cram, D. J.: J. Amer. Chem. Soc. (1974) *96*, 7097
206. Timko, J. M., Helgeson, R. C. and Cram, D. J.: J. Amer. Chem. Soc. (1978) *100*, 2828
207. Timko, J. M., Moore, S. S., Walba, D. M., Hiberty, P. C. and Cram, D. J.: J. Amer. Chem. Soc. (1977) *99*, 4207
208. Toke, L, Fenichel, L., Bako, P. and Szejtli, J.: Acta Chem. Acad. Sci. Hung. (1978) *98*, 357

209. Tomoi, M., Abe, O., Ikeda, M., Kihara, K. and Kakiuchi, H.: Tetrahedron Letters (1978) 3031
210. Trucks, R. O. and Steiner, E. C.: U.S. 4,113,739, 12 Sep. 1978
211. Tundo, P. and Fendler, J. H.: J. Amer. Chem. Soc. (1980) *102*, 1760
212. Tusek, L.: Com. Naz Energ Nucl. (1974) 74, Chem. Abstr. (1975) 57662g
213. Ungara, R., El Haj, B. and Smid, J.: J. Amer. Chem. Soc. (1976) *98*, 5198
214. Ueno, K,. Jpn. Priority: 18, JP: 060426.
215. van der Ven, R. H., Kellogg, R. M., Vos, A. and van Bergen, T. J.: J. C. S., Chem. Commun. (1978) 923
216. Vartanyan, S. A., Akopyan, T. R. and Paronikyan, E. G.: Arm. Khim Zh. (1979) *32*, 720
217. Van Keulen, B. J., Kellogg, R. M. and Piepers, O.: J. C. S., Chem. Commun. (1979) 285
218. Vögtle, F. and Dix, I.: Liebigs Ann. Chem. (1977) 1698
219. Vögtle, F. and Frensch, K.: Angew. Chem. Int. Ed. Engl. (1976) *15*, 685
220. Vögtle, F. and Jansen, B.: Tetrahedron Letters (1976) 4895
221. Vögtle, F., Muller, W. M. and Weber, E.: Chem. Ber. (1980) *113*, 1130
222. Vögtle, F. and Newmann, P.: Tetrahedron Letters (1970) 115
223. Vögtle, F. and Weber, E.: Angew. Chem. (1974) *86*, 126
224. Vögtle, F. and Zuber, M.: Tetrahedron Letters (1972) 561
225. Wada, F., Arata, R., Goto, T., Kibokawa, K. and Matsuda, T.: Bull. Chem. Soc. Jpn. (1980) *53*, 2061
226. Weber, E.: Angew. Chem. Int. Ed. Engl. (1979) *18*, 219
227. Weber, E. and Vögtle, F.: Liebigs Ann. Chem. (1976) 891
228. Wong, K. H., Bourgoin, M. and Smid, J.: J. C. S., Chem. Commun. (1974) 715
229. Wong, K. H., Konizer, G. and Smid, J.: J. Amer. Chem. Soc. (1970) *92*, 666
230. Wong, K. H. and Ng, H. L.: Synth. Commun. (1980) *10*, 385
231. Wudl, F. and Gaeta, F.: J. C. S., Chem. Commun. (1972) 107
232. Yamashita, I., Fujii, M., Kaneda, T., Misumi, S. and Otsubo, T.: Tetrahedron Letters (1980) *21*, 541
233. Yamashita, I, Fujii, M., Kaneda, T., Otsubo, T., Sakata, Y. and Misumi, S.: Kokagaku Toronkai Koen Yoshishu (1979) 64
234. Yamashita, T, Nakamura, H., Takagi, M. and Ueno, K.: Bull. Chem. Soc. Jpn. (1980) *53*, 1550

Chapter 4

Syntheses of Azacrowns

4.1 Introduction

Almost as soon as Pedersen announced his discovery of the crown ethers (see Chaps. 2 and 3) it was recognized by many that these species were similar to those prepared by Busch and coworkers for binding coinage and transition metals (see Sect. 2.1). The latter compounds contained all or a predominance of nitrogen and sulfur (see also Chap. 6) in accordance with their intended use. The crown ethers and the polyazamacrocycles represented two extremes in cation binding ability and preparation of the intermediate compounds quickly ensued. In the conceptual sense, monoazacrowns are the simplest variants of the macrocyclic polyethers and these will be discussed first.

4.2 Monoazacrowns

Greene[43a] reported the preparation of monoaza-18-crown-6 (*1*) in 1972. His approach was direct, involving condensation of N-trityldiethanolamine with tetraethylene glycol ditosylate. Removal of the protecting group could be achieved by acid cleavage. Unfortunately, neither details of the synthesis nor physical properties were included in this report.

In 1977, Gokel and Garcia[40] prepared *1* in the hope that it could be treated with KH to form the self-solvating and overall neutral base system, *2*.

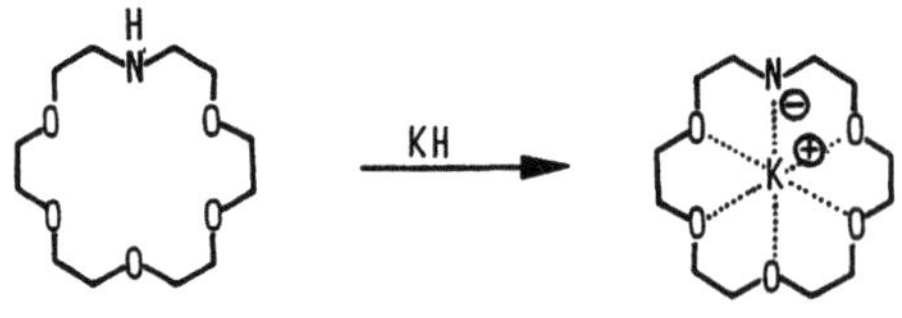

1 *2* (4.1)

The synthesis of *1* was accomplished by stirring N-benzyldiethanolamine in the presence of NaH/DMF and tetraethylene glycol ditosylate at 25 °C for 48 h. After chromatography, N-benzyl-*1* was isolated in 25% yield as a yellow oil. Debenzylation (H_2/Pd) gave the free ligand as a white solid, mp 48–51 °C. The synthesis is shown in Eq. (4.2), below.

$$C_6H_5CH_2N(CH_2CH_2OH)_2 + TsOCH_2(CH_2OCH_2)_3CH_2OTs \xrightarrow{NaH/DMF}$$

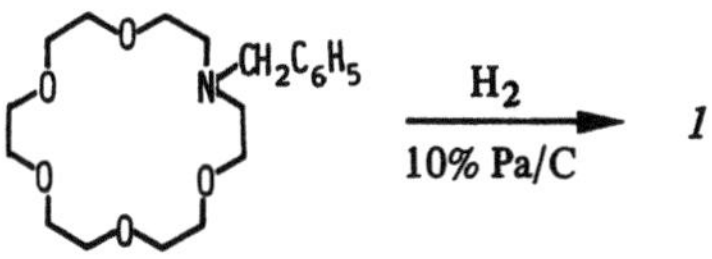

(4.2)

Treatment of *2* with KH did afford the self-solvating base system as suggested by Eq. (4.1), but its properties were somewhat more predictable than had been hoped. An interesting substance was isolated from the debenzylation step, however. When either dichloromethane or chloroform was used in work-up, prior to hydrogenolysis, a compound analyzing for *1* · H_3OCl was obtained. A crystal structure was determined for this compound and it was found to be the hydrochloride salt of *1*, which has a molecule of water in the hole and is hydrogen bonded to the macroring[40]. The compound can also be formed simply by titrating *1* with HCl as shown in Eq. (4.3).

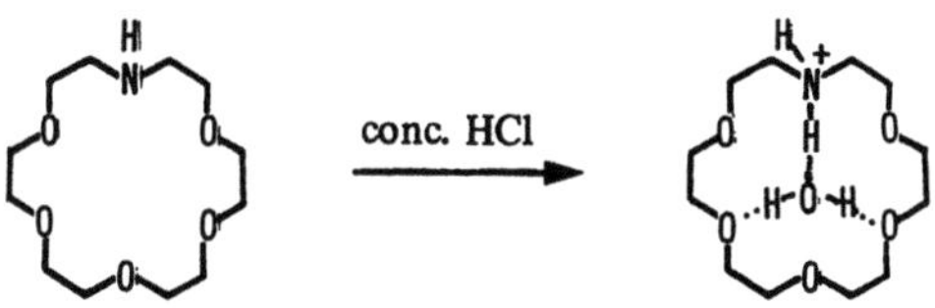

(4.3)

A number of azacrown compounds were reported by Lockhart and coworkers in 1973[75]. In a typical case, 2-aminophenol was heated at reflux with an equivalent of tetraethylene dichloride for 48 h. A mixture of the monoaza-15-crown-5 (*3*) and the monoaza-12-crown-4 (*4*) shown in Eq. (4.4) were obtained. Structural assignments were based on ^{1}H-NMR, IR and mass spectral data.

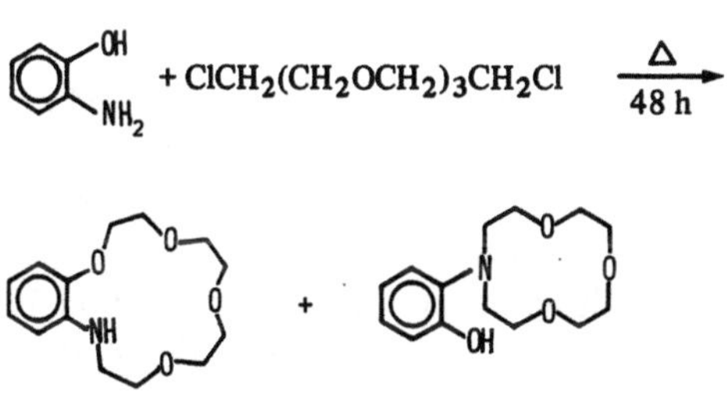

(4.4)

3 *4*

In more recent work, Lockhart and Thompson have formed derivatives of *3* by substituting the secondary nitrogen. Substituents include $CO{-}CH_2{-}CH_2{-}CH_2{-}COOH$ and $CH_2CH_2CH_2OEt$[74]. The latter compound should exhibit quite interesting binding properties for alkali metals, but binding constants for these compounds do not appear to have been determined.

Cram and Hogberg[47a] have prepared a number of compounds similar to *3*, utilizing mesyl and tosyl as protecting groups. These syntheses are generally unequivocal, but removal of the sulfonamide has proved troublesome (see also Sect. 4.5).

With respect to the monoazacrown ethers, Pedersen has utilized some interesting technology which is applicable to this class of compounds. For example, he has used benzylamine as a nitrogen transfer agent. This is accomplished by the sequence shown in Eq. (4.5)[99]. In the same report, Pedersen utilized the *t*-butyloxycarbonyl (tBOC)

$$R{-}OH + Ts{-}Cl \rightarrow R{-}O{-}Ts \xrightarrow{C_6H_5CH_2NH_2} R{-}NH{-}CH_2{-}C_6H_5 \xrightarrow[H_2]{Pd/C} R{-}NH_2 \qquad (4.5)$$

protecting group for differentiating the two nitrogens in ortho-phenylenediamine[99]. An alternative means of differentiating these nitrogens is by oxidation state. Thus chain extension using ethylene oxide can be effected as shown in Eq. (4.6), below[98,99].

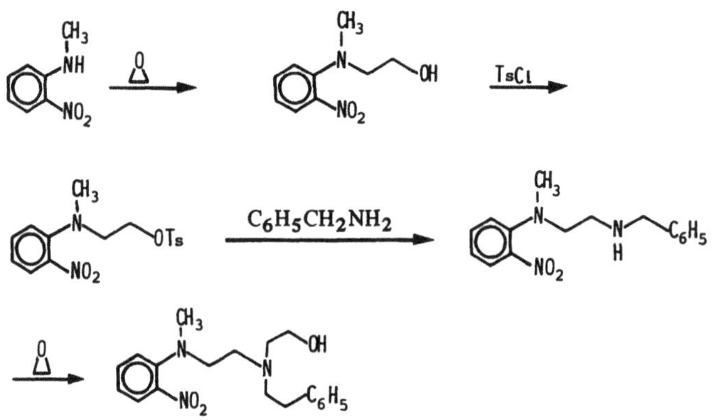

(4.6)

Okahara and coworkers have conducted considerable work on "one-pot" cyclization reactions which result in the formation of various crowns and related species[64]. These efforts are mentioned in Sects. 3.12 and 5.4. These techniques have been applied to the syntheses of azacrowns as well. The approach involves methodology similar to that developed by Pedersen and described above.

In a typical case, ethylamine is allowed to react with ethylene oxide to produce N-ethyldiethanolamine (*5*). The latter is then treated with additional ethylene oxide to afford N,N-di(polyoxyethylene)ethylamine (*6*) where the sum a + b is 3. This material is then stirred at room temperature for 3 h with toluenesulfonyl chloride and powdered sodium hydroxide in dioxane solution. After filtration and Kugelrohr distillation, N-ethylmonoaza-15-crown-5 is isolated in 75% yield[64] as illustrated below in Eq. (4.7).

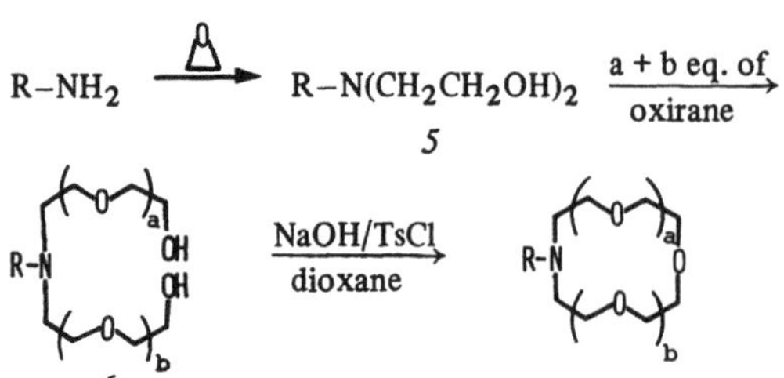

(4.7)

The economic advantages of such an approach are manifest. This is an especially appealing synthesis considering the ease with which one may functionalize commercially available diethanolamine[65]. Nevertheless, as with the synthesis of oxygen crowns by this method, a difficulty arises in purification of the intermediate diol. When a + b equivalents of oxirane are added, the desired product is formed, but other, closely related oligomers are also produced. Separation of these diols is possible by distillation, but this is often a tedious job. In large scale applications where time might be devoted to solving the engineering problems, such an approach is probably unexcelled for economy and convenience.

Quite recently, Okahara and his coworkers have extended their method to the formation of long chain N-alkylmonoazacrowns. It was expected that such compounds as N-decylmonoaza-18-crown-6 "may be useful as new surfactants with complexing ability with metal salts, phase transfer catalysts and selective ion carriers"[65].

A single nitrogen atom can be incorporated in a macrocyclic polyether as part of a subcyclic unit. Of the possible nitrogen containing heterocycles such as pyrrole, imidazole and pyridine, the latter has been incorporated most often. Certain compounds of this sort are mentioned in Sect. 5.5. Although the first report of monopyridyl-18-crown-6 was by Newcomb, Gokel and Cram[83], by far the majority of work on this class of compounds has been conducted by Newkome and his collaborators[85–97]. For the most part, these compounds have been prepared by Williamson ether syntheses, sometimes involving nucleophilic aromatic substitution. An appropriate appreciation for the range of structures now known can be best learned from a perusal of tables 4.14 to 4.19 at the end of this chapter.

4.3 Bridged Monoazacrowns

A number of "bridged" crown ethers have been prepared. Although the Simmons-Park "in-out" bicyclic amines (see Sect. 1.3.3) are the prototype, Lehn's cryptands (see Chap. 8) are probably better known. Intermediates between the cryptands (which Pedersen referred to as "lanterns"[98]) and the simple monoazacrowns are monoazacrowns bridged by a single hydrocarbon strand. Pedersen reports the synthesis of such a structure (see 7, below) which he referred to as a "clam" compound for the obvious reason[98]. Although Pedersen appears not to have explored the binding properties of his "clam" in any detail, he did attempt to complex Na^+ and Cs^+ ions. "A 0.0001 molar solution of the 'clam' compound is prepared in ethanol. The metal ions Na^+ and Cs^+ are added to the clam-ethanol solutions as salts. Ultraviolet spectra of these solutions indicate that a small amount of the Na^+ is complexed by the 'clam' compound but none of the Cs^+"[98].

More recently, Johnson and Sutherland have used bridged monoazacrowns as host molecules for bis-ammonium salts[49]. The intramolecular bis-crown diammonium complex formed when the size relationships are appropriate is shown below as structure *8*. It should be noted that the host structures were prepared from the simple monoazacrowns by acylating with a bis-acid chloride followed by hydride reduction.

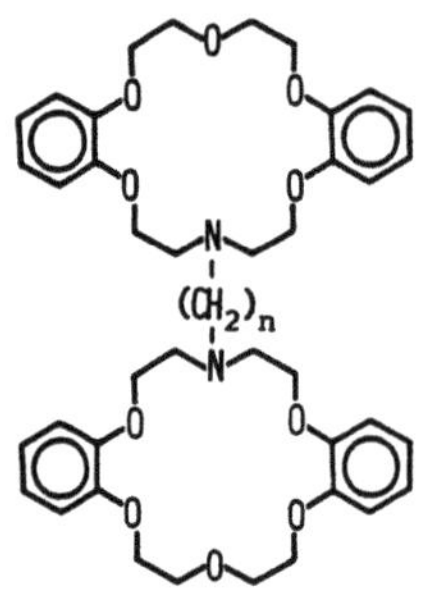

7

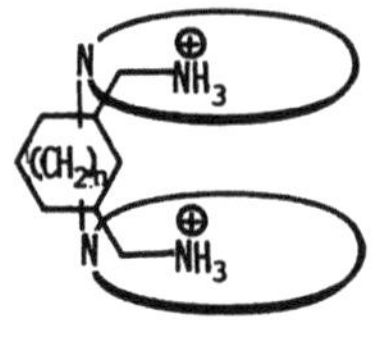

8

Although the "clam" structures are of interest as ammonium ion binders, they are by no means the only azacrown compounds of interest in this application. Sutherland and coworkers have examined a number of azacrowns as primary ammonium ion binders[49–53]. In addition, Metcalfe and Stoddart[78] have utilized bis-azacrowns to bind secondary ammonium cations.

4.4 Diazacrowns

The synthesis of 1,10-diaza-18-crown-6 (*9*) has been an important problem because this is the key starting material in the synthesis of numerous cryptands (see Chap. 8). Although first synthesized some years ago, the process has recently been patented[69]. Diazacrown *9* is prepared by a high dilution condensation of 1,8-diamino-3,7-dioxaoctane with ethylene glycol diacetyl chloride. The resulting diamide is then reduced with lithium aluminum hydride to give *9* in 56% overall yield from the open-chained diamine. The synthesis is illustrated in Eq. (4.8), below.

$H_2NCH_2(CH_2OCH_2)_2CH_2NH_2 + Cl{-}CO{-}CH_2OCH_2CH_2OCH_2{-}CO{-}Cl \longrightarrow$

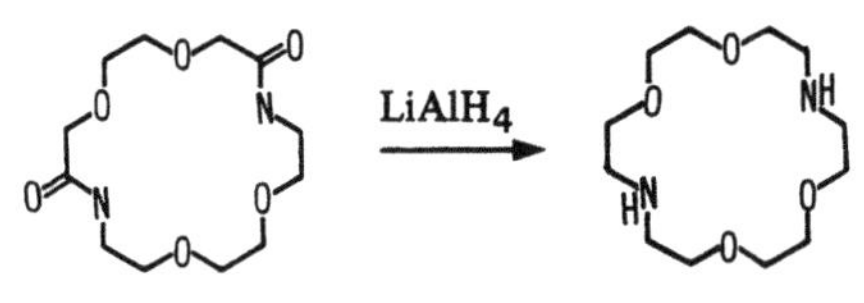

(4.8)

9

Interest in this compound has been high over the years and other methods have been attempted in the hope of improving yield or convenience in the synthesis of *9*. King and Krespan[55] noted that trifluoroacetamides could be hydrolytically removed from secondary nitrogen much more readily than is possible using either tosylate or mesylate as a protecting group. They therefore attempted to adapt the use of this group to the synthesis of diazacrown compounds, notably *9*. The scheme presented in their paper is shown below as Eq. 4.9. Unfortunately, the authors state: "Attempts to adapt this amine synthesis to the case in which cyclic diamine [*9*] is formed gave at best a 3% yield"[55].

$$Cl(CH_2CH_2O)_2CH_2CH_2Cl + \text{excess } NH_3 \rightarrow [H_2N(CH_2CH_2O)_2CH_2CH_2]_2NH +$$

$$H_2N(CH_2CH_2O)_2CH_2CH_2NH_2 + 9$$

CF_3CO_2Et (↓) ; (↑) 1. NaH 2. $TsO(CH_2CH_2O)_3Ts$ 3. H_2O (4.9)

$$CF_3CONH(CH_2CH_2O)_2CH_2CH_2NHCOCF_3$$

In the course of experiments directed at the synthesis of open-chained equivalents of macrocyclic polyethers (see Chap. 7), Rasshofer and Vögtle found that *9* or its derivatives were formed rather than the desired compounds[105]. Although the yields obtained by this method are modest, the facts that the reaction is conducted without the use of high dilution and in one-pot recommend it.

In a typical experiment, triethylene glycol was treated with two equivalents of sodium toluenesulfonamide in dry DMF solution. After 6 h at reflux, the solution was distilled and product obtained by a standard work-up procedure. By this procedure, *9* was obtained in about 10% yield. The transformation is illustrated below as Eq. (4.10). Note also that Vögtle and his coworkers have also utilized phthalimide as a source of nitrogen in the preparation of such azacrown precursors as $H_2N(CH_2CH_2O)_2CH_2CH_2NH_2$[105]. In such reactions, a standard hydrazine cleavage was used to remove the phthaloyl residue.

$$TsOCH_2(CH_2OCH_2)_2CH_2OTs + TsNH^-Na^+ \xrightarrow{DMF} 9 \qquad (4.10)$$

The one-pot synthesis of *9* described above appears to afford only modest yields of azacrowns. One might wonder why any crown at all would be formed under non-high dilution conditions intended to yield only open-chained material. Vögtle suggests that this "can be explained in terms of template, steric and entropy effects"[105]. These factors are of doubtless significance, but it is interesting to note that in the synthesis of polyazamacrocycles, Richman and Atkins[103] found that there was no significant template effect observed. The question of the template effect in the syntheses of *9* has recently been addressed by Kulstad and Malmsten[61,62]. They conclude that the formation of *9* is assisted by the presence of alkali metal cations.

Compound *9* has been prepared by the latter authors using triethylene glycol diiodide and triethylene glycol diamine (see Eq. 4.11) and an alkali metal carbonate in acetonitrile solution[63].

It is interesting to note that Kulstad and Malmsten have utilized yet another method for introducing nitrogen into the crown precursors. They utilize sodium azide in DMSO to displace halogen from triethylene glycol dichloride. The bis-azide is then reduced using hydrogen sulfide in ethanol[61].

$$H_2NCH_2(CH_2OCH_2)_2CH_2NH_2 + ICH_2(CH_2OCH_2)_2CH_2I \xrightarrow{M_2CO_3} 9 \qquad (4.11)$$

Although the principal application for *9* has been in the synthesis of cryptands (see Chap. 8), this material has also served as precursor to a number of nitrogen based lariat ethers[28,61,123], sometimes referred to as "crown complexanes"[117]. Binding constants for such compounds have been measured for a few examples in a few cases[28,123], but generalizations are not yet appropriate. These compounds should prove to have interesting cation binding properties.

Functionalization may be accomplished before cyclization by preparing the appropriate diethanolamine derivative first[28]. An alternative and more straightforward approach is simply to prepare *9* by any of the typical methods and then substitute the secondary nitrogens afterward. Reaction between *9* and chloroacetamide[61] and acrylonitrile[117] yielded compounds *10* and *11* in 61% and 100% yields respectively.

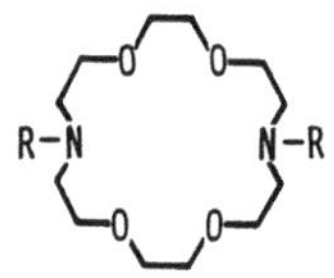

10: R = CH_2CONH_2

11: R = CH_2CH_2CN

1,10-Diaza-18-crown-6 has also been converted into polymeric materials by the reaction of *9* with toluenediisocyanate[77]. The polymeric materials were prepared by stirring commercially available *9* with TDI in dichloromethane solution. Reaction was rapid and exothermic, but the mixture was not worked up until the next day. The product was a white solid which softened between 170–190° and decomposed between 250–270°. A detailed spectral analysis was undertaken and mechanical properties are described, but no information is reported on the cation binding abilities of the various species thus produced[77].

Diazacrown compounds have been studied extensively by the groups of Sutherland[13,45,46,47,50,71,72] and Stoddart[78]. Particular emphasis has been placed on the interactions of such species with primary and secondary alkylammonium ions (see above). The syntheses of the compounds utilized in these studies generally parallel the methodology described above.

4.5 Cyclams

Although the all-nitrogen heteromacrocycles, i.e. "cyclams" have been of interest to the inorganic community for some time[11], it was the advent of crown ether chemistry which stimulated the preparation of many of these structures by modifications of technology which had proved useful in the latter context. Richman and Atkins, working at Dupont's Central Research Department[103], were able to prepare such molecules as tetraaza-12-crown-4 (cyclen, *12*) in high yield without resorting to high dilution conditions. In a typical preparation, the disodium salt of ethylenediamine tritosylate was allowed to react in DMF (100°, 1–2 h) with diethanolamine tritosylate to give the tetra-N-tosyl derivative of *12* (80% yield). Detosylation was effected by heating with 97% sulfuric acid at 100° for 48 h. After workup *12* was isolated in 90% yield (72% overall yield)[11]. The reaction is illustrated in Eq. (4.12), below.

A number of other, related compounds along with detailed experimental procedures are reported by Atkins, Richman and Oettle[5]. Rings of other sizes and different leaving groups distinguish the latter work from that previously cited. In this case, as in the pre-

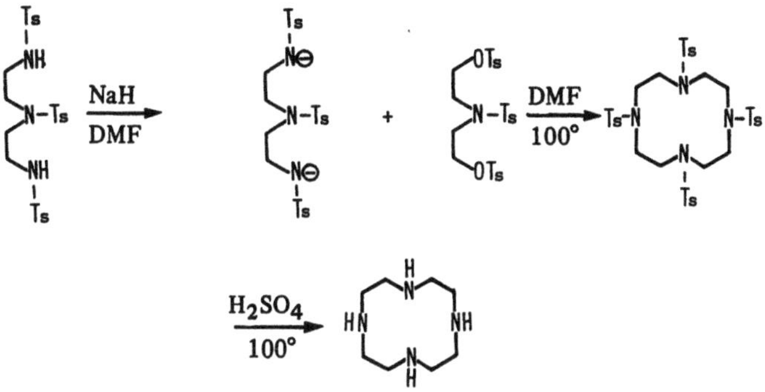

12

(4.12)

vious work, detosylation is accomplished using hot, concentrated sulfuric acid, for as long as 70 h in some cases.

Detosylation has been a singular problem in the development of polyazamacrocycle syntheses. Gokel and Garcia overcame this problem by utilizing the benzyl protecting group which could be removed by hydrogenolysis[40]. Greene used the trityl protecting group in his original synthesis of monoaza-18-crown-6[43a]. Both of these groups have been utilized by Hediger and Kaden in their syntheses of mono-N-substituted tetraaza-macrocycles[44].

Despite these alternatives, tosyl remains an attractive nitrogen protecting group because of its stability and crystallinity. Vögtle and coworkers have addressed the detosylation problem and have reported using lithium aluminum hydride, phenol and hydrobromic acid, and sodium bis(methoxyethoxy)aluminum hydride to this end[17]. Of these, lithium aluminum hydride in THF solution seemed most effective in deprotecting the di-N-tosyl derivative of *9*. More recently, Kossai, Simonet and Jeminet have reported quite high conversions for the electrochemical detosylation of several azamacrocycles[56]. The latter method requires a mercury electrode for the reductive detosylation. This requirement may be a disadvantage in laboratories not equipped to conduct such reactions. Kimura and Yatsunami have effected detosylation using 48% HBr after a high dilution synthesis in DMF. These authors refer to the cyclams as "cyclic spermidines"[52].

It is interesting to note that although the first examples of "template" effects were observed in nitrogen macrocycles (see chapter 2) no template effect appears to operate in the synthesis of *12*. Richman and Atkins note this in their original report[103]. The authors replaced the sodium cation with tetramethylammonium cations and still obtained greater than 50% yield of tetra-N-tosyl-*12*. Shaw considered this problem and suggested that because of the bulky N-tosyl groups, ". . . the loss of internal entropy on cyclization is small"[110]. He offered this as an explanation for the apparent lack of a template effect in the cyclization.

Tabushi and Fujiyoshi have prepared alkylated cyclam derivatives for the purpose of suspending these systems from a polymer backbone. They have utilized a malonic ester alkylation reaction followed by cyclization and reduction to accomplish this end[112]. The approach is illustrated below in Eq. (4.13).

$$\mathrm{EtOCOCH_2COOEt + RX \xrightarrow{Base} EtOCOCHRCOOEt + H_2NCH_2CH_2NH(CH_2)_3NHCH_2CH_2NH \xrightarrow{EtOH}}$$

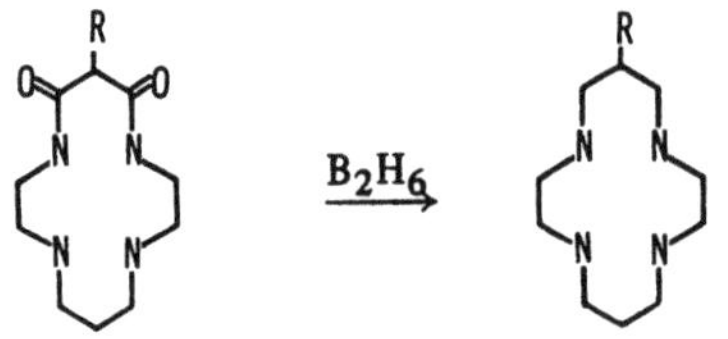
(4.13)

4.6 Cyclic Oximes

An obvious method for including both nitrogen and oxygen in the same heteromacrocycle is to prepare cyclic oxime derivatives. This has been accomplished by the groups of DeSarlo[25] and Vögtle[104]. The polymerization of acetonitrile oxide provides macrocycle *13*, shown below[25]. Cation binding by this compound was demonstrated using infrared analysis, but the binding was not reported in any quantitatively useful form.

Note that the acetonitrile oxide cyclooligomers (e.g. *13*) are not true oxime derivatives. Such derivatives have been prepared from biacetyl, however[104]. Derivatives related to *14*, below, were prepared and found not to be good complexing agents. They were, nevertheless, capable of phase transferring either sodium or potassium permanganate into dichloromethane.

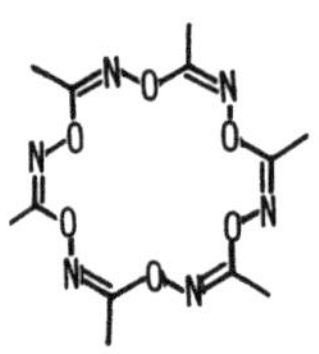
13

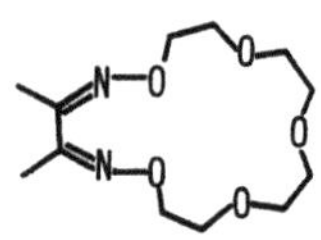
14

4.7 Macrocyclic Imines

Macrocycles have been prepared by formation of macrocyclic imines as well as by using variations of the Williamson ether synthesis[29,33,43]. Typically, a diamine or dialdehyde is treated with its counterpart to yield the Schiff's base. The saturated macrocycle may then be obtained by simple reduction, using sodium borohydride, for example[43]. The cyclization may be metal-ion templated[29]. In the special case of the all-nitrogen macrocycle, *15*, the condensation of diamine with glyoxal shown in Eq. (4.14), was unsuccessful[33].

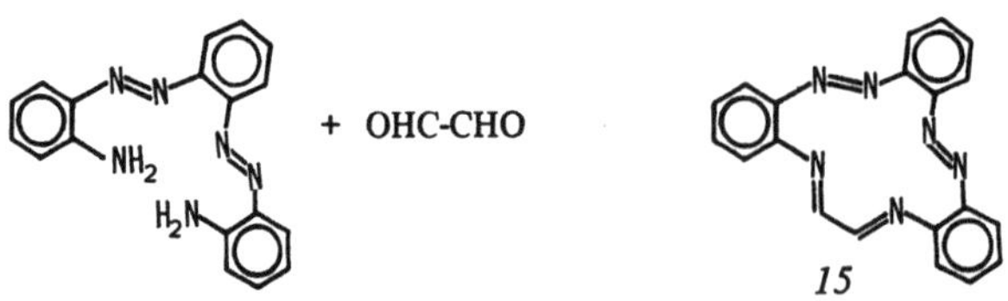

15
(4.14)

4.8 Thioazamacrocycles

A number of compounds containing both nitrogen and sulfur have been reported. For a survey of the structural types, see Chap. 6 and Table 4.6. Interest in these compounds seems to have arisen largely from curiosity concerning the effect a sulfur-oxygen interchange might have on cation binding. Some data are reported for thioazamacrocycles and open-chained relatives[4].

Although synthetic approaches to these systems vary enough that no method can really be called typical, variations on the approach illustrated below as Eq. (4.15) has been used by both Pelissard and Louis[100] and Tabushi, Okino, and Kuroda[115]. The example shown here is the work of the latter authors.

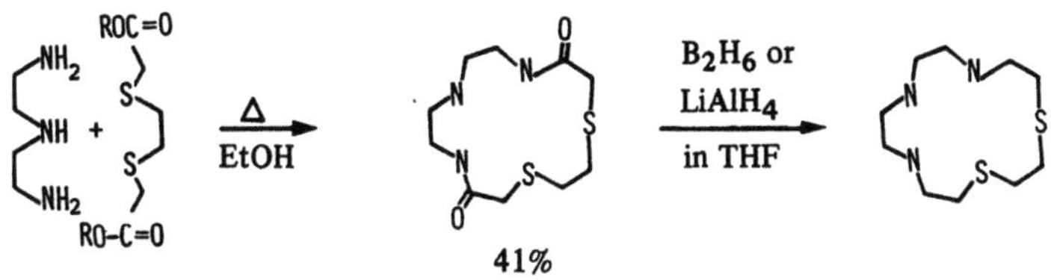

(4.15)

4.9 Biologically Active Azacrowns

One of the earliest reports of a nitrogen-containing compound derived from a biologically active precursor was reported by Wudl and Gaeta in 1972[130]. These authors converted both L-proline and D-ψ-ephedrine into the corresponding 20-membered macrocycles as illustrated for the synthesis of *16* in Eq. (4.16). Although active precursors were employed in the syntheses of these molecules, the results of biological testing, if any, do not appear to have been published.

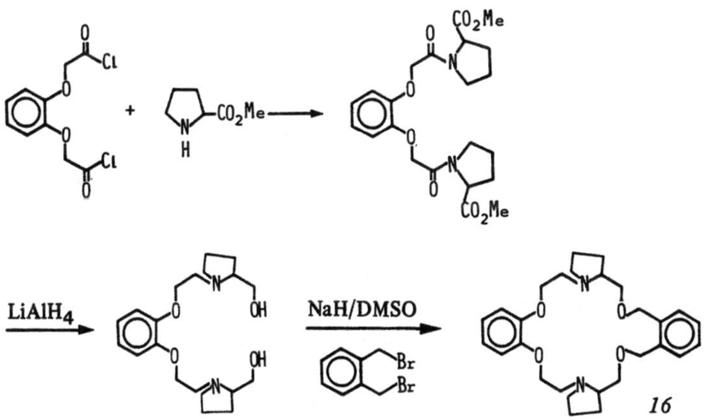

(4.16)

Vögtle and Elben have been particularly interested in attaching crown ethers to residues of known biological or pharmacological activity[36,125]. These efforts have been directed in large measure at azacrown compounds. For example, using actinomycin-D as a model, Elben and Vögtle[36] have replaced the macrocyclic depsipeptide rings with 1,10-diaza-18-crown-6 residues. The authors have dubbed these species "cryptomycins". Although the synthesis and properties (including allusion to x-ray structural analysis) are disclosed, it is not clear from this report whether or not these substances had the intended biological activity[36]. In other work, these same authors reported the synthesis of conceptual analogs of cinnarizine, procaine and dibenzosuberonimine ether utilizing diazacrown rings[125]. Although complexes of these new compounds with sodium, potassium, and calcium ions were prepared, it is not clear what biological testing was conducted on them.

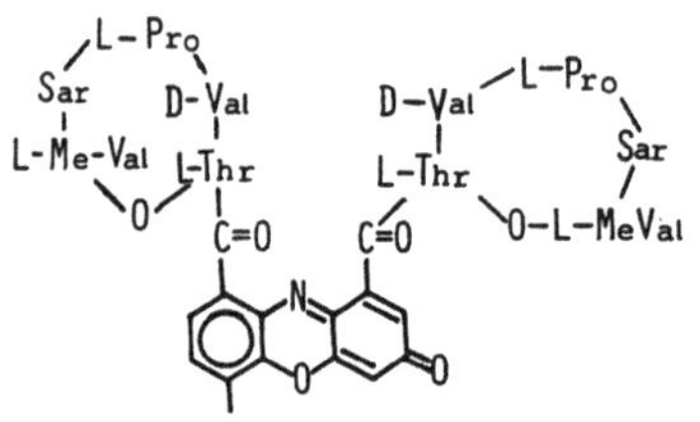

Actinomycin-D

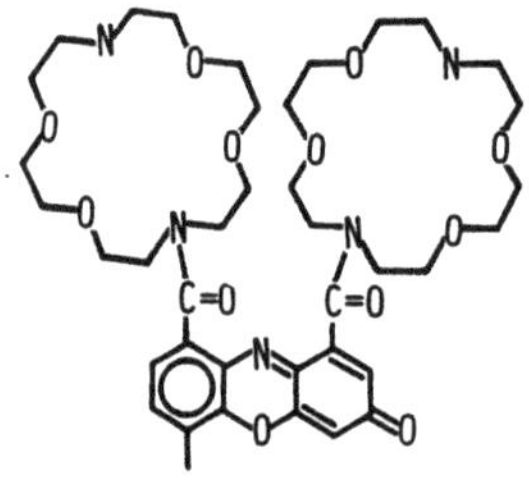

a "cryptomycin"

Kulstad and Malmsten[61] tested a number of derivatives of 1,10-diaza-18-crown-6 (*9*) for toxicity in mice (intravenous). Compound *9* and the bis-N-2-hydroxyethyl, bis-N-2-carboxamidomethyl, and bis-N-2-carboxymethyl derivatives of *9* were tested and all were found to be toxic in the range of 5–50 mg/kg. At a dose level of 5 mg/kg of *9*, the two mice tested were still alive after a week, but at a dose level of 25 mg/kg, the mice died in less than 24 h[61]. If these data could be translated directly to human beings, a 100 kg human would require something between 0.5–2.5 g of compound (i.v.) to produce a lethal effect.

4.10 Crown Ether Based Dyes

In Pedersen's early experiments, the relative binding of cations by crown ethers was assessed by extraction of alkali metal picrates into an organic phase. In these experiments, the crown ether served to draw into the organic phase a colored molecule which was ordinarily insoluble in this medium. An extension and elaboration of this notion has been developed by Dix and Vögtle[31,32] and Nakamura, Takagi, and Ueno[131]. In efforts by both of these groups, crown ether molecules were appended to chromophoric or colored residues. Ion-selective extraction and interaction with the crown and/or chromophore could produce changes in the absorption spectrum. Examples of molecules so constructed are illustrated below as *17* and *18* from refs. 32 and 131, respectively.

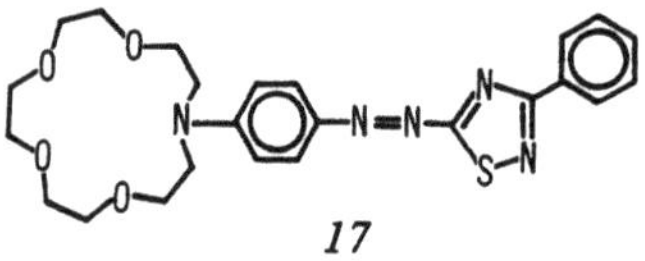

17

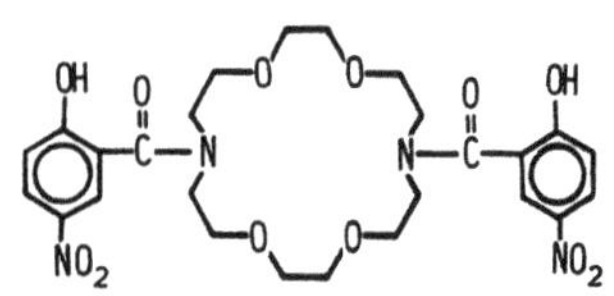

18

4.11 Newkome's Pyridine-containing Macrocycles

Many of the macrocycles reported by Newkome contain the pyridine subunit although in a number of these cases, the heterocycles have their nitrogens directed away from the binding cavity. Rather than discuss those having inward-directed nitrogen in this section and outward-directed nitrogen in Sect. 3.11, we have devoted the latter section to this discussion. The closely related pyridine, bipyridine and pyrimidine compounds are likewise difficult to classify and have been consigned to a variety of tables. Some of these may be found in this chapter, Tables 4.14, 4.16, and 4.18, and some are found in Chap. 3 (3.25, 3.26). In addition, the reader is referred to Sects. 5.5 and 6.3 and the corresponding tables for additional information.

4.12 Listing of Tables

4.13 Tables 4.1–4.21

Table 4.1. Monoazacrowns

Cpd. No.	No. C's	Ring size	Structure	Mp or (bp) in °C	Yield (%)	Refs.
1	10	15	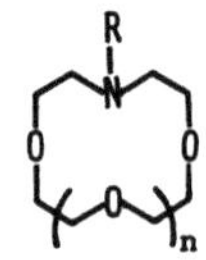 n = 2, R = H	27–30	40	51
2	11	15	n = 2, R = CH_3	nr nr	37 45	51 64
3	13	15	n = 2, R = $CH_2CH{=}CH_2$	nr	48	64
4	13	18	n = 3, R = CH_3	nr	55	64
5	13	15	n = 2, R = $CH_2CH_2OCH_3$	oil	69	28
6	14	15	n = 2, R = $(CH_2)_3CH_3$	nr	58	64
7	14	12	n = 1, R = 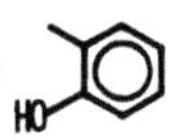	90	nr	75
8	15	15	n = 2, R = $(CH_2CH_2O)_2CH_3$	oil	47	28
9	15	18	n = 3, R = $CH_2CH{=}CH_2$	nr oil	40 38	64 41
10	15	18	n = 3, R = $CH_2CH_2OCH_3$	oil	53	28
11	15	18	n = 3, R = $CH_2CHOHCH_3$	oil	100	41
12	16	18	n = 3, R = $(CH_2)_3CH_3$	nr	67	64
13	16	15	n = 2, R = $(CH_2)_5CH_3$	(150–153/0.01)	63	65
14	17	18	n = 3, R = $(CH_2CH_2O)_2CH_3$	oil	50	28
15	18	18	n = 3, R = $(CH_2)_5CH_3$	(170–173/0.01)	39	65
16	18	15	n = 2, R = $(CH_2)_7CH_3$	(165–168/0.09)	62	65
17	20	15	n = 2, R = $(CH_2)_9CH_3$	(203–205/0.1)	64	65
18	20	18	n = 3, R = $(CH_2)_7CH_3$	(196–197/0.03)	63	65
19	20	21	n = 4, R = $(CH_2)_5CH_3$	(202–204/0.03)	51	65
20	22	18	n = 3, R = $(CH_2)_9CH_3$	(205–206/0.03)	57	65
21	22	21	n = 4, R = $(CH_2)_7CH_3$	(210–212/0.03)	44	65

Table 4.1 (continued)

Cpd. No.	No. C's	Ring size	Structure	Mp or (bp) in °C	Yield (%)	Refs.
22	22	15	n = 2, R = $(CH_2)_{11}CH_3$	(201–202/0.04)	44	65
23	24	18	n = 3, R = $(CH_2)_{11}CH_3$	(217–218/0.04)	63	65
24	24	21	n = 4, R = $(CH_2)_9CH_3$	(220–228/0.06)	40	65

Table 4.2. Diazacrowns containing secondary nitrogen

Cpd. No.	No. C's	Ring size	Structure	Mp or (bp) in °C	Yield (%)	Refs.
1	8	9	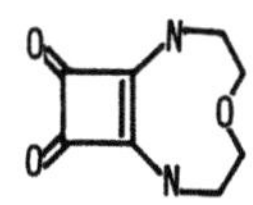	240 dec.	78	124
2	8	11	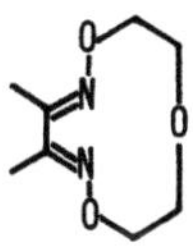	nr	nr	104
3	8	12		oil	41	106
4	8	12		83–84	46	69
5	10	15	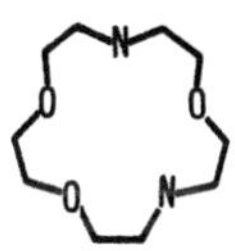	oil	67	69
6	10	12	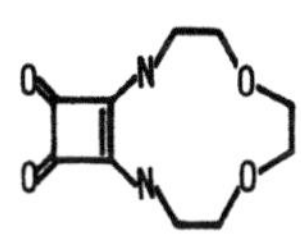	290–296 dec.	24	124
7	10	14	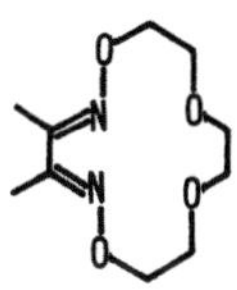	nr	nr	104

Table 4.2 (continued)

Cpd. No.	No. C's	Ring size	Structure	Mp or (bp) in °C	Yield (%)	Refs.
8	10	15		~0	~ 50	100
9	10	15		~5	<40	100
10	10	15	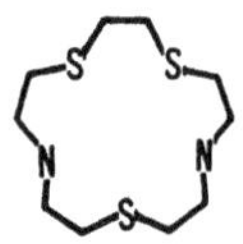	~31	<55	100
11	12	12	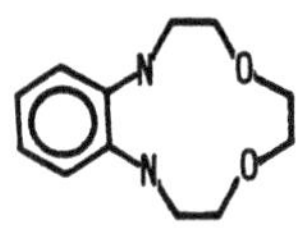	92–94	nr	75
12	12	15	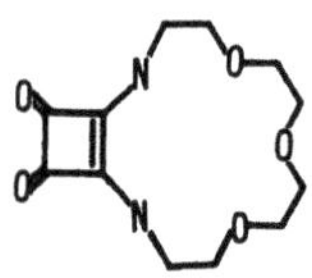	190–200	39	124
13	12	17	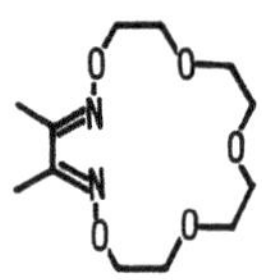	nr	nr	104
14	12	18	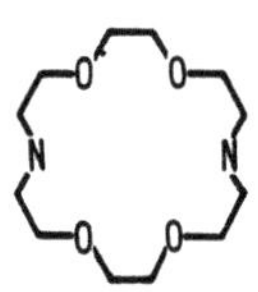	115–116 nr nr	60 30 nr	26 61 17
15	12	18	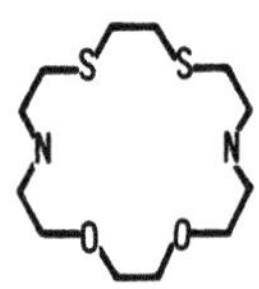	45	42	69
16	12	18	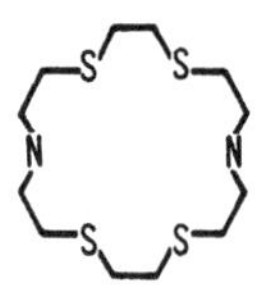	125 125	42 8	69 7

Table 4.2 (continued)

Cpd. No.	No. C's	Ring size	Structure	Mp or (bp) in °C	Yield (%)	Refs.
17	14	19	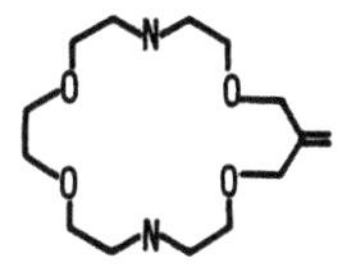	nr (140–142/ 5 x 10^{-3})	nr 35	80 118
18	14	21	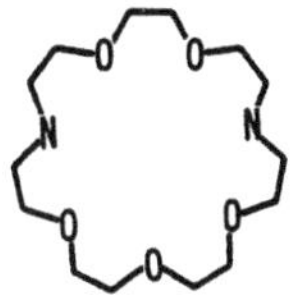	~0 ~15	52 nr	69 26
19	15	15	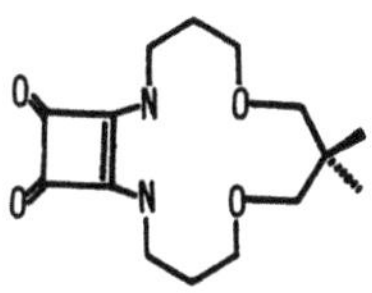	240–241	63.5	124
20	16	24	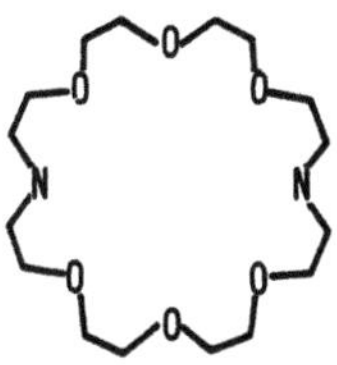	oil	40	26, 69
21	16	18		nr	nr	124
22	18	14	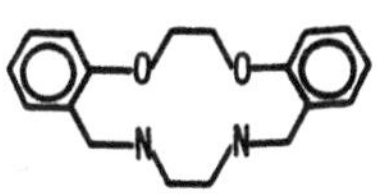	169	"low"	43
23	19	15	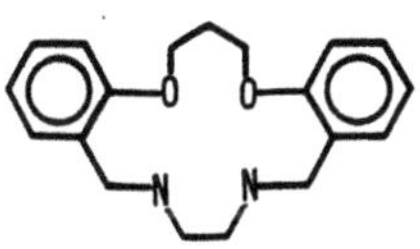	nr	46	43
24	19	15	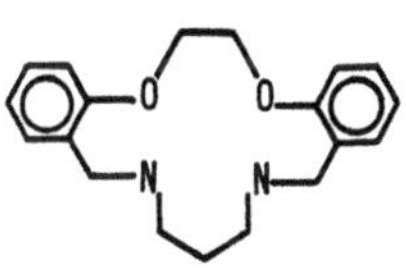	143 139–140	90 nr	43 56
25	19	15	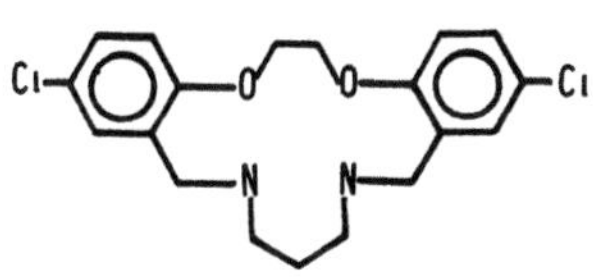	210	75	43

Table 4.2 (continued)

Cpd. No.	No. C's	Ring size	Structure	Mp or (bp) in °C	Yield (%)	Refs.
26	19	15	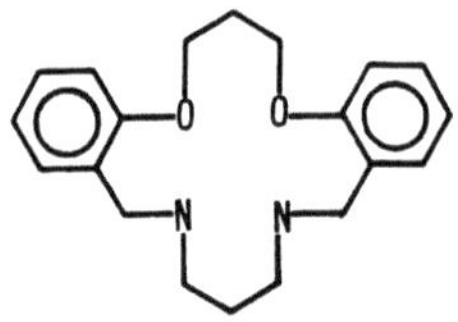	91	40	43
27	22	21	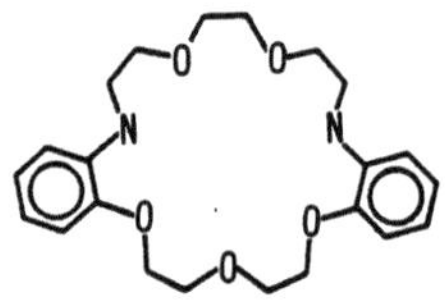	117	nr	74
28	26	18	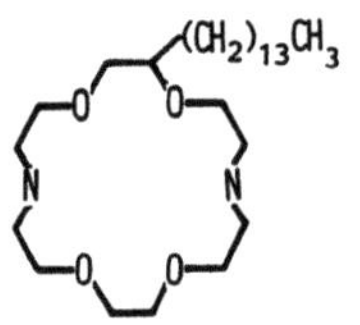	38–39	60.5	20
29	28	18	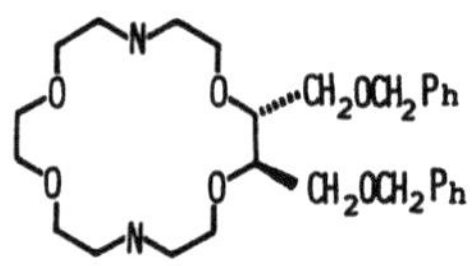	oil	21	3

Table 4.3. Diaza-18-crowns containing tertiary nitrogen

Cpd. No.	No. C's	Ring size	Structure	Mp or (bp) in °C	Yield (%)	Refs.
1	14	18	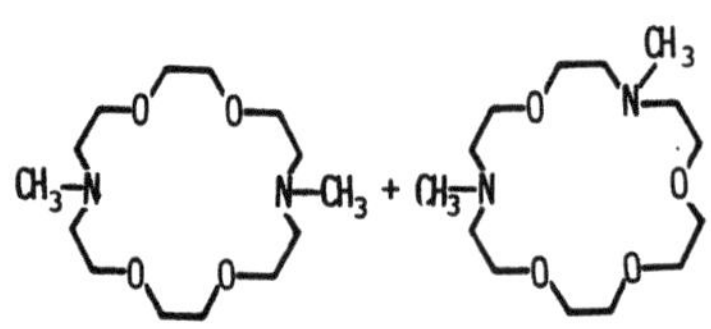	nr	24	64
			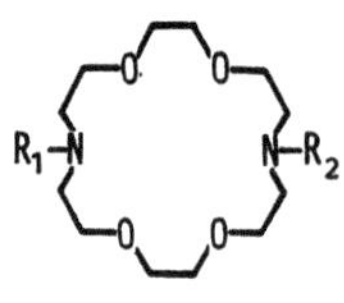			
2	14	18	$R_1 = R_2 = CH_3$	(127–132/0.01)	34	47

Table 4.3 (continued)

Cpd. No.	No. C's	Ring size	Structure	Mp or (bp) in °C	Yield (%)	Refs.
3	16	18	$R_1 = R_2 = CH_2CH_2OH$	(210–214/0.15)	58	61
4	16	18	$R_1 = R_2 = CH_2\overset{O}{\overset{\|}{C}}NH_2$	161	61	61
5	16	18	$R_1 = R_2 = CH_2CO_2H$	171–173	47	61
6	18	18	$R_1 = R_2 = CH_2CH_2OCH_3$	(197–199/0.1)	29	61
7	18	18	$R_1 = R_2 = CO_2CH_2CH_3$	89	86	125
8	18	18	$R_1 = R_2 = CH_2CH_2CO_2H$	128.5–130	65	117
9	20	18	$R_1 = R_2 = CH_2CO_2CH_2CH_3$	(218–220/0.2)	71	61
10	20	18	$R_1 = H, R_2 =$ 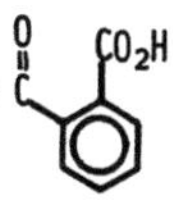	170–173	65	125
11	21	18	$R_1 = H, R_2 =$ 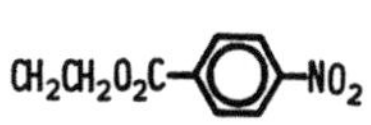	118–120	24	125
12	21	18	$R_1 = H, R_2 =$ 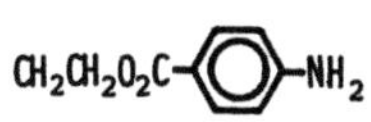	oil	60	125
13	24	18	$R_1 = R_2 =$ 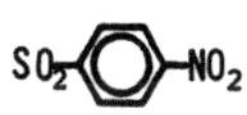	197–199	86	125
14	25	18	$R_1 = H, R_2 = CHPh_2$	88–89	22	125
15	26	18	$R_1 = R_2 =$ 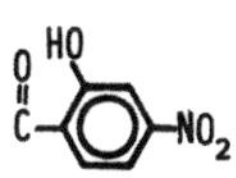	114.5–117	38	130
16	26	18	$R_1 = R_2 =$ 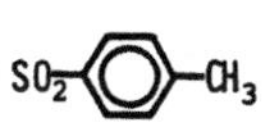	161	15	105
				nr	80	5
				163.5–164.5	80	103
17	26	18	$R_1 = R_2 =$ 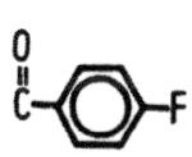	134	86	35
18	26	18	$R_1 = R_2 =$ 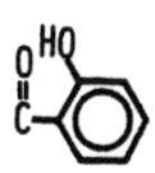	141–142	36	130
19	26	18	$R_1 = R_2 =$ 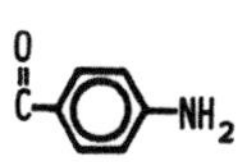	178–180	72	35

Table 4.3 (continued)

Cpd. No.	No. C's	Ring size	Structure	Mp or (bp) in °C	Yield (%)	Refs.
20	26	18	$R_1 = R_2 =$ 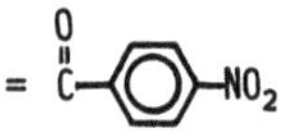	164–165	82	35
21	28	18	$R_1 = H, R_2 = (CH_2)_{15}CH_3$	90–92	94	20
22	30	18	$R_1 = R_2 =$ 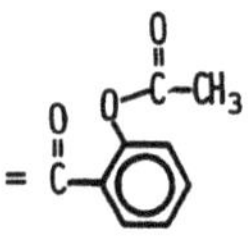	125–127	73	35
23	30	18	$R_1 = R_2 = \overset{O}{\overset{\|}{C}}CH_2CH_2Ph$	76–79	94	125
24	30	18	$R_1 = R_2 = \overset{O}{\overset{\|}{C}}-CH{=}CH-Ph$ (E)	148–150	72	125
25	30	18	$R_1 = R_2 =$ 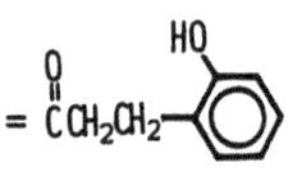	131–131.5	80	131
26	30	18	$R_1 = R_2 =$ 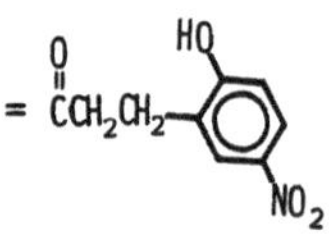	124–125	48	131
27	32	18	$R_1 = \overset{O}{\overset{\|}{C}}OCH_2Ph$, $R_2 =$ 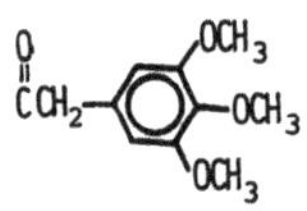	nr	34	35
28	34	18	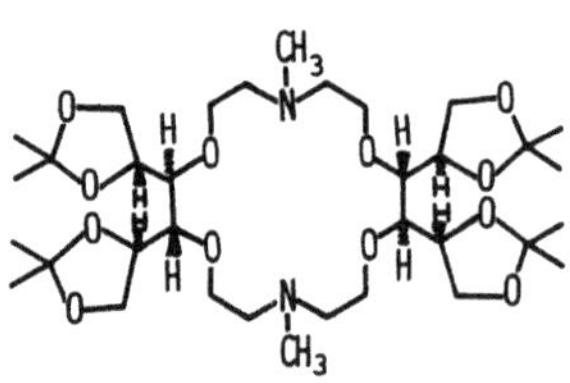	106–108	28	68
29	34	18	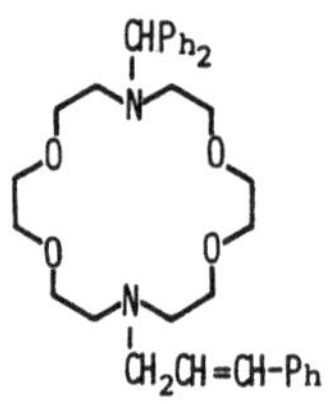 (E)	oil	87	125

Table 4.3 (continued)

Cpd. No.	No. C's	Ring size	Structure	Mp or (bp) in °C	Yield (%)	Refs.
30	34	18	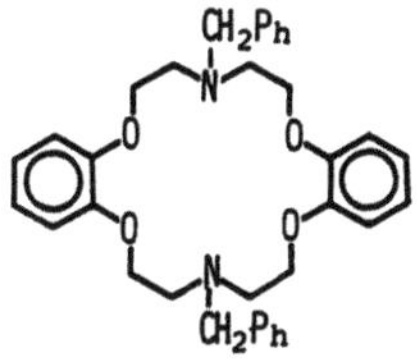	176	nr	99
			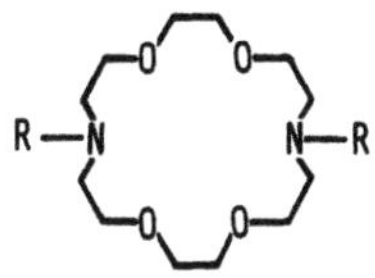			
31	36	18	R = 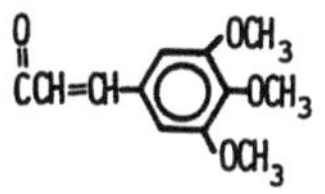	158	83	35
32	38	18	R = $CHPh_2$	109–109.5	86	125
33	38	18	R = 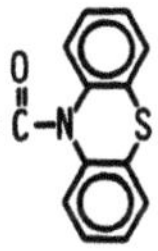	158–160	77	35
34	38	18	R = 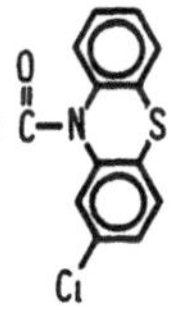	145–147	80	35
35	40	18	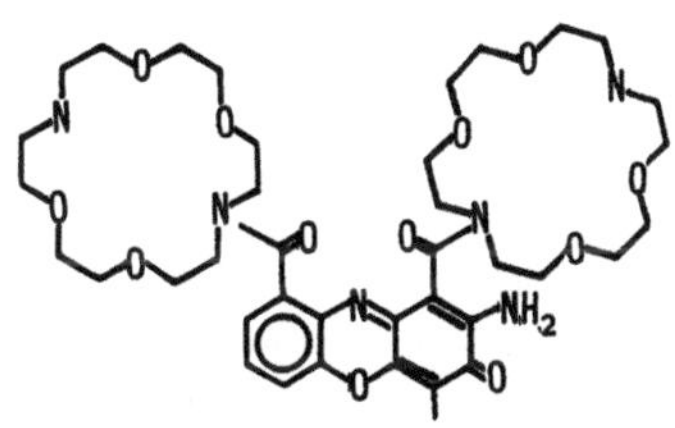	121–123	16	36, 37
36	42	18	R = 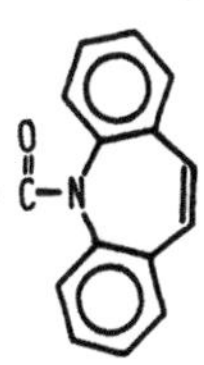	165–166	45	125

Table 4.3 (continued)

Cpd. No.	No. C's	Ring size	Structure	Mp or (bp) in °C	Yield (%)	Refs.
37	42	18	R = 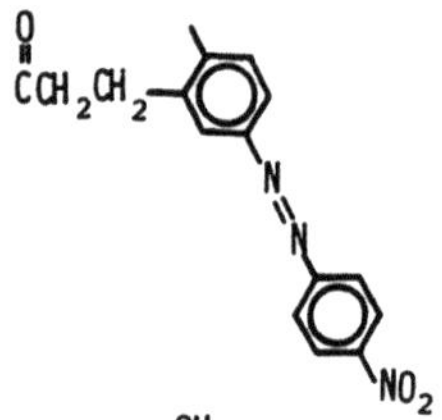	196–203 dec.	49	131
38	50	18	R = 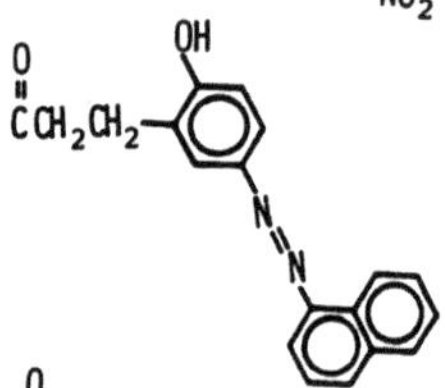	88–92	43	131
39	50	18	R = 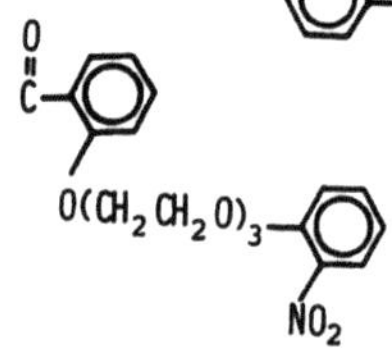	oil	78	111
40	54	18	R = 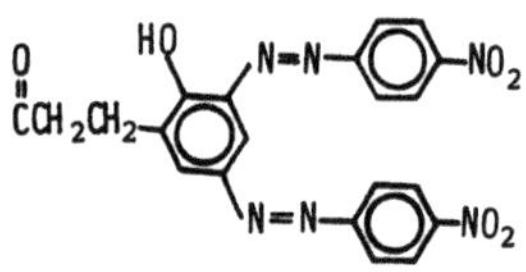	144 dec.	14	131
41	56	18	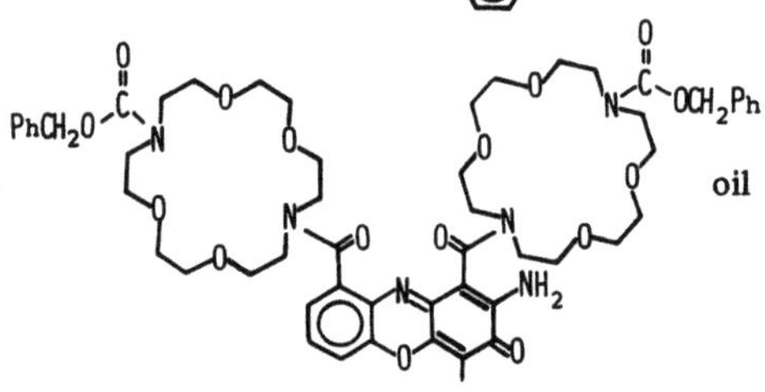	oil	18	36, 37

Table 4.4. Diazacrowns containing tertiary nitrogens

Cpd. No.	No. C's	Ring size	Structure	Mp or (bp) in °C	Yield (%)	Refs.
1	13	14	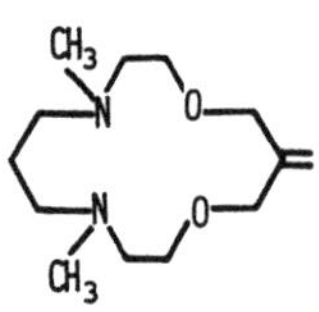	nr	nr	79

Table 4.4 (continued)

Cpd. No.	No. C's	Ring size	Structure	Mp or (bp) in °C	Yield (%)	Refs.
2	13	14	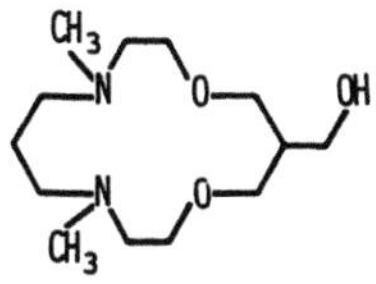	nr	nr	81
3	14	18	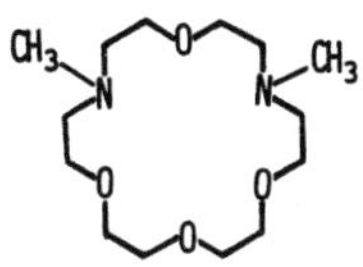	(160/0.03)	20	47
4	15	19	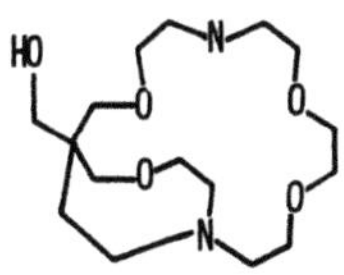	nr	27	59
5	18	18	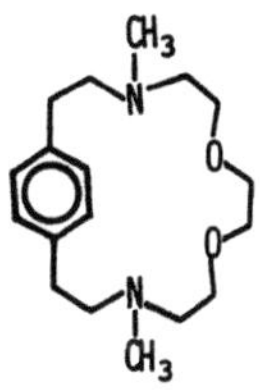	(175/0.005)	13.5	5a
6	20	9	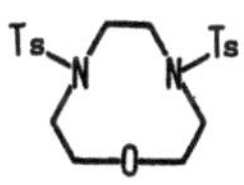	199–201 201 nr	32 41 32	107 105 5
7	21	10	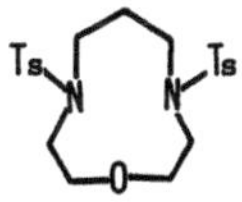	121–129 nr	66 47	107 5
8	22	11	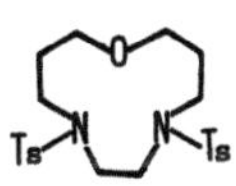	nr glass	63 63	5 107
9	22	12	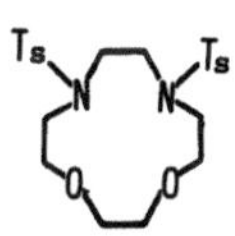	nr 220–226 223–224	52 43 52	5 105 107
10	22	12	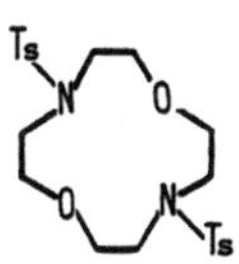	nr 203–204	80 80	5 103

Table 4.4 (continued)

Cpd. No.	No. C's	Ring size	Structure	Mp or (bp) in °C	Yield (%)	Refs.
11	22	20	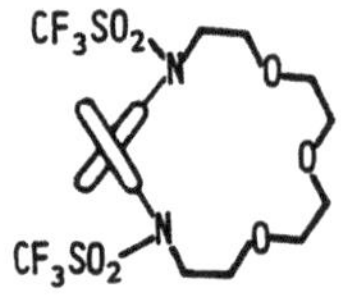	glass	73	22
12	12	12	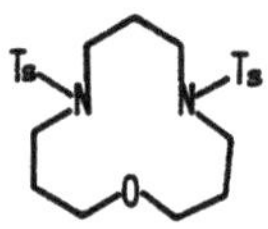	nr	38	5
13	23	12	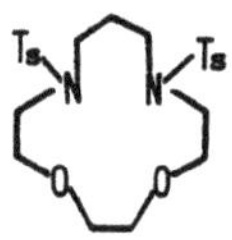	nr 197–198	28 28	5 107
14	24	15	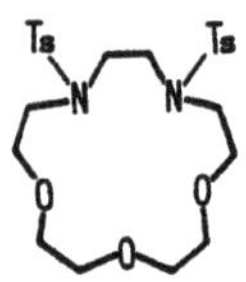	201–207 nr	23 23	107 5
15	24	15	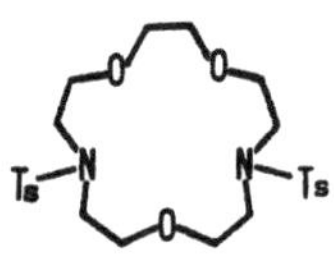	nr	74	5
16	25	16	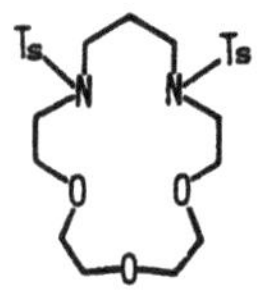	195–196 nr	7 7	107 5
17	26	12	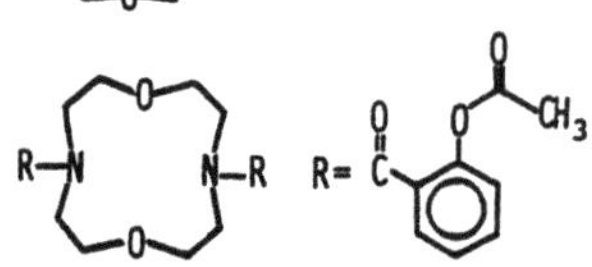	oil	81	35
18	26	18	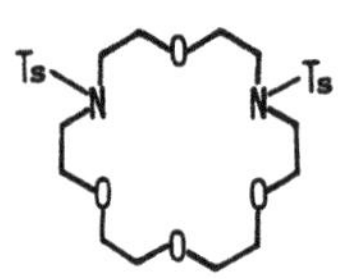	nr	69	5
19	28	15	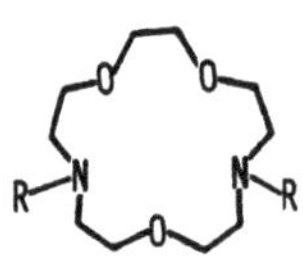 R= C(=O)CH=CHPh (E)	120–122	64	35

Table 4.4 (continued)

Cpd. No.	No. C's	Ring size	Structure		Mp or (bp) in °C	Yield (%)	Refs.
20	28	15	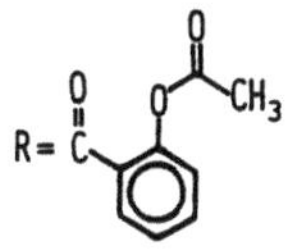		oil	88	35
21	28	20	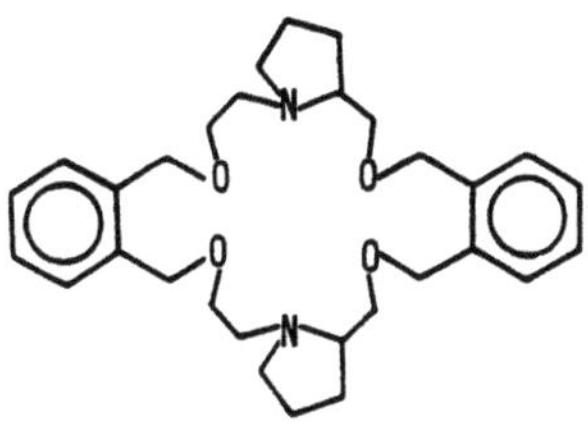		98–99	nr	130
22	30	9	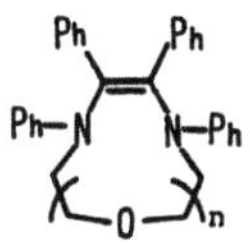	n = 1	nr	40	12
23	32	12		n = 2	150–151	51	12
24	38	20	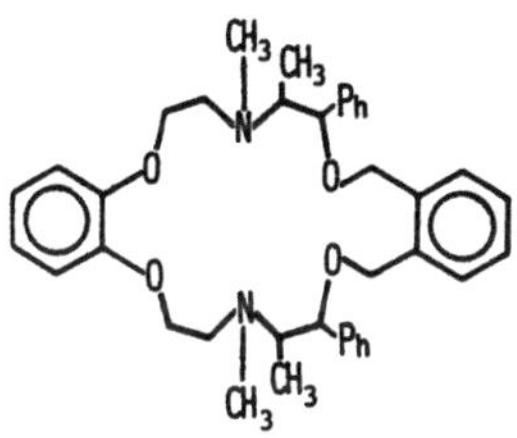		94–95	40	130
25	44	20	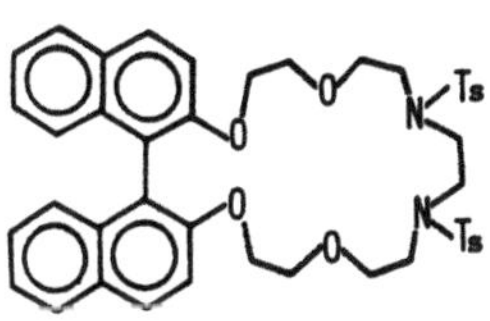	(racemic)	130–202	20	22
26	62	22	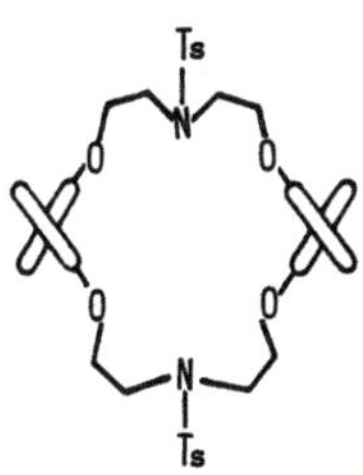	(RR)	130–200	11.3	22

Table 4.5. Azacrowns containing two substituted nitrogens

Cpd. No.	No. C's	Ring size	Structure	Mp or (bp) in °C	Yield (%)	Refs.
1	13	15	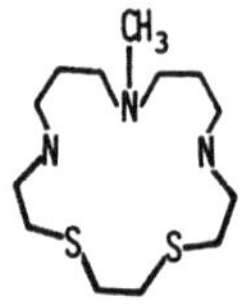	~5	<50	100
2	14	15	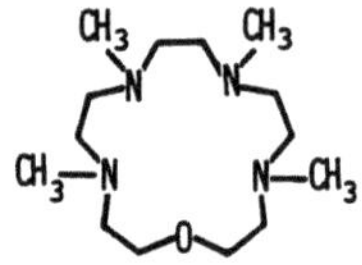	oil	~55	100
3	15	18	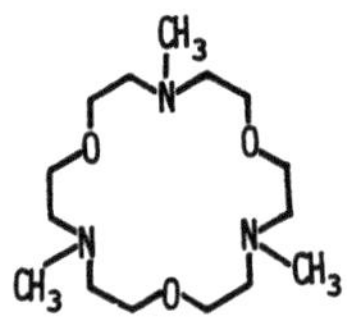	nr	nr	70
4	29	12	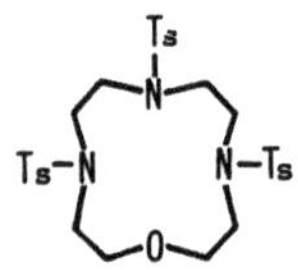	nr 200–204	90 63	5 107
5	29	18	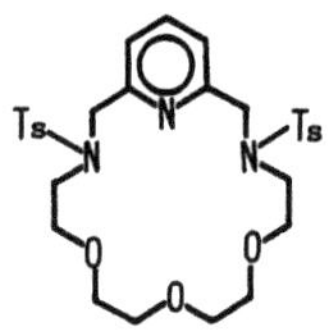	163–165	27	126
6	30	13	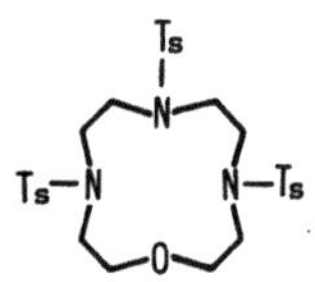	nr	29	5
7	31	14	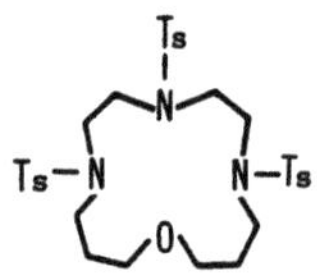	204–207	29	107
8	31	14	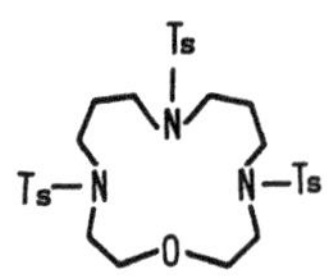	nr	58	5

Table 4.5 (continued)

Cpd. No.	No. C's	Ring size	Structure	Mp or (bp) in °C	Yield (%)	Refs.
9	31	15	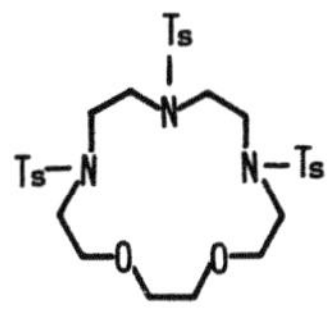	nr 198–199	88 71	5 107
10	33	16	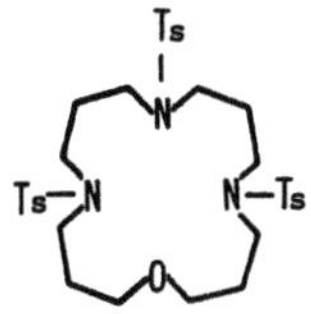	nr	69	5
11	33	17	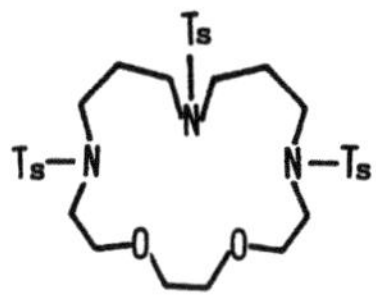	102 104 nr	85 85	107 5
12	33	18	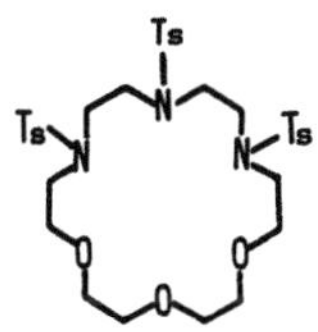	nr 196–198	44 14	5 107
13	35	20	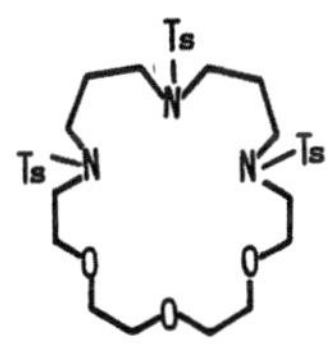	nr glass	83 83	5 107
14	38	15	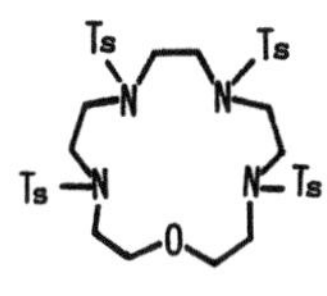	253–255 nr	52 65	107 5
15	40	17	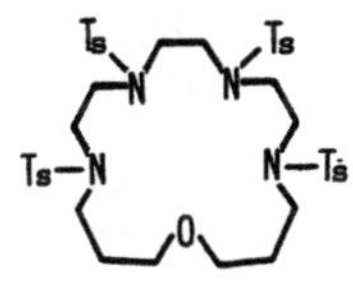	glass nr	44 nr	107 5
16	40	18	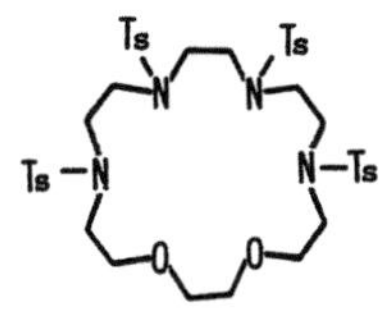	nr 93–96	72 25	5 107

Table 4.5 (continued)

Cpd. No.	No. C's	Ring size	Structure	Mp or (bp) in °C	Yield (%)	Refs.
17	42	21	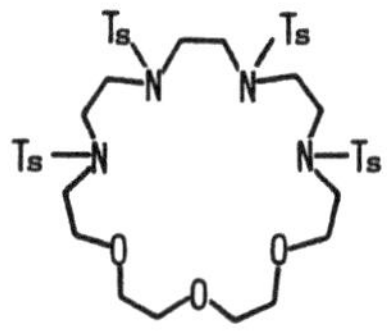	nr 192–193	35 35	5 107
18	46	27	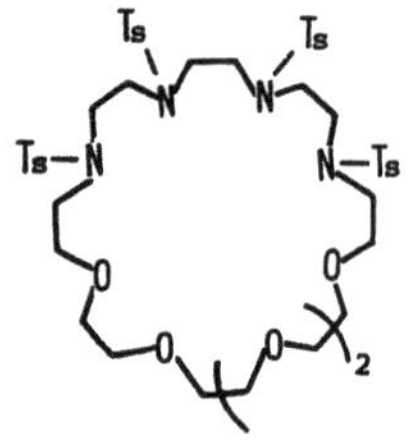	192–194	12	107

Table 4.6. Azacrowns containing more than two substituted nitrogens

Cpd. No.	No. C's	Ring size	Structure	Mp or (bp) in °C	Yield (%)	Refs.
1	8	12		83–85 238–240 (tripicrate) 235–239 (tripicrate)	61 nr 100	106 56 107
2	10	15	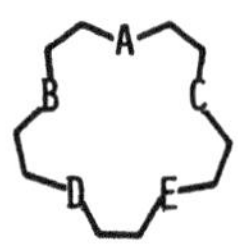 A = B = C = N D = E = O	paste nr oil 215–218 (tripicrate)	<25 48 19 nr	115 114 106 56
3	10	15	A = B = C = D = N, E = O	66–69 54–56	64 <10	106 115
4	10	15	A = B = C = N, D = E = S	nr 62.5–65	nr <32	114 115
5	10	15	A = B = C = D = N, E = S	nr paste	nr <20	114 115

Table 4.6 (continued)

Cpd. No.	No. C's	Ring size	Structure	Mp or (bp) in °C	Yield (%)	Refs.
6	12	18	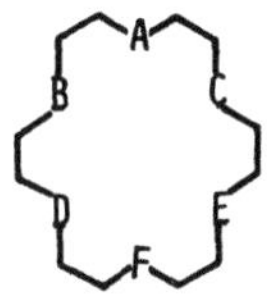 A = B = C = D = N E = F = O	nr oil paste	nr 74 <50	114 106 115
7	12	18	A = B = C = D = E = N, F = O	paste	<10	115
8	12	18	A = B = C = D = N, E = F = S	paste nr	<23 nr	115 114
9	12	18	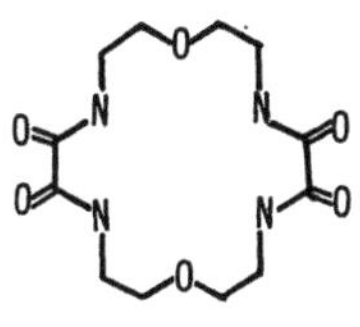	200 dec.	75	124
10	14	18	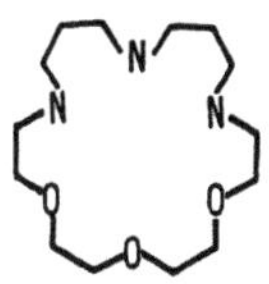	oil	35	106
11	16	24	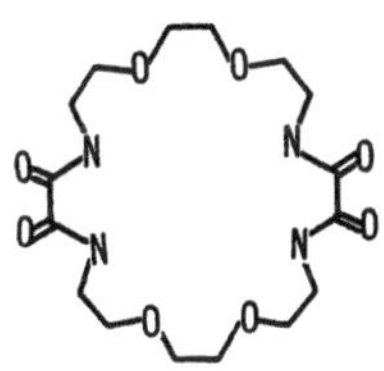	186–194	60	124
12	16	22	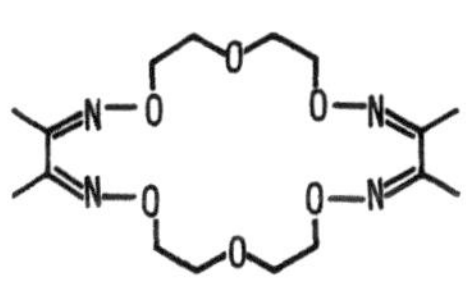	123	1.7	104
13	24	22	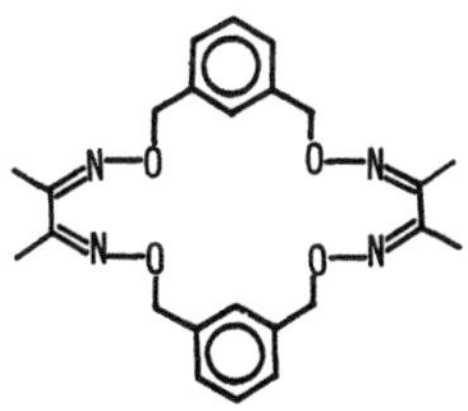	nr	nr	104

Table 4.6 (continued)

Cpd. No.	No. C's	Ring size	Structure	Mp or (bp) in °C	Yield (%)	Refs.
14	26	30	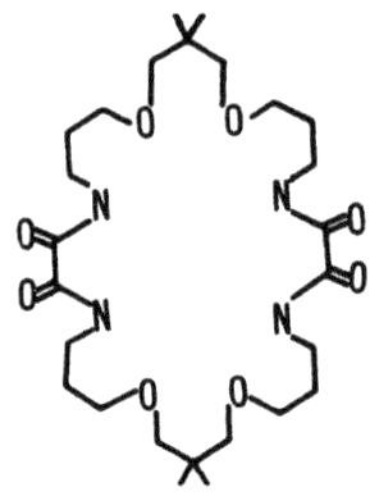	189–213	63	124

Table 4.7. All-nitrogen crowns having the nitrogens predominantly unsubstituted

Cpd. No.	No. C's	Ring size	Structure	Mp or (bp) in °C	Yield (%)	Refs.
1	8	12	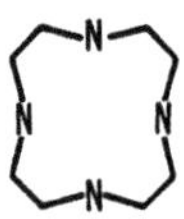	34–35	nr	56
			· 4HCl	nr	72	103
2	10	14	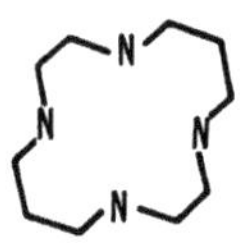	185–186	nr	56
				173 (subl. at 125)	nr	11
3	10	15	 · 3HBr	nr	29	52
4	11	11	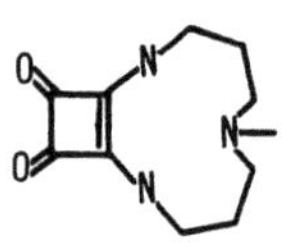	230	70	124
5	12	14	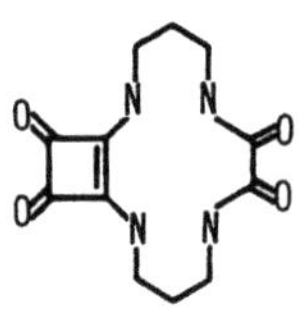	250 dec.	92	124

Table 4.7 (continued)

Cpd. No.	No. C's	Ring size	Structure	Mp or (bp) in °C	Yield (%)	Refs.
6	12	18	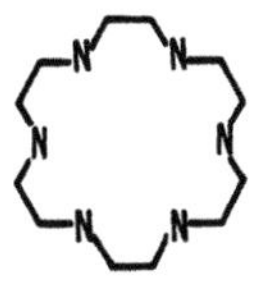	147–150	49–50	5
7	12	16	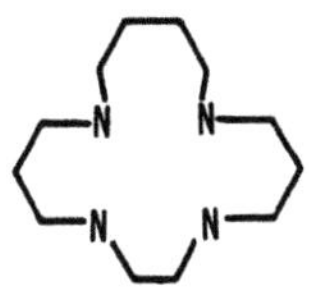 · 4HBr	256.5–257 dec.	27	52
8	13	17	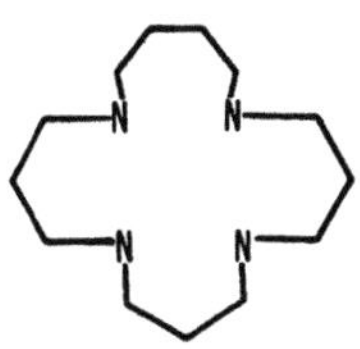 · 4HBr	278–279 dec.	37	52
9	15	14	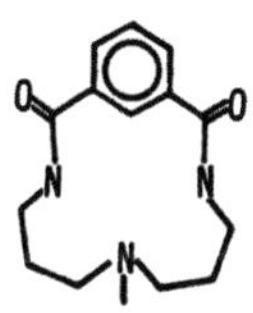	189–191	26	126
10	18	24	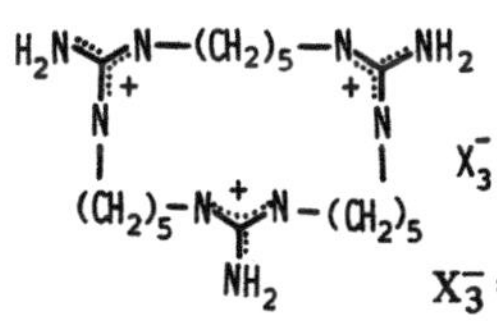 X_3^- = Cl or Br	170 dec. tripicrate	36	27
11	18	17	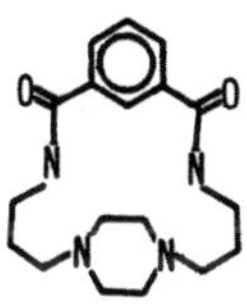	253–256	31	126
12	22	22	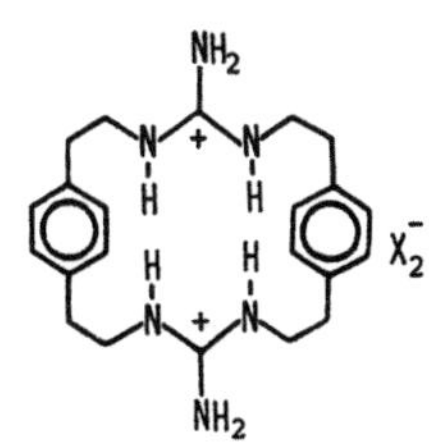 · $2Br^-$ $X_2 = Br_2$ or Cl_2	>300	56	27

Table 4.7 (continued)

Cpd. No.	No. C's	Ring size	Structure		Mp or (bp) in °C	Yield (%)	Refs.
13	24	18	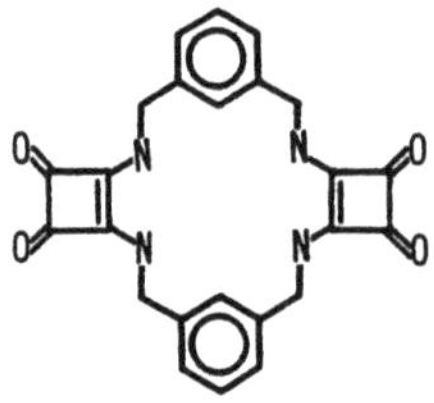		230 dec.	70	124
14	26	14	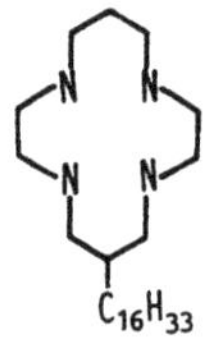		132–133	10	112
15	26	14	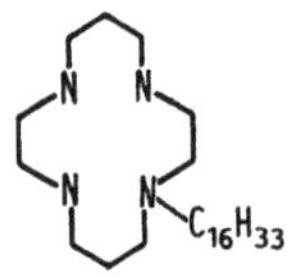	· 4HCl	197–198	10	112
16	26	30	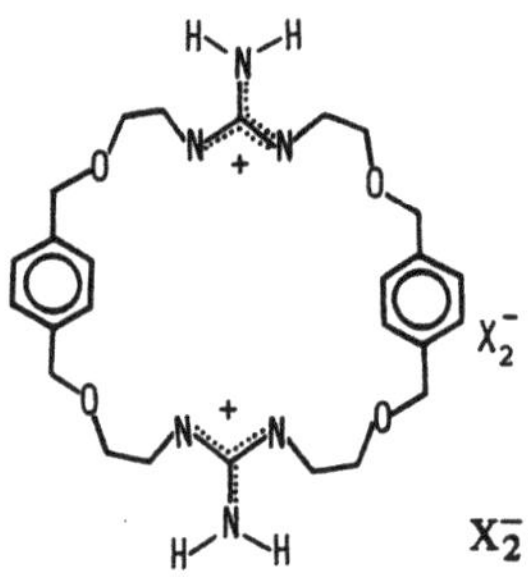	· 2Br⁻ X_2^- = Br or Cl	209–210	64	27
17	42	14	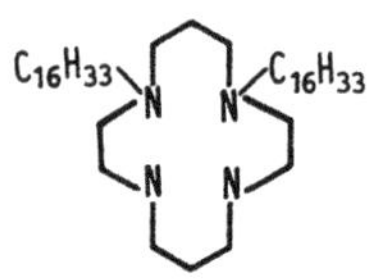	· 4HCl	180–181	16	112
18	42	53	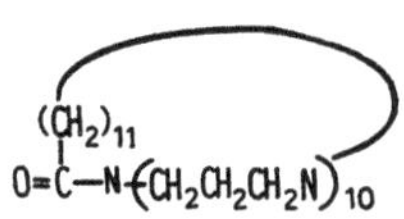		nr	nr	58
19	48	18	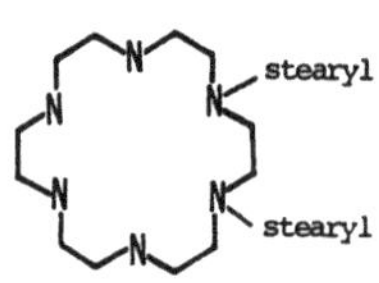		nr	8	113

Table 4.8. All-nitrogen crowns having the nitrogens predominantly substituted

Cpd. No.	No. C's	Ring size	Structure		Mp or (bp) in °C	Yield (%)	Refs.
			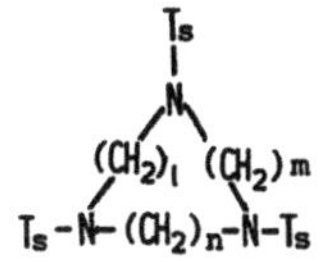				
1	27	9	l = m = n = 2		nr	71	5
					224–225	57	105
					222–223	71	103
2	28	10	l = m = 2, n = 3		nr	84	5
					234–236	84	103
3	29	11	l = m = 2, n = 4		nr	81	5
4	29	11	l = m = 3, n = 2		nr	58	5
5	30	12	l = m = 2, n = 5		172–173	55	5, 103
6	30	12	l = m = n = 3		nr	67	5
7	31	13	l = m = 2, n = 6		205–206	50	103
					nr	40–50	5
8	31	9	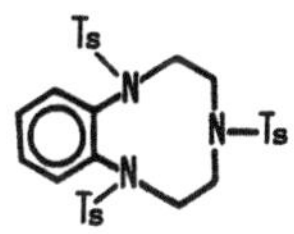		nr	24	5
9	31	14	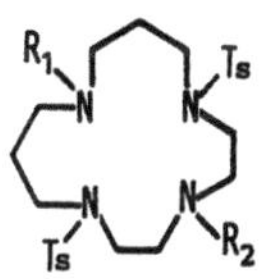	R_1 = H, R_2 = Ts	179–181	27.5	44
10	31	14		R_1 = Ts, R_2 = H	160 dec.	nr	44
			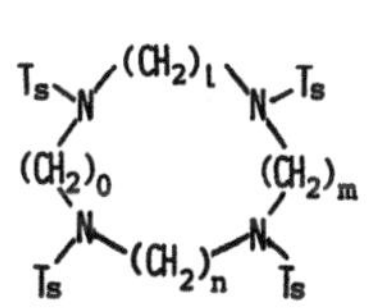				
11	37	13	l = m = n = 2, o = 3		nr	77	5
12	38	14	l = m = 2, n = o = 3		234–236	58	5, 103
					nr	80	5
13	38	14	l = n = 2, m = o = 3		nr	70	5
14	40	16	l = m = n = o = 3		nr	90	5

Table 4.8 (continued)

Cpd. No.	No. C's	Ring size	Structure	Mp or (bp) in °C	Yield (%)	Refs.
15	45	15	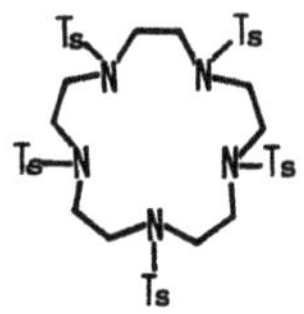	278–280	83	5, 103
16	46	16	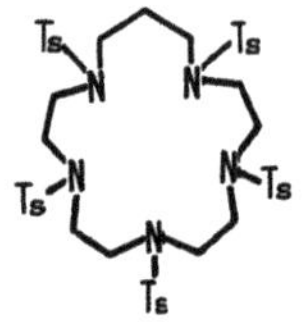	nr	51	5
17	54	18	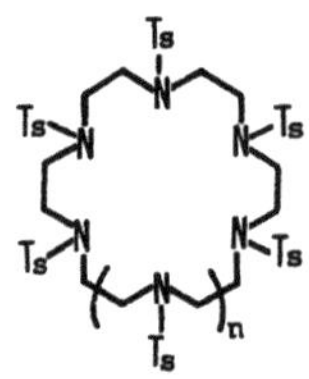 n = 1	260–290 311–313	70–77 75	5 103
18	56	20	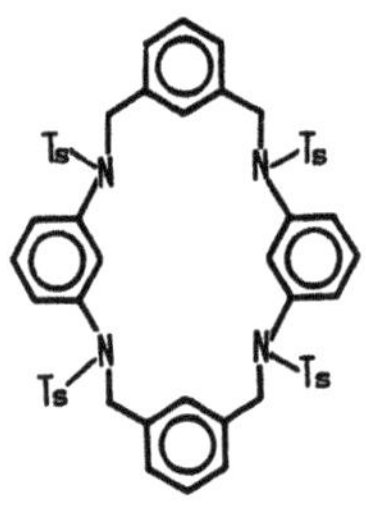	248–250	4.5	121
19	63	21	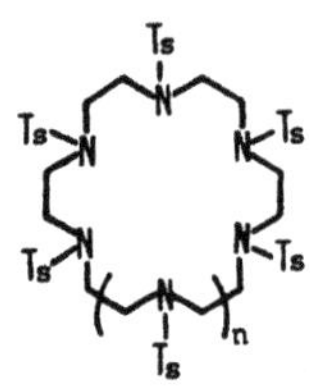 n = 2	nr 183–184	45 45	5 103
20	72	24	n = 3	nr	73	5

Table 4.9. Nitrogen crowns containing amide functions

Cpd. No.	No. C's	Ring size	Structure	Mp or (bp) in °C	Yield (%)	Refs.
1	10	16	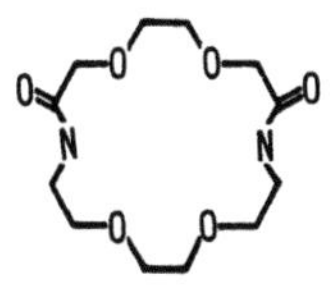	122	84	16
2	10	16	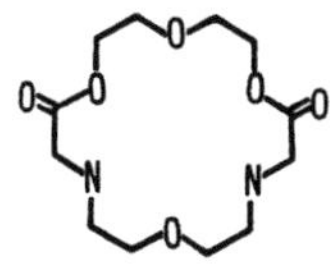	126	89	16
3	26	15	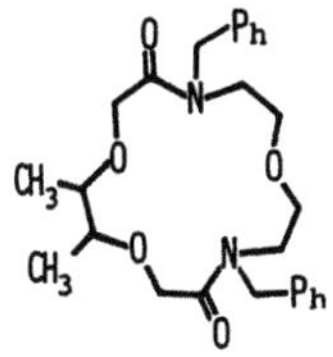	oil	48	101
4	28	18	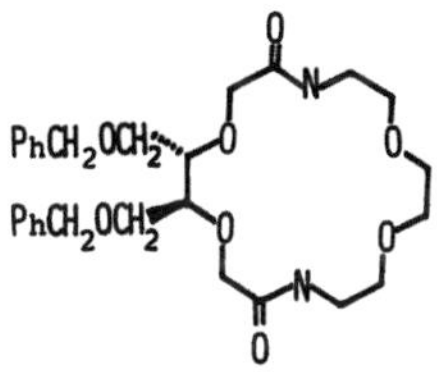	oil	23	3
5	28	18	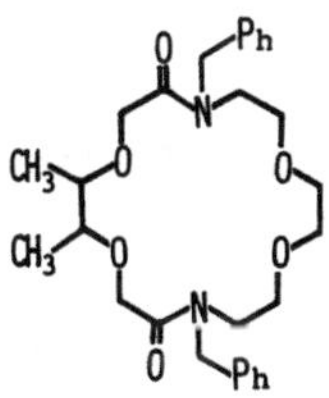	oil	50	101
6	30	21	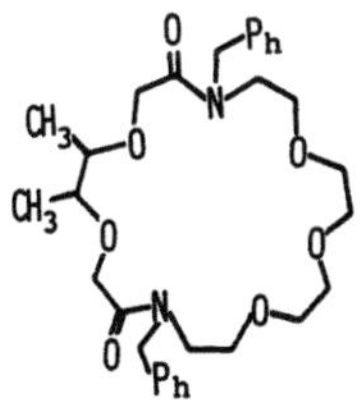	oil	59	101

Table 4.10. Miscellaneous benzocrowns containing nitrogen

Cpd. No.	No. C's	Ring size	Structure		Mp or (bp) in °C	Yield (%)	Refs.
1	12	12	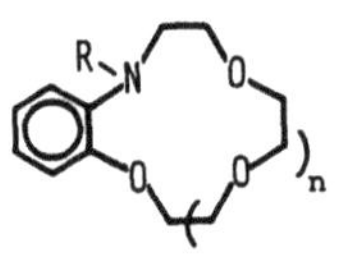	n = 1; R = H	80	nr	75
2	14	15		n = 2, R = H	nr	nr	74
					101.5	nr	75
					100–101	nr	99
3	14	15	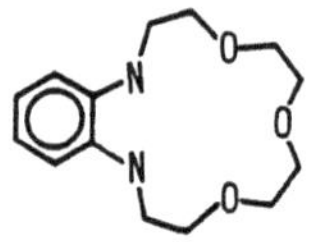		110	nr	75
4	16	15	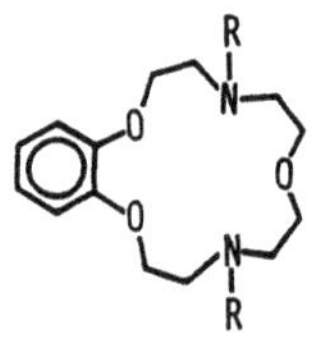	R = CH_3	nr	nr	45
					(210/0.4)	42	46
5	16	18	See 1, n = 3, R = H		oil	nr	98
6	17	15	See 1, n = 2, R = $CH_2CH_2CH_3$		"brown liq"	15	74
7	17	15	See 1, n = 2, R = $(CH_2)_3Br$		"solid"	86	74
8	17	15	See 1, n = 2, R = $COCH_2CH_2Br$		"brown liq"	95	74
9	18	15	See 1, n = 2, R = $COCH_2CH_2COOH$		"solid"	59	74
10	18	18	See 12, X = NMe		(220/0.5)	48	47
11	19	15	See 1, n = 2, R = $(CH_2)_3OEt$		"brown liq"	75	74
12	20	18	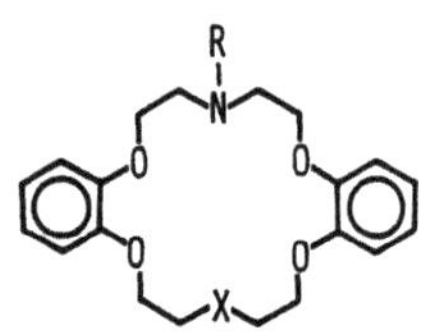	X = O, R = H	152–153	22	98, 99
13	20	18	See 12, X = NH, R = H		175–176	80	98
14	22	18	See 12, X = NMe, R = Me		158–160	22	47
15	28	18	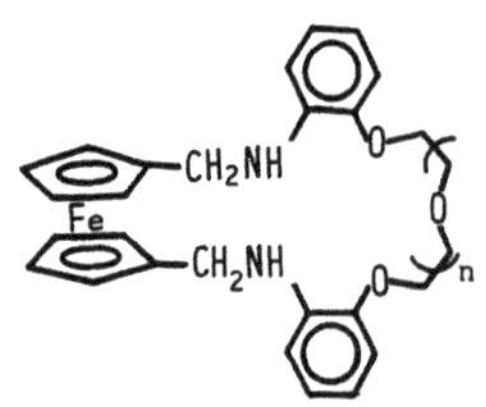	n = 1	nr	60	6

Table 4.10 (continued)

Cpd. No.	No. C's	Ring size	Structure	Mp or (bp) in °C	Yield (%)	Refs.
16	28	20	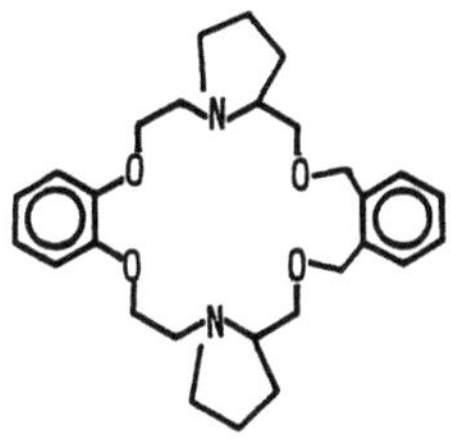	98–99	nr	130
17	30	15	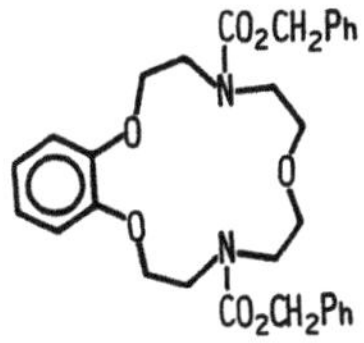	(320/0.2)	55	27, 46
18	30	21	See 15, n = 2	nr	60	6
19	38	20	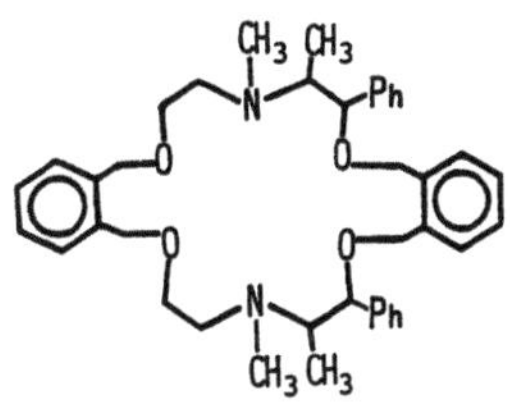	94–95	40	130

Table 4.11. Monoazacrowns containing a 4-substituted phenyl on nitrogen

Cpd. No.	No. C's	Ring size	Structure		Mp or (bp) in °C	Yield (%)	Refs.
1	16	15	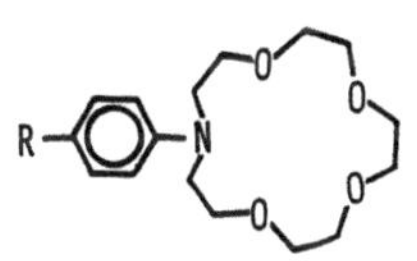	R = H	44–45 44–45	nr 16	30 32
2	16	15		R = NH_2	46	nr	30
3	16	15		R = N=O	127	83	32
4	16	15		R = NO_2	127	nr	30
5	17	15		R = CHO	80 87–88	nr 75	30 32

Table 4.11 (continued)

Cpd. No.	No. C's	Ring size	Structure	Mp or (bp) in °C	Yield (%)	Refs.
6	18	18	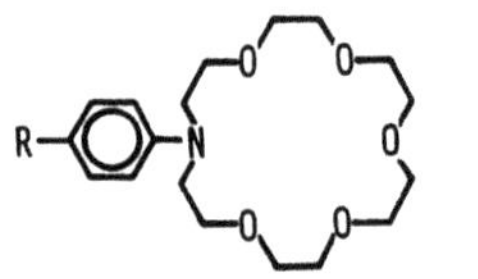 R = H	30–35	nr	30
				30–35	13	32
7	18	18	R = NH_2	52–53	86	32
8	19	18	R = CHO	43–45	65	32
9	19	15	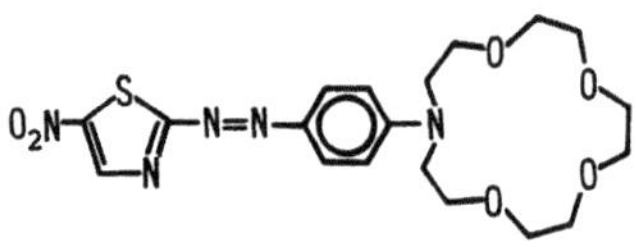 (E)	196–202	21	32
10	20	21	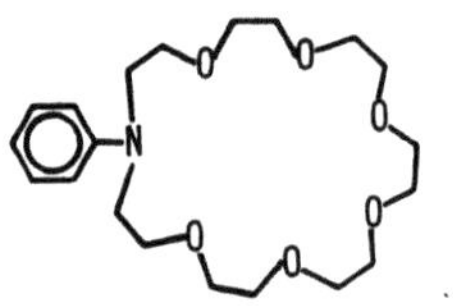	oil	nr	30
11	22	15	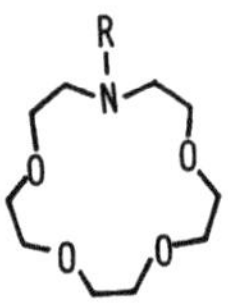 R= 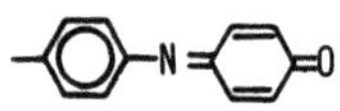	50–55	91	32
12	22	15	R= 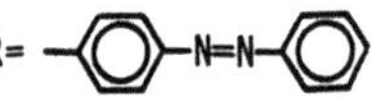 (E)	nr	nr	30
13	22	15	R= 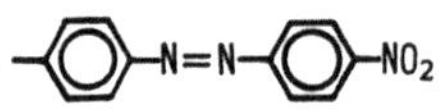 (E)	132–134	nr	30
14	22	15	R= 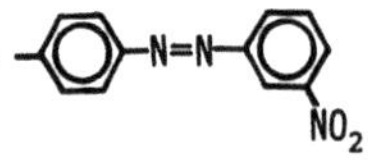 (E)	110	nr	30
15	22	15	R= 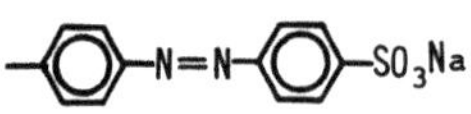 (E)	138–145	nr	30
16	23	15	R= 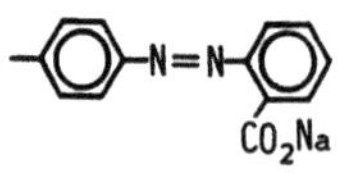 (E)	139–149	nr	30
17	23	15	R= 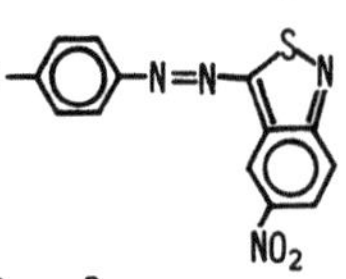 (E)	196–204	15	32
18	24	9	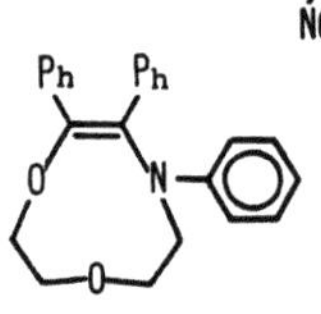	nr	10	12

Table 4.11 (continued)

Cpd. No.	No. C's	Ring size	Structure	Mp or (bp) in °C	Yield (%)	Refs.
19	24	18	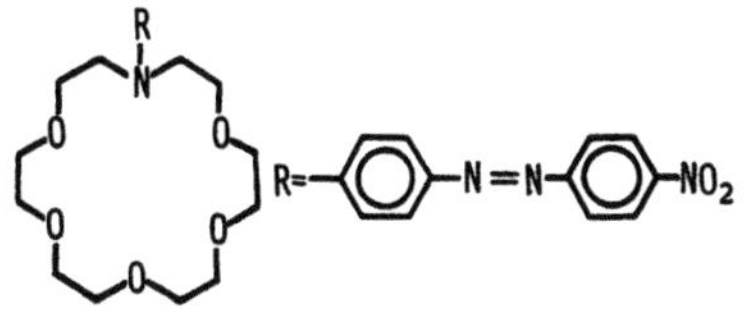 (E)	116 106–109	nr 56	30 32
20	24	15	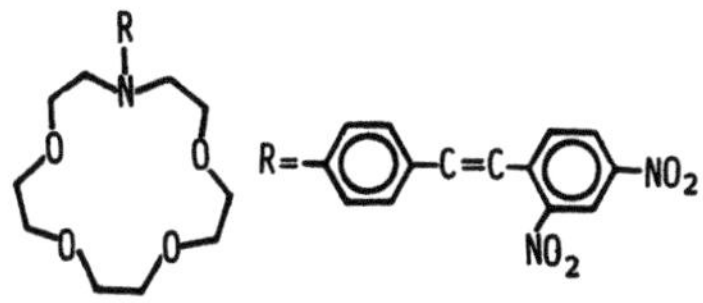 (E)	143–146	46	32
21	24	15	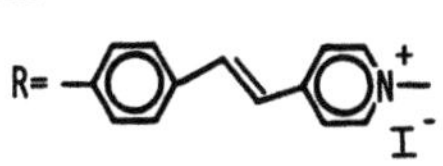 (E)	168–170	78	32
22	24	15	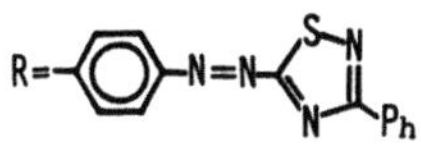 (E)	158–160	71	32
23	25	15	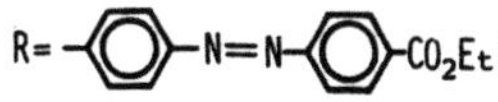 (E)	115	nr	30
24	26	15	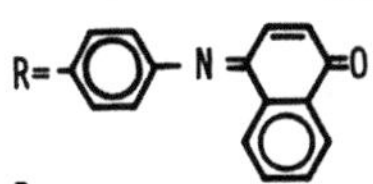	124–125	65	32
25	26	18	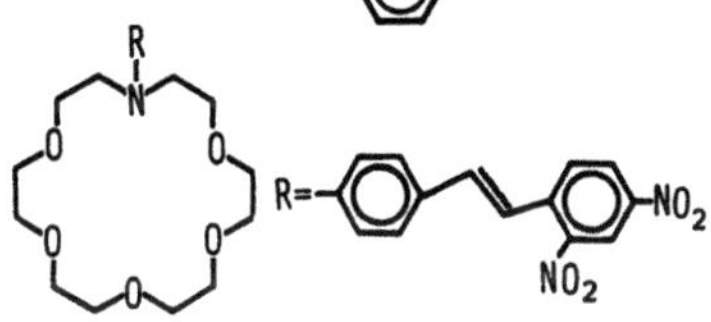 (E)	90–91	39	32
26	26	18	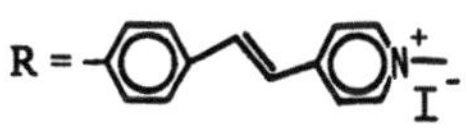 (E)	89–90	58	32
27	26	21	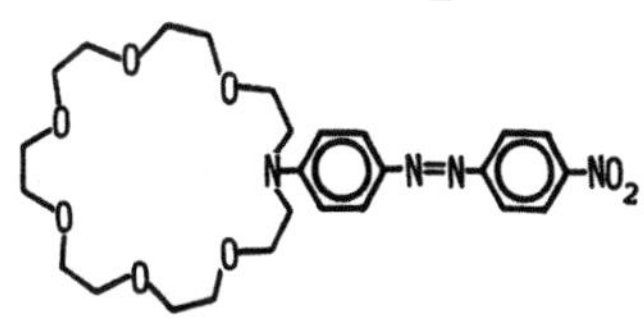 (E)	95–97	nr	30
28	31	15	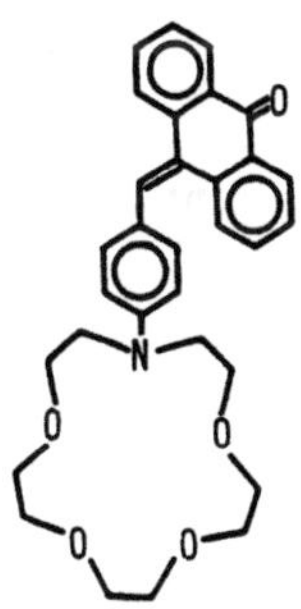	oil	40	32

Table 4.12. Azacrowns containing the 3,3-bis(methyleneoxy)oxetane subunit

Cpd. No.	No. C's	Ring size	Structure	Mp or (bp) in °C	Yield (%)	Refs.
1	13	16	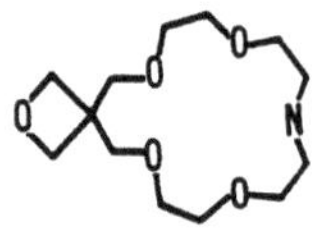	79–81	67	59, 60
2	13	16	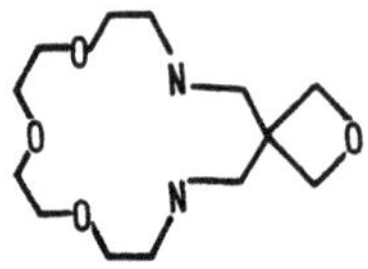	(108–110/ 1 x 10^{-4})	17	59, 60
3	15	16	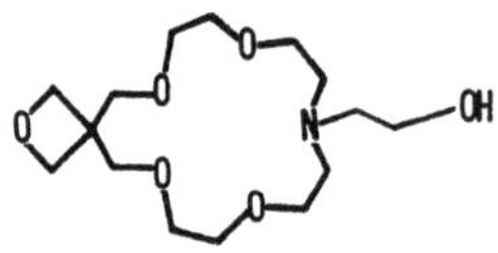	oil	87	59
4	15	19	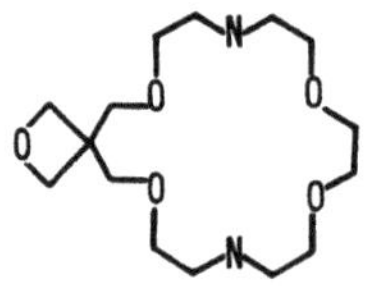	oil	71	59
5	18	20	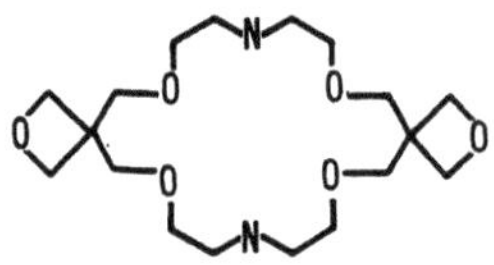	118.5–119	17	57

Table 4.13. Azacrowns containing the furan subunit

Cpd. No.	No. C's	Ring size	Structure	Mp or (bp) in °C	Yield (%)	Refs.
1	14	18	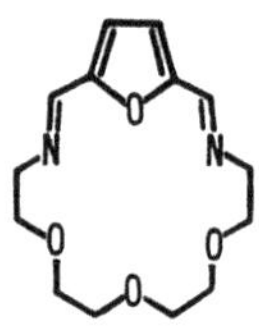	nr	nr	38
2	14	15	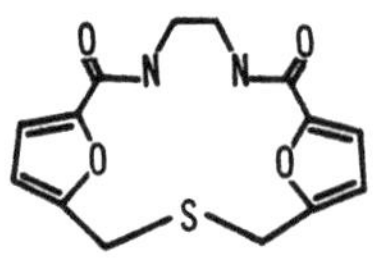	210–212	50	119

Table 4.13 (continued)

Cpd. No.	No. C's	Ring size	Structure	Mp or (bp) in °C	Yield (%)	Refs.
3	15	16	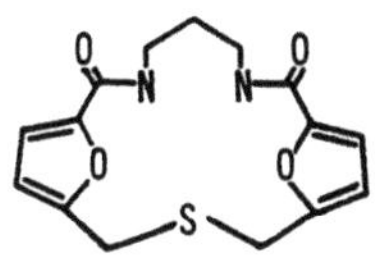	72–73	22	119
4	16	15	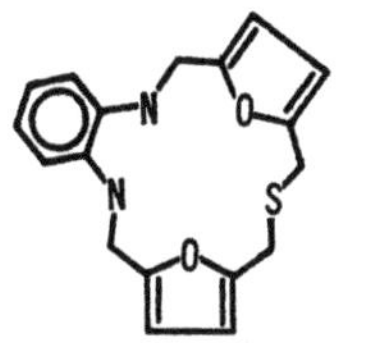	197–198	79	119
5	16	18	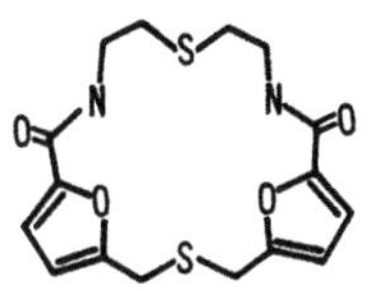	45–46	93	119
6	18	16	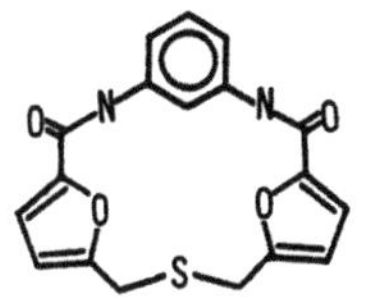	204–205	80	119
7	18	17	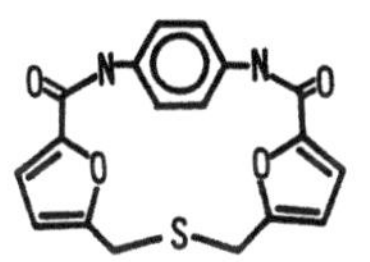	224–225	91	119
8	18	19	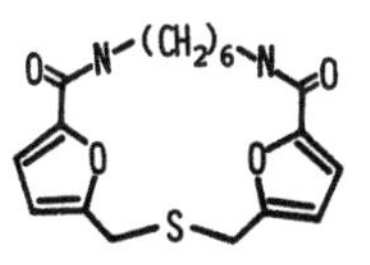	87–88	50	119
9	18	12	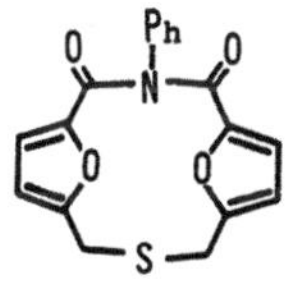	184–185	97	119
10	20	12	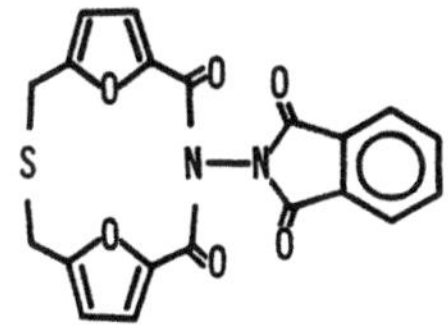	88–89	62	119

Table 4.14. Azacrowns containing the pyridine-2,6-diheteroatom subunit

Cpd. No.	No. C's	Ring size	Structure	Mp or (bp) in °C	Yield (%)	Refs.
1	9	10	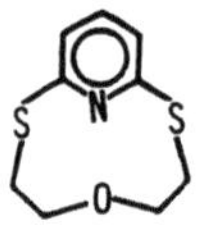	92–94	13	126
2	10	13	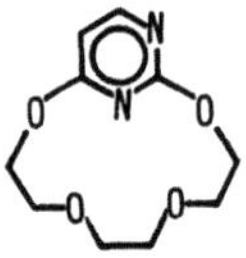	118–121	2	95
3	11	13		83–84	16	93
4	11	13	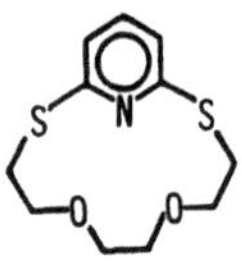	75–77	18	126
5	12	13	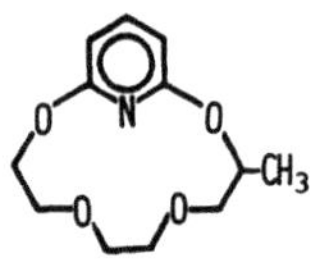	55	3	94
6	13	16	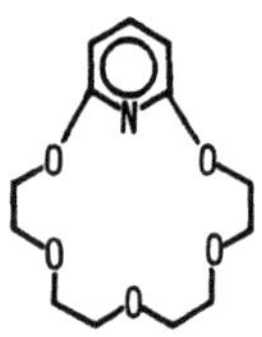	76–78	4	93, 94
7	13	16	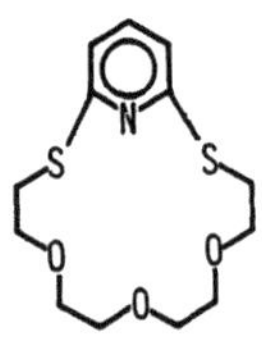	74–76	28	126
8	14	14	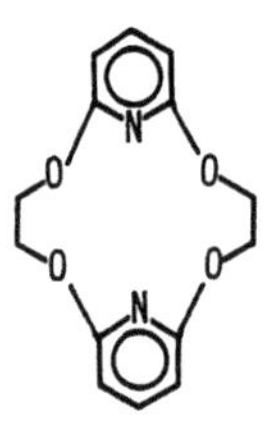	215–216	16	93

Table 4.14 (continued)

Cpd. No.	No. C's	Ring size	Structure		Mp or (bp) in °C	Yield (%)	Refs.
9	14	16	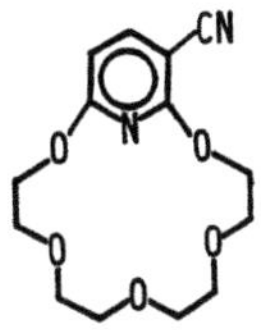		nr	nr	86
10	14	19	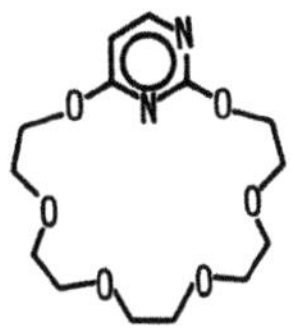		(178–182/1.0)	16	95
11	15	19	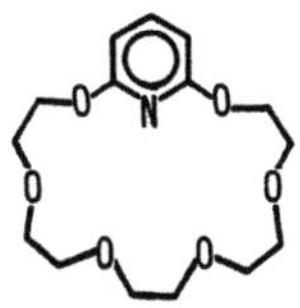		(155–160/0.15)	12	94
12	16	18	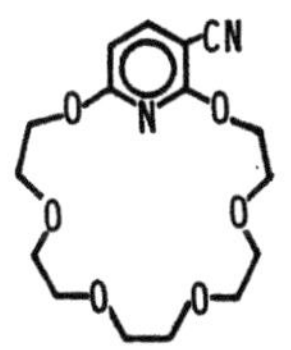		nr	48	86
13	16	16			(205/0.3)	20	88, 89
14	16	19	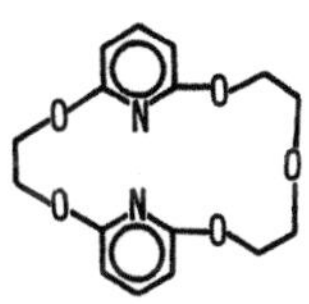		94.5–95.5	5	93, 94
15	17	22	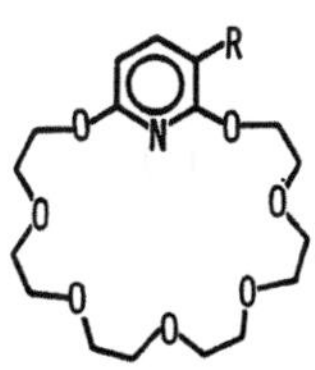	R = H	(190–193/0.1)	48	94
16	18	22		R = CN	nr	7	86

Table 4.14 (continued)

Cpd. No.	No. C's	Ring size	Structure	Mp or (bp) in °C	Yield (%)	Refs.
17	18	20	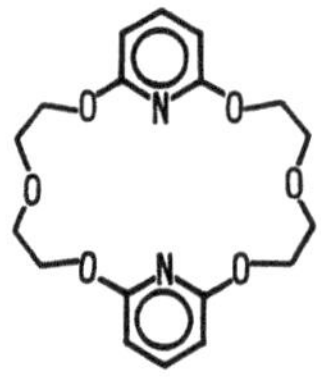	111–112	18	93, 94
18	18	19	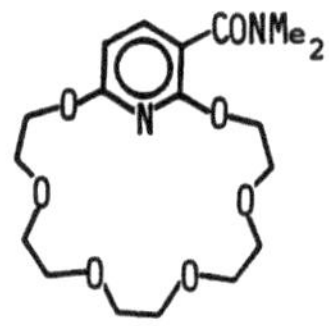	(220/0.4)	32	88, 89
19	20	14	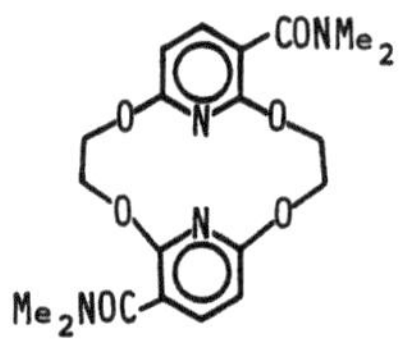	270 dec.	2	87
20	20	22	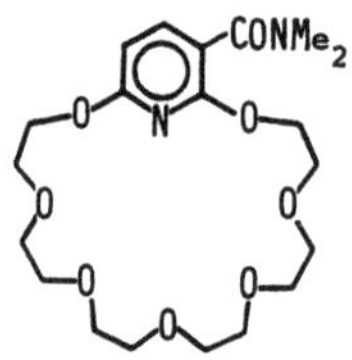	(240/0.3)	17	88, 89
21	22	26	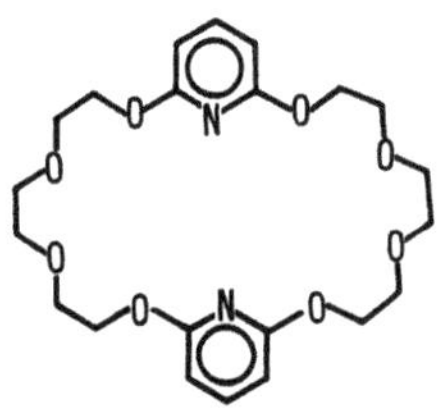	117–120	2	93, 94
22	24	26	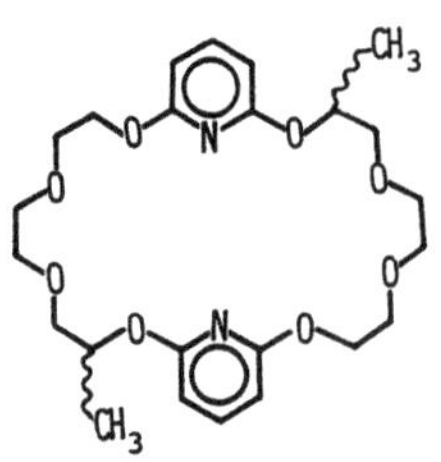	111	15	94

Table 4.14 (continued)

Cpd. No.	No. C's	Ring size	Structure	Mp or (bp) in °C	Yield (%)	Refs.
23	26	3	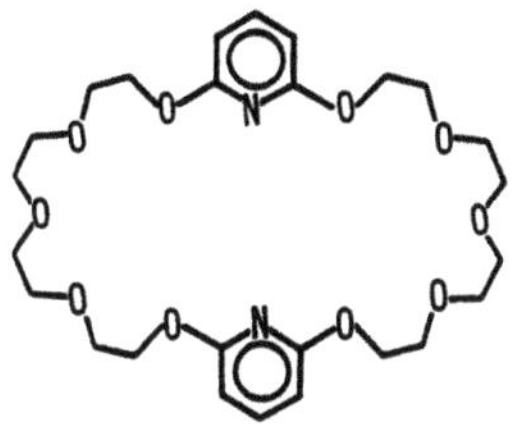	83–84	5	94
24	27	30	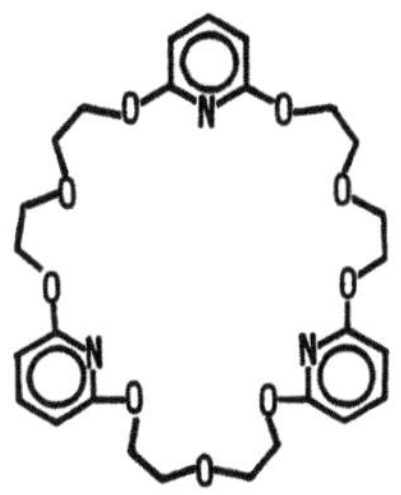	120.5–121.5	3	93, 94
25	20	38	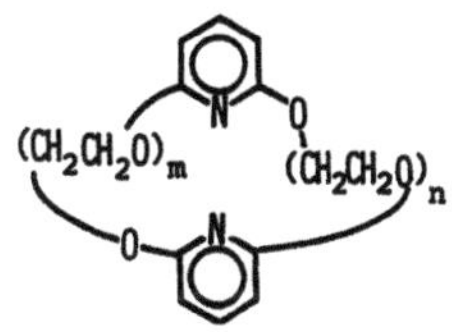 m = n = 5	91	8	94
26	32	41	m = 6, n = 5	nr	4	94

Table 4.15. Azacrowns containing the 2,6-picolinoyl subunit

Cpd. No.	No. C's	Ring size	Structure	Mp or (bp) in °C	Yield (%)	Refs.
1	11	12	 n = 1	228–230	14	126
2	13	15	n = 2	200–201	60	126
3	14	14	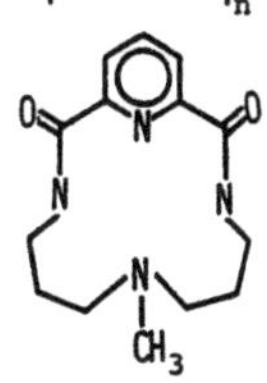	164–166	59	126

Table 4.15 (continued)

Cpd. No.	No. C's	Ring size	Structure	Mp or (bp) in °C	Yield (%)	Refs.
4	15	18	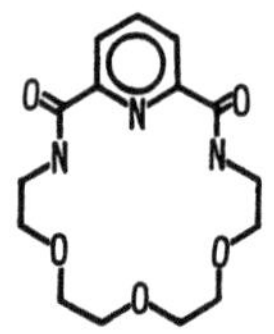	126–129	72	126
5	17	17	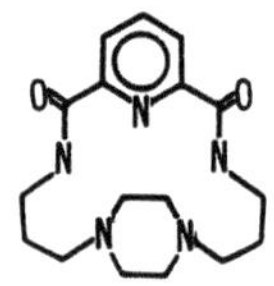	226–228	37	126
6	18	12	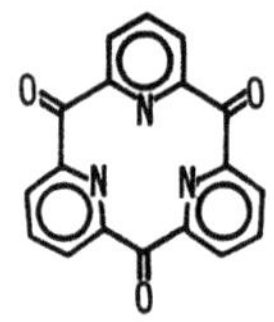	134–135	<2	97
7	18	18	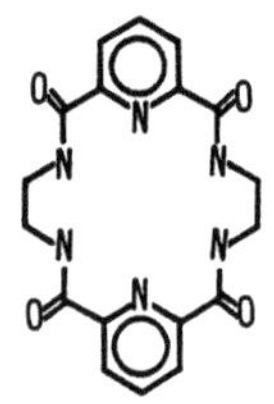	>250 dec.	36	126
8	22	24	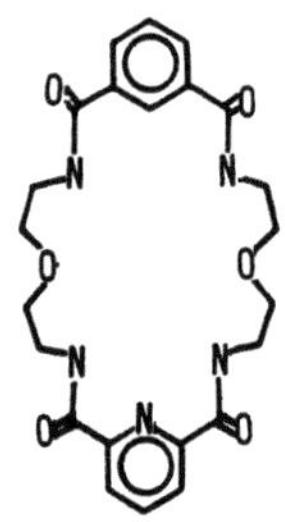	338–340	3	126, 128

Table 4.16. Azacrowns containing the pyridine-2,6-dimethylene subunit

Cpd. No.	No. C's	Ring size	Structure		Mp or (bp) in °C	Yield (%)	Refs.
1	11	12		X = O	133–135	nr	122
2	11	12		X = S	162–163	50	129
3	11	13		X = S–S	151–153 subl.	13	129
4	12	12		X = N–CH_3	67–69	11	129
5	13	15		X = SCH_2CH_2S	131–133	38	129
6	13	15		X = OCH_2CH_2O	90–91	nr	122
7	14	12	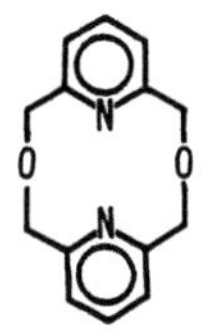		nr	6	83, 84
8	15	18	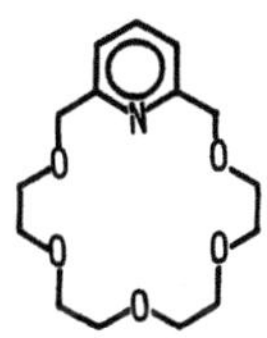		40–41	29	83, 84
9	15	18	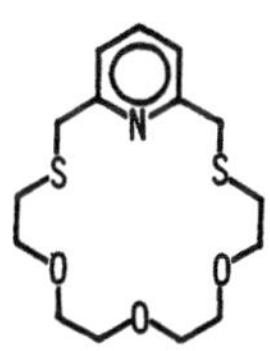		58–59	nr	122
10	15	18	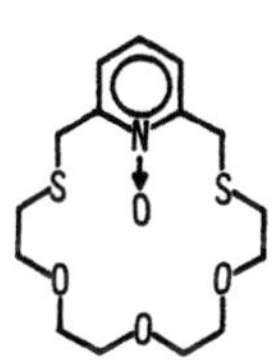		oil	nr	122
11	17	21			oil	nr	122

Table 4.16 (continued)

Cpd. No.	No. C's	Ring size	Structure	Mp or (bp) in °C	Yield (%)	Refs.
12	18	18	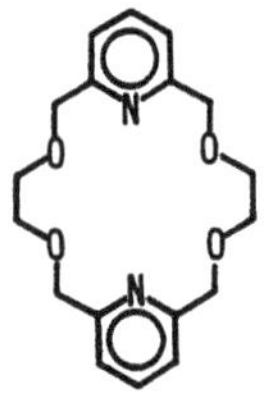	147–148	18	83, 84
13	18	18	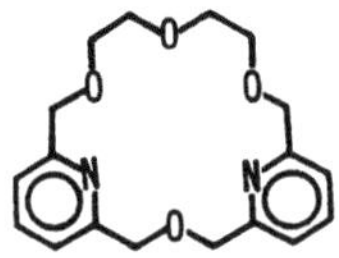	nr	nr	83, 84
14	18	18	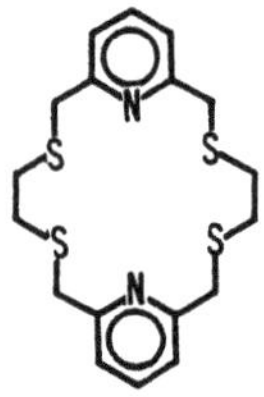	213–216 dec.	19	129
15	21	15	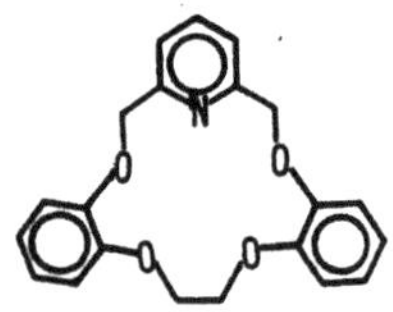	123–124	56	120
16	21	18	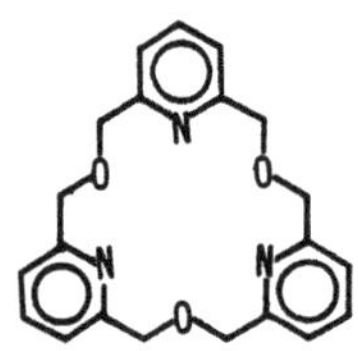	125–128	32	83, 84
17	22	15	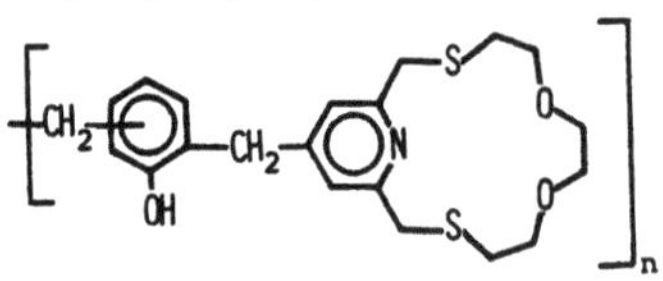	nr	nr	10
18	22	22	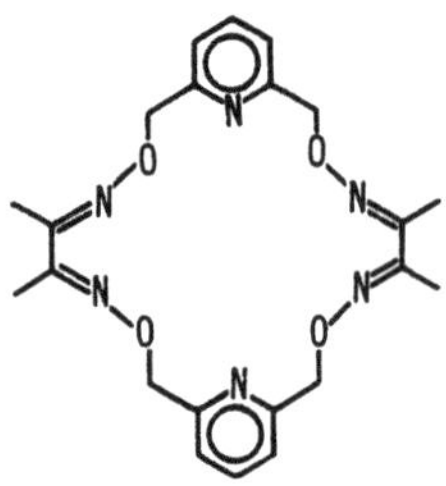	nr	nr	104

Table 4.16 (continued)

Cpd. No.	No. C's	Ring size	Structure	Mp or (bp) in °C	Yield (%)	Refs.
19	22	18	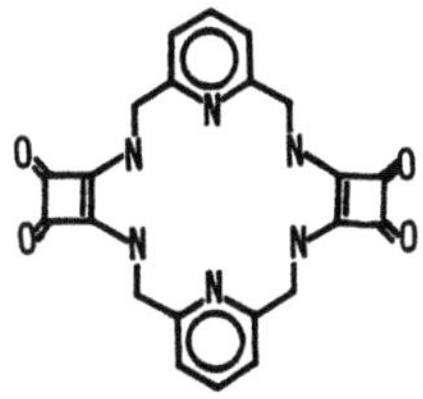	nr	nr	124
20	23	15	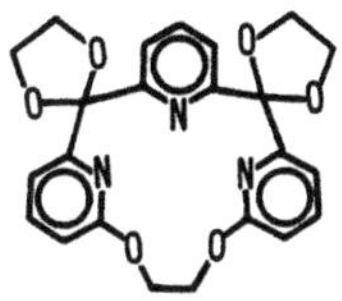	209–211	46	92
21	23	18	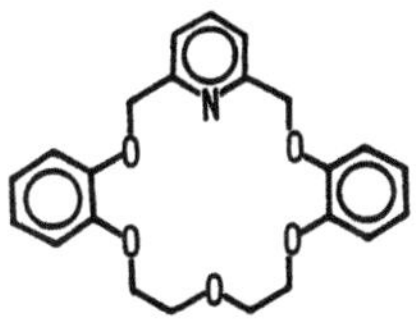	132 dec.	36	128
22	23	18	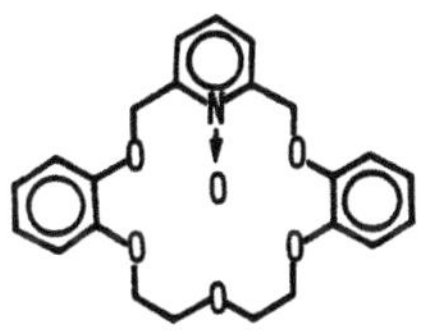	159 dec.	25	128
23	23	19	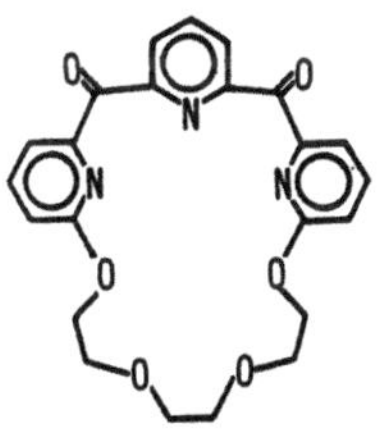	145–146	26	92
24	23	19	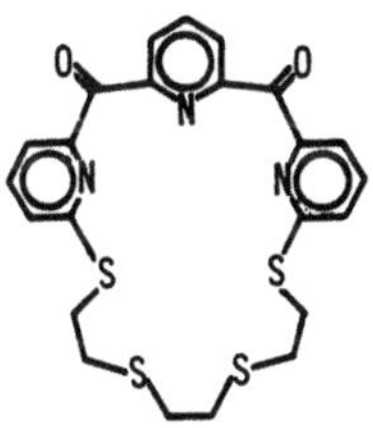	129–130	1.2	92
25	25	12	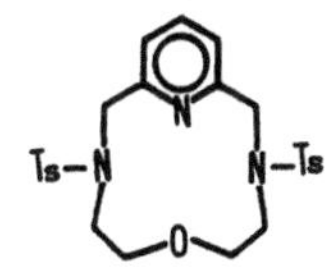	175–177	34	128

Table 4.16 (continued)

Cpd. No.	No. C's	Ring size	Structure	Mp or (bp) in °C	Yield (%)	Refs.
26	25	21		99–100	65	120
27	25	21	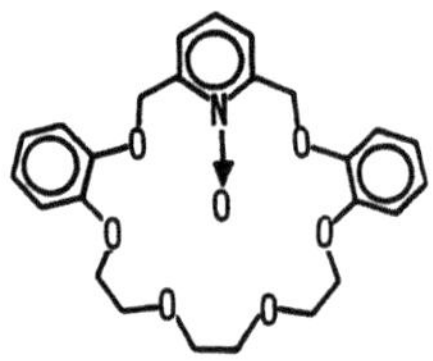	168–170	93	120
28	25	24	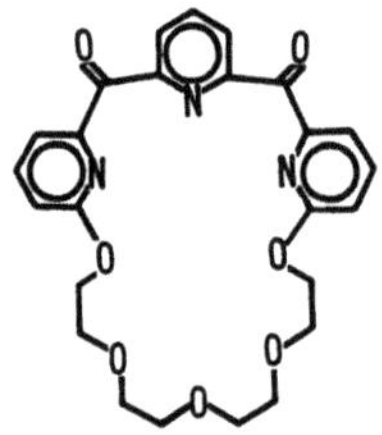	118–120	nr	92
29	26	18	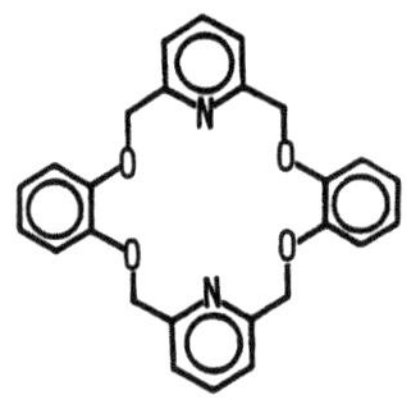	184–186	9	83, 84
30	27	15	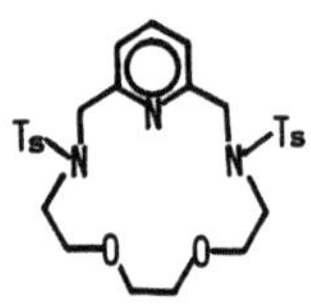	184–185	32	128
31	27	27	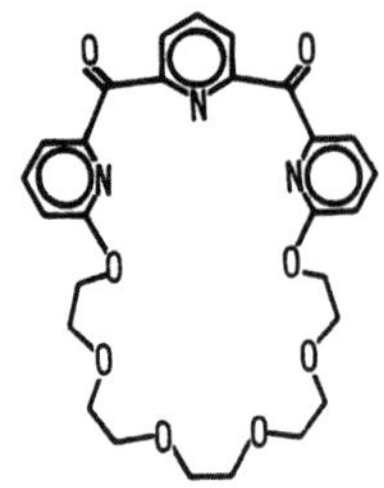	76–77	nr	92

Table 4.16 (continued)

Cpd. No.	No. C's	Ring size	Structure	Mp or (bp) in °C	Yield (%)	Refs.
32	28	24	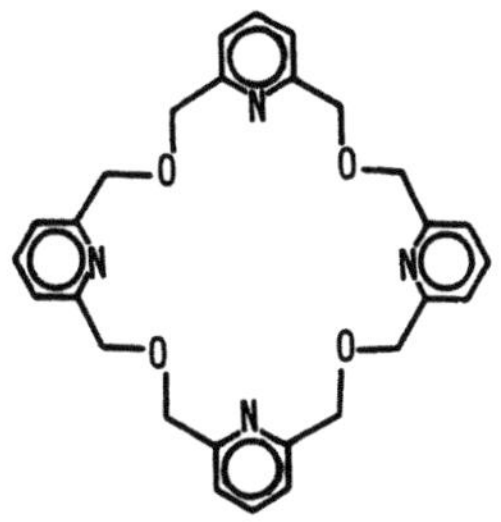	nr	6	83, 84
33	29	18	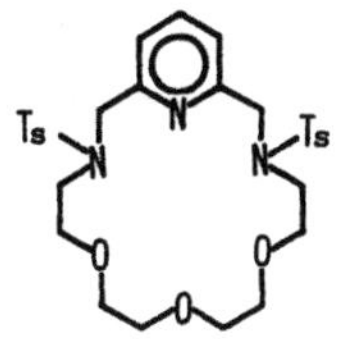	163–165	27	128
34	29	30	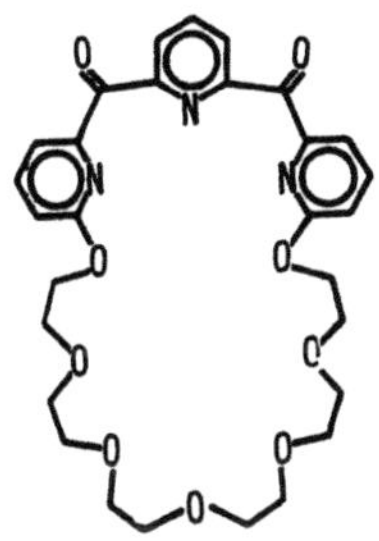	62–63	nr	92
35	30	18	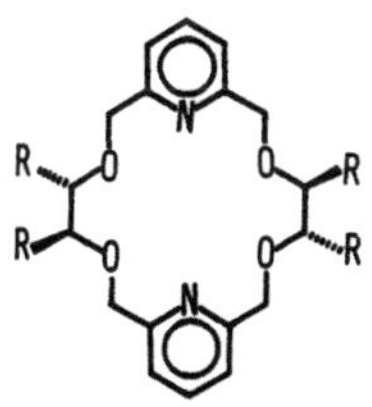 R = C(=O)N(CH$_3$)$_2$	224	15	39
36	30	22	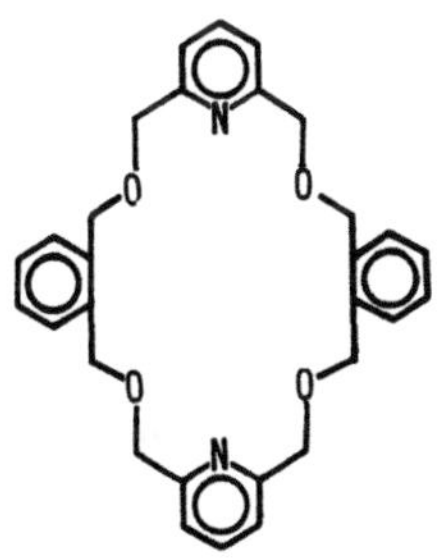	142–143	40	96

Table 4.16 (continued)

Cpd. No.	No. C's	Ring size	Structure	Mp or (bp) in °C	Yield (%)	Refs.
37	38	18	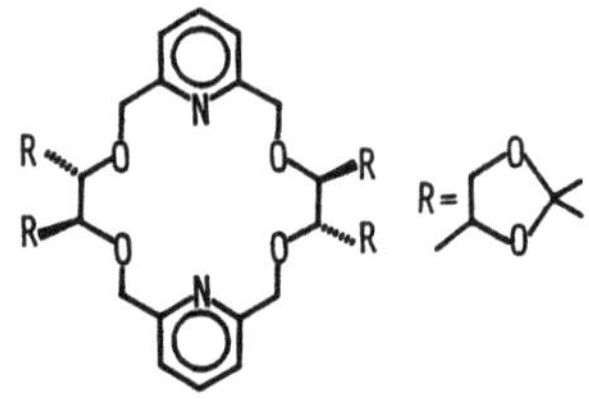	147–149	7.5	68
38	44	20	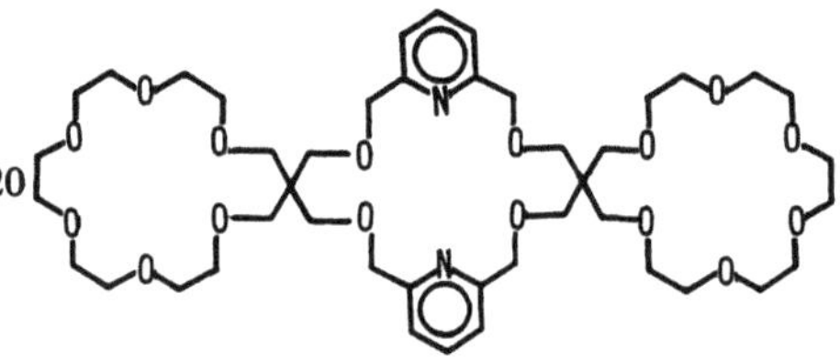	123–125	nr	127
39	45	33	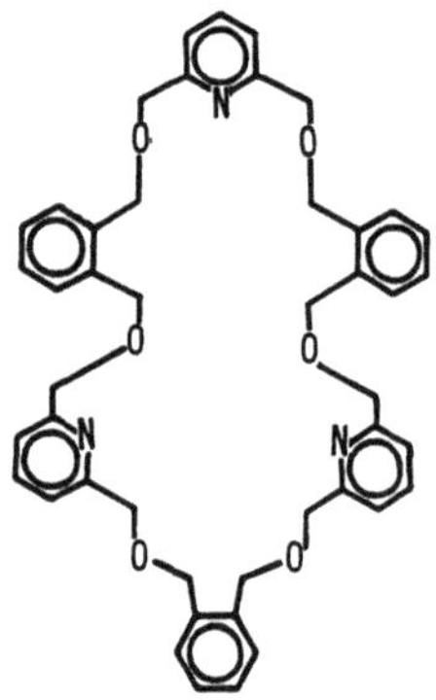	129–130	9	96
40	51	22	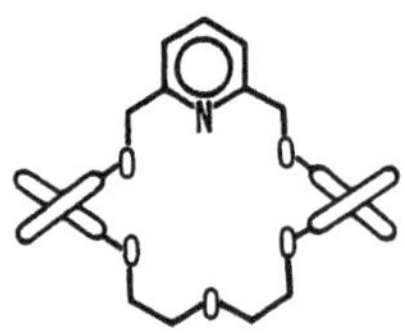 (SS)	foam	43	67
41	52	22	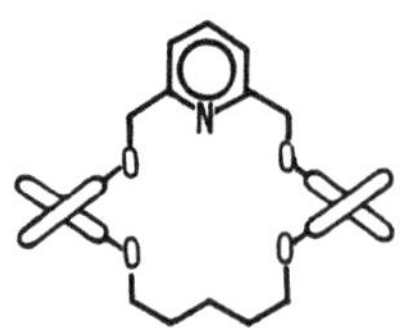 (SS)	foam	29	67

Table 4.16 (continued)

Cpd. No.	No. C's	Ring size	Structure		Mp or (bp) in °C	Yield (%)	Refs.
42	54	22		(SS)	288–292 dec.	26	67, 83
43	54	22	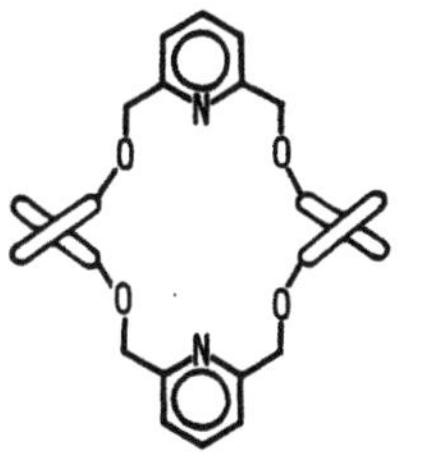		nr	nr	42
44	55	22	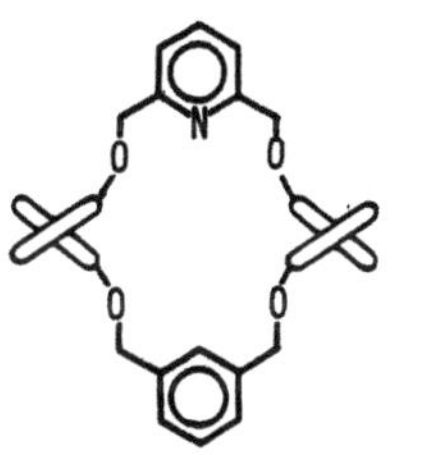	(SS)	foam	43	67
45	60	44	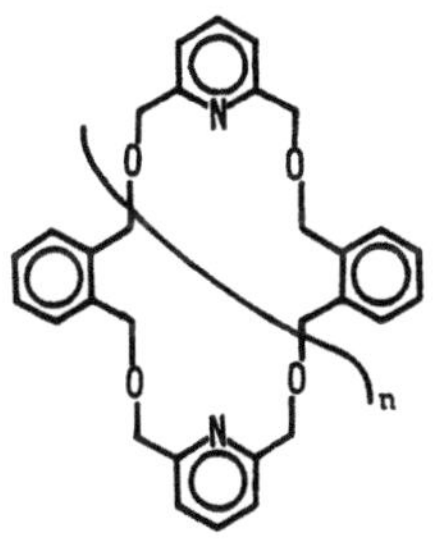	n = 3	108–109	nr	96
46	75	55		n = 4	104–105	nr	69

Table 4.17. Azacrowns containing the benzene-1,3-bis(aminomethylene) subunit

Cpd. No.	No. C's	Ring size	Structure	Mp or (bp) in °C	Yield (%)	Refs.
1	14	12	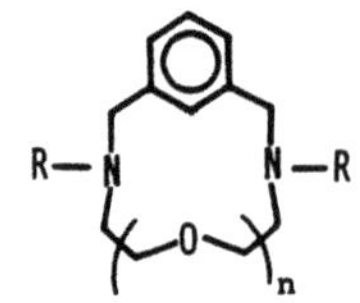 n = 1, R = CH_3	oil	44	71
2	14	15	n = 2, R = H	oil	28.5	72
3	16	15	n = 2, R = CH_3	oil	40	71
4	18	15	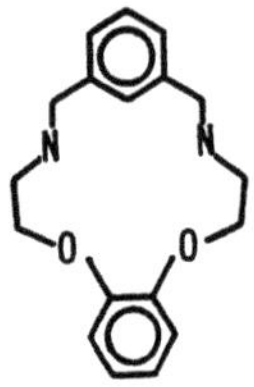	119–122	45	46
5	18	15	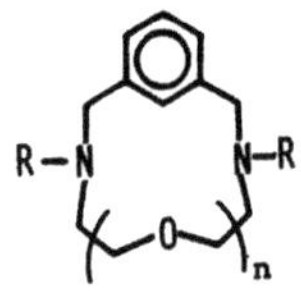 n = 2, R = CH_2CH_2OH	(200/0.04)	96	72
6	18	18	n = 3, R = CH_3	oil	22	71
7	20	15	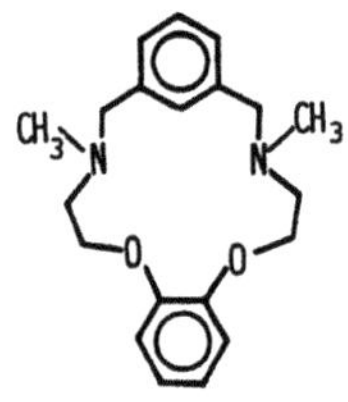	76–78	32	46
8	20	15	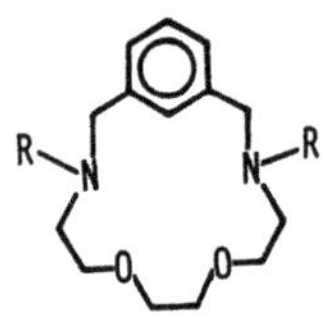 R = $CH_2C(=O)NHCH_3$	124–128	97	72
9	22	15	R = $CH_2C(=O)OCH_2CH_3$	(220/0.02)	85	72
10	22	15	R = $CH_2C(=O)N(CH_3)_2$	oil	58	72
11	26	15	R = CH_2– 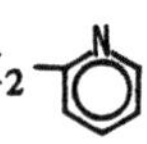	oil	55	72
12	30	15	R = $CHC(=O)NHCH(CO_2CH_3)CH(CH_3)_2$	76–77	80	72
13	32	15	R = $(CH_2)_8CH_3$	oil	84	72
14	32	15	R = $C(=O)(CH_2)_7CH_3$	oil	92	72

Table 4.18. Azacrowns containing the 1,3-pyridazine or 1,3,5-triazine subunit

Cpd. No.	No. C's	Ring size	Structure	Mp or (bp) in °C	Yield (%)	Refs.
1	8	11	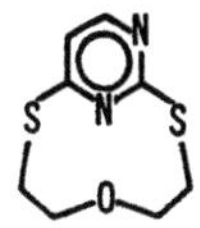	123–125	6	95
2	10	13	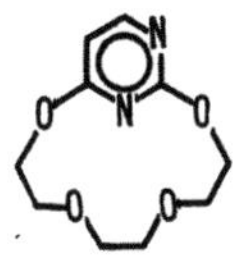	118–121	2	95
3	12	16	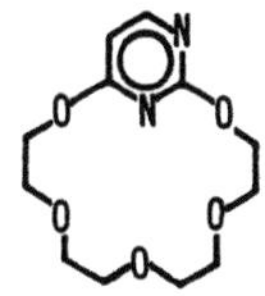	65–67	8	95
4	12	16	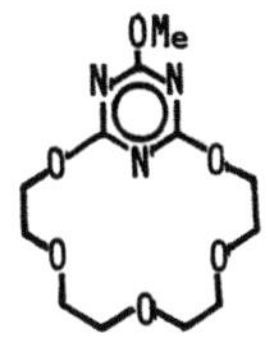	nr	nr	66
5	13	16	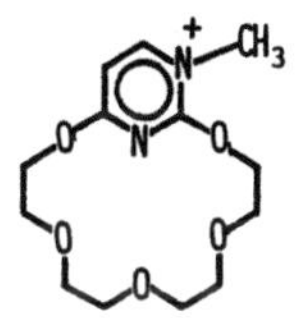	nr	nr	95
6	14	19	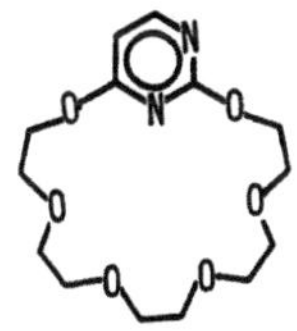	(178–182/1.0)	16	95
7	14	17	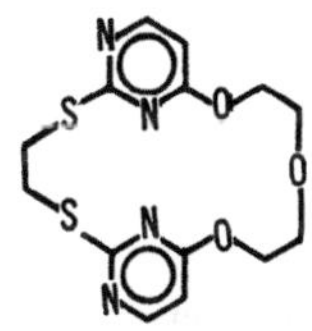	169–172	55	95
8	16	19	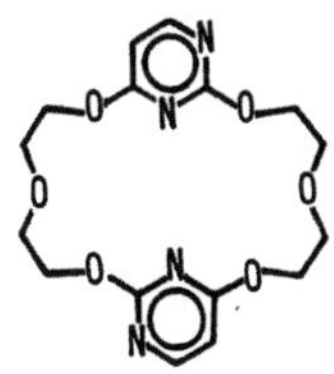	171–173	11	95

Table 4.18 (continued)

Cpd. No.	No. C's	Ring size	Structure		Mp or (bp) in °C	Yield (%)	Refs.
9	16	19	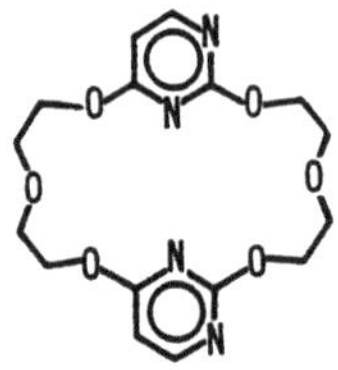		163–165	7	95
10	16	20	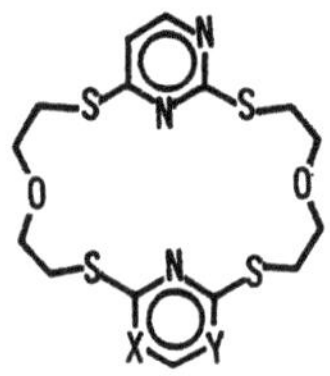	X ≠ Y = CH or N	159–160	8	95
11	16	22	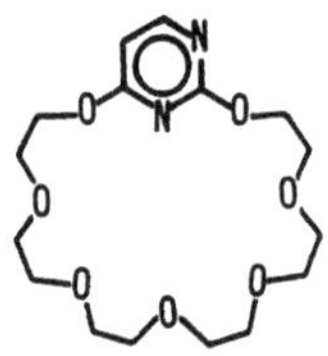		(165–169/0.5)	18	95
12	16	22	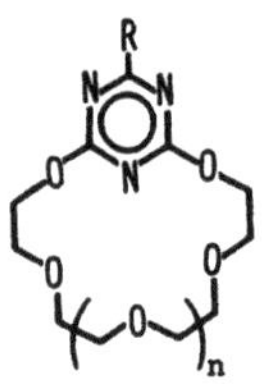	R = OCH_3 n = 3	nr	nr	66
13	17	16		R = OC_6H_{13} n = 1	nr	nr	60
14	19	16		R = C_8H_{17} n = 1	nr	nr	66
15	20	26	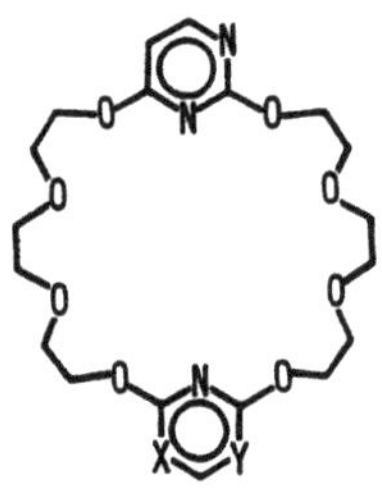	X ≠ Y = CH or N	131–133	5	95
16	21	22	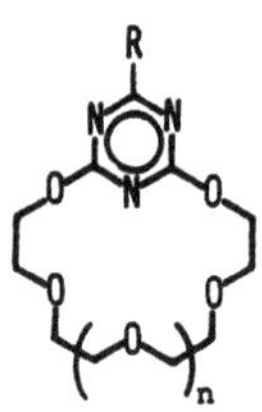	R = OC_6H_{13} n = 3	nr	nr	66
17	23	16		R = $C_{12}H_{25}$ n = 1	nr	nr	66
18	23	22		R = C_8H_{17} n = 3	nr	nr	66

Table 4.18 (continued)

Cpd. No.	No. C's	Ring size	Structure		Mp or (bp) in °C	Yield (%)	Refs.
19	24	30	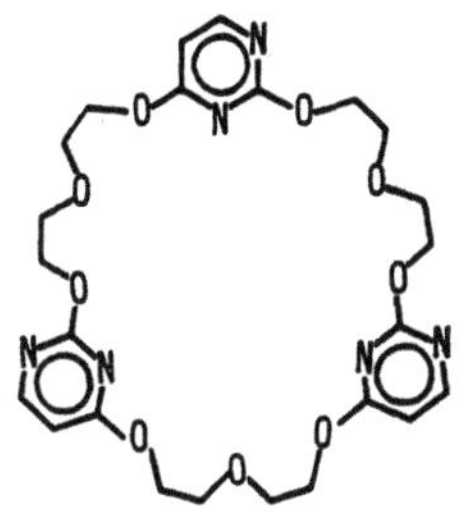		(151–156/0.1)	8	95
20	27	16	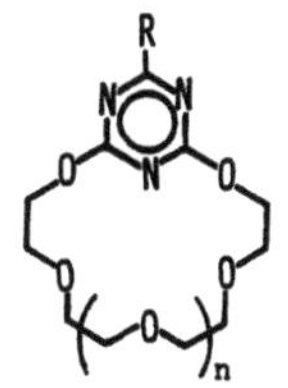	$R = C_{16}H_{33}$ n = 1	nr	nr	66
21	27	22		$R = C_{12}H_{25}$ n = 3	nr	nr	66
22	28	38		n = 4 X ≠ Y = CH or N	(200–203/1.0)	8	95
23	31	22	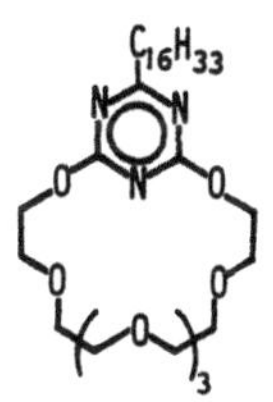		nr	nr	66

Table 4.19. Azacrowns containing the 2,2′-bipyridyl subunit

Cpd. No.	No. C's	Ring size	Structure	Mp or (bp) in °C	Yield (%)	Refs.
1	16	16	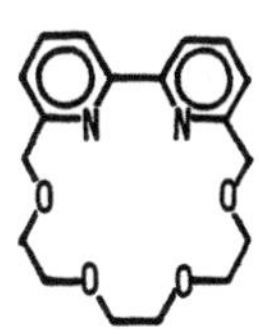	106–108	2	91

Table 4.19 (continued)

Cpd. No.	No. C's	Ring size	Structure	Mp or (bp) in °C	Yield (%)	Refs.
2	18	18	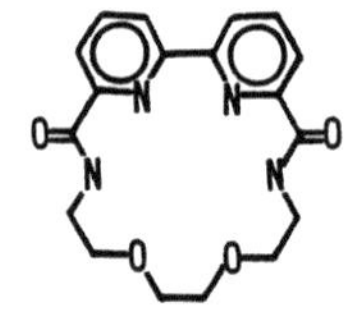	227–228	59	18
3	18	19	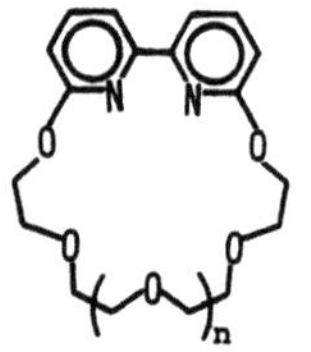 n = 1	99–101	24	91
4	20	22	n = 2	90–92	28	91
5	22	25	n = 3	41–43	30	91
				41–43	nr	85
6	24	27	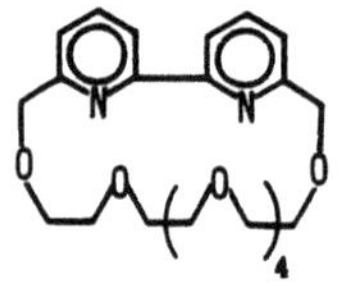	oil	21	91
7	28	26	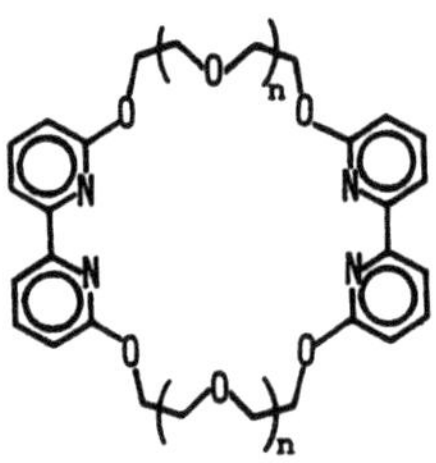 n = 1	103–105	7	91
8	32	32	n = 2	146–148	13	91
9	36	38	n = 3	121–122	15	91
10	40	44	n = 4	99–101	16	91
11	44	50	n = 5	70–72	20	91
12	48	54	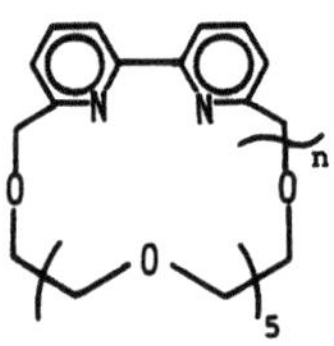 n = 2 (dimer)	nr	10	90
13	72	81	n = 3 (trimer)	nr	3	89

Table 4.20. Azacrowns containing two or more macrorings

Cpd. No.	No. C's	Ring size	Structure	Mp or (bp) in °C	Yield (%)	Refs.
1	24	15	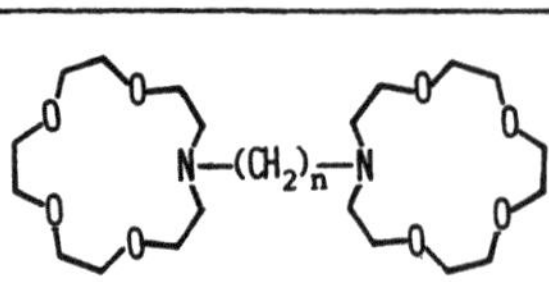 n = 4	(229–231/0.01)	80.5	49
2	25	15	n = 5	(235–242/0.01)	84	49
3	26	15	n = 6	(210–220/0.02)	81	49

Table 4.20 (continued)

Cpd. No.	No. C's	Ring size	Structure	Mp or (bp) in °C	Yield (%)	Refs.
4	27	15	n = 7	(245–248/0.05)	62	49
5	32	15	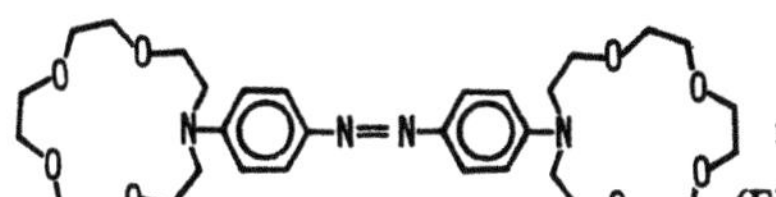 (E)	nr	65	32
6	32	15	monoazoxy derivative of 5 (above) (E)	145–151	nr	30
7	33	15	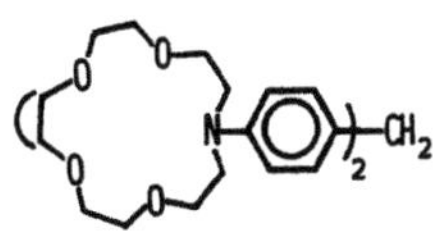	80–85	nr	30
8	35	15	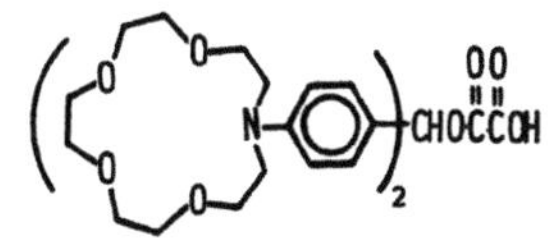	nr	nr	30
9	37	15	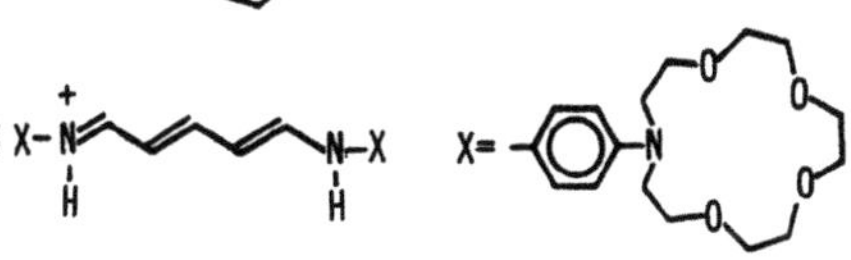	106	89	32
10	39	15	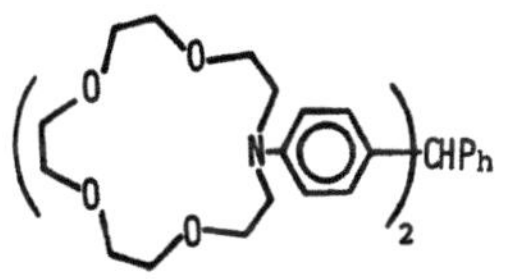	117–121	nr	30
11	39	15	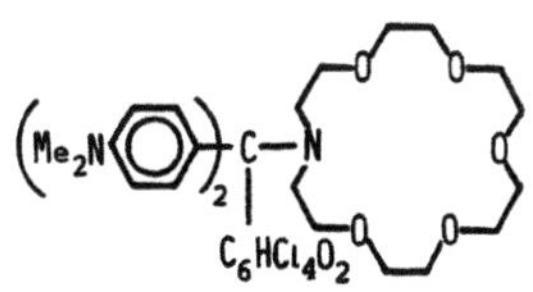	nr	nr	30
12	41	15	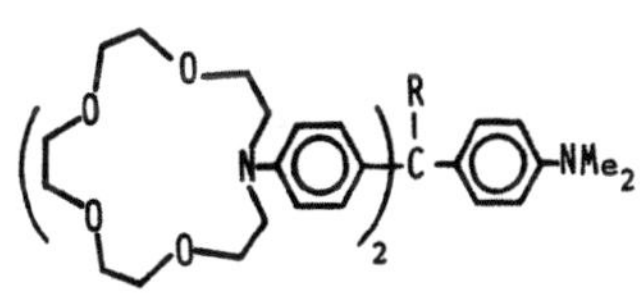 R = H	oil	nr	30
13	41	15	R = Cl	nr	nr	30
14	45	15	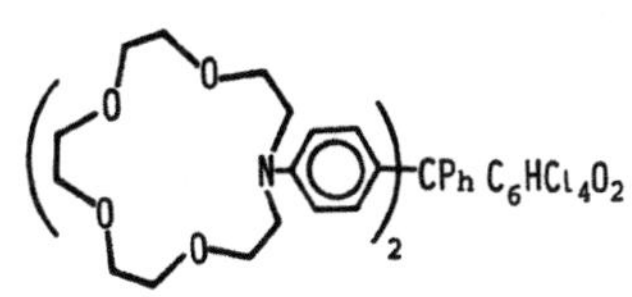	nr	nr	30

Table 4.20 (continued)

Cpd. No.	No. C's	Ring size	Structure	Mp or (bp) in °C	Yield (%)	Refs.
15	49	15	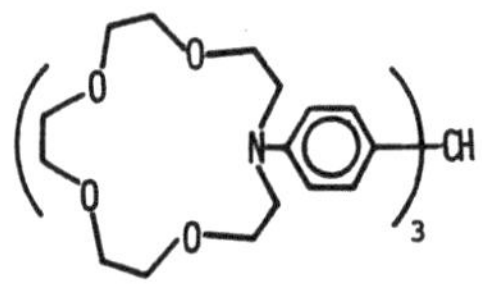	oil	nr	30
16	50	18	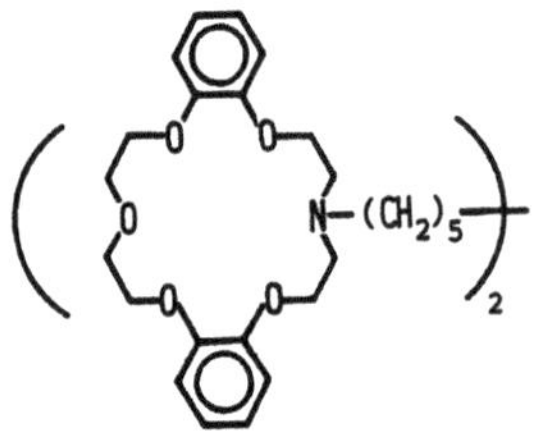	122–124	nr	98, 99
17	55	15	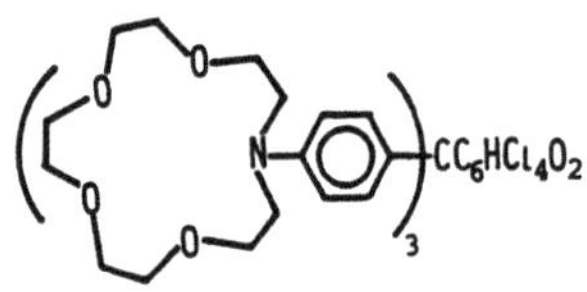	nr	nr	30

Table 4.21. Miscellaneous azacrowns

Cpd. No.	No. C's	Ring size	Structure	Mp or (bp) in °C	Yield (%)	Refs.
1	17	14	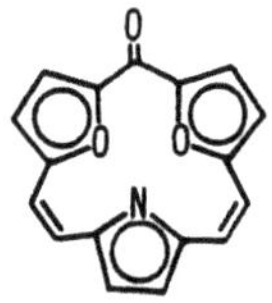	299–300	13.7	23
2	17	20	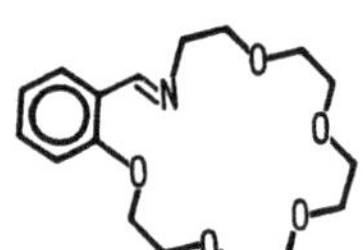	nr	nr	104
3	19	18	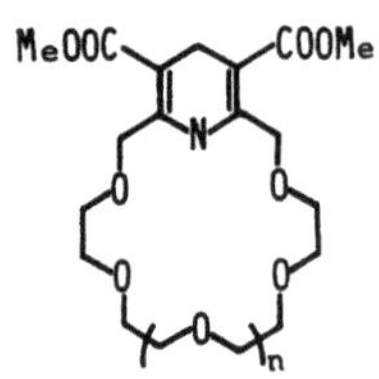 n = 1	115.5–116.5	11	54

Table 4.21 (continued)

Cpd. No.	No. C's	Ring size	Structure	Mp or (bp) in °C	Yield (%)	Refs.
4	21	18	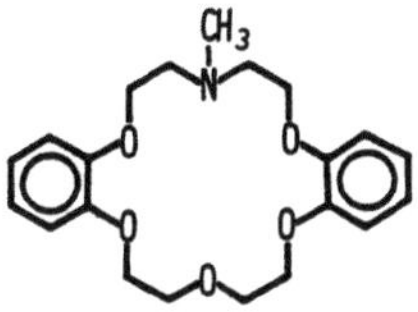	132–133	2	128
5	21	21	see structure 3, n = 2	92.5–93.5	10	54
6	22	20	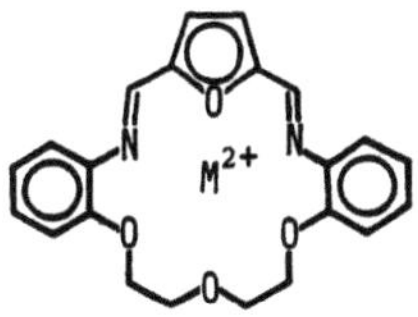 M = Ca, Sr, Ba	nr	nr	38
7	23	24	see structure 3, n = 3	nr	1.7	54
8	26	22	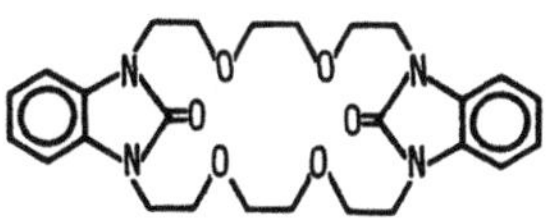	114	12	48
9	28	16	 A = N, B = O	230.5	15	14
10	28	16	see structure 9, A = O, B = N	260	69	14
11	33	11	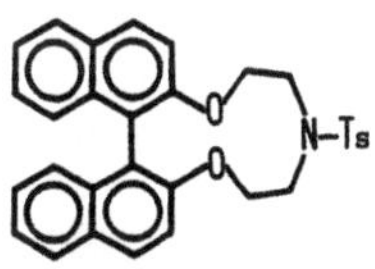	239–240	42	22
12	37	20	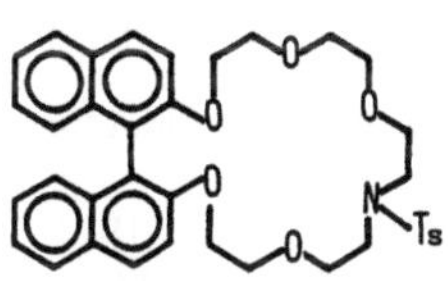	119–120	31	22
13	40	16	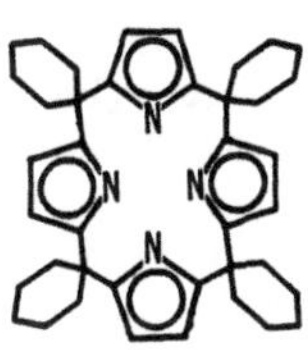	272–272.5	9	15

Table 4.21 (continued)

Cpd. No.	No. C's	Ring size	Structure	Mp or (bp) in °C	Yield (%)	Refs.
14	58	20	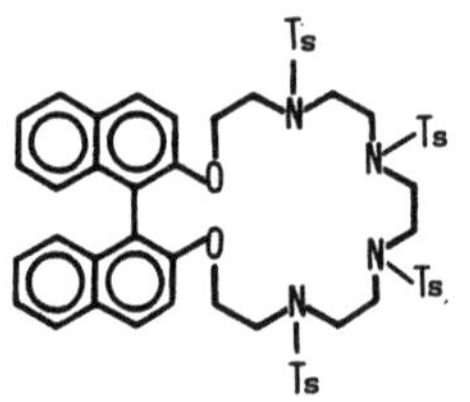	211–213	38	22

4.14 References

1. Adam, K. R., Lindoy, L. F., Smith, R. J., Anderegg, G., Henrick, K., McPartlin, M. and Tasker, P. A.: J. Chem. Soc., Chem. Commun. (1979) 812
2. Alcock, N. W. and Tasker, P. A.: J. Chem. Soc., Chem. Commun. (1972) 1239
3. Ando, N., Yamamoto, Y., Oda, J. and Inouye, Y.: Synthesis (1978) 688
4. Arnaud-Neu, F., Schwing-Weill, M. J., Louis, R. and Weiss, R.: Inorg. Chem. (1979) *18*, 2956
5. Atkins, T. J., Richman, J. E. and Oettle, W. F.: Org. Synth. (1978) 58
5a. Beckford, H. F., King, R. M. and Stoddart, J. F., Tetrahedron Letter (1978) 171
6. Bell, A. P. and Hall, C. D.: J. Chem. Soc., Chem. Commun. (1980) 163
7. Black, D. St. C. and McLean, I. A.: Chem. Commun. (1968) 1004
8. Black, D. St. C. and McLean, I. A.: Tetrahedron Letters (1969) 3961
9. Black, D. St. C. and McLean, I. A.: Aust. J. Chem. (1971) *24*, 1401
10. Blasius, E., Janzen, K. P. and Newmann, W.: Mikrochimica Acta (1977) *II*, 279
11. Bosnich, B., Poon, C. K. and Tobe, M. L.: Inorg. Chem. (1965) *4*, 1102
12. Boujlel, K. and Simonet, J.: Tetrahedron Letters (1979) 1497
13. Bovill, M. J., Chadwick, D. J. Johnson, M. R., Jones, N. F., Sutherland, I. O. and Newton, R. F.. J. Chem. Soc., Chem. Commun. (1979) 1065
14. Brown, W. H. and French, W. N : Can. J. Chem. (1958) *36*, 371
15. Brown, W. H., Hutchinson, B. J. and MacKinnan, M. H.: Can. J. Chem. (1971) *49*, 4017
16. Buhleier, E., Frensch, K., Luppertz, F. and Vögtle, F.: Liebigs Ann. Chem. (1978) 1586
17. Buhleier, E., Rasshofer, W., Wehner, W., Luppertz, F. and Vögtle, F.: Liebigs Ann. Chem. (1977) 1344
18. Buhleier, E., Wehner, W. and Vögtle, F.: Chem. Ber. (1978) *111*, 200
19. Cinquini, M., Montanari, F. and Tundo, P.: J. Chem. Soc., Chem. Commun. (1975) 393
20. Cinquini, M. and Tundo, P.: Synthesis (1976) 516
21. Clay, R. M., Micheloni, M., Paoletti, P. and Steele, W. V.: J. Am. Chem. Soc. (1979) *101*, 4120
22. Cram, D. J.: Ger. Offen. 2,414,188, 21 Nov. 1974
23. Cresp, T. M. and Sargent, M. V.: J. Chem. Soc. Perkin Trans. I (1973) 1786
24. Curry, J. D. and Busch, D. H.: J. Am. Chem. Soc. (1964) *86*, 592
25. DeSarlo, F., Guarno, A. and Speroni, G. P.: J. Chem. Soc., Chem. Commun. (1977) 549
26. Dietrich, B., Lehn, J. M. and Sauvage, J. P.: Tetrahedron Letters (1969) 2885
27. Dietrich, B., Fyles, T. M., Lehn, J. M., Pease, L. G. and Fyles, D. L.: J. Chem. Soc., Chem. Commun. (1978) 934
28. Dishong, D. M., Schultz, R. A. and Gokel, G. W.: unpublished (1980)
29. Drew, M. G. B., bin Othman, A. H., Mellroy, P. D. A. and Nelson, S. M.: J. Chem. Soc., Dalton (1975) 2507

30. Dix, J. P. and Vögtle, F.: Angew. Chem. Int. Ed. Engl. (1978) *17*, 857
31. Dix, J. P. and Vögtle, F.: Angew. Chem. (1978) *90*, 893
32. Dix, J. P. and Vögtle, F.: Chem. Ber. (1980) *113*, 457
33. Ehrensperger, C. P., Heberlein, M. and Skrabal, P.: Helv. Chim. Acta (1978) *61*, 2813
34. Ekstrom, A., Lindoy, L. F. and Smith, R. J.: J. Am. Chem. Soc. (1979) *101*, 4014
35. Elben, U., Fuchs, H. B., Frensch, K. and Vögtle, F.: Liebigs Ann. Chem. (1979) 1102
36. Elben, U. and Vögtle, F.: J. Chem. Res. (S) (1978) 316
37. Elben, U. and Vögtle, F : J. Chem. Res. (M) (1979) 3791
38. Fenton, D. E., Cook, D. H., Nowell, I. W. and Walker, P. E.: J. Chem. Soc., Chem. Commun. (1977) 623
39. Girodeau, J. M., Lehn, J. M. and Sauvage, J. P.: Angew. Chem. Int. Ed. Engl. (1975) *14*, 764
40. Gokel, G. W. and Garcia, B. J.: Tetrahedron Letters (1977) 317
41. Gokel, G. W. and Diamond, C. J.: unpublished (1980)
42. Gokel, G. W., Timko, J. M. and Cram, D. J.: J. Chem. Soc., Chem. Commun. (1975) 444
43. Grimsley, P. G., Lindoy, L. F., Lip, H. C., Smith, R. J. and Baker, J. T.: Aust. J. Chem. (1977) *30*, 2095
43a. Greene, R. N.: Tetrahedron Letters (1972) 1793
44. Hediger, M. and Kaden, T. A.: J. Chem. Soc., Chem. Commun. (1978) 14
45. Hodgkinson, L. C., Leigh, S. J. and Sutherland, I. O.: J. Chem. Soc., Chem. Commun. (1976) 639
46. Hodgkinson, L. C. and Sutherland, I. O.: J. Chem. Soc. Perkin. I (1979) 1908
47. Hodgkinson, L. C., Johnson, M. R., Leigh, S. J., Spencer, N., Sutherland, I. O. and Newton, R. F.: J. Chem. Soc. Perkin. I (1979) 2193
47a. Hogberg, S. A. G. and Cram, D. J.: J. Org. Chem. (1975) *40*, 151
48. Htay, M. M. and Meth-Cohn, O.: Tetrahedron Letters (1976) 79
49. Johnson, M. R., Sutherland, I. O. and Newton, R. F.: J. Chem. Soc., Chem. Commun. (1979) 306
50. Johnson, M. R., Sutherland, I. O. and Newton, R. F.: J. Chem. Soc., Chem. Commun. (1979) 309
51. Johnson, M. R., Sutherland, I. O. and Newton, R. F.: J. Chem. Soc. Perkin I (1979) 357
52. Kimura, E., and Yatsunami, T.: Chem. Pharm. Bull. (1980) *28*, 994
53. Johnson, M. R., Sutherland, I. O. and Newton, R. F.: J. Chem. Soc. Perkin. Trans. I (1980) 586
54. Kellogg, R. M., Van Bergen, T. J., Van Doren, H., Hedstrand, D., Kool, J., Kruizinga, W. H. and Troostwijk, C. B.: J. Org. Chem. (1980) *45*, 2854
55. King, A. P. and Krespan, C. G.: J. Org. Chem. (1974) *39*, 1315
56. Kossai, R., Simonet, J. and Jeminet, G.: Tetrahedron Letters (1979) 1059
57. Kotzyba-Hibert, F., Lehn, J. M. and Vierling P.: Tetrahedron Letters (1980) *21*, 941
58. Kramer, U., Guggisberg, A. and Hesse, M.: Helv. Chim. Acta (1979) *62*, 2317
59. Krespan, C. G.: J. Org. Chem. (1975) *40*, 1205
60. Krespan, C. G.: U.S. Patent 3,952,015, 20 Apr. 1976
61. Kulstad, S. and Malmsten, L. A.: Acta Chem. Scand., Ser. B (1979) *B33*, 469
62. Kulstad, S. and Malmsten, L. A.: Tetrahedron Letters (1980) *36*, 521
63. Kulstad, S. and Malmsten, L. A.: Tetrahedron Letters (1980) *21*, 643
64. Kuo, P. L., Miki, M., Ikeda, I. and Okahara, M.: Tetrahedron Letters (1978) 4273
65. Kuo, P. L., Miki, M., Ikeda, I. and Okahara, M.: J. Am. Oil Chem. Soc. (1980) 227
66. Kuwamura, T. and Yoshida, S.: Nippon Kagaku Kaishi (1980) 427
67. Kyba, E. P., Gokel, G. W., deJong, F., Koga, K., Sousa, L. R., Siegel, M. G., Kaplan, L., Sogah, G. D. Y. and Cram, D. J.: J. Org. Chem. (1977) *42*, 4173
68. Laidler, D. A. and Stoddart, J. F.: J. Chem. Soc., Chem. Commun. (1976) 979
69. Lehn, J. M.: U.S. Patent 3,888,877, 10 June 1980
70. Lehn, J. M. and Vierling, P.: Tetrahedron Letters (1980) 1323
71. Leigh, S. J. and Sutherland, I. O.: J. Chem. Soc., Chem. Commun. (1975) 414
72. Leigh, S. J. and Sutherland, I. O.: J. Chem. Soc. Perkin I (1979) 1089
73. Liang, B. F., Margerum, D. W. and Chung, C. S.: Inorg. Chem. (1979) *18*, 2001
74. Lockhart, J. C. and Thompson, M. E.: J. Chem. Soc. Perkin Trans. I (1977) 202
75. Lockhart, J. C., Robson, A. C., Thompson, M. E., Furtado, D., Kaura, C. K. and Allan, A. R.: J. Chem. Soc. Perkin I (1973) 577

76. Margulis, T. N. and Zompa, L. J.: J. Chem. Soc., Chem. Commun. (1979) 430
77. Mathias, L. J. Al-Jumah, K.: Organic Coatings and Plastics Chemistry, Vol. 42, ACS (1980) 262
78. Metcalfe, J. C., Stoddart, J. F. and Jones, G.: J. Am. Chem. Soc. (1977) *99*, 8317
79. Mitsubishi Chemical Industries, Japan Kokai 78:91,327, 26 July 1978
80. Mitsubishi Chemical Industries, Japan Kokai 78:91,328, 9 Feb. 1980
81. Mitsubishi Chemical Industries, Japan Kokai 78:91,331, 26 July 1978
82. Okahara, M.: Japan Kokai 78:33,319, 22 Mar. 1978
83. Newcomb, M., Gokel, G. W. and Cram, D. J.: J. Am. Chem. Soc. (1974) *96*, 6810
84. Newcomb, M., Timko, J. M., Walba, D. M. and Cram, D. J.: J. Am. Chem. Soc. (1977) *99*, 6392
85. Newkome, G. R. and Hager, D. C.: J. Am. Chem. Soc. (1978) *100*, 5567
86. Newkome, G. R. and Kawato, T.: J. Org. Chem. (1979) *44*, 2693
87. Newkome, G. R. and Kawato, T.: J. Am. Chem. Soc. (1979) *101*, 7088
88. Newkome, G. R. and Kawato, T.: Tetrahedron Letters (1978) 4643
89. Newkome, G. R., Kawato, T. and Nayak, A.: J. Org. Chem. (1979) *44*, 2697
90. Newkome, G. R., Kohli, D. K. and Fronczek, F.: J. Chem. Soc., Chem. Commun. (1980) 9
91. Newkome, G. R., Nayak, A., Fronczek, F., Kawato, T., Taylor, H. C. R., Meade, L. and Mattice, W.: J. Am. Chem. Soc. (1979) *101*, 4472
92. Newkome, G. R., Nayak, A., Sauer, J. D., Mattschei, P. K., Watkins, S. F., Fronczek, F. and Benton, W. H.: J. Org. Chem. (1979) *44*, 3816
93. Newkome, G. R., McClure, G. L., Simpson, J. B. and Danesh-Khoshboo, F.: J. Am. Chem. Soc. (1975) *97*, 3232
94. Newkome, G. R., Nayak, A., McClure, G. L., Danesh-Khoshboo, F. and Broussard-Simpson, J.: J. Org. Chem. (1976) *42*, 1500
95. Newkome, G. R., Nayak, A., Otemaa, J., Van, D. A. and Benton, W. H.: J. Org. Chem. (1978) *43*, 3362
96. Newkome, G. R. and Robinson, J. M.: J. Chem. Soc., Chem. Commun. (1973) 831
97. Newkome, G. R., Sauer, J. D., Mattschei, P. K. and Nayak, A.: Heterocycles (1978) *9*, 1555
98. Pedersen, C. J. and Bromels, M. H.: U.S. Patent 3,847,949, 12 Nov. 1974
99. Pedersen, C. J. and Bromels, M. H.: U.S. Patent 4,031,111, 21 June 1977
100. Pelissard, D. and Louis, R.: Tetrahedron Letters (1972) 4589
101. Petranek, J. and Ryba, O.: Tetrahedron Letters (1977) 4249
102. Richman, J. E.: Tetrahedron Letters (1977) 559
103. Richman, J. E. and Atkins, T. J.: J. Am. Chem. Soc. (1974) *96*, 2268
104. Rasshofer, W., Müller, W. M., Oepen, G. and Vögtle, F.: J. Chem. Research (S) (1978) 72; J. Chem. Research (M) (1978) 1001
105. Rasshofer, W. and Vögtle, F.: Liebigs Ann. Chem. (1978) 552
106. Rasshofer, W. and Vögtle, F.: Liebigs Ann. Chem. (1977) 1340
107. Rasshofer, W., Wehner, W. and Vögtle, F.: Liebigs Ann. Chem. (1976) 916
108. Sabatini, L. and Fabbrizzi, L.: Inorg. Chem. (1979) *18*, 438
109. Samuels, G. J. and Espenson, J. H.: Inorg. Chem (1979) *18*, 2587
110. Shaw, B. L.: J. Am. Chem. Soc. (1975) *97*, 3856
111. Sieger, H. Vögtle, F.: Liebigs Ann. Chem. (1980) 425
112. Tabushi, I. and Fujiyoshi, M.: Heterocycles (1977) 7, 851
113. Tabushi, I., Kobuke, Y. and Kato, H.: Jpn. Kokai Tokkyo Koho: 79,130,593, 9 Oct. 1979
114. Tabushi, I., Kuroda, H., Kato, H., Hirai, T. and Okino, H.: Japan Kokai 77: 106,882, 07 Sep. 1977
115. Tabushi, I., Okino, H. and Kuroda, Y.: Tetrahedron Letters (1976) 4339
116. Takagi, M., Tazaki, M. and Ueno, K.: Chem. Letters (1978) 1179
117. Takagi, M., Tazaki, M. and Ueno, K.: Chem. Letters (1978) 1179
118. Tomoi, M., Abe, O., Ikeda, M., Kihara, K. and Kakiuchi, H.: Tetrahedron Letters (1978) 3031
119. Vartanyan, S. A., Akopyan, T. R. and Paronikyan, E. G.: Arm. Khim. Zh. (1978) *31*, 349
120. Vögtle, F., Muller, W. M. and Weber, E.: Chem. Ber. (1980) *113*, 1130
121. Vögtle, F. and Newmann, P.: Tetrahedron Letters (1970) 115
122. Vögtle, F. and Weber, E.: Angew. Chem. (1974) *86*, 126

123. Von Deuten, K., Knoechel, A., Kopf, J., Oehler, J. and Rudolph, G.: J. Chem. Res., Synop. (1979) 358
124. Vögtle, F. and Dix, P.: Liebigs Ann. Chem. (1977) 1698
125. Vögtle, F. and Elben, U.: Chem. Ber. (1978) *111*, 1434
126. Vögtle, F., Weber, E., Wehner, W., Nätscher, R. and Grütze, J.: Chem. Zeitung (1974) *98*, 562
127. Weber, E.: Angew. Chem. Int. Ed. Engl. (1979) *18*, 219
128. Weber, E. and Vögtle, F.: Chem. Ber. (1976) *109*, 1803
129. Weber, E. and Vögtle, F.: Liebigs Ann. Chem. (1976) 891
130. Wudl, F. and Gaeta, F.: J. Chem. Soc., Chem. Commun. (1972) 107
131. Yamashita, T., Nakamura, H., Takagi, M. and Ueno, K.: Bull. Chem. Soc. Jpn. (1980) *53(6)*, 1550–4
132. Zeigerson, E., Ginzburg, G., Schwartz, N., Luv, Z. and Meyerstein, D.: J. Chem. Soc. (1979) 241

Chapter 5

Crown Esters and Macrocyclic Polyether Lactones

5.1 Introduction

In 1930, in one of many pioneering studies, Carothers showed that certain condensation polymers could be cyclic or macrocyclic rather than exclusively linear[21]. In addition to this very important observation, he showed that thermolysis *in vacuo* of certain polymers could also yield macrocyclic materials. Quite a number of papers have been published on this subject since that time, especially dealing with the chemistry of phthalate[15,22,24,25,26,27,30], isophthalate[5,10,15,44], and terephthalate esters[15,32,33,34,49,62,64,77]. Many of these structures are tabulated at the end of this chapter.

5.2 General Ester Synthesis

Despite this considerable literature noted above, it was really Bradshaw and coworkers who pioneered the synthesis of macrocyclic ester analogs of crown ethers[8–15,31,36–38,44,47,69]. Bradshaw and his coworkers have listed several motivations for their explorations in this area. "One objective of [the] research program is to prepare and study a series of multidentate compounds which resemble naturally occurring macrocyclic compounds"[3]. Further, Bradshaw and his coworkers have said that "it is our hope that we can prepare macrocycles to mimic the selectivities of the naturally occurring cyclic antibiotics and thereby make available models for the investigation of biological cation transportation and selectivity processes"[31]. These workers have presented a number of comparisons with valinomycin[37]. The other expressly stated goal of their research "is to prepare molecules which will allow us to systematically examine the parameters which affect complex stability and to understand that stability in terms of ΔH and $T\Delta S$ values for complex formation".

These authors have prepared quite a variety of crown-related esters and have incorporated numerous heterocyclic subunits in the macrorings. The structural variety can be gleaned from a perusal of the tables at the end of this chapter. Despite this variety, one basic approach has been utilized in most of the syntheses thus far presented. This method involves simultaneous addition of separate benzene solutions of the diacyl halide and diol to an additional several volumes of benzene heated at 45–60°. After the addition is complete, the solution is maintained at that temperature for 2–3 days. In what appears to be the preponderance of work from this group, the time is 48 hours and the temperature is kept between 45–50°. An exception to this is found in the synthesis of the pyridyl-diester, preparation of which is illustrated in Eq. (5.1), below. In this reaction, a 1 : 1 mixture of THF and benzene were used rather than pure benzene, although the import of this alteration is not obvious[35].

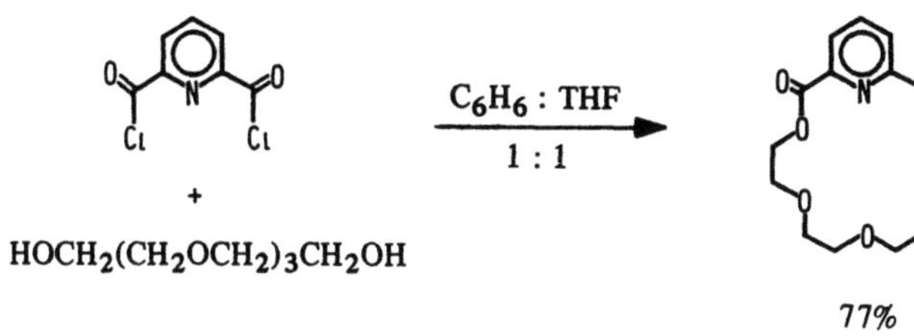

(5.1)

Frensch and Vögtle[32,33] have utilized a somewhat more complex apparatus in the preparation of certain compounds identical to those prepared by Bradshaw and coworkers[15]. In fact it was the report of one such similar compound which produced a minor controversy. Vögtle and Frensch condensed terephthaloyl chloride and tetraethylene glycol and reported the preparation of compound *1*, shown below. This is the 1 : 1 adduct of the two starting materials, sometimes referred to as "monomer". This is to be contrasted with the 2:2 compound, often referred to as "dimer" (*2*, below). When Bradshaw and Thompson[15] treated terephthaloyl chloride with tetraethylene glycol by the approach described above, they reported isolation of the dimer, mp 93.5–95.0° in 1% yield. The identity of this material as dimer was confirmed by molecular weight determination. The dimer has a melting point essentially identical to that reported by Frensch and Vögtle[32] for the monomer. Recently, Frensch and Vögtle have reported the conduct of this reaction at high dilution from which they were able to isolate both the monomer and the dimer[33]. The molecular weights of the two compounds were confirmed by vapor phase osmometry. It was found that the respective monomer and dimer melting points were 98–99° and 96°. The respective yields of these two compounds were 13% and 20% when conducted on a 10 mmole scale. The ^{1}H-NMR spectra of the two compounds were significantly different. The reaction is illustrated in Eq. (5.2).

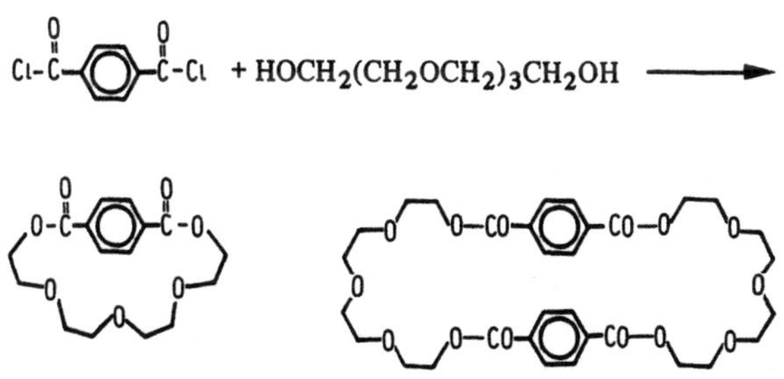

1 *2*

(5.2)

5.3 Macrocyclic Ester Properties

Certain observations concerning the compounds themselves rather than the syntheses *per se* are interesting and merit mention here. In particular, it is know from Cram's extensive work (see Sect. 3.12) that complexation of primary ammonium ions is favored by 18-crown-6 systems. Bradshaw and coworkers found that when a heterocycle incorporated in the ring system was a poor donor group, ammonium ion complexation was favored by larger ring sizes[10]. The implication of this is that the ammonium ion is utilizing the aliphatic ring oxygens and is not binding in the centrosymmetric fashion normally associated with 18-crown-6 systems[10]. Such poor donor groups include the 2,5-furano-(3) and 1,3-benzo-(4) units shown below.

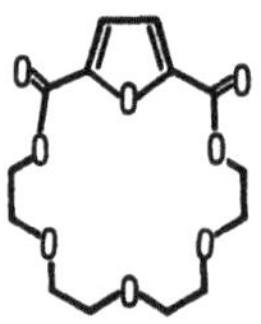

3

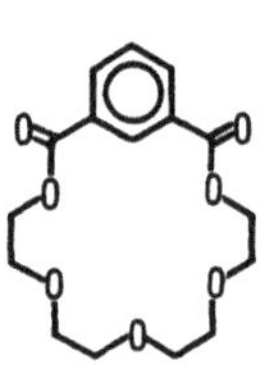

4

One heterocyclic subunit which is an excellent donor group is the 2,6-pyrido-substitutent[52–54]. However, when 4-chloro-2,6-dicarboxypyridine was utilized as precursor to the macrocycle, the product apparently was less stable than was the corresponding nonchlorinated ester[13]. The decomposition was apparently initiated by complexation of either Sr^{2+} or Ba^{2+} but the mechanism by which this degeneration occurred was not suggested nor is it obvious to the present authors, especially in the absence of any commentary concerning the products of decomposition[13]. The synthesis of the apparently unstable system is illustrated below.

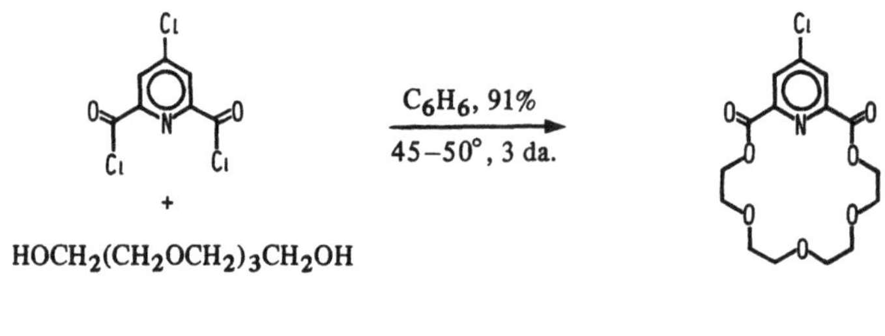

5.4 One-pot Macrocyclic Lactone Synthesis

Okahara and his coworkers have made a number of contributions to the synthesis of crown ethers using a "one-pot" method (see Sect. 3.13). These methods have been applied largely to the preparation of simple aliphatic crown ether systems. In addition, this group has prepared macrocyclic ester compounds using a one-pot procedure. Although this approach is somewhat limited in applicability, it is certainly worthy of note for the preparation of macrocyclic lactones[48].

In this method, an oligoethylene glycol is treated with 0.3 eq. of sodium bromoacetate in methanol. The resulting monocarboxylic acid salt is then treated with toluenesulfonyl chloride, sodium carbonate and dioxane. After heating the reaction mixture at 50° for an hour, the product was obtained either by Kugelrohr distillation or extraction. The two compounds produced in the reaction below (i.e., n = 1 and n = 2) were formed in 38% and 42% yields respectively[48].

These authors suggest that the reaction "may proceed through the formation of their mixed anhydrides", and illustrate a metal ion templated, mixed carboxylic-sulfonic anhydride.

$$HOCH_2(CH_2OCH_2)_{n+3}OH + BrCH_2CO_2Na \xrightarrow{MeOH} HOCH_2(CH_2OCH_2)_{n+3}OCH_2CO_2Na$$

$$\xrightarrow[\text{dioxane, 1 h, 50°}]{\text{TsCl, Na}_2\text{CO}_3}$$

(5.4)

5.5 The Hantzsch Condensation

Van Bergen and Kellogg[70] have reported utilization of the Hantzsch condensation as a route to certain pyridine-containing esters which have been of interest in the latter's group. The approach involves forming the bis-acetoacetate ester of an oligoethylene glycol followed by treatment with ammonium carbonate and aqueous formaldehyde solution. In a typical reaction, illustrated below (Eq. 5.5), the diester of triethylene glycol was converted into the dilactone-crown in 25% yield, after dehydrogenation of the intermediate using N_2O_3. The "dimers" are reported to form in these reactions in "13–32% yields". The authors "consider it possible that the excess of ammonium ion acts as a template in the condensation to bring the acetoacetate into closer proximity to enhance intramolecular ring closure". It is not clear exactly when the template effect occurs, however, in this reaction which these workers refer to as "mechanistically complex"[70].

It should be noted that compounds of this type[52–54] have been of interest as models for hydride transfer reductions in biological systems. Van Bergen and Kellogg state clearly their hope, if not expectation, that when ions such as H^+, Zn^{2+}, Mg^{2+}, etc. are complexed "in the vicinity of the donor, beneficial catalytic effects might be forthcoming"[70].

$$CH_3\overset{O}{\overset{\|}{C}}CH_2\overset{O}{\overset{\|}{C}}OCH_2(CH_2OCH_2)_nCH_2O\overset{O}{\overset{\|}{C}}CH_2\overset{O}{\overset{\|}{C}}CH_3 \xrightarrow[\text{2. } N_2O_3]{\text{1. } (NH_4)_2CO_3,\ CH_2O}$$

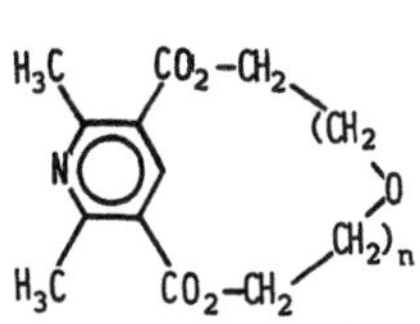

+ dimer

(5.5)

Similar compounds to those described above have been prepared by the same group using the bis-cesium salts of pyridine-3,5-dicarboxylic acids. In a number of experiments Piepers and Kellogg showed that "Cs^{2+} plays an irreplacable role as compared with Na^+, K^+ and Rb^+ for the formation of [the macrocycle shown below]"[60].

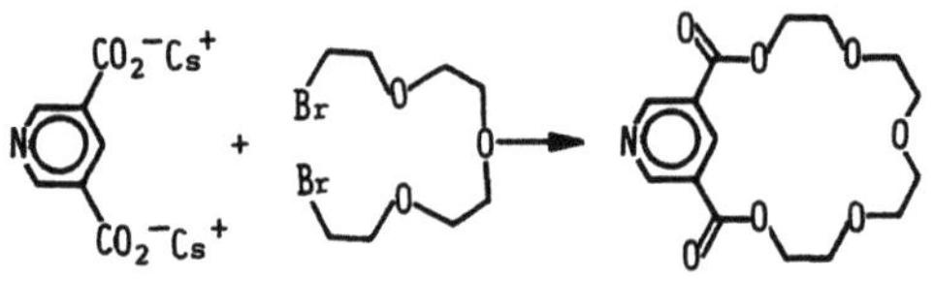

85%

(5.6)

It should be stressed that there has been considerable interest in macrocycles containing the pyridine subcyclic unit, especially by Newkome and coworkers[52–54] (see also Chap. 4). Most of his interest appears to have been focussed on polyethers rather than ester compounds. Bradshaw has also reported syntheses of compounds containing the pyridine subcyclic unit[13].

5.6 Antibiotic Models

In light of the considerable recent interest in the macrocyclic polyether-polyester antibiotics, Samat and coworkers have prepared a large-ring macrocycle as a model for the nonactine series[65]. The synthetic approach is more or less standard, involving a double nucleophilic substitution. Unlike some of the more recent approaches to macrocyclic ester compounds, these reactions were conducted in DMF rather than benzene for 22 h rather than 48 h, and at 70–80° rather than 45–50°. The overall yield is reported to be 30%. Unfortunately, it is not clear whether the reaction was conducted under high dilution conditions[65]. No report of the potential antibiotic activity of these compounds has yet appeared.

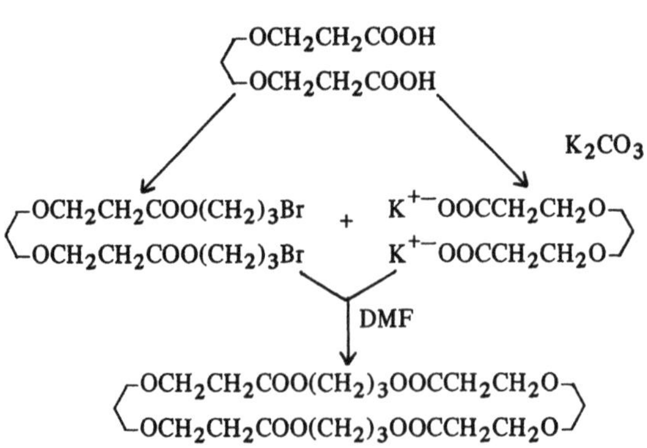

30%

(5.7)

5.7 Stannoxanes as Covalent Templates

Shanzer and Mayer-Shochet[66] have utilized disubstituted stannoxanes as "covalent templates" for the formation of tetralactones. In this procedure, a dialkyl stannoxane is condensed with a glycol to form a cyclic-distannoxane as illustrated below. In a typical preparation, the distannoxane (*5*) was dissolved in CCl_4 (ca. 0.04 *M*) at reflux and sebacoyl chloride (2 eq) was added dropwise at reflux. After heating for 20 h, the solution was concentrated to a solid. Chromatography afforded the tetralactone in 30% yield[66].

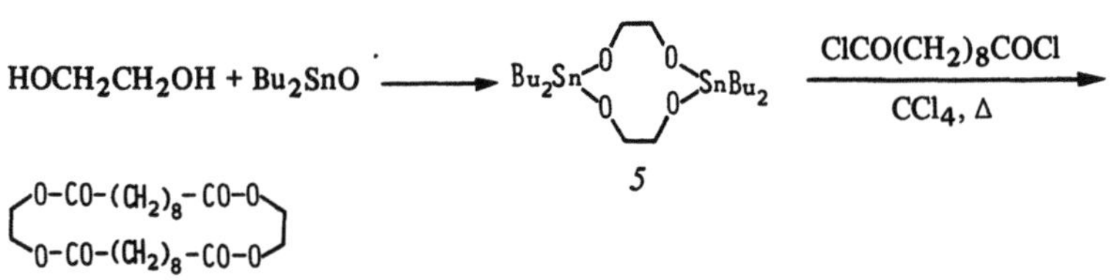

5

(5.8)

Although several examples are presented in this paper, the method is too new for much literature to have yet accumulated.

5.8 Macrocyclic Esters by Ozonolysis

From their structures, it appears that the hydrolytic stability of macrocyclic lactones must necessarily be inferior to macrocyclic polyethers. Ease of synthesis of the cyclic esters is therefore one of the aspects which commend them to interest. It is probably for this reason that such lactones have not been made more often by the interesting approach of Kögel and Schröder[40]. These workers report the ozonolysis of dibenzo-18-crown-6 in a mixture of methanol and dichloromethane at –20°. Reduction of the ozonide at –75° using dimethylsulfide followed by warming and addition of acetone led to formation of *6* in 14% yield. The bis-oxalate had mp 164–165° from acetone, very similar to that of the starting crown. The transformation is illustrated below in Eq. (5.9).

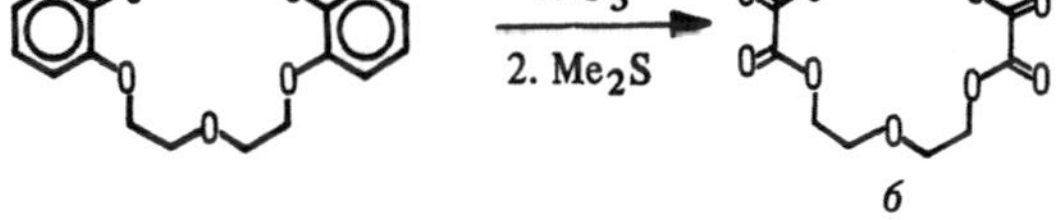

6

(5.9)

5.9 Photochemical Macrocyclization

An interesting footnote to the general synthesis of macrocyclic crown ether-type esters is found in the work of Ors and Srinivasan[55,56]. These workers used cinnamate esters and diesters as precursors to large ring lactones. Depending on orientation, either truxillic or truxinic acids may result, and these authors report a variety of products. The compound resulting from this work which most closely resembles a crown ether is shown below.

Such a structure might well exhibit significant cation binding but this possibility does not appear to have been investigated.

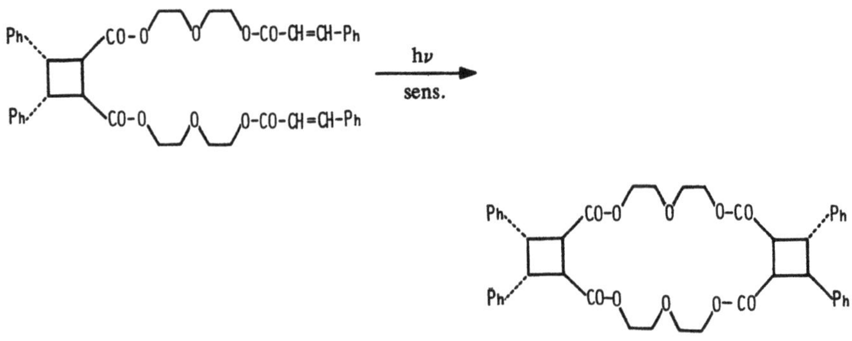

5.10 Listing of Tables

5.11 Tables 5.1–5.14

Table 5.1. Macrocyclic oxalate esters

Cpd. No.	No. C's	Ring size	Structure		Mp or (bp) in °C	Yield (%)	Refs.
1	10	15	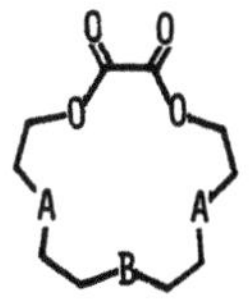	A = B = O	45–46	28	31
2	10	15		A = O, B = S	88–90	6	31
3	10	15		A = S, B = O	45–47	8	31
4	12	18	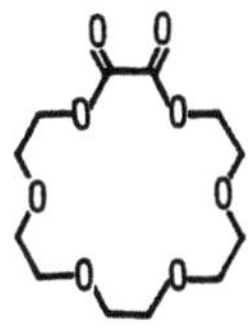		84–85	55	31
5	12	18	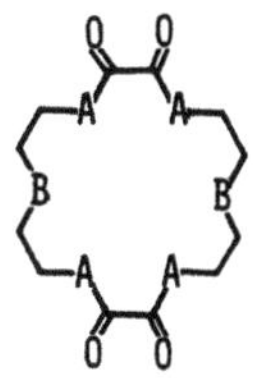	A = B = O	162.5–164 164–165	4 14	31 40
6	12	18		A = O, B = S	195–203	14	31
7	12	18		A = S, B = O	208–210	67	31
8	12	18		A = B = S	108–109 dec.	62	31
9	14	21	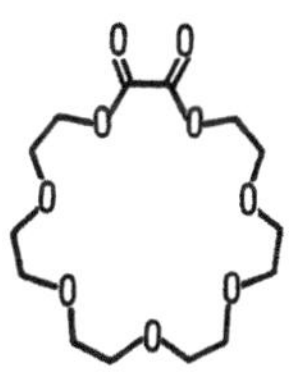		nr	nr	31
10	16	24	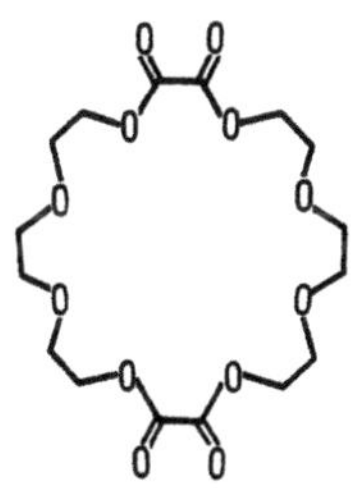		68–72	2.5	31
11	20	16	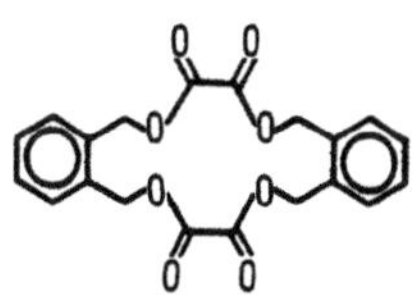		233–235	35	26

Table 5.2. Macrocyclic malonate esters

Cpd. No.	No. C's	Ring size	Structure		Mp or (bp) in °C	Yield (%)	Refs.
1	9	13	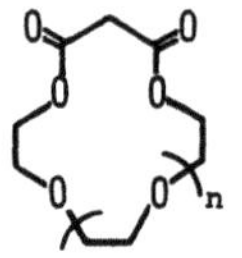	n = 1	51.5–52.5	34	11, 12
2	11	16		n = 2	67–68	27	11, 12
3	11	16	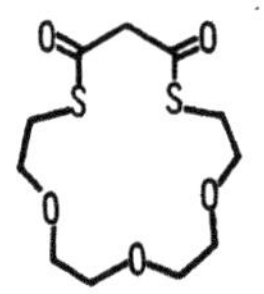		nr	nr	11
4	13	16	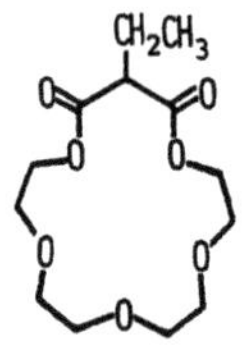		(146–148/0.17)	28	69
5	13	19	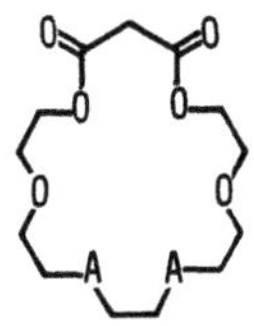	A = O	57–58	38	11, 12, 37
6	13	19		A = S	nr	85	11
7	17	16	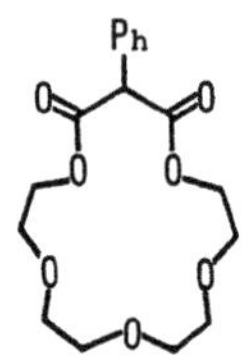		83.5–85	28	69
8	22	18	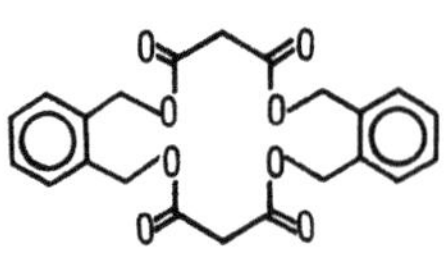		215–218	0.6	26

Table 5.3. Macrocyclic butanoate and butenoate esters

Cpd. No.	No. S's	Ring size	Structure		Mp or (bp) in °C	Yield (%)	Refs.
1	12	16	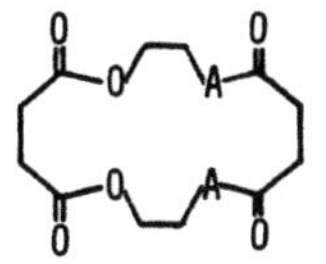	A = O	nr	nr	3, 21, 74
2	12	16		A = N	190–192	46	3
3	12	17	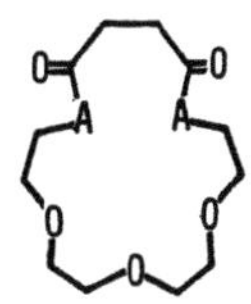	A = O	oil	20	11
4	12	17		A = S	oil	~ 5	11
5	13	17	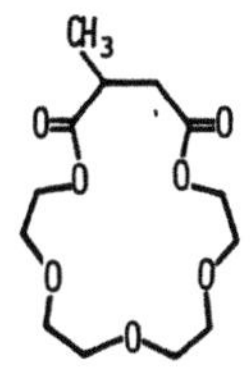		(146–154/ 0.1–0.2)	18	69
6	14	19	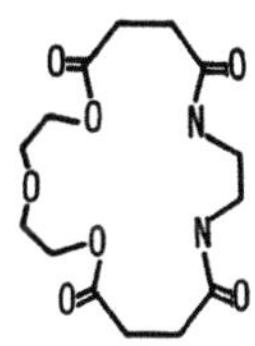		167–168	8	3
7	14	20	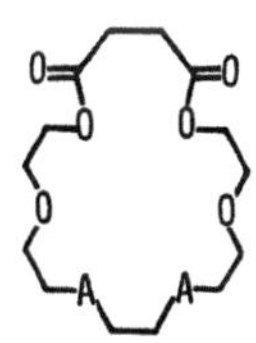	A = O	oil	90	11
8	14	20		A = S	oil	nr	11
9	16	22	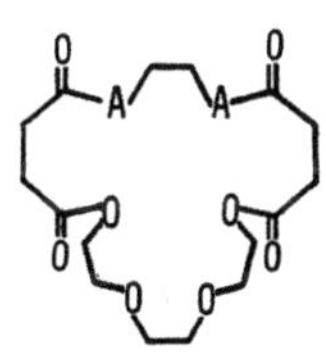	A = O	nr	0.5	3
10	16	22		A = N	159–161	10	3
11	18	17	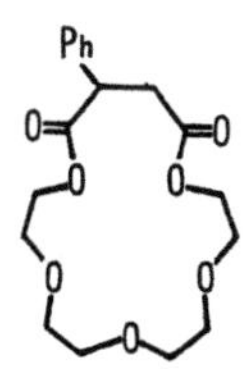		oil	1	69

Table 5.3 (continued)

Cpd. No.	No. C's	Ring size	Structure	Mp or (bp) in °C	Yield (%)	Refs.
12	21	19	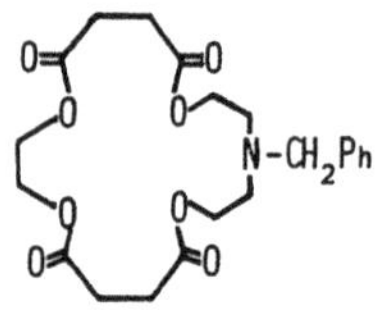	nr	5	3
13	24	20	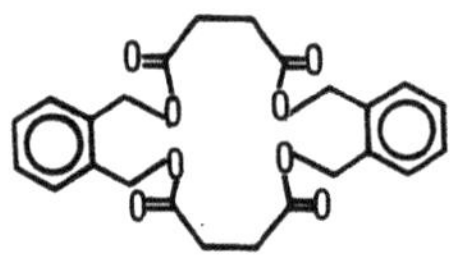	217	16	26
14	16	20	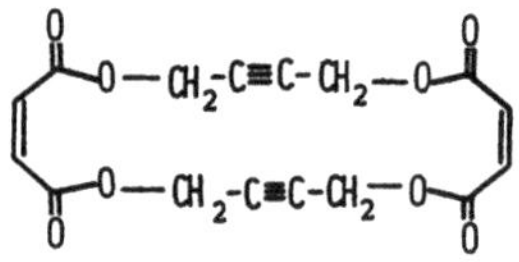	184	nr	27
15	16	22	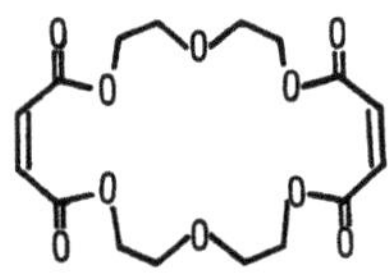	127.5–128	16	3
16	18	20	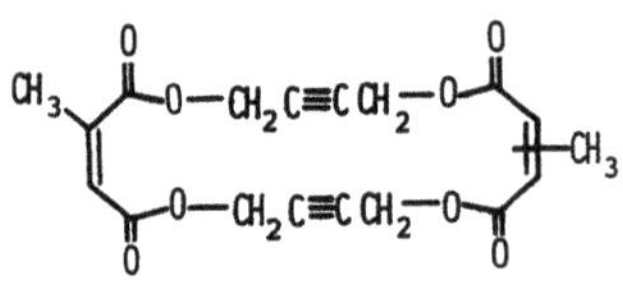	254	nr	27
17	18	22	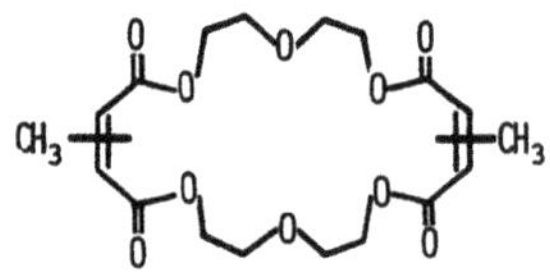	nr	1	3
18	20	24	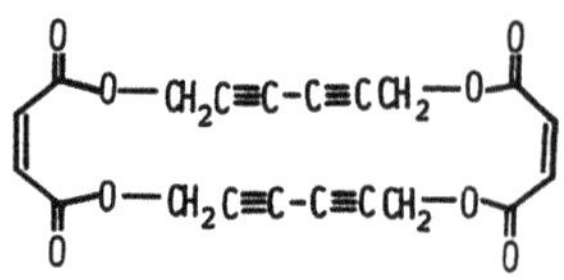	160 dec.	nr	27
19	22	24	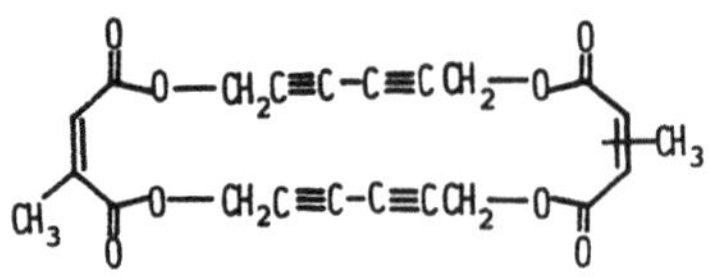	≥160 dec.	nr	27

Table 5.3 (continued)

Cpd. No.	No. C's	Ring size	Structure		Mp or (bp) in °C	Yield (%)	Refs.
20	24	35	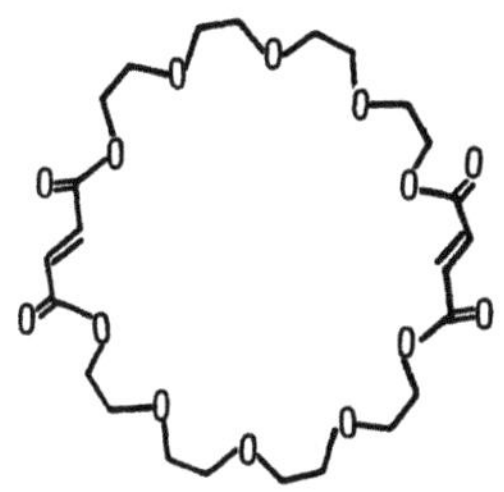	(E, E)	67–71	2	69

Table 5.4. Macrocyclic diglycolate esters

Cpd. No.	No. C's	Ring size	Structure		Mp or (bp) in °C	Yield (%)	Refs.
1	10	15	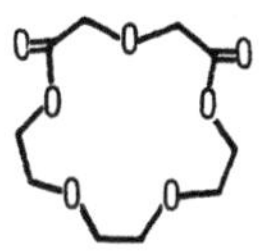		100.5–101.5	26	47
2	11	12			209–212	12	20
3	12	18	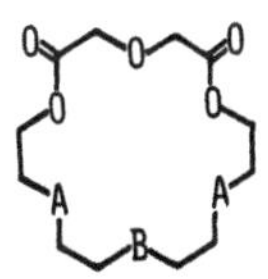	A = B = O	78.5–79.5	35	37, 47
4	12	18		A = O, B = S	113.5–115	36	47
5	12	18		A = S, B = O	36–36.5	19.5	47
6	14	15	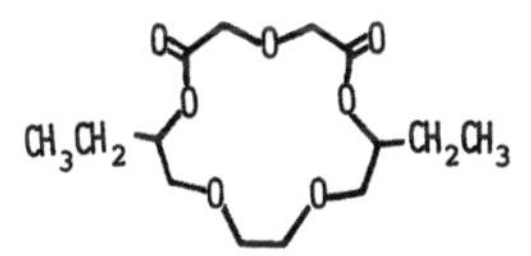		102.5–104	13	38
7	14	21	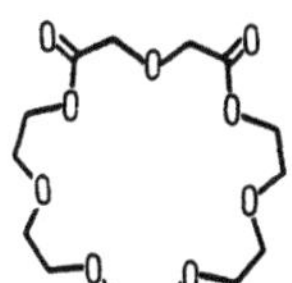		(200/0.2)	25.6	47

Table 5.4 (continued)

Cpd. No.	No. C's	Ring size	Structure		Mp or (bp) in °C	Yield (%)	Refs.
8	16	18	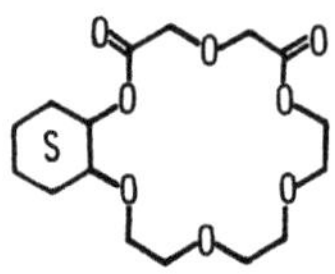	(E)	50.5–51.5	33	38
9	16	18	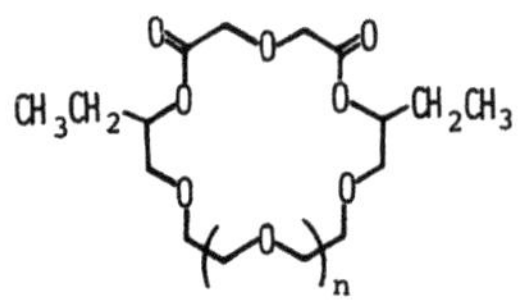	n = 1	61.5–63	46	38
10	18	21		n = 2	(188–190/0.3)	34	38
11	20	18	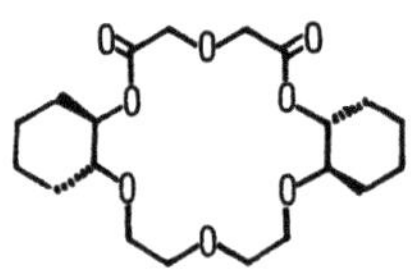	(trans-anti-trans)	124.5–126.5	nr	38
12	20	18	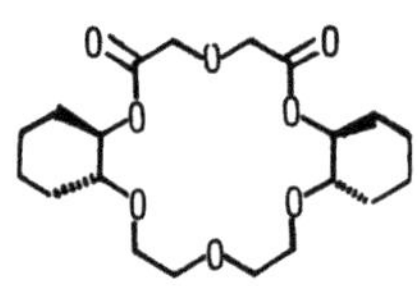	(trans-syn-trans)	180–181	nr	38
13	28	18	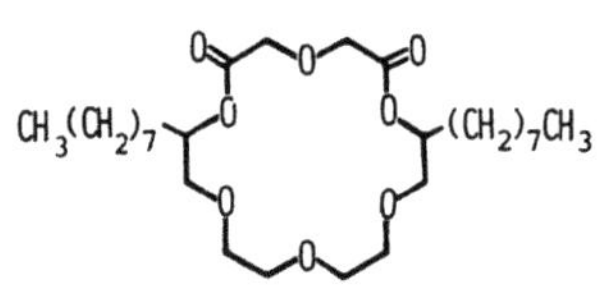		68.5–69	35	38

Table 5.5. Macrocyclic thiodiglycolate esters

Cpd. No.	No. C's	Ring size	Structure		Mp or (bp) in °C	Yield (%)	Refs.
1	8	12	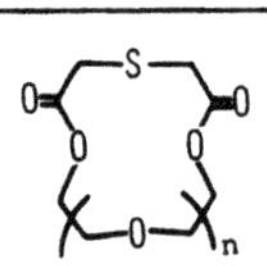	n = 1	93.5–94.5	5	47
2	10	15		n = 2	85.5–86.5	23	47
3	12	12	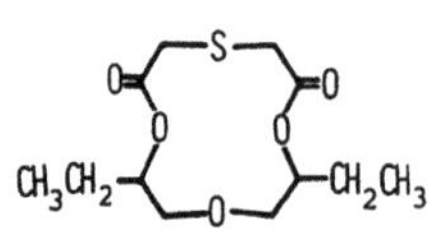		96.5–97.5	6	38

Table 5.5 (continued)

Cpd. No.	No. C's	Ring size	Structure		Mp or (bp) in °C	Yield (%)	Refs.
4	12	18	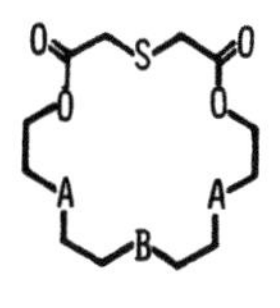	A = B = O	43.5–44.5	20	36, 47
5	12	18		A = O, B = S	106–107	11.5	47
6	12	18		A = S, B = O	nr	31	47
7	14	15	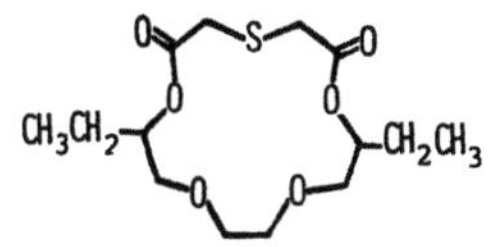		121.5–122.5	20	38
8	14	21	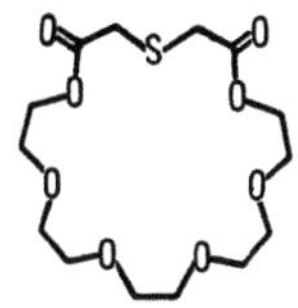		(204/0.2)	23	47
9	16	18	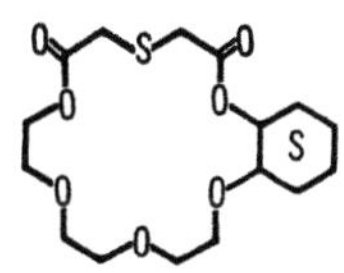	(E)	(165–170/1)	30	38
10	16	18	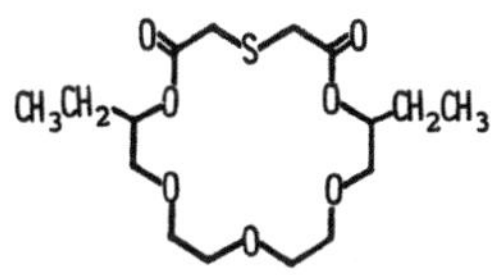		(194–195/1.3)	34	38
11	18	21	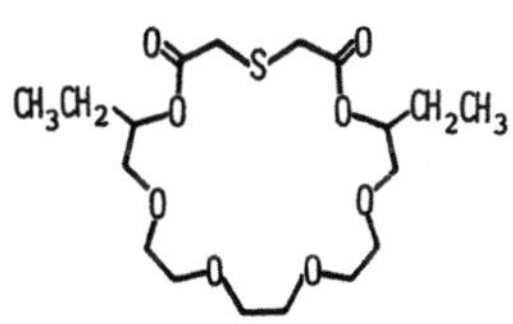		(180–181/1.5)	53	38
12	20	18	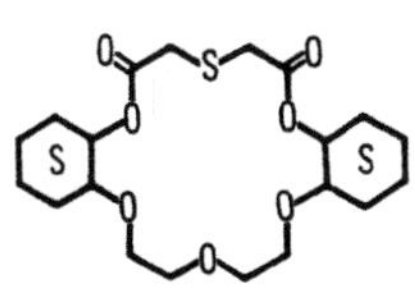		120.5–121	28	38

Table 5.6. Macrocyclic phthalate esters

Cpd. No.	No. C's	Ring size	Structure		Mp or (bp) in °C	Yield (%)	Refs.
1	10	8	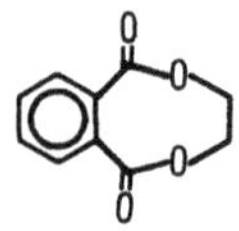		59.5–61.8	3	29
2	12	9	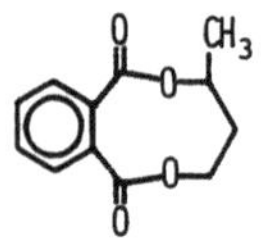		78.6–78.8	5	29
3	12	10	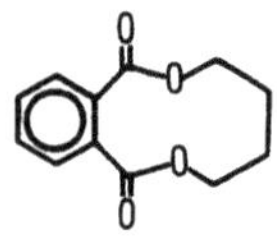		111	nr	22
					106	13	24
4	12	11	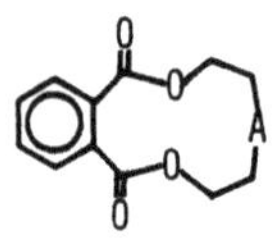	A = O	55–57	nr	22
						89	29
					61.8–63.2	89	30
5	12	11		A = S	107.2–107.7	2	29
6	13	11	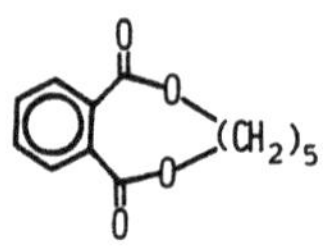		101	15.5	24
					106	nr	22
					nr	94	29
7	14	11	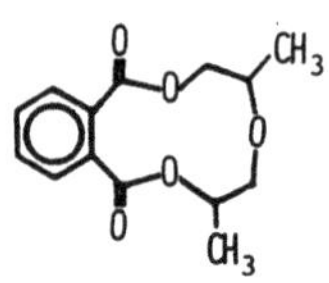	2 isomers	A) 130.5–131.5	15	29, 30
					B) 105.8–107.1		
8	14	11	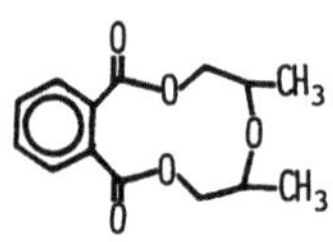	2 isomers	nr	nr	29
9	14	11	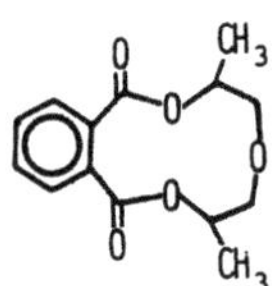	2 isomers	nr	nr	29
10	14	11	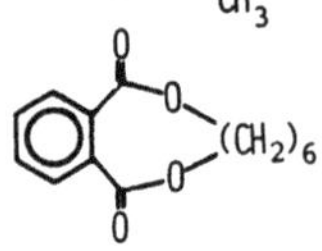		63.7–64.5	88	29
					64	8	24
11	14	14	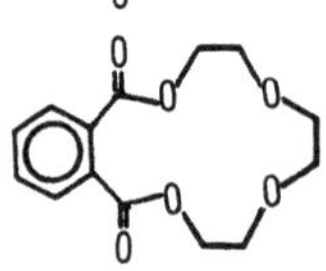		119.5–121	90	29
					118.5–120	66	30
					91.5–93.5	20	60

Table 5.6 (continued)

Cpd. No.	No. C's	Ring size	Structure	Mp or (bp) in °C	Yield (%)	Refs.
12	16	10	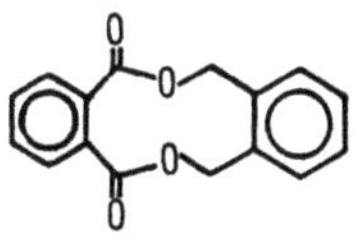	148	12.7	26
13	16	14	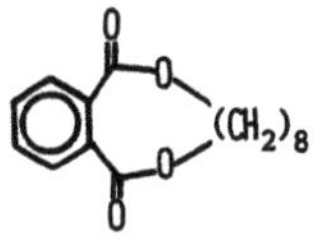	nr	3	24
14	16	17	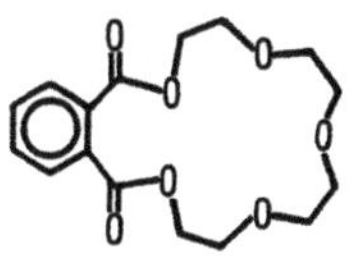	70–75	5	69
15	18	16	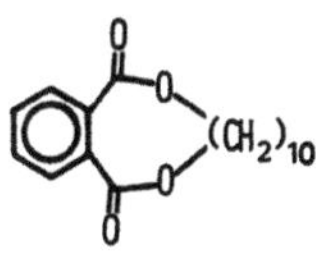	nr	3	24
16	18	20	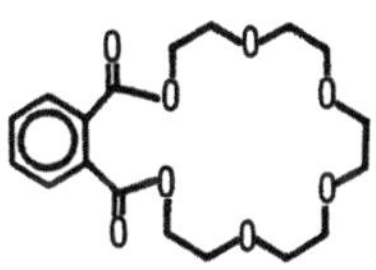	oil	<1	15
17	20	16	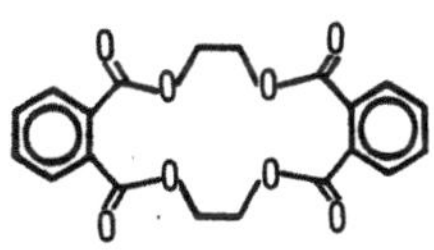	189 nr	10 0.5	24 29
18	20	18	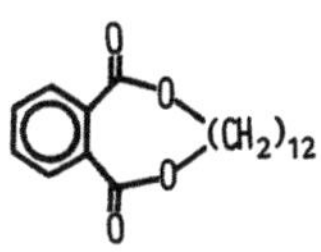	nr	4	24
19	20	23	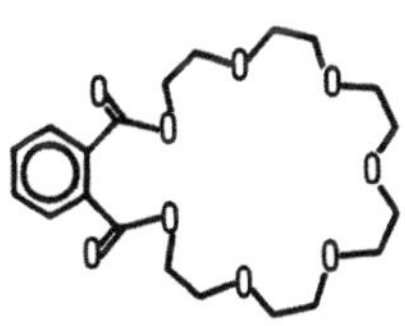	oil	0.4	15
20	22	16	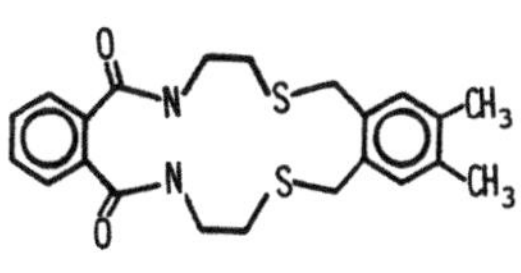	nr	50	39

Table 5.6 (continued)

Cpd. No.	No. C's	Ring size	Structure	Mp or (bp) in °C	Yield (%)	Refs.
21	22	18	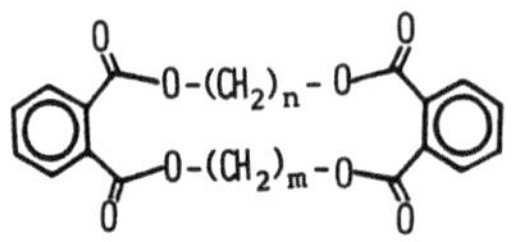 n = 2, m = 4	104	10	25
22	22	18	n = m = 3	191	5	22
23	23	19	n = 3, m = 4	131	12	25
24	24	20	n = 2, m = 6	147	20	25
25	24	20	n = 3, m = 5	165	7.5	25
26	24	20	n = m = 4	193 140	9.4 6.3	24 25
27	24	20	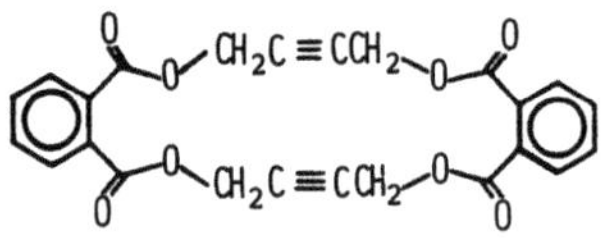	280	nr	27
28	24	20	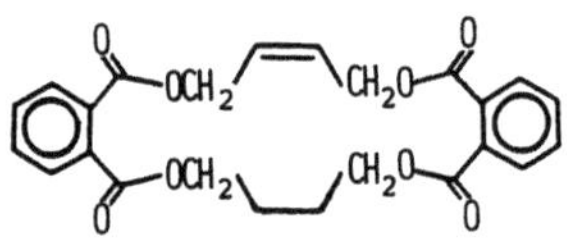	149	31.7	25
29	25	21	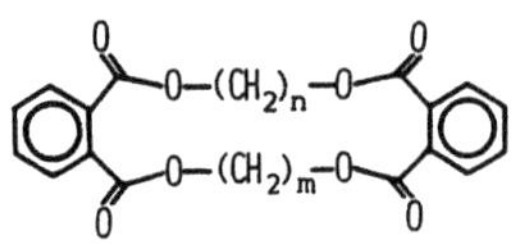 n = 4, m = 5	100	7	25
30	26	22	n = 4, m = 6	108	14	25
31	26	22	n = m = 5	139	9	25
32	27	23	n = 4, m = 7	115–117	9.6	25
33	28	24	n = 4, m = 8	138	7.3	25
34	28	24	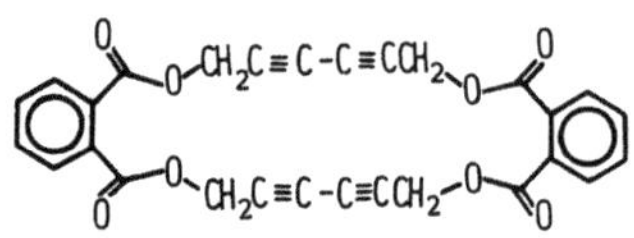	160 dec.	nr	27
35	28	24	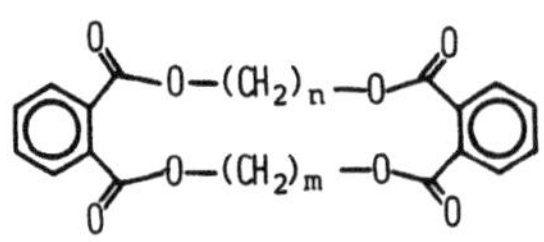 n = m = 6	66	9	25
36	29	25	n = 4, m = 9	112	8	25
37	29	25	n = 6, m = 7	60	6.7	25
38	30	16	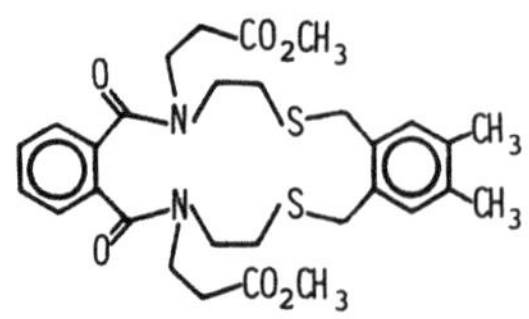	nr	nr	39

Table 5.6 (continued)

Cpd. No.	No. C's	Ring size	Structure		Mp or (bp) in °C	Yield (%)	Refs.
39	30	26	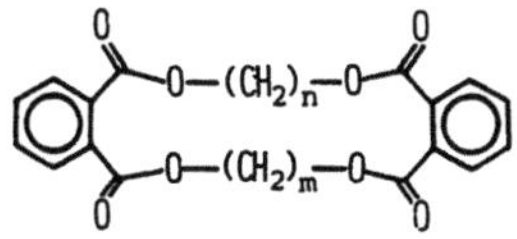	n = 6, m = 8	93	8	25
40	30	26		n = m = 7	163	6.5	24
41	31	27		n = 7, m = 8	89	9.6	25
42	32	28	n = m = 8		64–65	8.4	25
43	33	29	n = 8, m = 9		75	6.4	25
44	34	30	n = m = 9		140–141	4.2	24
45	35	31	n = 9, m = 10		56	3.7	25
46	35	32	n = 8, m = 12		95	9.4	25
47	36	32	n = m = 10		78–79	3.6	25
48	37	33	n = 10, m = 11		64	3.5	25
49	38	34	n = m = 11		138–139	2.7	24
50	39	35	n = 11, m = 12		75	7	25
51	40	36	n = m = 12		84	8.6	25

Table 5.7. Macrocyclic isophthalate esters

Cpd. No.	No. C's	Ring size	Structure		Mp or (bp) in °C	Yield (%)	Refs.
1	14	15	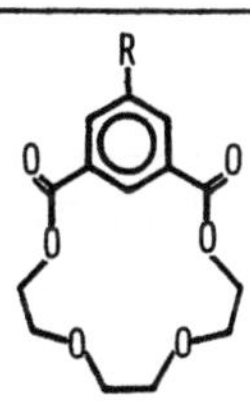	R = H	138–140	2	15
2	14	15		R = NO_2	161–163	2.6	15
3	16	18	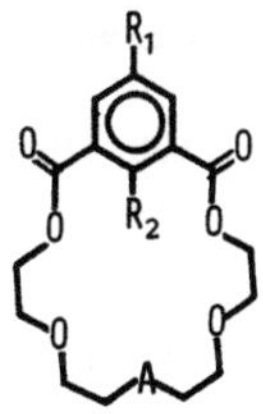	A = O, $R_1 = R_2$ = H	95.5–97 95.5–96 93.5	60 33 31	60 15 32
4	16	18		A = S, $R_1 = R_2$ = H	nr (170/0.1)	1.5 21	15 15

Table 5.7 (continued)

Cpd. No.	No. C's	Ring size	Structure		Mp or (bp) in °C	Yield (%)	Refs.
5	16	18	$A = O, R_1 = H, R_2 = NO_2$		124–126	26	32
6	16	18	$A = O, R_1 = NO_2, R_2 = H$		157–158	17	15
7	18	21	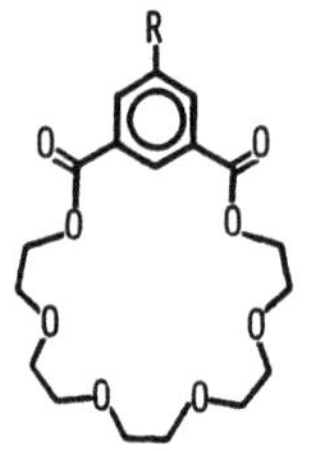	R = H	103.5–104.5	28	15
8	18	21		$R = NO_2$	104–105	30	15
9	20	24		R = H	106.5–108.5	31	15
10	20	24		$R = NO_2$	90–92	6	15
11	20	18	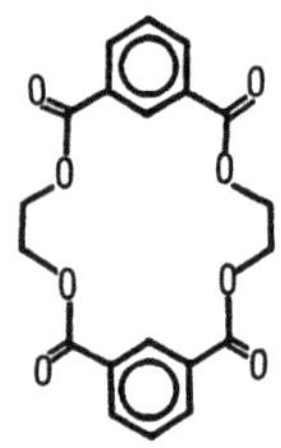		325–327	10–15	5

Table 5.8. Macrocyclic terephthalate esters

Cpd. No.	No. C's	Ring size	Structure		Mp or (bp) in °C	Yield (%)	Refs.
1	16	19	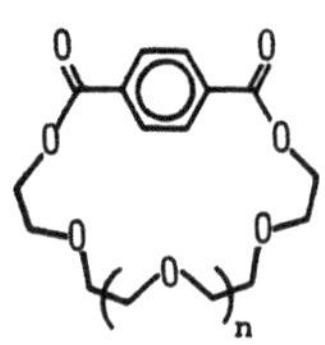	n = 1	96	23	32
					98–99	13	33
2	18	22		n = 2	108–109.5	<1	15
					110–111	10	60
3	20	25		n = 3	oil	20	60
					69–70	2	15

Table 5.8 (continued)

Cpd. No.	No. C's	Ring size	Structure	Mp or (bp) in °C	Yield (%)	Refs.
4	20	20	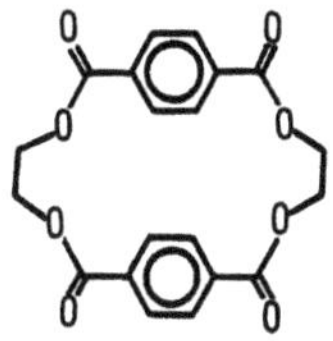	229	17	62
5	22	23	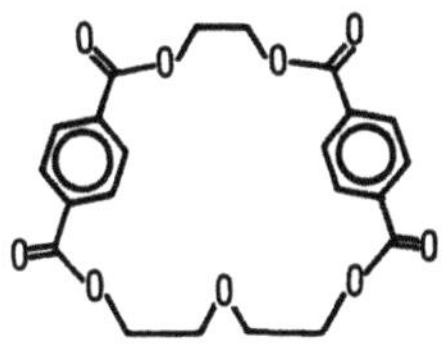	173–179 165–167	nr nr	49 34
6	30	30	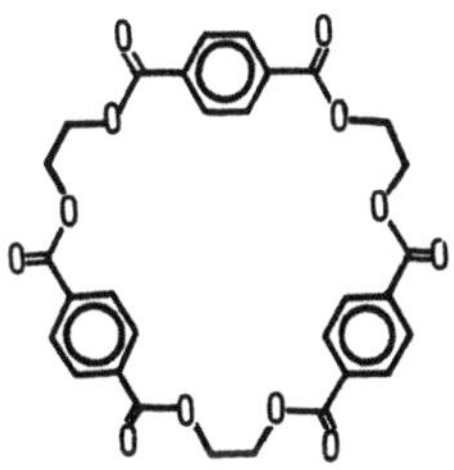	315–316 nr	8 48	49 77
7	32	38	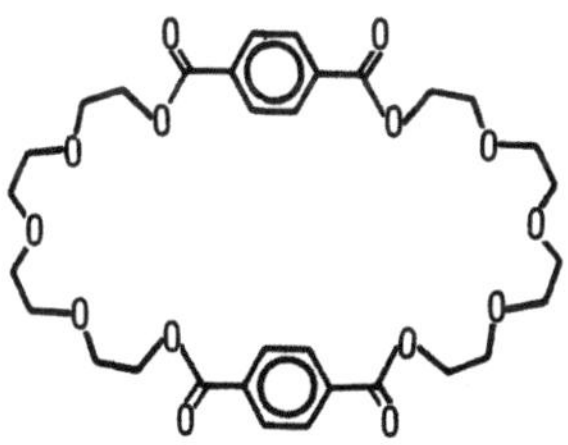	93.5–95 96	1 20	15 33
8	40	40	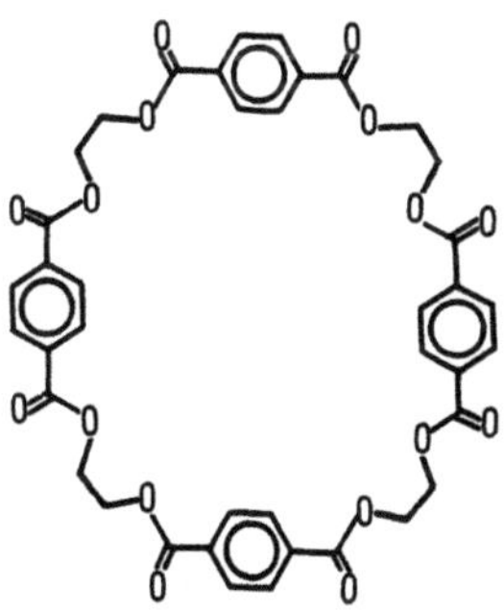	328–331 324–325	nr 40	49 77

Table 5.8 (continued)

Cpd. No.	No. C's	Ring size	Structure		Mp or (bp) in °C	Yield (%)	Refs.
9	48	48	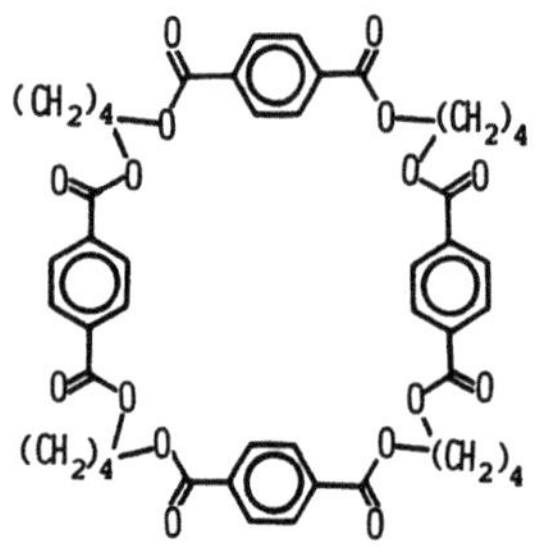		247–249	nr	49
10	50	50	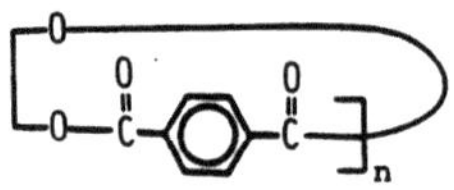	n = 5	264	45	77
11	60	60		n = 6	304–305	35	77
12	70	70		n = 7	238–240	25	77

Table 5.9. Macrocyclic esters of furan-2,5-dicarboxylic acid

Cpd. No.	No. C's	Ring size	Structure		Mp or (bp) in °C	Yield (%)	Refs.
1	14	18	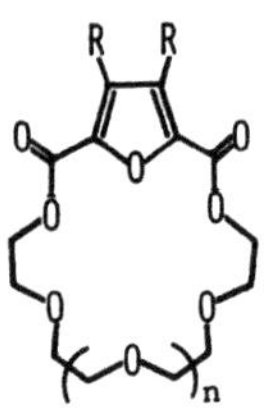	n = 1, R = H	117–118	60	10
2	16	18		n = 1, R = OCH_3	167–167.5	67	10
3	16	21		n = 2, R = H	133–134	27	10
4	18	21		n = 2, R = OCH_3	84–85	31	10
5	18	24		n = 3, R = H	93.5–95	30	10
6	20	24		n = 3, R = OCH_3	67–67.5	14	10

Table 5.10. Macrocyclic amides and esters containing the 2,6-dipicolinate subunit

Cpd. No.	No. C's	Ring size	Structure		Mp or (bp) in °C	Yield (%)	Refs.
1	11	12	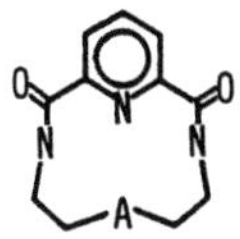	A = O	228–230	11	75
2	11	12		A = S	242–243	20	76
3	11	13	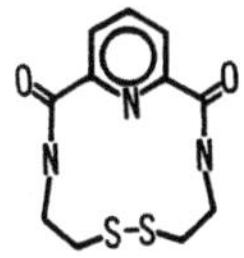		234–236	22	76
4	13	15	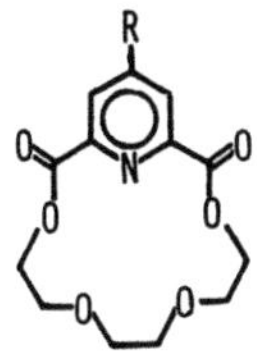	R = H	139–140	9.6	8
5	13	15		R = Cl	126–127	9	14
6	13	15	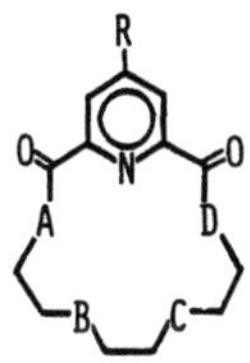	A = D = N, B = C = O, R = H	200–201	60	75
7	13	15		A = B = C = D = S, R = H	135–138	<10	32
8	14	15		A = B = C = D = O, R = OCH_3	154–154.5	1	14
9	15	18	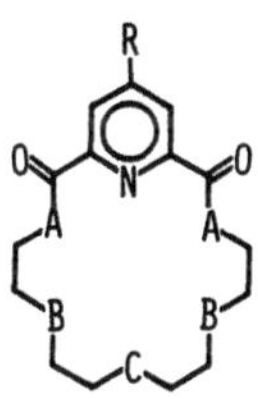	A = B = C = O, R = H	86.5–87.5 86.5 83.5–85 86.5–87.5	70–78 70 77 78	8 32 36 8
10	15	18		A = B = C = O, R = Cl	104–105	91	13, 14
11	15	18	A = N, B = C = O, R = H		127–129	71	75
12	15	18	A = B = O, C = S, R = H		96–97.5	10.7	8
13	15	18	A = S, B = C = O, R = H		146–148	46	32
14	15	18	A = C = O, B = S, R = H		127–129	4.2	8
15	16	18	A = B = C = O, R = OCH_3		116–117	54	13, 14
16	17	17	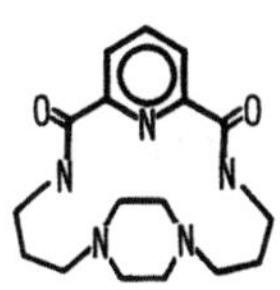		226–228	37	73

Table 5.10 (continued)

Cpd. No.	No. C's	Ring size	Structure		Mp or (bp) in °C	Yield (%)	Refs.
17	17	15	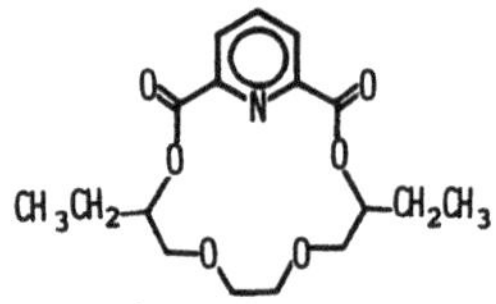		(170–175/1.5)	31	38
18	17	18	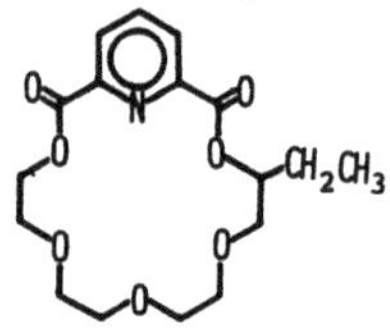		87–88.5	nr	38
19	17	21	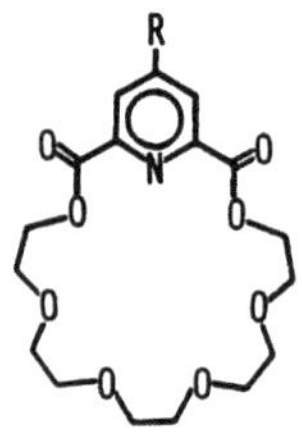	R = H	143–144.5	24.7	8
20	17	21		R = Cl	70–71 nr	71 72.3	13 14
21	18	21		R = OCH_3	122–123	19	13, 14
22	18	18	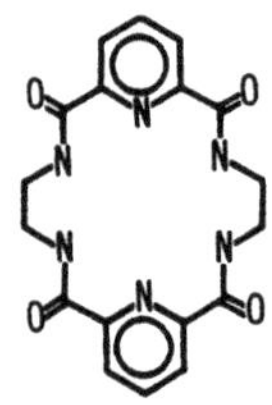		~250 dec.	37	76
23	19	15	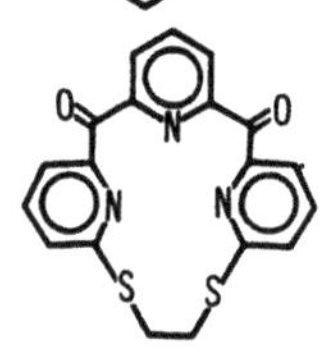		199–201	3.6	53
24	19	24	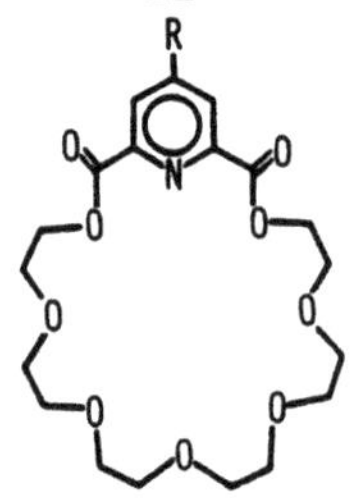	R = H	110–111	28.5	8
25	19	24		R = Cl	65–66	38	14
26	19	18	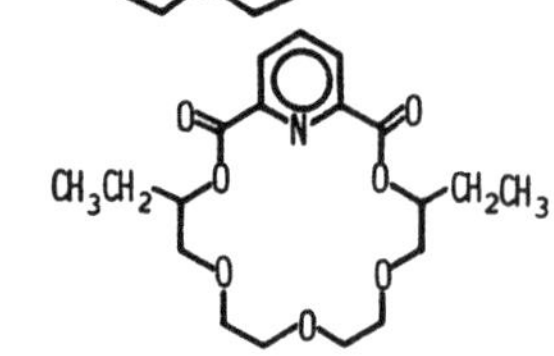		103.5–106.5	3	38

Table 5.10 (continued)

Cpd. No.	No. C's	Ring size	Structure	Mp or (bp) in °C	Yield (%)	Refs.
27	19	18	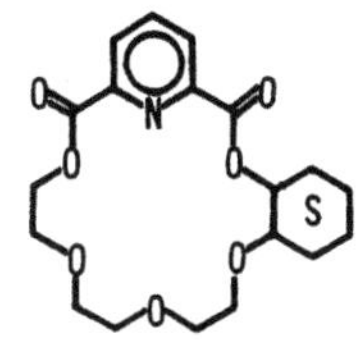 (E)	95–96.5	42	38
28	20	24	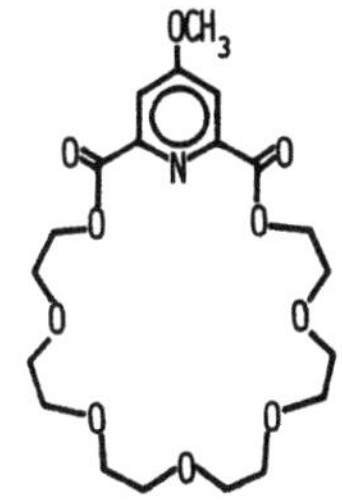	72–73	43	14
29	21	17	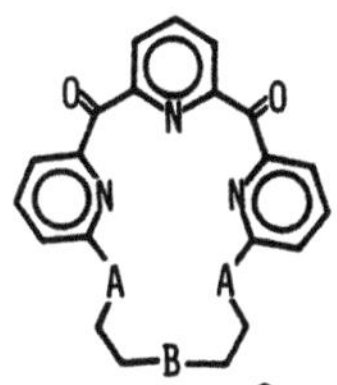 A = B = O	119–121	26.5	53
30	21	17	A = S, B = O	198–200	28	53
31	21	17	A = B = S	158–160	15	53
32	21	21	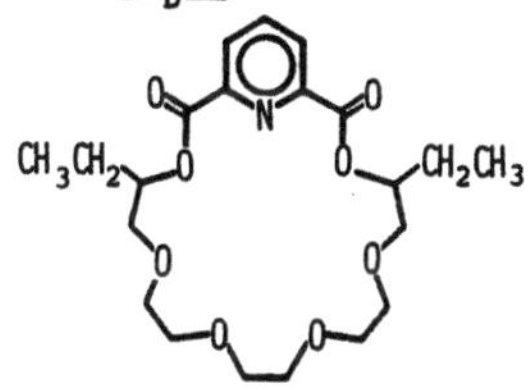	(201/1)	31	38
33	21	27		52–53	45	14
34	22	24	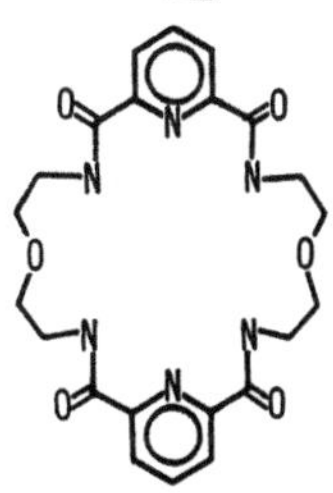	338–340 subl.	3	75

Table 5.10 (continued)

Cpd. No.	No. C's	Ring size	Structure		Mp or (bp) in °C	Yield (%)	Refs.
35	22	27	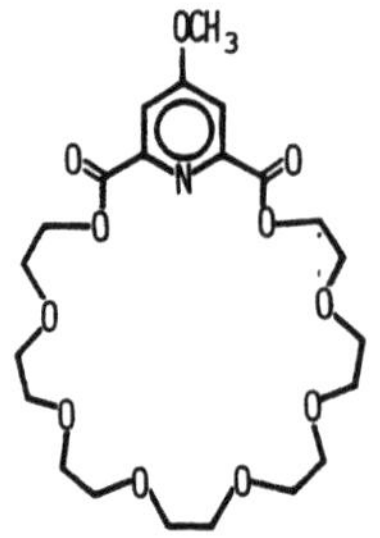		oil	36.5	14
36	23	18	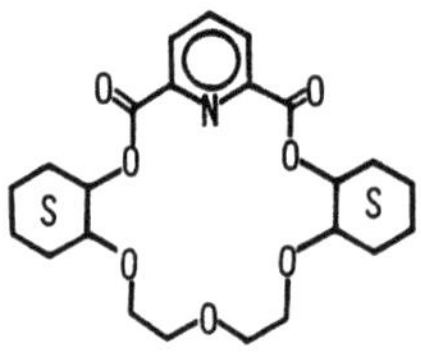	(E)	137–138.5	6	38
37	24	20	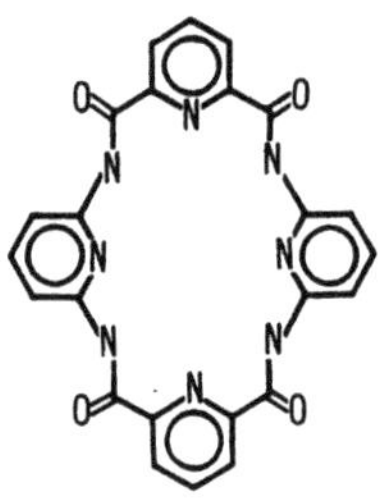		>360	8	76
38	28	24	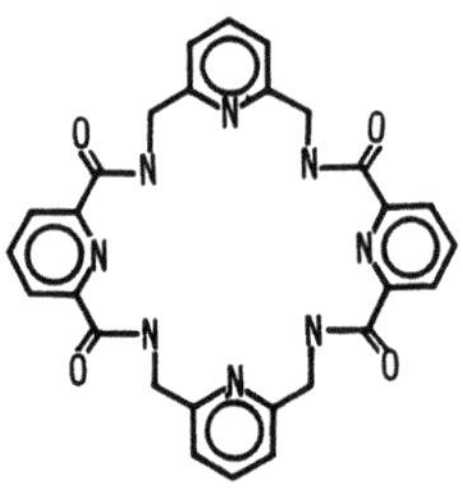		320–322	18	20
39	30	24	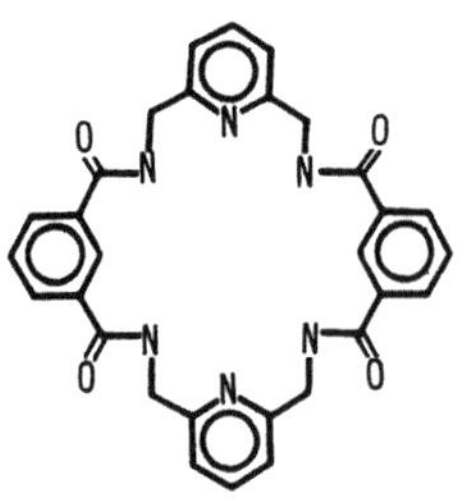		>300	4	20

Table 5.10 (continued)

Cpd. No.	No. C's	Ring size	Structure		Mp or (bp) in °C	Yield (%)	Refs.
40	34	34	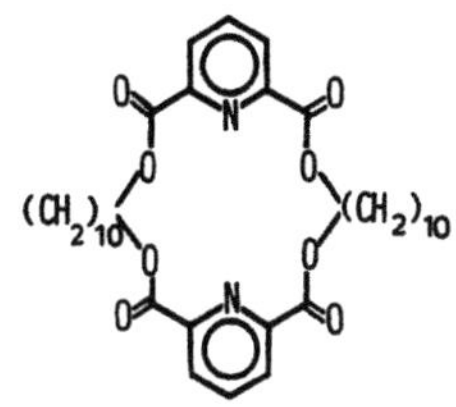		156–158	48	32

Table 5.11. Macrocyclic esters containing the 3,5-dipicolinate subunit

Cpd. No.	No. C's	Ring size	Structure		Mp or (bp) in °C	Yield (%)	Refs.
1	13	15	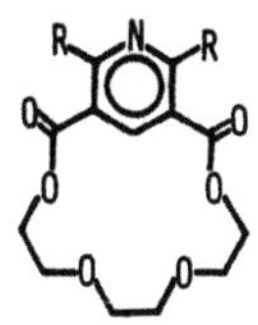	R = H	137–137.5 128–130	1 20	8 60
2	15	15		$R = CH_3$	167–169	25	70
3	15	18	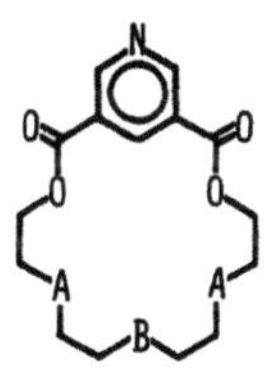	A = B = O	94.5–96.5 112–113	75 32.7	60 8
4	15	18		A = O, B = S	108–109	9.3	8
5	15	18		A = S, B = O	103–104	8.3	8
6	17	18	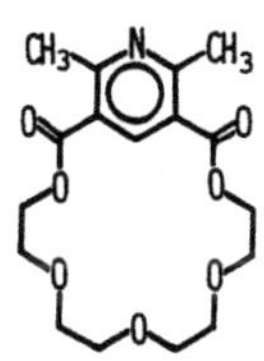		90–92 90–92 90–92	70 15 20	60 70 38a
7	17	18	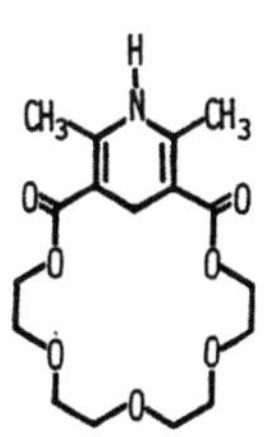		114–117 $\cdot H_2O$	nr	38a

Table 5.11 (continued)

Cpd. No.	No. C's	Ring size	Structure	Mp or (bp) in °C	Yield (%)	Refs.
8	17	21	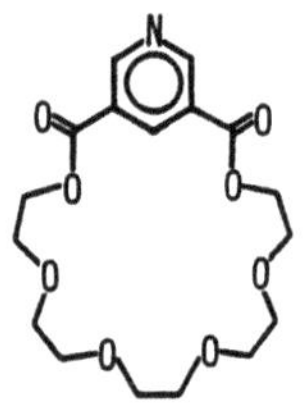	53–57 54.5-55.5	90 6.9	60 8
9	18	18	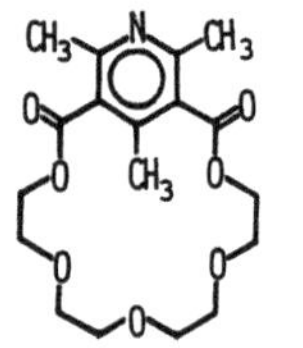	186.5–187.5	25	38a
10	18	18	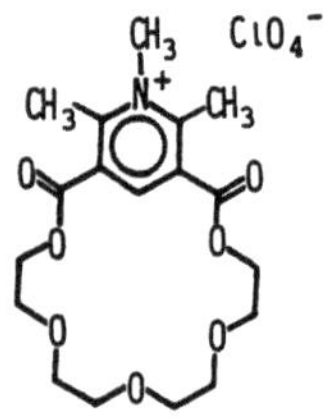	190–193	nr	70
11	18	18	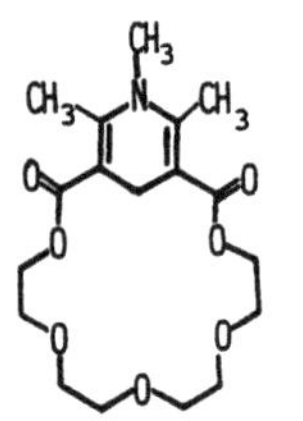	110–113 110–113	nr 86	70 38a
12	18	18	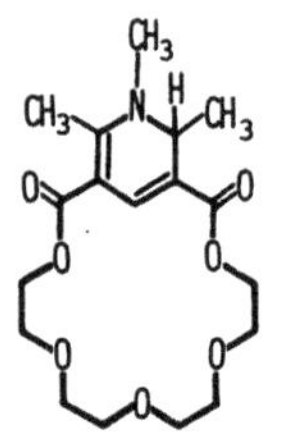	131.5–133 131.5–133	nr 41	70 38a
13	19	18	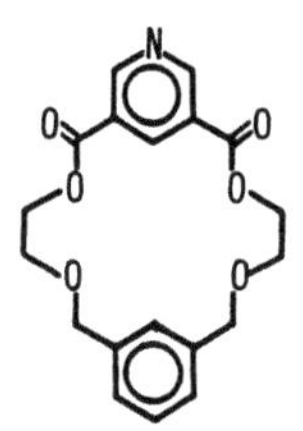	127–128	25	38a

Table 5.11 (continued)

Cpd. No.	No. C's	Ring size	Structure	Mp or (bp) in °C	Yield (%)	Refs.
14	19	24	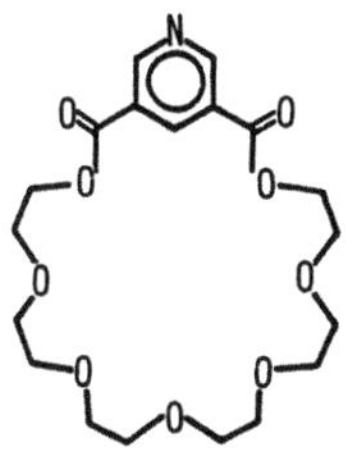	oil 50–51	30 7.5	60 8
15	20	18	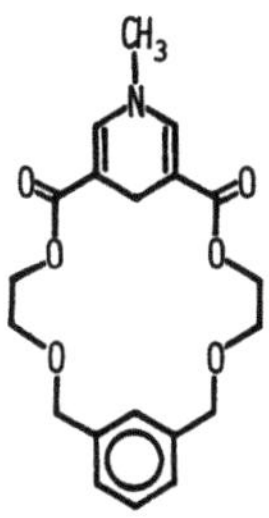	134–136	86	38a
16	21	18	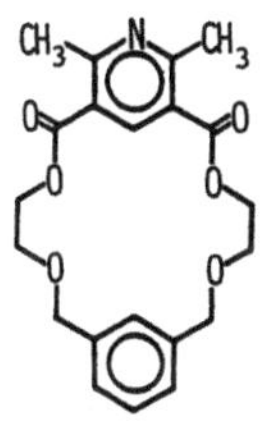	158.5–160	15	38a
17	21	18	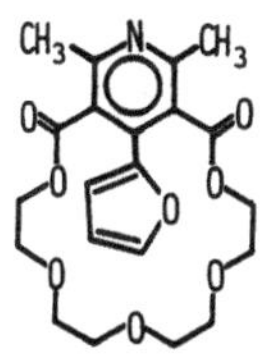	179–180	nr	38a
18	22	18	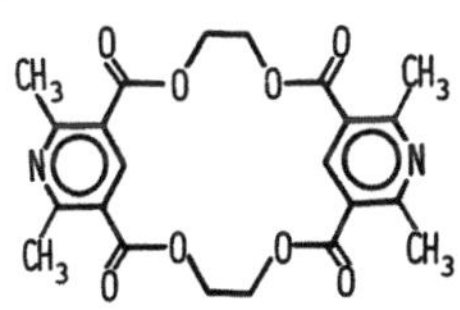	196–198	nr	70
19	23	18	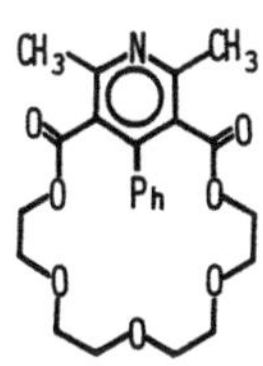	179–181	25	38a

Table 5.11 (continued)

Cpd. No.	No. C's	Ring size	Structure		Mp or (bp) in °C	Yield (%)	Refs.
20	26	24	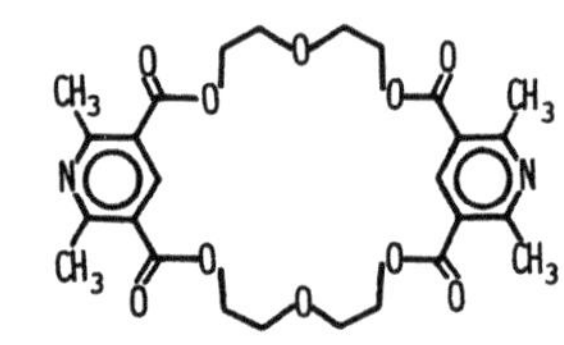		196–198	1	38a
21	–	–	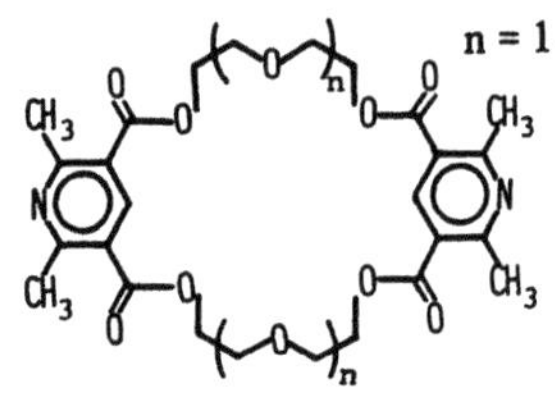	n = 1–3 (mixture)	nr	nr	70
22	30	30		n = 2	oil	12	38a
23	34	36		n = 3	oil	19	38a
24	38	42		n = 4	oil	22	38a

Table 5.12. Macrocyclic diphenyldisulfides

Cpd. No.	No. C's	Ring size	Structure		Mp or (bp) in °C	Yield (%)	Refs.
1	18	14	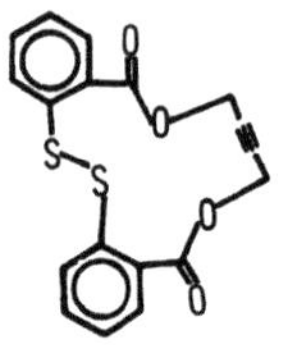		148–150	51.4	28
2	18	14	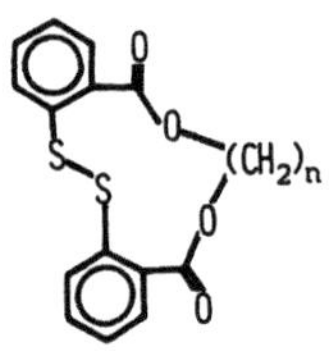	n = 4	157	95	28
3	19	15		n = 5	85–87	9	28
4	20	16		n = 6	147	53	28
5	20	18	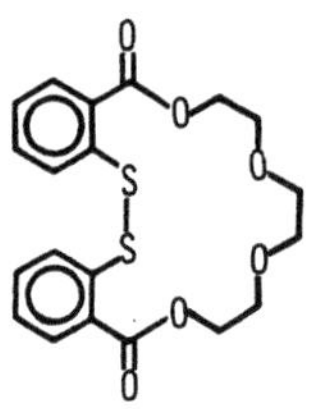		193	5.4	28

Table 5.12 (continued)

Cpd. No.	No. C's	Ring size	Structure	Mp or (bp) in °C	Yield (%)	Refs.
6	21	17	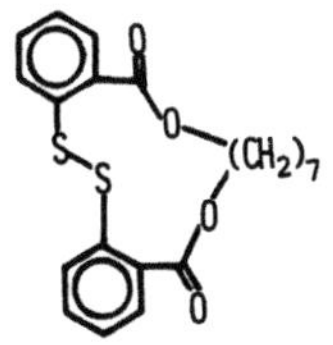	117	57	28
7	22	21	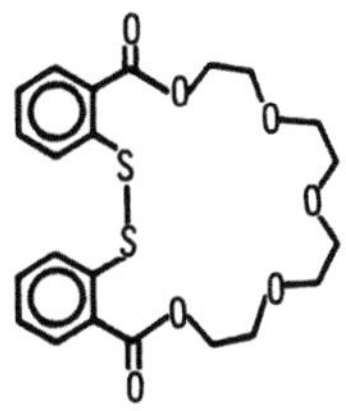	141–142	4.9	28
8	22	18	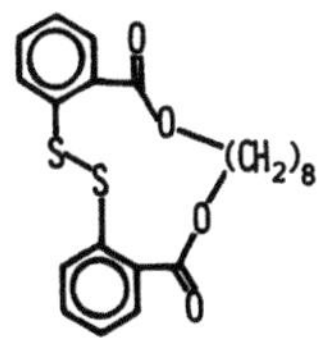	186	5.5	28

Table 5.13. Miscellaneous mono- and diesters

Cpd. No.	No. C's	Ring size	Structure	Mp or (bp) in °C	Yield (%)	Refs.
1	10	15	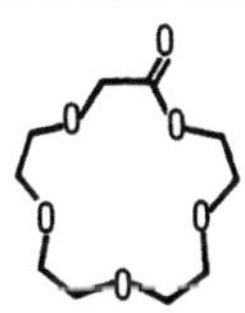	(99–102/0.04)	38	48
2	12	16	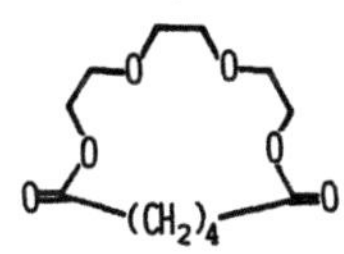	57–58	31	69
3	12	18	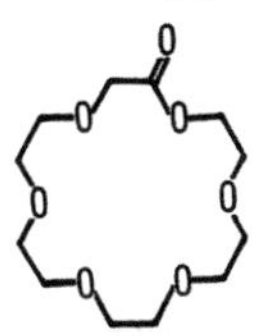	(116–118/0.04)	42	48

Table 5.13 (continued)

Cpd. No.	No. C's	Ring size	Structure		Mp or (bp) in °C	Yield (%)	Refs.
4	13	18	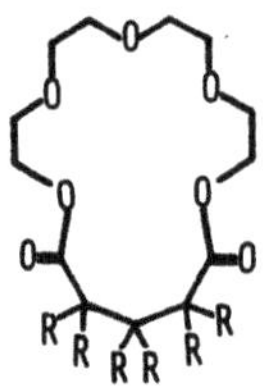	R = H	(155–157/0.65)	31	69
5	13	18		R = F	(135–137/ 0.2–0.3)	23	69
6	14	15	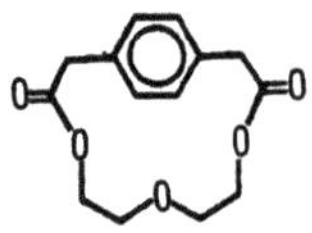		158.5–159.5	nr	64
7	14	18	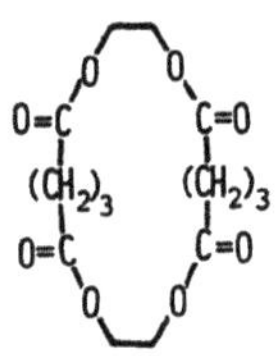		144–145	20	66
8	14	18	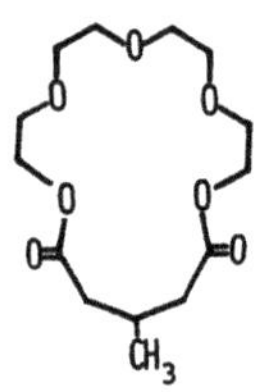		(130–142/0.12)	37	59
9	14	19	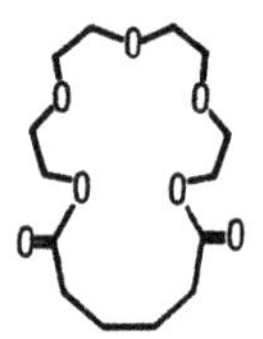		(154–155/ 0.06–0.07)	20	69
10	15	16	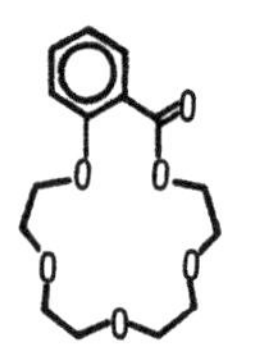		oil	36	71
11	15	21	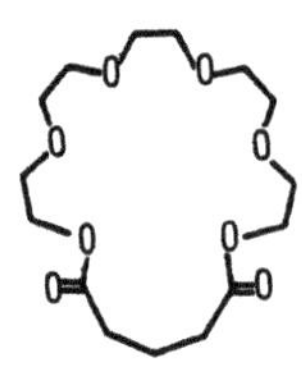		nr	80	11

Table 5.13 (continued)

Cpd. No.	No. C's	Ring size	Structure	Mp or (bp) in °C	Yield (%)	Refs.
12	16	15	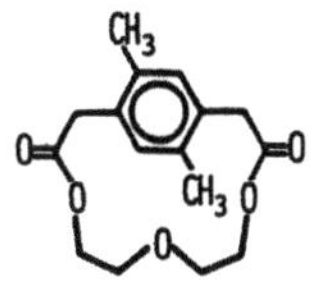	96–97	nr	65
13	16	17	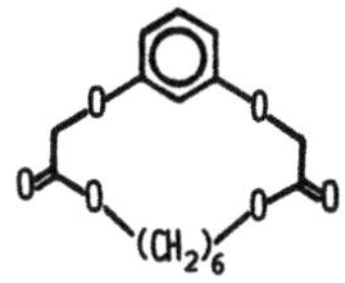	nr	nr	50
14	16	17	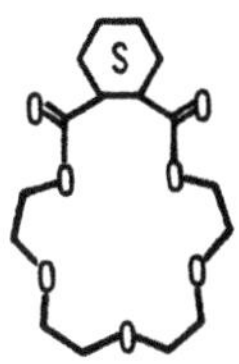 (Z)	oil	<1	69
15	16	18	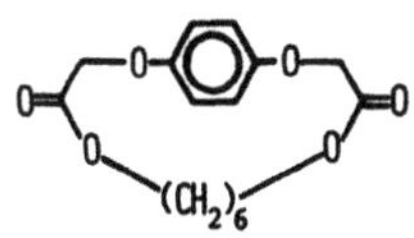	nr	nr	50
16	16	18	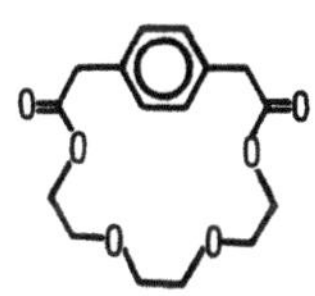	39–40	nr	64
17	16	19	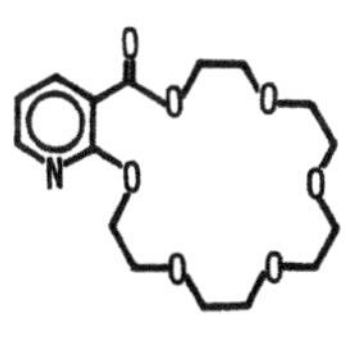	oil	48	52
18	17	19	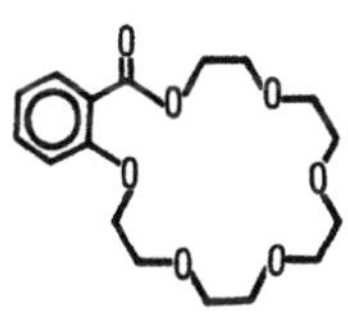	oil	54	71
19	18	18	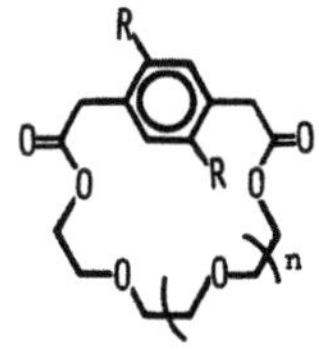 n = 1, R = CH_3	70.5–71	nr	64
20	18	21	n = 2, R = H	95.5–96.5	nr	64

Table 5.13 (continued)

Cpd. No.	No. C's	Ring size	Structure	Mp or (bp) in °C	Yield (%)	Refs.
21	18	22	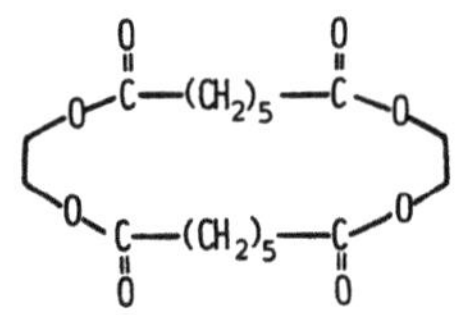	147–148	15	66
22	19	22	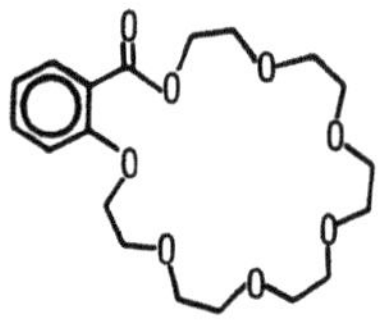	oil	68	71
23	20	20	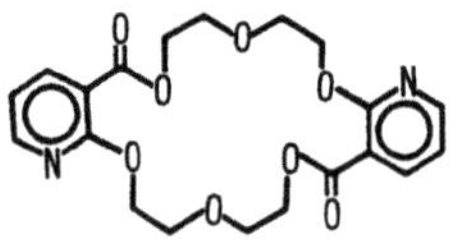	169.5–170	6.9	52
24	20	20	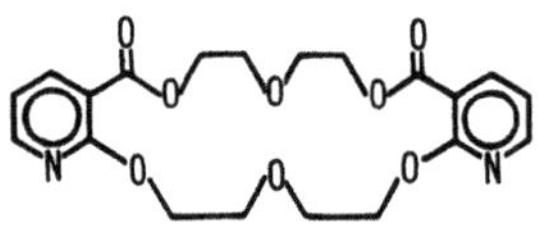	112.5–113	7.4	52
25	20	21	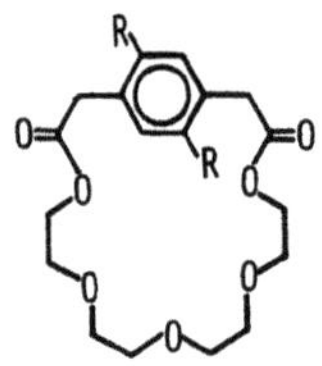 R = CH_3	85.5–86.5	nr	64
26	20	21	R = OCH_3	94.5–95	nr	64
27	22	21	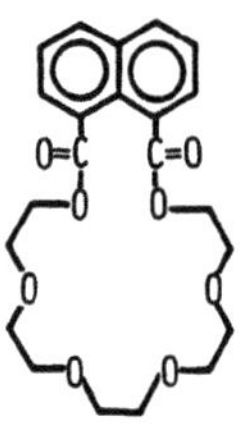	122–125	<1	15
28	22	26	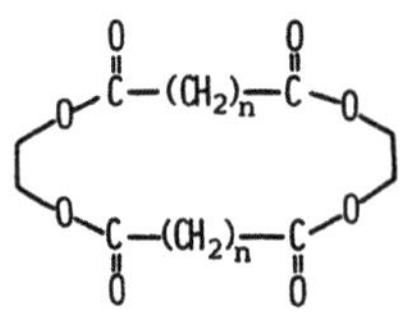 n = 7	147–148	53	66
29	24	28	n = 8	79–81	30	66
30	24	32	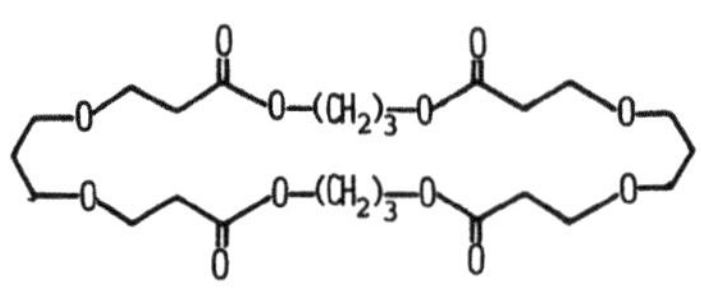	nr	30	65

Table 5.13 (continued)

Cpd. No.	No. C's	Ring size	Structure	Mp or (bp) in °C	Yield (%)	Refs.
31	26	22	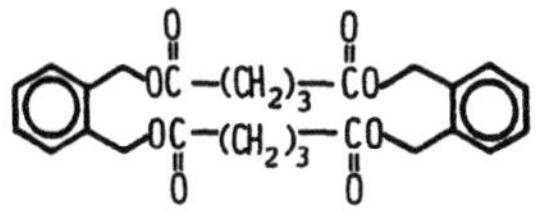	173	3	26
32	28	26	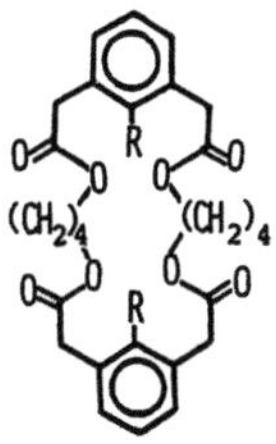 R = H	92–93	70	72
33	28	26	R = Cl	163–164	75	72
34	28	21	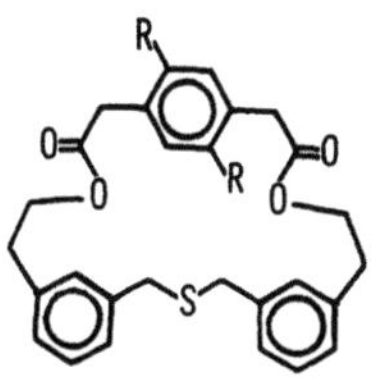 R = H	125–125.5	~10	63
35	30	21	R = CH_3	128.5–129	~10	63
36	28	24	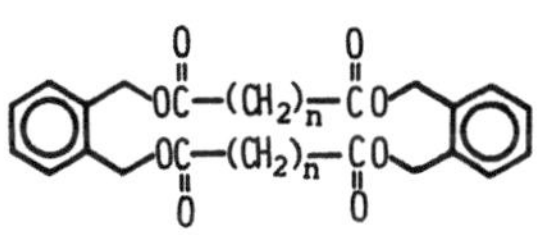 n = 4	144	3	26
37	30	26	n = 5	121	0.9	26
38	32	28	n = 6	134	7.3	26
39	34	30	n = 7	119	0.3	26
40	36	32	n = 8	135	3.1	26
41	42	18	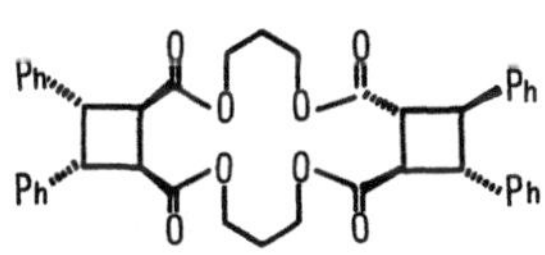	159–161	32	55
42	44	22	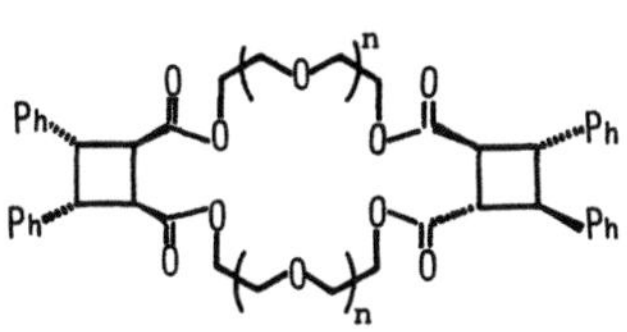 n = 1	nr	nr	56
43	48	28	n = 2	nr	nr	56

Table 5.14. Miscellaneous lactams, cyclic carbamates, carbonates, ureas, etc.

Cpd. No.	No. C's	Ring size	Structure		Mp or (bp) in °C	Yield (%)	Refs.
1	5	8	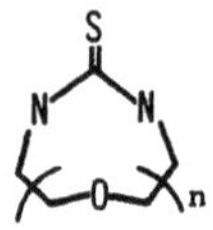	n = 1	216	41	7
2	7	11		n = 2	141	27	7
3	8	13	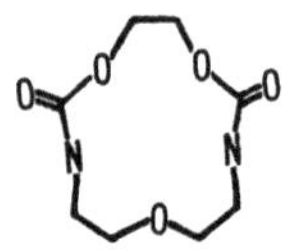		185	37	18
4	9	11	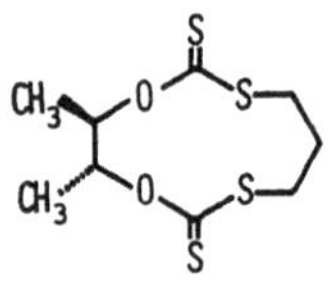	(E)	125–126	14	45
5	10	13	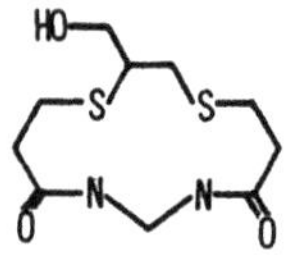		214–215	nr	16
6	10	16	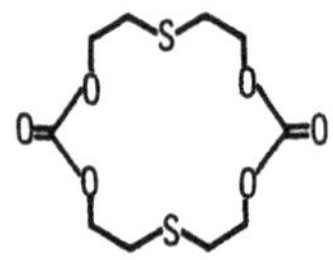		nr	nr	16
7	10	16	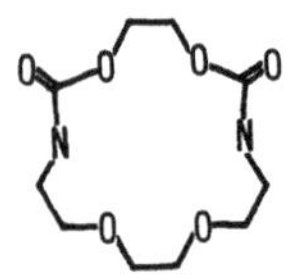		122	84	18
8	10	16	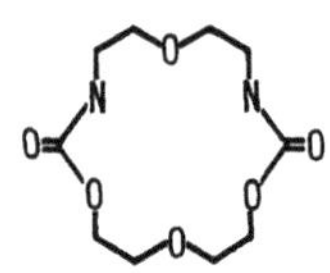		126	89	18
9	10	16	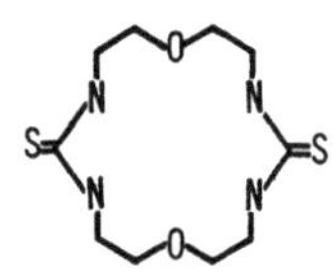		233	18	7
10	10	16	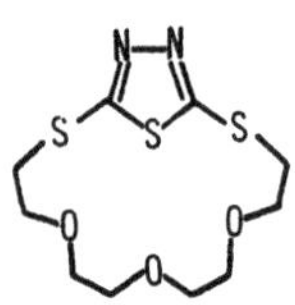		84–86	nr	16

Table 5.14 (continued)

Cpd. No.	No. C's	Ring size	Structure		Mp or (bp) in °C	Yield (%)	Refs.
11	11	11	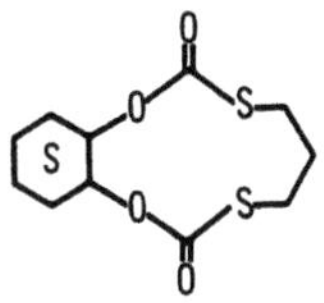	(Z) (E)	125–126 179–180	23 51 (E + Z)	45
12	11	16	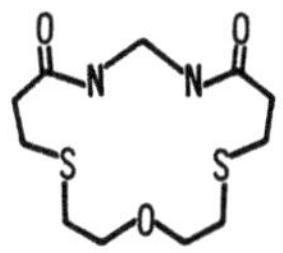		nr	30	16
13	11	17	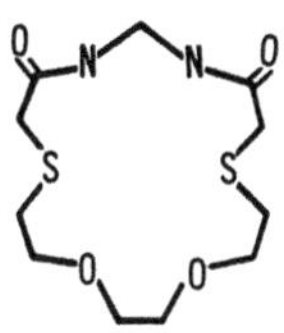		209–210	78	16
14	12	12	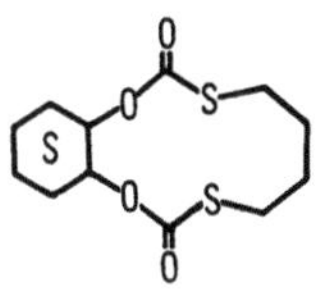	(Z) (E)	126–127 140–141	22 46 (E + Z)	45
15	12	13	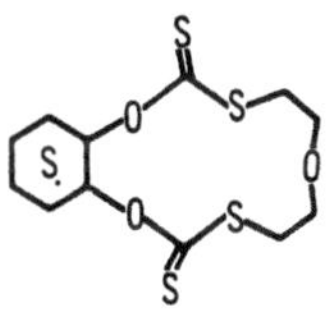	(Z) (E)	160–161 240–241	47 (E + Z)	45
16	12	13	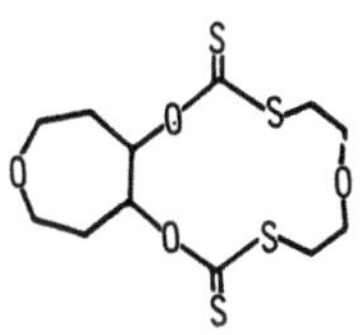		90–91	61	45
17	12	14	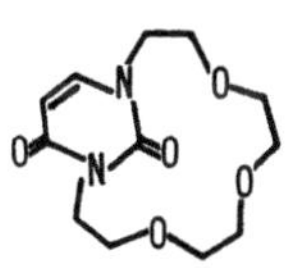		128–129	28	54
18	12	18	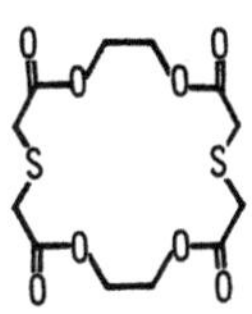		150.5–151.5	nr	16

Table 5.14 (continued)

Cpd. No.	No. C's	Ring size	Structure	Mp or (bp) in °C	Yield (%)	Refs.
19	12	19	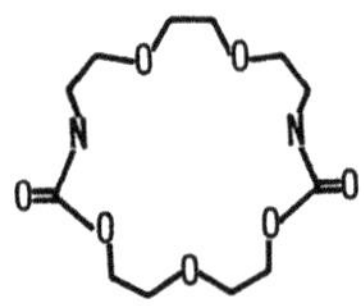	112	77	18
20	12	19	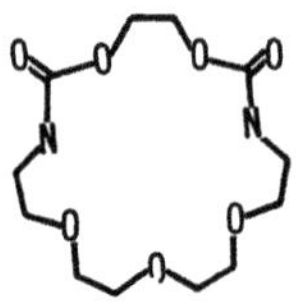	88	84	18
21	13	13	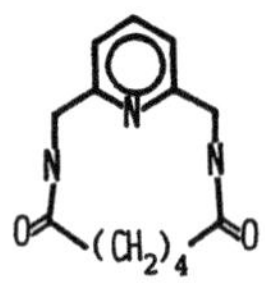	152	17	20
22	13	13	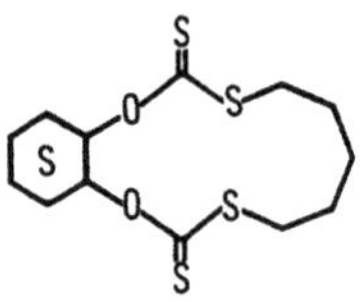 (E) mixture	148–149 nr	20 30	45
23	13	13	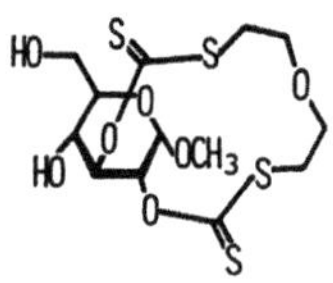	90–92	90	45
24	13	15	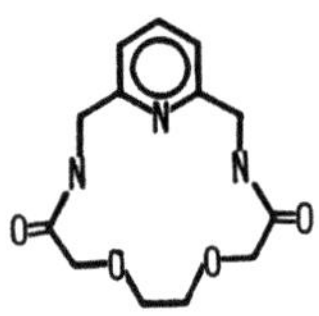	249–251	25	20
25	13	15	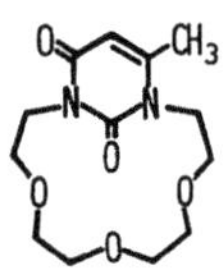	92–94	0.2	35
26	13	16		135	59	18

Table 5.14 (continued)

Cpd. No.	No. C's	Ring size	Structure		Mp or (bp) in °C	Yield (%)	Refs.
27	13	16	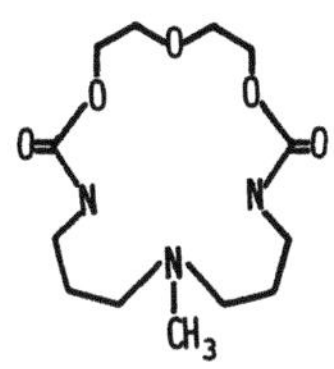		93	62	18
28	13	19	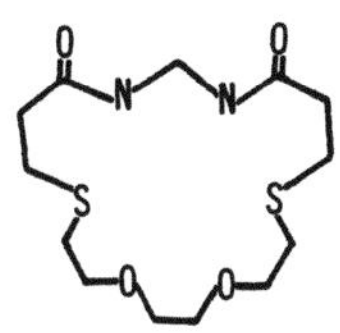		197–199	nr	16
29	13	19	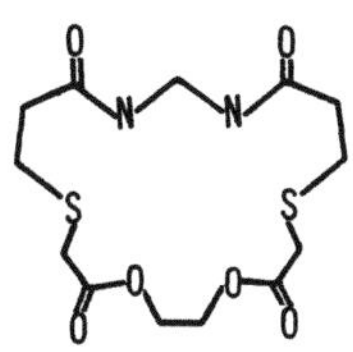		177–178	nr	16
30	14	14	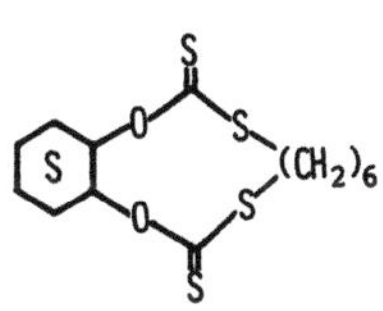	(Z) (E) mixture	95–96 120–121 –	26 – 32	45
31	14	16	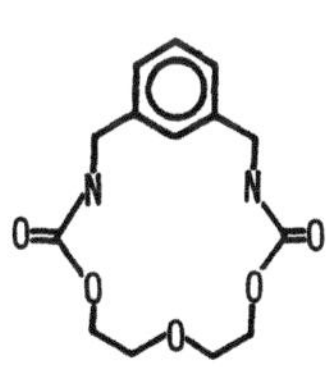		145–147	66	18
32	14	16	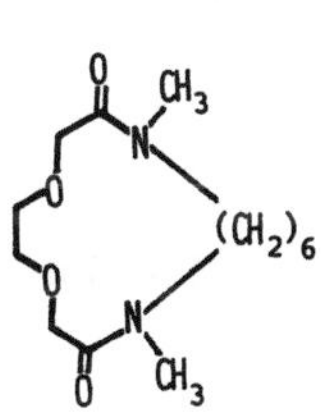		nr	nr	2
33	14	20	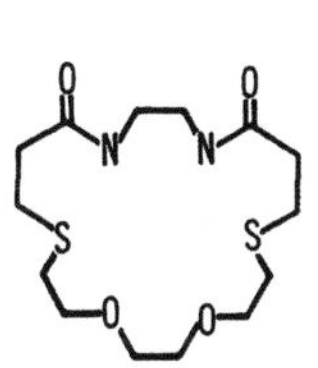		nr	nr	16

Table 5.14 (continued)

Cpd. No.	No. C's	Ring size	Structure	Mp or (bp) in °C	Yield (%)	Refs.
34	14	22	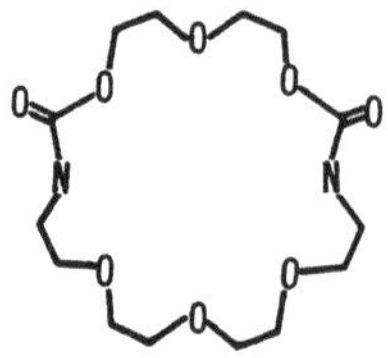	62	90	18
35	14	22	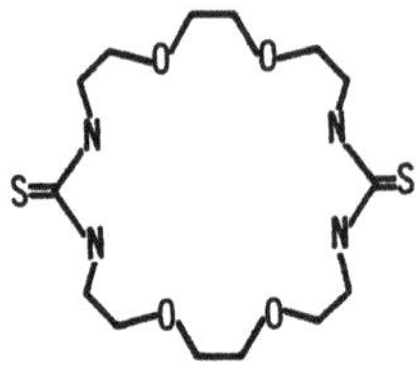	223	8	7
36	15	14	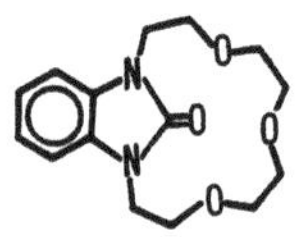	117–118	14	35
37	16	15	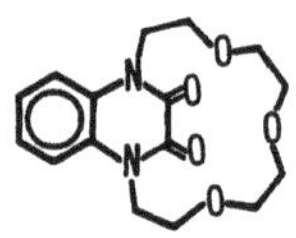	184	7	35
38	16	19	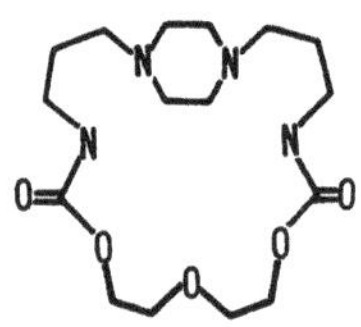	89	59	18
39	17	14	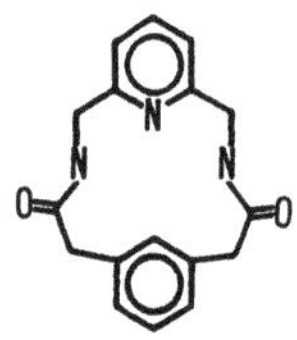	243	31	20
40	17	14	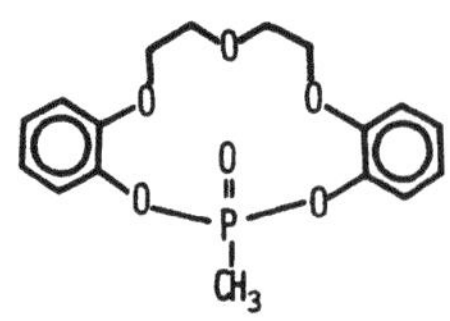	104–106	17	41
41	18	14	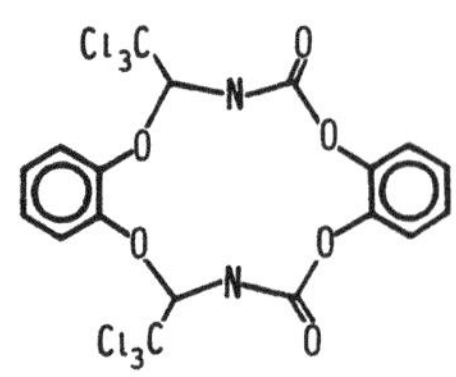	83–85	9	43

Table 5.14 (continued)

Cpd. No.	No. C's	Ring size	Structure	Mp or (bp) in °C	Yield (%)	Refs.
42	18	14	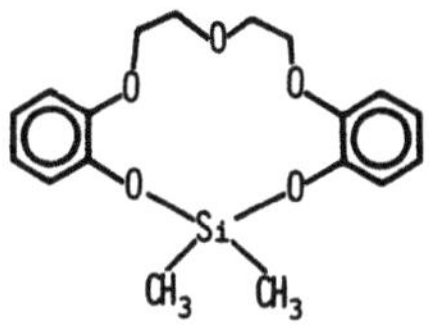	88–92	15	46
43	18	16	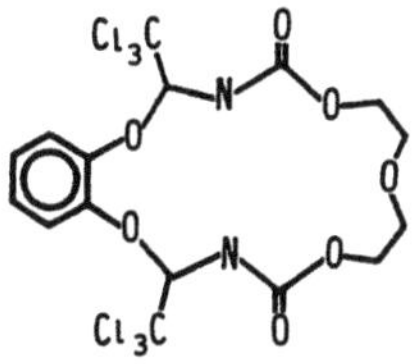	143–146	6	43
44	18	17	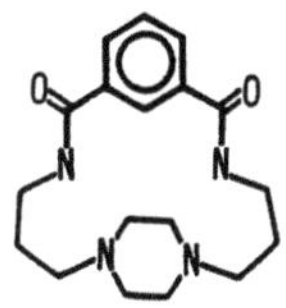	253–256	31	73
45	18	20	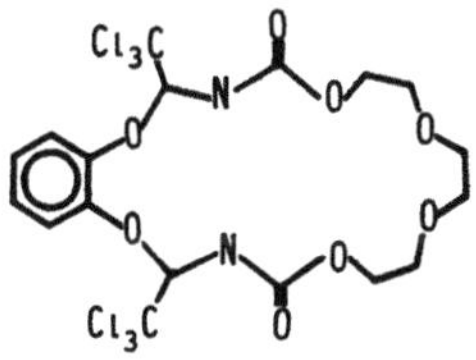	62–65	16	43
46	19	20	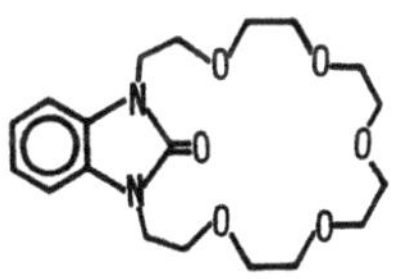	(95–100/0.05)	78.5	35
47	20	13	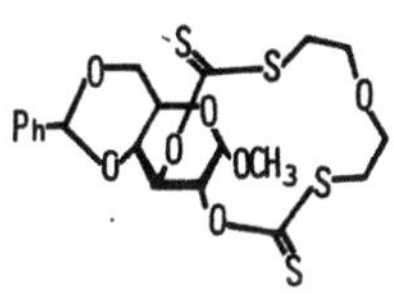	190–222	40	45
48	20	18	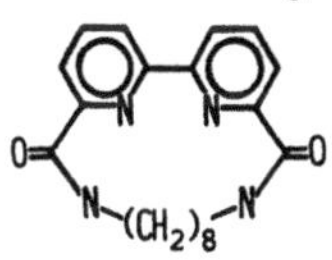	184–185	41	19
49	20	20	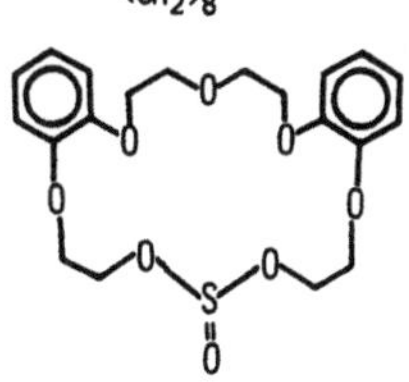	133 133	33 nr	58 59

Table 5.14 (continued)

Cpd. No.	No. C's	Ring size	Structure	Mp or (bp) in °C	Yield (%)	Refs.
50	20	21		192–193	24	19
51	20	23	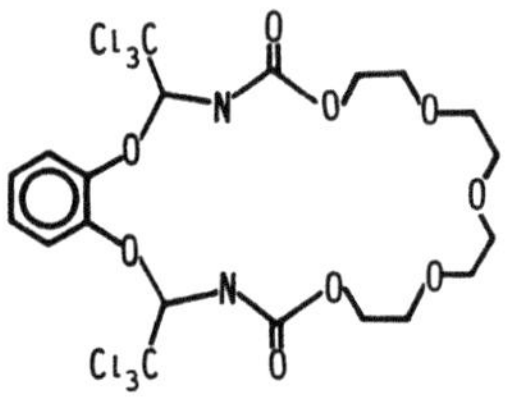	132–135	9	43
52	20	26	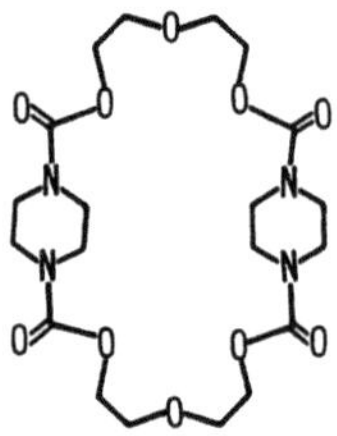	108–112	35	18
53	21	20	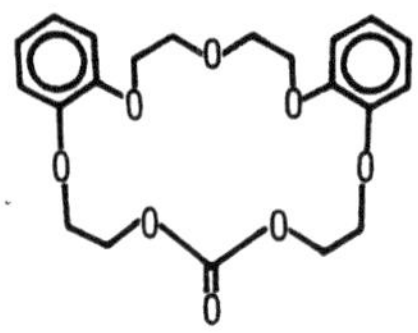	122–125	48	57
54	22	16	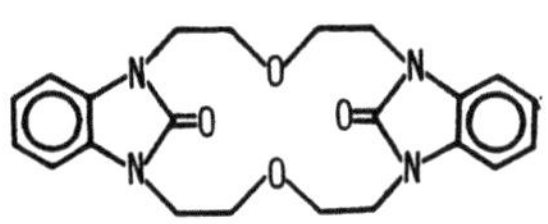	197–199	15	35
55	22	18	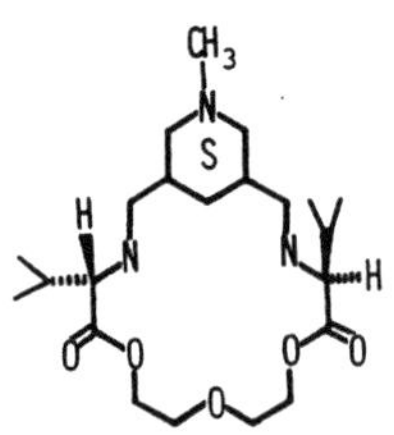	140.5–142.5	25	23
56	22	20	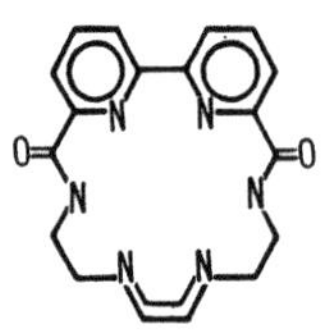	207	19	19

Table 5.14 (continued)

Cpd. No.	No. C's	Ring size	Structure	Mp or (bp) in °C	Yield (%)	Refs.
57	22	21	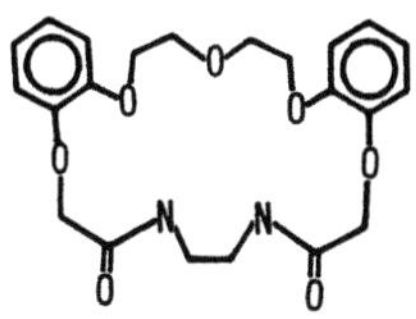	168–170	15	4
58	22	22	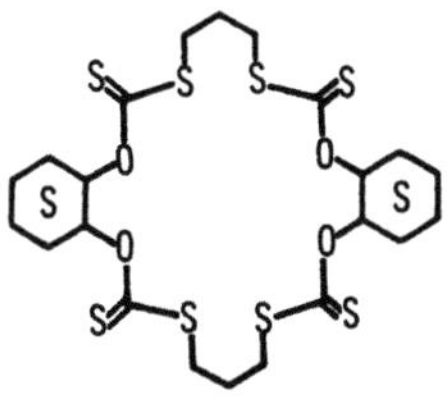	180–181	4	45
59	23	21	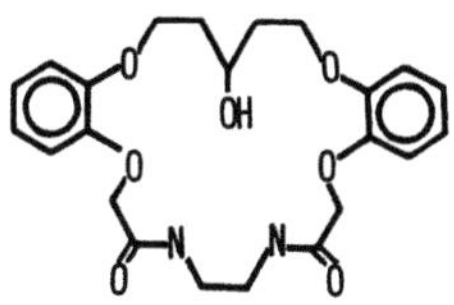	202	12	4
60	24	24	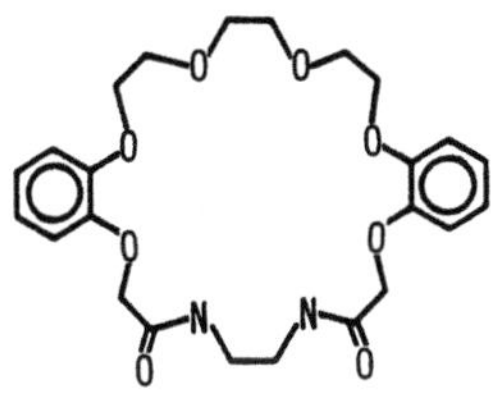	176–177	16	4
61	25	18	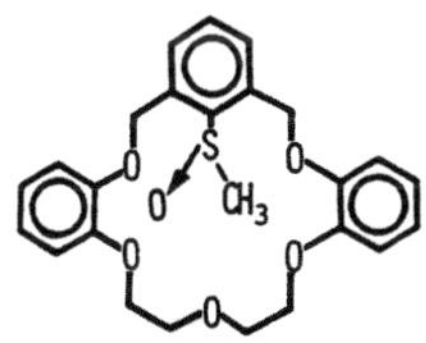	273–274	8	75
62	26	26	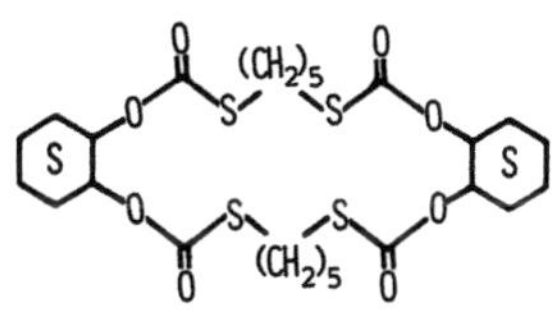	109–111	7	45
63	28	17	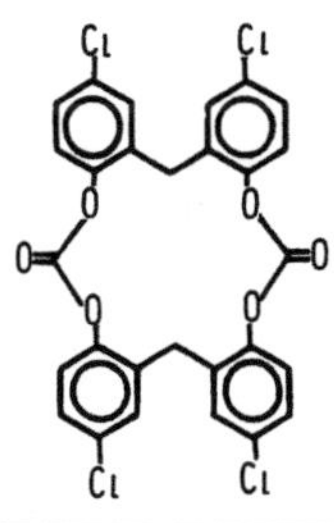	365–368	nr	17

Table 5.14 (continued)

Cpd. No.	No. C's	Ring size	Structure	Mp or (bp) in °C	Yield (%)	Refs.
64	28	23	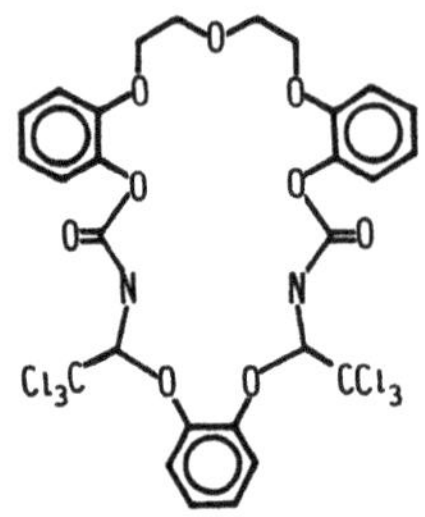	163–165	15	43
65	28	23	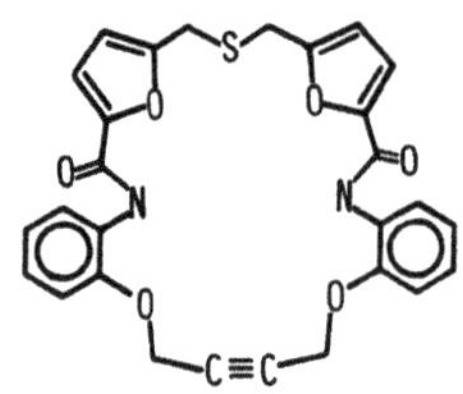	nr	nr	1
66	28	24	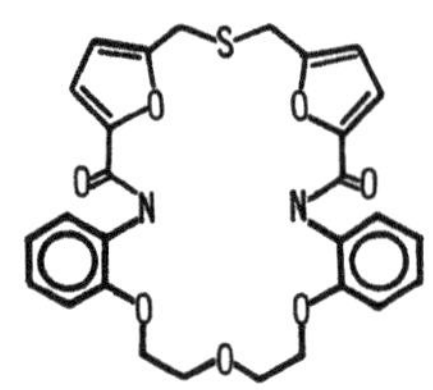	nr	nr	1
67	30	17	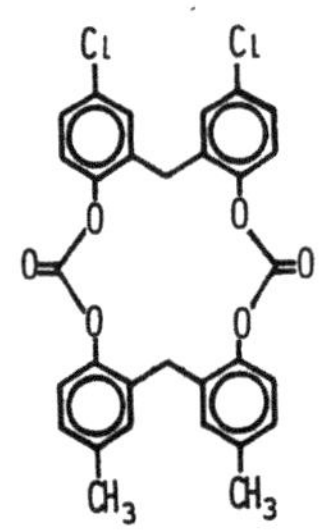	330–333		17
68	30	27	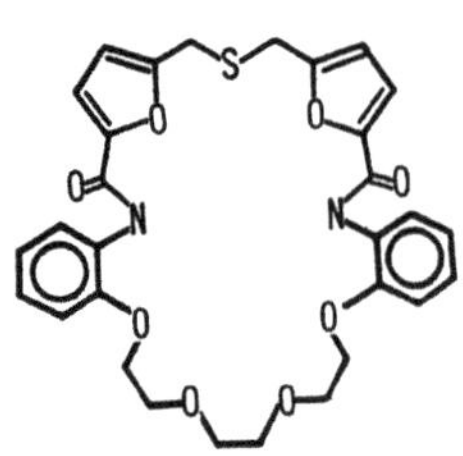	nr	nr	1

Table 5.14 (continued)

Cpd. No.	No. C's	Ring size	Structure	Mp or (bp) in °C	Yield (%)	Refs.
69	32	17	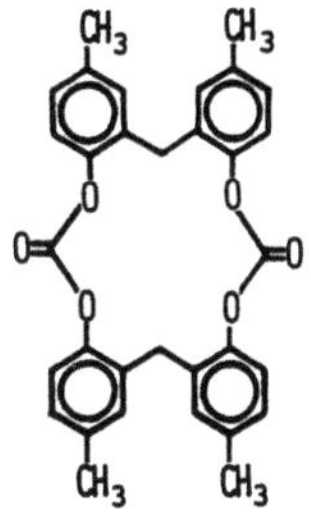	276–279 276–279 274.5–275	nr nr nr	17 51 61
70	34	28	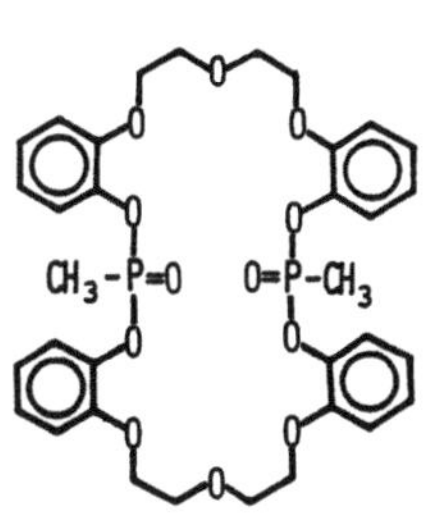	216–218	11	41
71	36	28	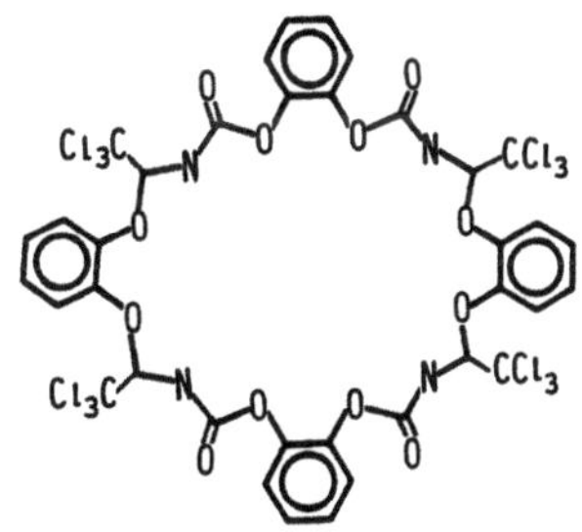	160–162	12	43
72	38	34	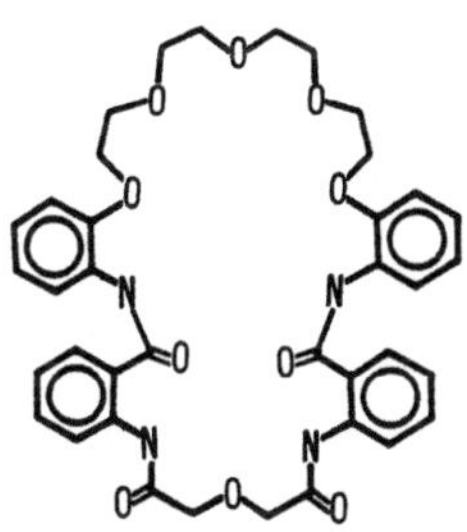	170	68	67

Table 5.14 (continued)

Cpd. No.	No. C's	Ring size	Structure	Mp or (bp) in °C	Yield (%)	Refs.
73	48	36	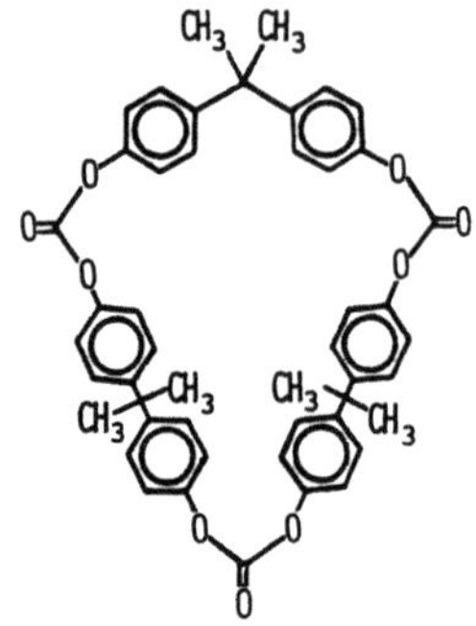	280	nr	51

5.12 References

1. Akopyan, T. R., Paronikyan, E. G., Sarkisyan, T. P., Avakimyan, D. A.: Arm. Khim. Zh. (1979) *32*(9), 716
2. Ammann, D., Pretsch, E., Simon, W.: Helv. Chim. Acta (1973) *56*, 1780
3. Asay, R. E., Bradshaw, J. S., Nielsen, S. F., Thompson, M. D., Snow, J. W., Masihdas, D. R. K., Izatt, R. M., Christensen, J. J.: J. Heterocyclic Chem. (1977) *14*, 85
4. Ashby, J., Hull, R., Cooper, M. J., Ramage, E. M.: Synth. Commun. (1974) *4*, 113
5. Berr, C. E.: J. Polymer Science (1955) *15*, 591
6. Bobranski, B., Sladowska, H.: Rocz. Chem. Ann. Soc. Chem. Polon. (1972) *46*, 459
7. Bogatsky, A. V., Lukyanenko, N. G., Kirichenko, T. I.: Tetrahedron Letters (1980) 313
8. Bradshaw, J. S., Asay, R. E., Maas, G. E., Izatt, R. M., Christensen, J. J.: J. Heterocyclic Chem. (1978) *15*, 825
9. Bradshaw, J. S., Asay, R. E., Baxter, S. L., Fore, P. E., Jolley, S. T., Lamb, J. D., Maas, G. E., Thompson, M. D., Izatt, R. M., Christensen, J. J.: Ind. Eng. Chem. Prod. Res. Dev. (1980) *19*, 86
10. Bradshaw, J. S., Baxter, S. L., Scott, D. C., Lamb, J. D., Izatt, R. M., Christensen, J. J.: Tetrahedron Letters (1979) 3383
11. Bradshaw, J. S., Bishop, C. T., Nielsen, S. F., Asay, R. E., Masihdas, D. R. K., Flanders, E. D., Hansen, L. D., Izatt, R. M., Christensen, J. J.: J. Chem. Soc. Perkin Trans. I (1976) 2505
12. Bradshaw, J. S., Hansen, L. D., Nielsen, S. F., Thompson, M. D., Reeder, R. A., Izatt, R. M., Christensen, J. J.: J. C. S. Chem. Comm. (1975) 874
13. Bradshaw, J. S., Maas, G. E., Izatt, R. M., Lamb, J. D., Christensen, J. J.: Tetrahedron Letters (1979) 635
14. Bradshaw, J. S., Maas, G. E., Lamb, J. D., Izatt, R. M., Christensen, J. J.: J. Am. Chem. Soc. (1980) *102*, 467
15. Bradshaw, J. S., Thompson, M. D.: J. Org. Chem. (1978) *43*, 2456
16. British Pat. 942,865, 27 Nov. 1963
17. British Pat. 1,047,767a, 9 Nov. 1966
18. Buhleier, E., Frensch, K., Luppertz, F., Vögtle, F.: Liebigs Ann. Chem. (1978) 1586
19. Buhleier, E., Wehner, W., Vögtle, F.: Chem. Ber. (1978) *111*, 200
20. Buhleier, E., Wehner, W., Vögtle, F.: Liebigs Ann. Chem. (1978) 537

21. Carothers, W. H., Dorough, G. L.: J. Am. Chem. Soc. (1930) *52*, 711
22. Crawford, L. M. R., Drewes, S. E., Sutton, D. A.: Chem. & Ind. (1970) 1351
23. DeVries, J. G., Kellogg, R. M.: J. Am. Chem. Soc. (1979) *101*, 2759
24. Drewes, S. E., Coleman, P. C.: J. C. S. Perkin Trans. I (1972) 2148
25. Drewes, S. E., Coleman, P. C.: J. C. S. Perkin Trans. I (1974) 2578
26. Drewes, S. E., Riphagen, B. G.: J. C. S. Perkin Trans. I (1974) 323
27. Drewes, S. E., Riphagen, B. G.: J. C. S. Perkin Trans. I (1974) 1908
28. Drewes, S. E., Riphagen, B. G.: J. C. S. Perkin Trans. I (1976) 2574
29. Ehrhart, W. A.. J. Org. Chem. (1968) *43*, 2930
30. Ehrhart, W. A.: U.S. Patent 3,431,279 4 Mar. 1969
31. Fore, P. E., Bradshaw, J. S., Nielsen, S. F.: J. Heterocycl. Chem. (1978) *15*, 269
32. Frensch, K., Vögtle, F.: Tetrahedron Letters (1977) 2573
33. Frensch, K., Vögtle, F.: J. Org. Chem. (1979) *44*, 884
34. Goodman, I., Nesbitt, B. F.: Brit. Patent 851,369, 12 Oct. 1960
35. Htay, M. M., Meth-Cohn, O.: Tetrahedron Letters (1976) 79
36. Izatt, R. M., Lamb, J. D., Asay, R. E., Mass, G. E., Bradshaw, J. S., Christensen, J. J.: J. Am. Chem. Soc. (1977) *99*, 6134
37. Izatt, R. M., Lamb, J. D., Maas, G. E., Asay, R. E., Bradshaw, J. S., Christensen, J. J.: J. Am. Chem. Soc. (1977) *99*, 2365
38. Jolly, S. T., Bradshaw, J. S.: J. Org. Chem. (1980) *45*, 3554
38a. Kellogg, R. M., Van Bergen, T. J., Van Doren, H., Hedstrand, D., Kool, J., Kruizinga, W. M., Troostwijk, C. B.: J. Org. Chem. (1980) *45*, 2854
39. Knutov, V. I., Butin, M. K., Voronkov, M. G.: Khim. Geterotsikl. Soedin. (1980) 123
40. Koegel, W., Schroeder, G.: Tetrahedron Letters (1978) 623
41. Kudrya, T. N., Shtepanek, A. S., Kirsanov A. V.: Zh. Obshch. Khim. (1978) *48*, 844
42. Kudrya, T. N., Shtepanek, A. S., Kirsanov, A. V.: Zh. Obshch. Khim. (1978) *48*, 927
43. Kudrya, T. N., Talanova, G. G., Bonadyk, S. V., Sinitsa, A. D.: Dopov. Akad. Nauk., UKr. RSR.: Geol., Khim. Biol. Nauki (1979) *6*, 434 [Chem. Abstr. (1980) *91*, 175313k]
44. Lamb, J. D., Izatt, R. M., Swain, C. S., Bradshaw, J. S., Christensen, J. J.: J. Am. Chem. Soc. (1980) *102*, 479
45. Lesiak, J., Boucher, M. N., Descotes, G.: Synthesis (1980) 240
46. Liptuga, N. I., Irodionova, L. F., Lozinskii, M. O.: Zh. Obshch. Khim. (1978) *48*, 1085
47. Maas, G. E., Bradshaw, J. S., Izatt, R. M., Christensen, J. J.: J. Org. Chem. (1977) *42*, 3937
48. Matsushima, K., Kawamura, N., Okahara, M.: Tetrahedron Letters (1979) 3445
49. Meraskentis, E., Zahn, H., Chem. Ber. (1970) *103*, 3034
50. Mirabella, F. M., Sheean, D. H.: Ger. Offen. 1,929,550, 2 Jan. 1970
51. Moody, L. S.: U.S. Patent 3,155,683, 3 Nov. 1964
52. Newkome, G. R., Kawato, T., Benton, W. H.: J. Org. Chem. (1980) *45*, 626
53. Newkome, G. R., Nayak, A., Sauer, J. D., Mattschei, P. K., Watkins, S. F., Fronczek, F., Benton, W. H.: J. Org. Chem. (1979) *44*, 3816
54. Newkome, G. R., Nayak, A., Otemaa, J., Van, D. A., Benton, W. H.: J. Org. Chem. (1978) *43*, 3362
55. Ors, J. A., Srinivasan, R.: J. Am. Chem. Soc. (1978) *100*, 315
56. Ors, J. A., Srinivasan, R.: J. C. S. Chem. Commun. (1978) 400
57. Pedersen, C. J.: J. Am. Chem. Soc. (1970) *92*, 391
58. Pedersen, C. J.: J. Org. Chem. (1971) *36*, 254
59. Pedersen, C. J.: U.S. Patent 3,987,061, 19 Oct. 1976
60. Piepers, O., Kellogg, R. M.: J.C.S. Chem. Commun. (1978) 383
61. Prochaska, R. J.: U.S. Patent 3,274,214, 20 Sept. 1966
62. Repin, H., Papanikolan, E.: J. Polymer Sci. (1967) 7, A-1, 3426
63. Sakamoto, K., Oki, M.: Chem. Lett. (1975) 615
64. Sakamoto, K., Oki, M.: Bull. Chem. Soc. Japan (1976) *49*, 3159
65. Samat, A., Elguero, J., Metzger, J.: J. C. S. Chem. Commun. (1979) 1182
66. Shanzer, A., Mayer-Shochet, N.: J. C. S., Chem. Commun. (1980) 176
67. Sieger, H., Vögtle, F.: Liebigs. Ann. Chem. (1980) 425

68. Stueben, K. C.: J. Polymer Sci. (1966) *4*, A-1, 829
69. Thompson, M. D., Bradshaw, J. S., Nielsen, S. F., Bishop, C. T., Cox, F. T., Fore, P. E., Maas, G. E., Izatt, R. M., Christensen, J. J.: Tetrahedron (1977) 3317
70. Van Bergen, T. J., Kellogg, R. M.: J. C. S. Chem. Commun. (1976) 964; Van der Veen, R. H., Kellogg, R. M., Vos, A., Van Bergen, T. J., J. C. S. Chem. Commun. (1978), 923
71. Van Keulen, B. J., Kellogg, R. M., Piepers, O.: J. C. S. Chem. Commun. (1979) 285
72. Vögtle, F., Newmann, P.: Tetrahedron Letters (1970) 115
73. Vögtle, F., Weber, E., Wehner, W., Nätscher, R., Grütze, J.: Chem. Zeitung (1974) *98*, 562
74. Vorländer, D.: Liebigs Ann. Chem. (1894) *280*, 167
75. Weber, E., Vögtle, F.: Chem. Ber. (1976) *109*, 1803
76. Weber, E., Vögtle, F.: Liebigs Ann. Chem. (1976) 891
77. Zahn, H., Repin, J. F.: Chem. Ber. (1970) *103*, 3041

Chapter 6

Miscellaneous Macrocycles

6.1 Macrocyclic Polyether Acetals

Although it has been known for many years that ethylene oxide and formaldehyde can form statistical copolymers, it was not until the interest in crown ethers developed that the potential of macrocyclic formals as complexing agents was recognized.

Rentsch and his coworkers[2] prepared 1,3,6,9,12-pentaoxacyclotetradecane (*1*) in 47% yield by heating tetraethylene glycol and paraformaldehyde in benzene solution. In this synthesis, the cationic resin "Amberlite IR 122" was used as catalyst. The product was isolated by high vacuum distillation at 175°, a temperature which may have also depolymerized some of the open-chained oligomers present[69].

$HO(CH_2CH_2O)_4H + (CH_2O)_n \longrightarrow$ 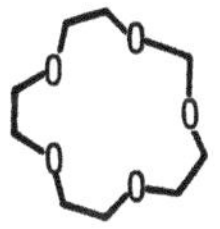 (6.1)

1

Kawakami, Suzuki and Yamashita showed that compound *1*, among many others, could be polymerized to derivatives of the corresponding open-chained species by treatment with boron trifluoride ether complex[36]. Yamashita and Kawakami[69] formed these same sorts of materials by heating the glycols and paraformaldehyde in the presence of toluenesulfonic acid. This led to "prepolymers" which were then thermally depolymerized to afford the cyclic oligomers which were separated by fractional distillation.

Although a number of such compounds have now been prepared, the chemistry of these species has not been extensively explored. Yamashita and Kawakami state: "Although they are easily hydrolyzed in acidic solution, they are fairly stable in basic solution and can be used as accelerating agents in nucleophilic synthetic reactions"[69]. They also demonstrated that the cyclic formals are less effective than crowns in the formation of n-butyl acetate from butyl bromide and a metallic acetate. In these experiments, equimolar amounts of nucleophile, substrate and ligand were used. In the same paper, it was stated that "the equilibrium constant for chelate formation was smaller for these cyclic formals than (for) crown ethers which have the same number of oxygen atoms"[69]. In later work, the same authors[35] expanded their observation as follows: "These macrocyclic formals were not as good complexing agents as crown ethers, but were much better than open chain polyethers which had the same number of oxygen atoms"[35]. In the S_N2 reaction described above, the cyclic formals were characterized as being "several times more effective" than the corresponding open-chained species[35].

Despite the statement above concerning the acid lability of cyclic formals, Gold and Sghibartz have shown that the acid catalyzed hydrolysis of these compounds is "markedly depressed by some metal ions"[30]. Although the smaller cyclic formals did not exhibit a substantial rate reduction even in the presence of small cations like lithium, in certain larger systems the rate reduction was more than an order of magnitude.

6.2 Macrocyclic Thioethers

6.2.1 Simple Sulfur Macrocycles

Although the polysulfur macrocycles became more interesting to the scientific community after Pedersen's report of the crown compounds (see Chap. 1), examples of such structures may be found in the literature dating back some time. The first report of a macrocyclic polysulfide appears to be that by Ray[64] who found in 1920 that when 1,2-dimercaptoethane and 1,2-dibromoethane were heated with alcoholic potassium hydrogen sulfide as shown in Eq. (6.2), triethylene trisulfide (*2*) and a substance presumed to be triethylene tetrasulfide (*3*) could be isolated.

$HSCH_2CH_2SH + BrCH_2CH_2Br \xrightarrow{KSH}$ 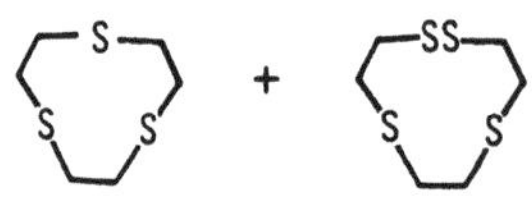 (6.2)

2 *3*

In 1960, Dann, Chiesa and Gates[25] of the Kodak Research Laboratories found that the disulfur analog of then unknown 18-crown-6 resulted from the reaction shown in Eq. (6.3). Since the complexation properties of such substances were not yet recognized, especially as they might apply to transition metals, this early and pioneering work has not had the recognition it probably deserved.

$Na_2S + ClCH_2(CH_2OCH_2)_2CH_2Cl \longrightarrow$ 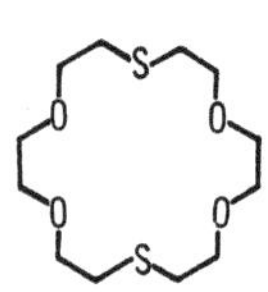 (6.3)

4

The first elaboration of crowns deliberately into sulfur-containing analogs seems to have been conducted by Pedersen[62,63]. The first of these efforts was directed at incorporating sulfur in complexes of macrocyclic polyethers[63]. Inclusion of sulfur as a portion of ring system *5* was also accomplished as shown in Eq. (6.4) and is reported in a patent [62] which issued before the latter[63], but was actually filed after it. We note here that the work of Busch[66] (see below) and the incorporation of sulfur in cryptands (see Chap. 8) predates these reports.

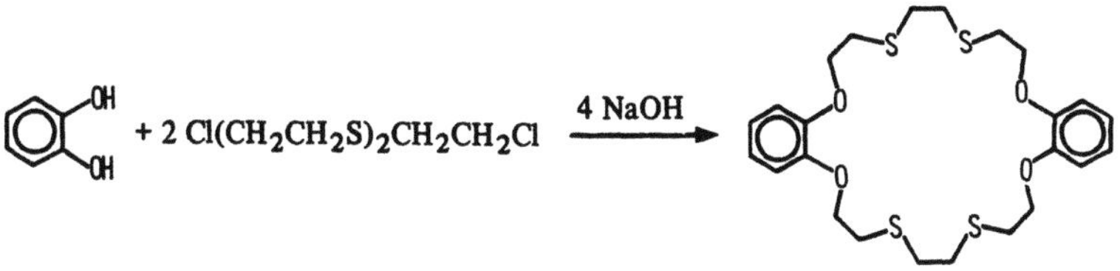

(6.4)

5

At about the same time Pedersen's work was underway, Bradshaw and his coworkers had undertaken the systematic preparation of numerous mono- and polysulfur macrocycles[15–17]. The syntheses were carried out in a fashion similar to that shown in Eq. (6.3). The two principal sources of sulfur in the first paper from this group[16] were sodium sulfide and 1,2-ethanedithiol. The latter was utilized in the presence of sodium hydroxide base. Ethanol was generally used as solvent and the reactions were conducted at high dilution. Products were purified either by vacuum distillation or recrystallization as appropriate. The yields were generally in the 5–30% range.

The example shown in Eq. (6.5) was conducted using 0.5 mole of bis-(2-mercaptoethyl) ether and 0.5 mole of 1,8-dichloro-3,6-dioxaoctane. The base was sodium hydroxide (1 mole) and between 900 and 2,000 mL of ethanol were used. The exact quantity of ethanol used in this synthesis is not clear from the experimental description. The reaction time was somewhere between 7 and 14 hours (again, unspecified), after which time the product (*6*) was isolated by vacuum distillation, as a viscous oil, in 27% yield. Similar approaches were utilized by this group for the synthesis of other analogs[15,17].

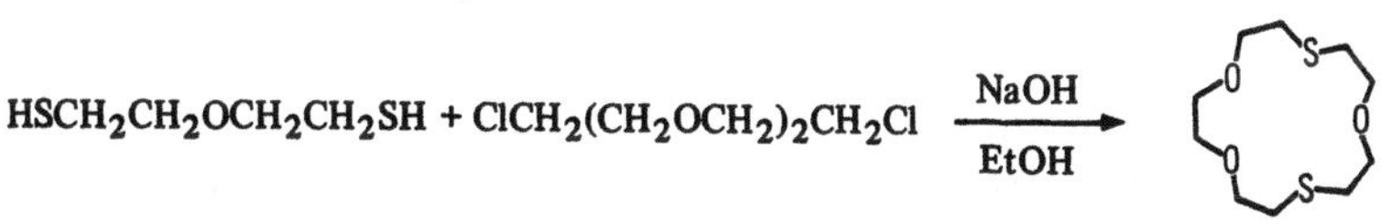

(6.5)

6

6.2.2 Macrocycles with Sulfur-containing Subunits

There are quite a number of crown compounds known which contain a thiophene or other sulfur-containing subunit. The reader is directed to the tables at the end of each chapter in order to locate these species. Because of the expense of including each structure under each possible category, some sulfur-containing compounds may not be listed in this chapter. It is hoped that the tables are well enough organized that the desired compound can be located if it is known by considering other structural features.

There are, however, some crown type compounds which contain no structural feature except the thiophene subunit, and these deserve some comment here. This is especially true since one of these compounds was prepared very early in the history of crown compounds. Ahmed and Meth-Cohn were interested in sulfur analogs of the porphyrin ring system and prepared compound 7 in 1969 by the method shown in Eq.

(6.6)[1]. The yield of 7 might have been higher had a transition metal template been used, but to our knowledge, this has not been tried.

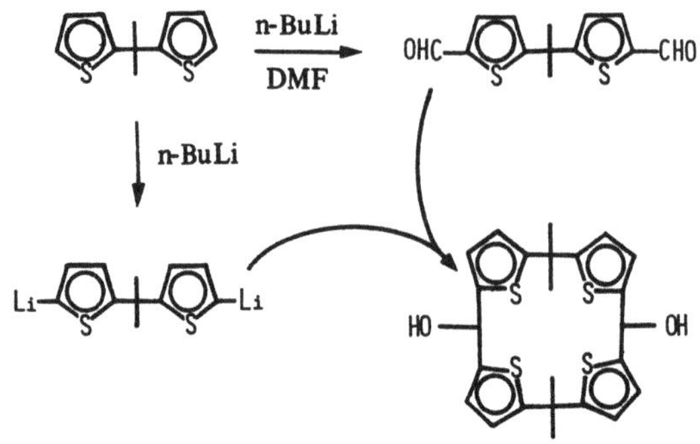

7

(6.6)

More recently, Bhattacharjya and Hortmann have condensed the dilithium salt of 1,3-propanedithiol with 2,5-bischloromethylthiophene at high dilution (aqueous THF). The product (*8*) shown in Eq. (6.7) was obtained in only 2.5% yield[6].

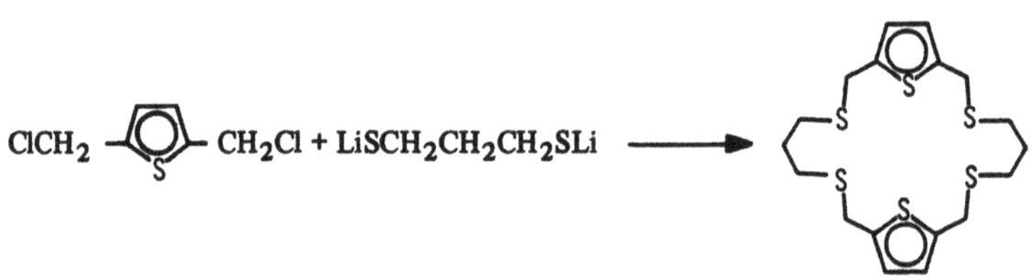

8

(6.7)

6.2.3 Macrocyclic Polysulfides

Although the first all-sulfur macrocycles were prepared many years ago[25,64] the first systematic study of such compounds was initiated by Busch and his coworkers[66], who were interested in the cation binding properties of such ligands. A sequential synthesis was utilized to produce 1,4,8,11-tetrathiacyclotetradecane [tetrathia-14-crown-4 (*10*)][67]. In the first step, 1,3-propanedithiol is metallated using sodium and alkylated with 2-chloroethanol. The diol was then treated with thiourea to form the dimercapto-dithioether compound *9*. The latter was once again metallated with sodium and allowed to react with 1,3-dibromopropane. The yield of *10* in the ring closure step, conducted at high dilution in absolute ethanol, was 7.5% after recrystallization. The entire sequence is illustrated in Eq. (6.8)[67].

$$NaSCH_2CH_2CH_2SNa + ClCH_2CH_2OH \xrightarrow[\text{(abs)}]{\text{EtOH}} CH_2(CH_2SCH_2CH_2OH)_2$$

(6.8)

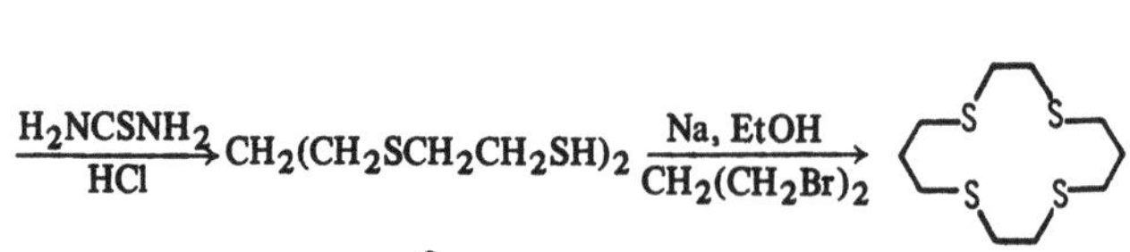

9 10

Somewhat later, Travis and Busch reported that extraction of the residue from Eq. (6.8) with hot ethanol afforded the dimer of *10* in 25% yield[78]. The latter, 1,4,8,11,15,18,22,25-octathiooctacosane, was characterized by mass spectrometric analysis as well as the customary analytical methods. It was found "that by dilution of reactants prior to mixing, the yield of [monomer] is greatly increased (50–60%) while the yield of [dimer] is lowered substantially"[78]. One might have expected the larger rather than the smaller ring to be more favored at higher dilution, but there is no further comment on this point.

It is interesting to note that a comparison of binding constants between polysulfur and polynitrogen macrocycles with their corresponding open-chained analogs has been made. Smith and Margerum[73] have shown that the difference in stability constants for Ni^{II} and *10* or its open-chained analog is relatively small, whereas the difference is large for the corresponding nitrogen compounds. The authors note that this "lends support to the hypothesis that (in addition to contributions from configurational entropy) the differing degree of ligand solvation by hydrogen bonding in water is a major source of the macrocyclic effect"[73].

Ochrymowycz and his coworkers have also prepared a number of polysulfur macrocycles for use in biological or biological model systems[28,29,59], The synthetic methodology is essentially similar to that described above except that certain of the sulfur containing fragments were prepared by addition reactions to ethylene. Two examples of this approach, taken from ref. 59, are shown in Eq. (6.9).

$$\text{a)}\quad SCl_2 + CH_2{=}CH_2 \longrightarrow ClCH_2CH_2SCH_2CH_2Cl$$

$$\text{b)}\quad (HSCH_2CH)_2S + SO_2Cl_2 \longrightarrow (ClSCH_2CH_2)_2 \xrightarrow{CH_2=CH_2} ClCH_2CH_2SCH_2CH_2SCH_2CH_2SCH_2CH_2Cl \qquad (6.9)$$

6.3 Mixed Sulfur-Nitrogen Macrocycles

Although a large number of such compounds have been prepared, many of the mixed nitrogen-sulfur macrocycles are included in other tables, most notably in Chap. 4. Some have also been recorded in Chap. 8 as monocyclic precursors to bi- or polycyclic ligands (i.e., cryptands). The reader is directed to the tables of other chapters where the desired compound, if known, may be reported.

Vögtle and his coworkers have prepared a large number of compounds containing sulfur and nitrogen[81–87,90–93]. An especially interesting type of compound is *11* which contains a pyridine subcyclic unit. This compound is prepared from one equivalent each of 2,2-dibromodiethyl ether and 2,6-dimercaptopyridine in a mixture of ethanol and 2-butanone. An excess (3 equivalents) of KOH served as base. Crystalline *11* (mp 168–170°) was obtained in 9% yield[91]. The approach is illustrated in Eq. (6.10).

It is interesting to compare this approach with the method used by Newkome and his coworkers for the preparation of a similar material containing the pyrimidine subcyclic unit[58]. In this case, 2,4-dichloropyrimidine was allowed to react with 1,2-eth-

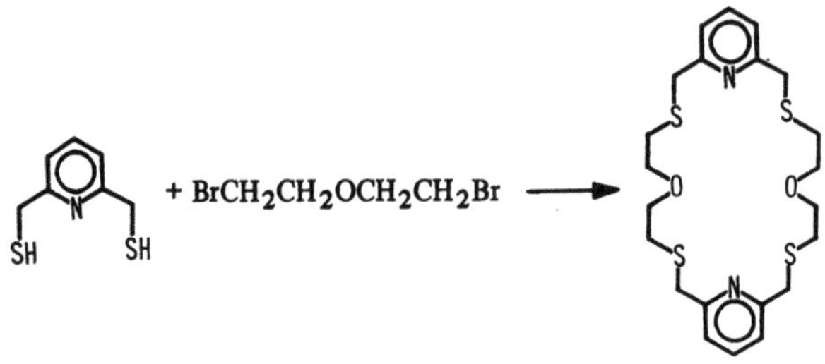

11

(6.10)

anedithiol and sodium hydride. Several products were formed, including the bridged bis-pyrimidine, *12*. Treatment of *12* with diethylene glycol and sodium hydride afforded macrocycle *13*, as a crystalline solid, mp 169–172°, in 55% yield.

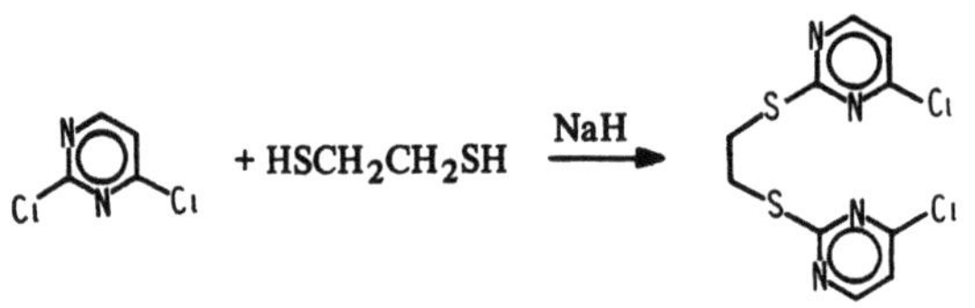

12

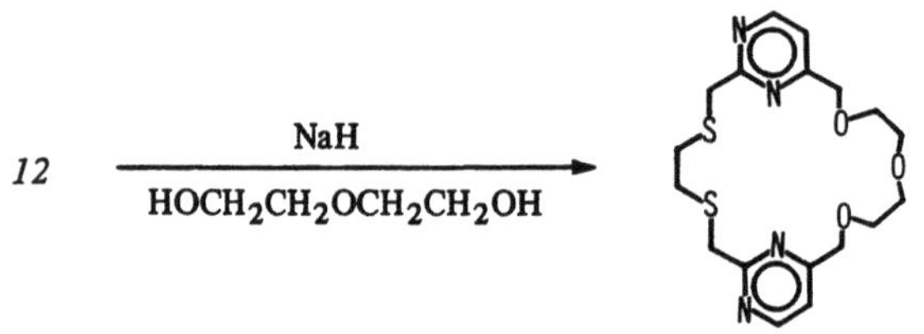

13

(6.11)

Quite a number of mixed sulfur-nitrogen macrocycles have been prepared, but these have largely been by the methods outlined in Chaps. 4 and 5 for the respective heteroatoms. An alternative method, involves the formation of a Schiff base, followed by reduction to the fully saturated system, if desired. An interesting example of the Schiff base formation is found in the reaction formulated in (6.12). Dialdehyde *14* is added to ethylenediamine in a solution containing ferrous ions. Although fully characterized, the yield for the reaction is not recorded[7]. To avoid confusion with the original literature, we note the claim that the "dialdehyde [*14*] was readily prepared in good yield by reaction of the disodium salt of 3-thiapentane-1,5-diol"[7]. The latter must be the dithiol rather than the diol.

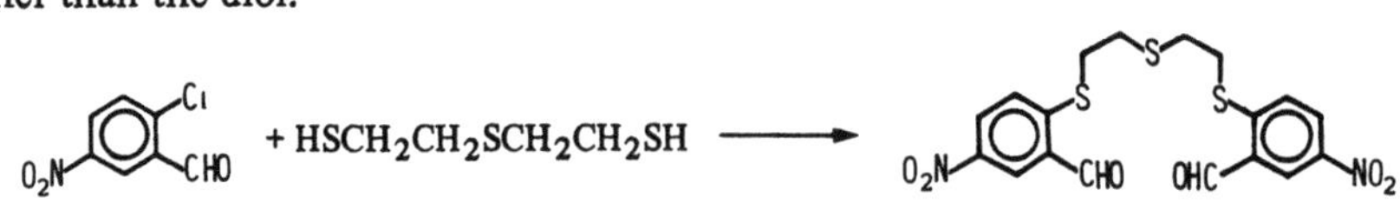

14

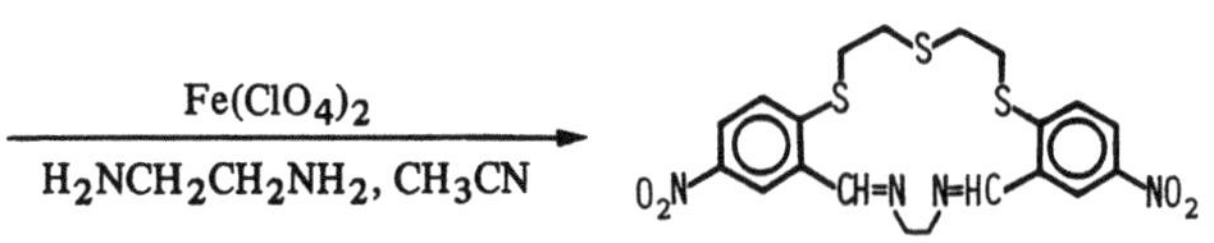

15

(6.12)

6.4 Phosphorus and Arsenic-containing Macrocycles

Although most of the macrocycles that contain phosphorus or arsenic which have thus far been prepared, are primarily transition metals binders, two compounds have been prepared which are essentially crown ethers containing phosphorus. Kudrya, Shtepanek and Kirsanov[43] have prepared two compounds which are essentially polyoxygen macrocycles but which contain one or two methylphosphonic acid esters as part of the ring. These two macrocycles are shown below as *16* and *17* and are both prepared by the reaction of 2,2′[oxybis(ethyleneoxy)] bisphenolate with methylphosphonic dichloride in a mixture of acetonitrile and benzene. The crystalline monomer (*16*) and dimer (*17*) were isolated in 17% and 11% yields respectively as indicated in Eq. (6.13).

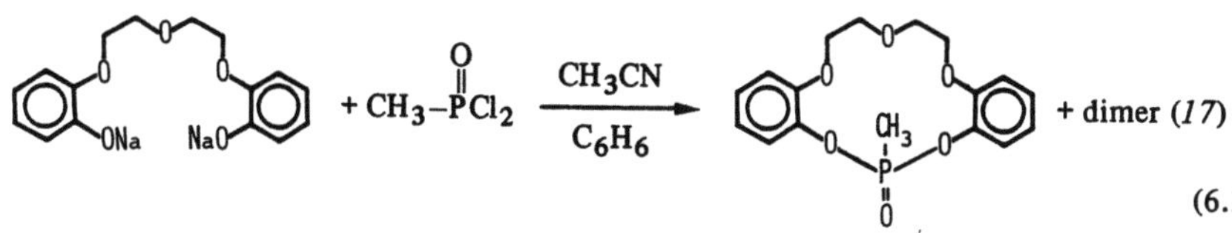

(6.13)

16

Quite a number of cyclic phosphonium salts have been prepared over the years. We have not generally included compounds here which contain fewer than three potential binding sites in a ring unless they are mentioned as by-products in another reaction. Horner, Kunz and Walach[32] have utilized the well-known alkylation approach to prepare cyclic phosphonium salts containing four phosphorus atoms. The formation of the cyclic tetraphosphonium salts is shown below in Eq. (6.14).

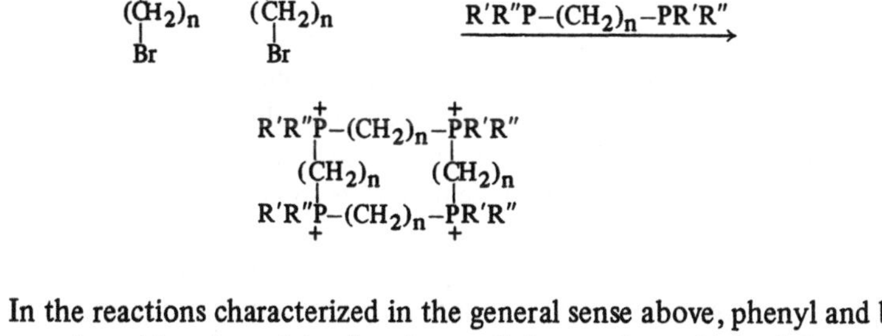

(6.14)

In the reactions characterized in the general sense above, phenyl and benzyl were the organic residues bound to phosphorus. It is not clear which of these groups is removed in the LAH dealkylation step. It should also be noted that as tetraquaternary phosphonium salts, these species do not qualify as crown analogs in the strictest sense because of the absence of lone pair electrons on phosphorus. Furthermore, the quaternary phosphorus can resist oxidation much better than secondary or tertiary phosphorus (see below) and these materials are therefore easier to prepare, at least in principle. The cyclic systems do not appear to have been dealkylated to afford tetraphosphorus systems which might be very interesting cation binders, and this remains an interesting problem.

The first polyphosphino macrocycles designed specifically for use as transition metal binders were reported in 1977 in back-to-back communications by Rosen[27] and Kyba[46] and their coworkers. The macrocycles reported in these papers were quite similar in some respects, but the synthetic approaches were markedly different. DelDonno and Rosen[27] began with bis-phosphinate *18*. Treatment of the latter with "Vitride" reducing agent and phosphinate *19*, led to the tris-phosphine, *20*. Formation of the nickel (II) complex of *20* followed by double alkylation (cyclization) and then removal of Ni^{++} by treatment of the complex with cyanide, led to *21* as illustrated in Eq. (6.15). The overall yield for this sequence is about 10%.

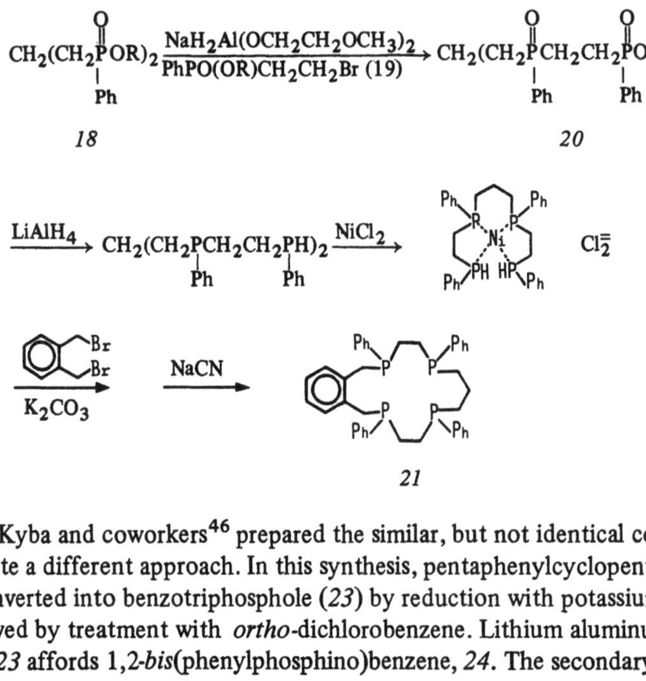

(6.15)

21

Kyba and coworkers[46] prepared the similar, but not identical compound, *26*, using quite a different approach. In this synthesis, pentaphenylcyclopentaphosphine (*22*) is converted into benzotriphosphole (*23*) by reduction with potassium metal in THF, followed by treatment with *ortho*-dichlorobenzene. Lithium aluminum hydride reduction of *23* affords 1,2-*bis*(phenylphosphino)benzene, *24*. The secondary phosphine may be deprotonated with n-butyllithium and alkylated with 3-chlorobromopropane. The two-armed bis-phosphine (*25*) which results may be treated with the dianion of *24* at high dilution to yield macrocycle *26*. The overall yield of *26* is about 4%. The synthetic approach is illustrated in Eq. (6.16), below.

Kyba and his coworkers[46] were also able to prepare a number of smaller, phosphorus-containing rings and several macrocycles containing both sulfur and phosphorus atoms.

Although the synthetic approaches taken by these two groups appear more or less straightforward, there are two difficulties associated with the synthesis of these phosphorus macrocycles which make these preparations all the more substantial. The first of these problems is the air sensitivity of phosphines, especially primary and secondary

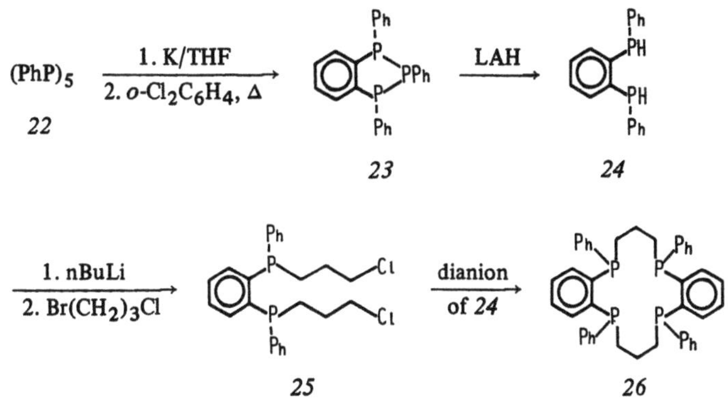

(6.16)

ones. In his preparation of these phosphorus-macrocycles, Kyba[46] refers to "anaerobic chromatography". Rosen[27] also notes that compound *21*, illustrated in Eq. (6.15) is an "air-sensitive oil".

The second problem concerns isomers of these various systems. As illustrated, *26* might be presumed to be planar. Unlike nitrogen, the inversion at phosphorus is very slow and there are, in fact, five possible isomers of it (three *meso* and two *dl* pairs)[46]. The interconversion of these isomers is reported in a subsequent paper by Davis, Hudson and Kyba[26] along with x-ray crystallographic data on the structure of the so-called "isomer B".

Kyba and his coworkers[44] have made numerous phosphorus-containing macrocycles and many of these are recorded in the tables at the end of this chapter. In addition, this group has prepared and reported quite a number of arsinocycles[44]. The observations concerning phosphorus macrocycles apply in general to the arsinocycles as well, but the inversion barrier is even higher for the latter than the former[44].

Quite recently, Ciampolini and coworkers[19] have reported the synthesis of two isomeric mixed oxygen-phosphorus macrocycles and the crystal structures of their cobalt complexes. Synthesis of macrocycle *27* was accomplished by condensation of 1,2-bis-(phenylphosphino)ethane dianion with 2,2′-dichlorodiethyl ether in THF. The two isomers of *27* were isolated in 1.5% and 2% yield. The synthesis is formulated in Eq. (6.17), below.

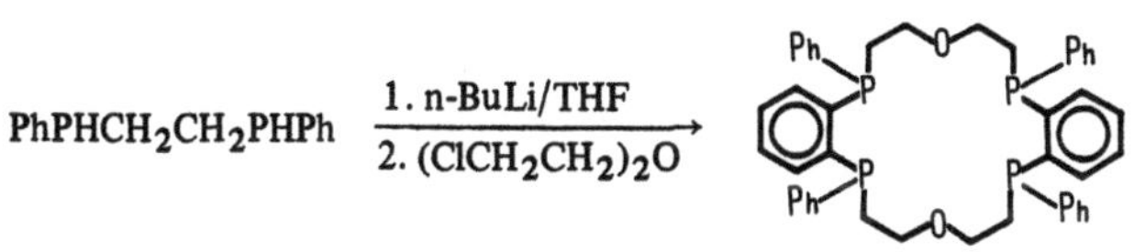

(6.17)

27

6.5 Silicon-containing Macrocycles

Examples of silicon-containing macrocycles are rare and are perhaps incidental to the report of crown-type compounds. Obviously, silicon is more like carbon than a heteroatom in its ability to complex a cation. Nevertheless, there is a recent report of a crown compound which contains silicon as part of the macroring. Liptuga, Irodionova and Lozinskii[50] have treated 2,2'[oxybis-(ethyleneoxy)]bisphenolate with dichlorodimethylsilane[50] in analogy to the phosphocycle synthesis illustrated in Eq. (6.13)[43]. The sila-cycle was obtained in 15% yield as a crystalline solid (mp 88–92°) as shown in Eq. (6.18), below.

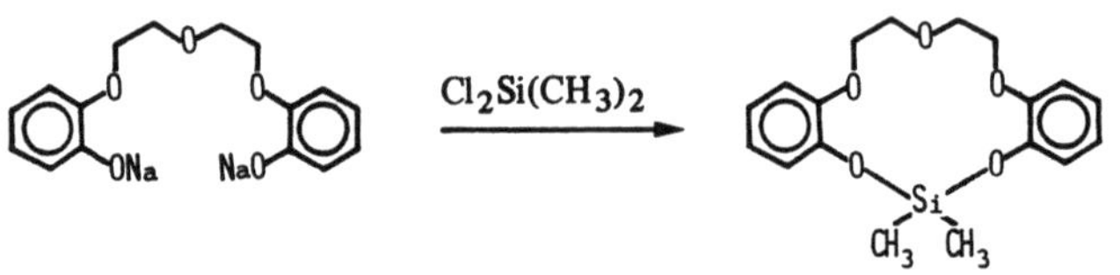

28

(6.18)

Bradshaw and Hui[14], in their review of sulfur-containing macrocycles, have included the analog of compound 7 (this chapter) in which each carbon bridge is replaced by a dimethylsilyl bridge. Nevertheless, silicon-containing crown relatives remain rare.

6.6 Polymer-bound Macrocycles

A good deal of work has been done on polymeric crown ethers during the last decade. Hogen Esch and Smid have been major contributors from the point of view of cation binding properties, and Blasius and coworkers have been especially interested in the cation selectivity of such species. Montanari and coworkers have developed a number of polymer-anchored crowns for use as phase transfer catalysts. Manecke and Storck[52] have recently published a review titled "Polymeric Catalysts", which may be useful to the reader in gaining additional perspective.

The polymeric crown ether species which have been prepared to date can be categorized very roughly into four major groups. The first of these groups comprises cation-binding species (crowns, cryptands) which are grafted to an existing backbone. A typical backbone is the chloromethylated divinylbenzene-styrene copolymer sometimes referred to as "Merrifield's polymer". The second group is constituted of crown ethers which have been copolymerized with other species and the crown forms an integral part or the backbone, rather than being appended to it as in the group above. The third group is probably the simplest in concept. These are polymers derived from polymerization of a crown ether containing a pendant polymerizable function. The final, and newest class, are the molecules referred to as "pseudo-crowns". These groups are discussed in turn, below.

6.6.1 Pendant Crown Polymers

Montanari and coworkers[20,44] have been particularly active in this area. They have generally utilized crowns or cryptands having long arms attached to them. These lipophilic arms are typically terminated in a primary or secondary amino function which may serve as a nucleophile in the reaction with a chloromethylated polystyrene residue.

A typical system is a chloromethylated polystyrene resin "cross-linked with 2 or 4% *p*-divinylbenzene and different amounts of chloromethylated sites (0.7–3.7 mequiv. of Cl per g of polymer)"[20]. The reaction is shown schematically in Eq. (6.19) and additional information may be found in Sects. 8.3 and 8.8.

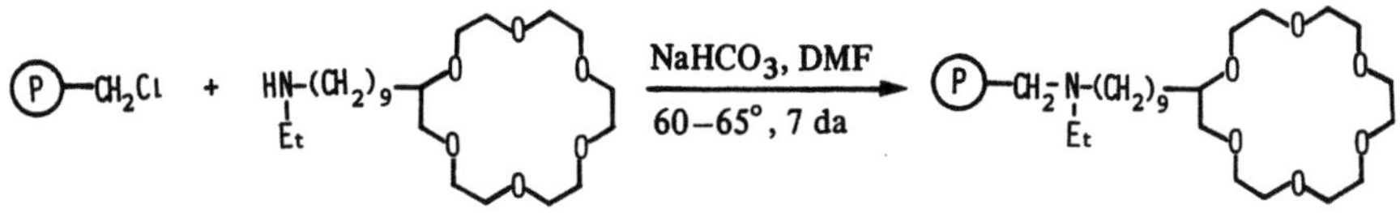

(6.19)

Blasius and coworkers have offered a somewhat different approach to systems of this general type[9]. In the first of these, shown in Eq. (6.20), he utilizes a hydroxymethyl-substituted 15-crown-5 residue as the nucleophile. This essentially similar to the Montanari method. The second approach is a variant also, but more different in the sense that covalent bond formation is effected by a Friedel-Crafts alkylation. In the reaction

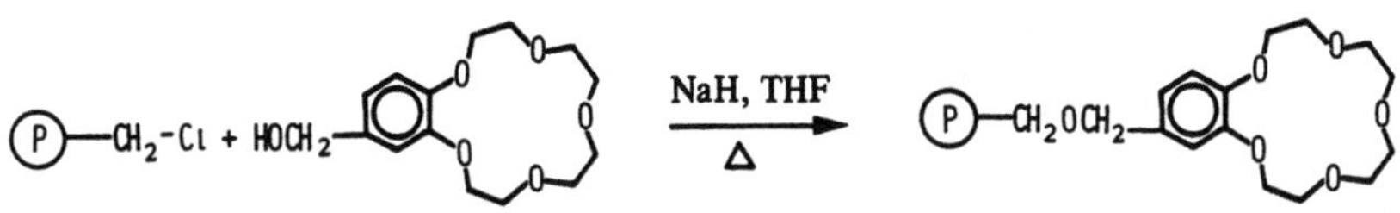

(6.20)

shown in Eq. (6.21), benzo-15-crown-5 is alkylated by the polystyrylmethyl carbonium ion, formed using $TiCl_4$ as a catalyst. The products formed by these two reactions are similar, but not identical. The obvious difference is that there is an ether link between the crown and polymer in the product of (6.20). In the product from (6.21), not only is there no ether linkage, but there exists the possibility of having the *ortho* as well as the *meta*-isomer as product.

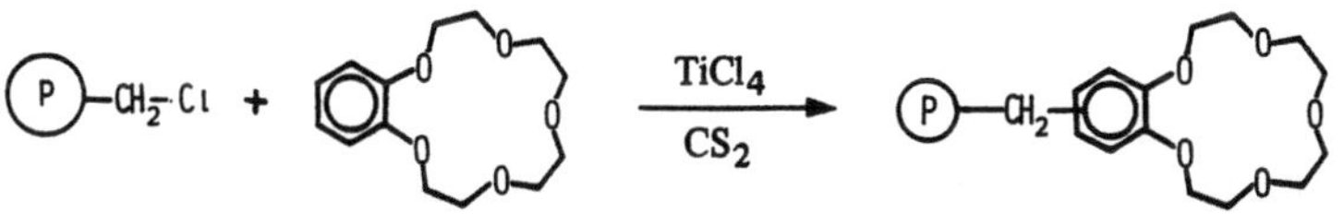

(6.21)

Before leaving consideration of this group of compounds, mention should be made of one unusual species of this general class. Tabushi, Kobuke and Nishiya[75] have reported a polymer-bound hexaketone, illustrated below as compound *29*, which is apparently of use in binding uranyl (UO_2^{++}) ions. The polymer has, in fact, been utilized in extraction of the uranyl ion directly from seawater[75].

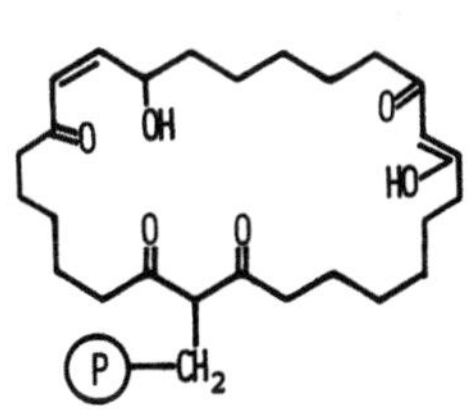

29

6.6.2 Crown Ether Copolymers

Most of the compounds in this class have been prepared from preexisting crown ether units. By far, the most common approach is to use a benzo-substituted crown and an electrophilic condensation polymerization. A patent issued to Takekoshi, Scotia and Webb (General Electric)[77] in 1974 which covered the formation of glyoxal and chloral type copolymers with dibenzo-18-crown-6. The latter were prepared by stirring the crown with an equivalent of chloral in chloroform solution. Boron trifluoride was catalyst in this reaction. The polymer which resulted was obtained in about 95% yield. The reaction is illustrated in Eq. (6.22).

(6.22)

An alternative copolymerization is illustrated by the method of Blasius[9]. In this preparation, a phenol-formaldehyde (novolac) type system is formed. Monobenzo-18-crown-6, for example, is treated with a phenol (or alkylated aromatic like xylene) and formaldehyde in the presence of acid. As expected for this type of reaction, a highly crosslinked resin results. The method is illustrated in Eq. (6.23). It should also be noted that the additional aromatic can be left out and a crown-formaldehyde copolymer can be prepared in analogy to (6.22)[9].

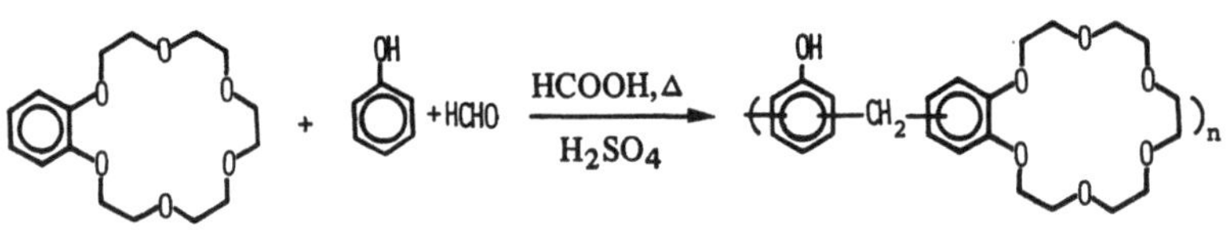

(6.23)

A different variety of copolymer has been prepared by Gramain and Frere[31] who treated 1,10-diaza-18-crown-6 with the bisglycidyl ether of bisphenol A. The reaction was conducted at reflux in a mixture of THF and methanol. The polymer, illustrated in Eq. (6.24) was formed 83% yield. The polymer was appparently quite stable, surviving aging tests conducted over a two-year period.

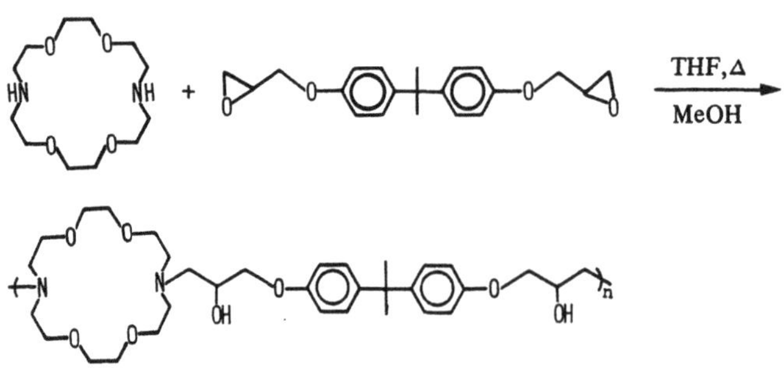

(6.24)

6.6.3 Polycrowns

For the purpose of our discussion, a polycrown is here defined as a polymer system arising by polymerization of a crown monomer unit. Extensive work has been done in this field by Kopolow, Hogen Esch and Smid and the examples presented here are taken from a paper by all three of these workers[39]. A typical preparation of vinylbenzo-15-crown-5 is accomplished according to the scheme shown as Eq. (6.25)[39]. One might also have utilized a formylation/Wittig sequence on benzo-15-crown-5 to accomplish the same end.

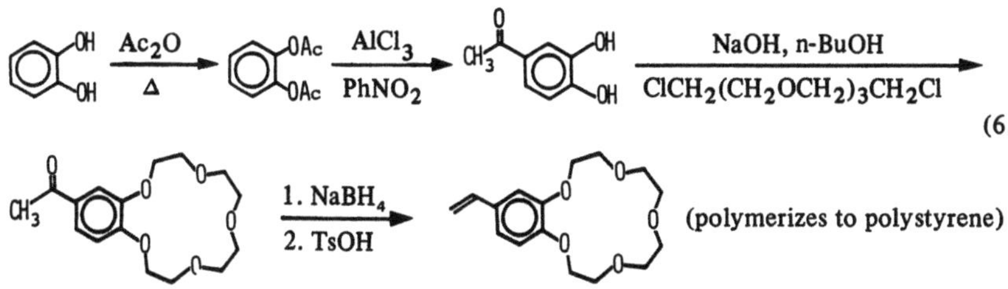

(6.25)

Monovinyldibenzo-18-crown-6 may be prepared by utilizing a preformed two-armed, monobenzo precursor as illustrated in Eq. (6.26)[39]. Such precursors are discussed in Chap. 3. With the two-armed precursor in hand, crown formation can be accomplished using acetylcatechol just as illustrated in Eq. (6.25), above.

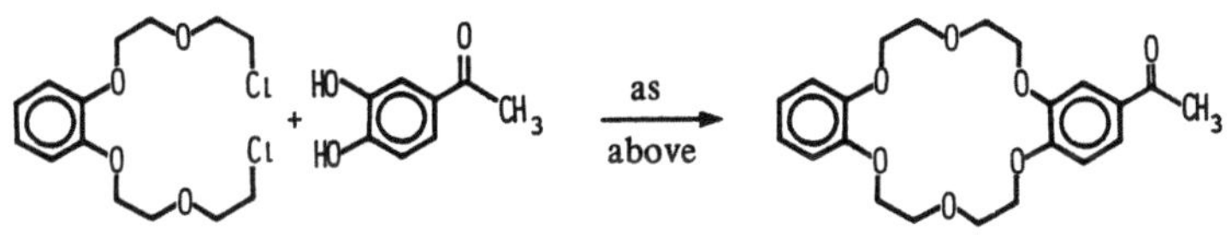

(6.26)

6.6.4 Pseudocrowns

Warshawsky and coworkers have recently reported the synthesis of a class of compounds which they call "polymeric pseudocrown ethers"[89]. A chloromethylated polystyrene matrix is used here as in 6.6.2, but instead of adding a crown to the backbone, a strand of ethyleneoxy units is allowed to react at two different positions on the chain, thus forming a crown. Such systems must necessarily be statistical, and the possibility exists for forming interchain bridges as well as intrachain species. Nevertheless, polymers which could be successfully characterized in a variety of ways were formed. A schematic representation of such structures is illustrated below as compound *30*[89].

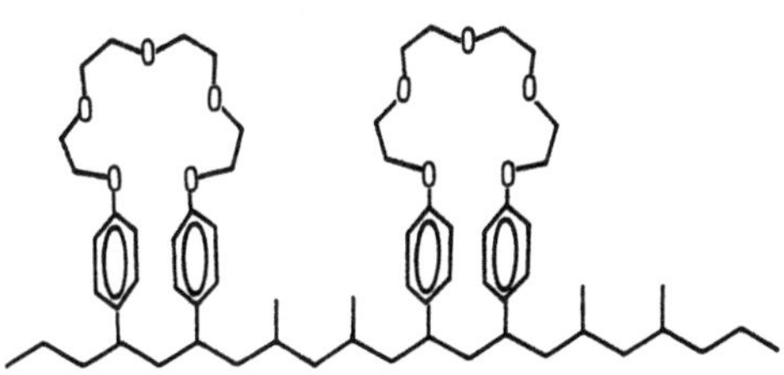

30

6.7 Listing of Tables

6.8 Tables 6.1–6.4

Table 6.1. Macrocyclic acetals and thioacetals

Cpd. No.	No. C's	Ring size	Structure		Mp or (bp) in °C	Yield (%)	Refs.
1	6	8	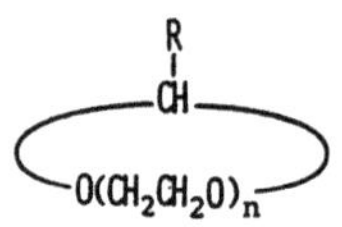	n = 2, R = CH_3	(32/1)	29	30
2	7	11		n = 3, R = H	27	65	2
3	8	11		n = 3, R = CH_3	nr	nr	35
					(60/0.2)	5	30
4	9	14		n = 4, R = H	23.5	47	2
					nr	nr	65
					20.5–21.5	nr	36
5	10	14		n = 4, R = CH_3	nr	nr	35
					(82/0.15)	8	30
6	12	16	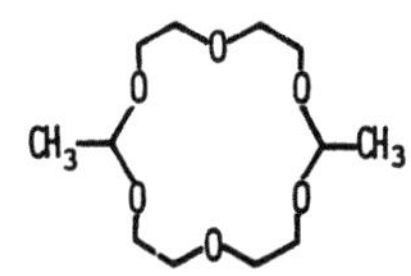		36–37	1.5	30
7	12	17	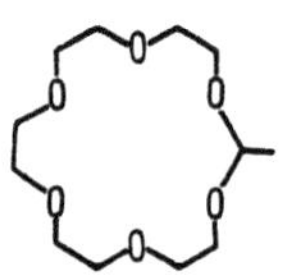		nr	nr	35
					(120/0.1)	13	30
8	14	9	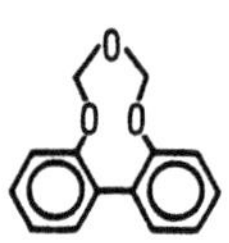		144–145	14	3
9	14	9			150–152	59	3

Table 6.1 (continued)

Cpd. No.	No. C's	Ring size	Structure	Mp or (bp) in °C	Yield (%)	Refs.
10	14	10	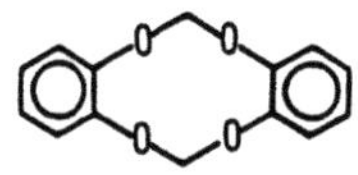	261–262	nr	13
11	14	20	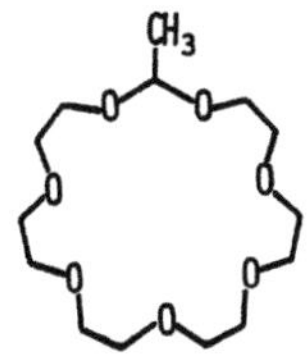	(142–144/0.2) (150/0.05)	4.5 20	35 30
12	14	22	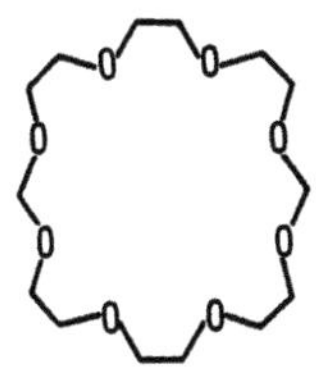	88 nr	18 nr	2 65
13	16	22	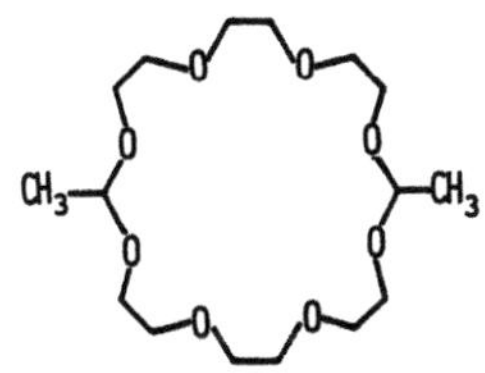	61–62	4	30
14	17	14	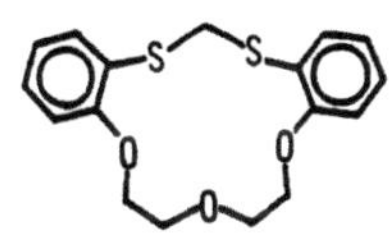	150–153	3	61, 62
15	18	14	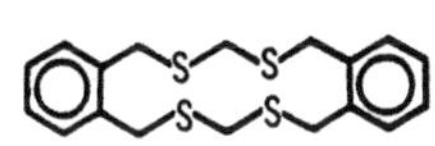	198–200	3	3
16	20	28	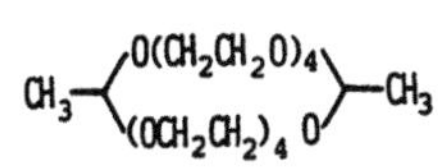	33–34	1	30
17	21	33	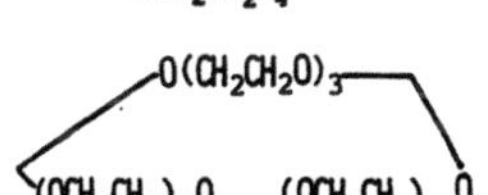	27	nr	65
18	22	10	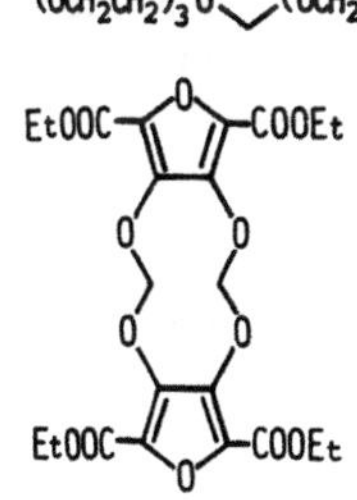	213	25	24

Table 6.1 (continued)

Cpd. No.	No. C's	Ring size	Structure	Mp or (bp) in °C	Yield (%)	Refs.
19	22	10	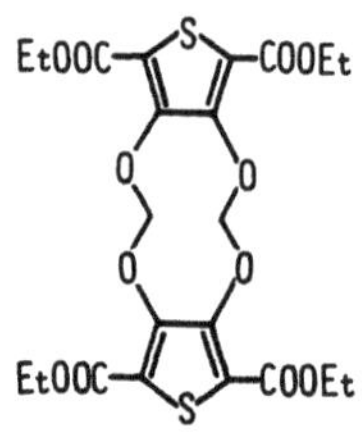	209–210	nr	24
20	24	34	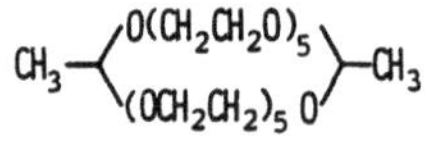	52–53.5	1	30
21	26	20	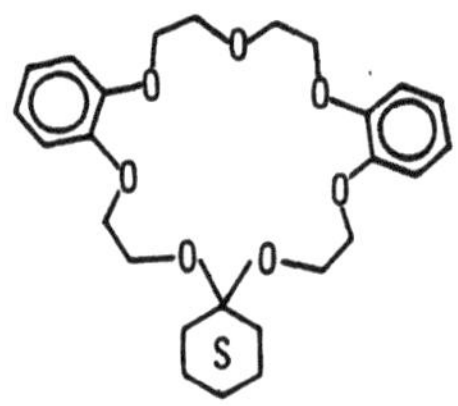	162	16	60
22	30	10	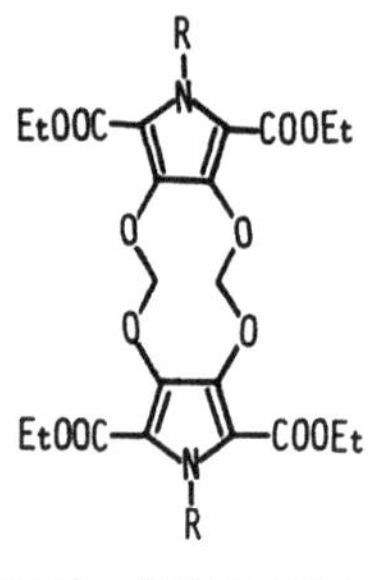 R = CH_2COOEt	192	72	24
23	34	10	R = Ph	214	74	24

Table 6.2. Macrocycles containing phosphorus or arsenic

Cpd. No.	No. C's	Ring size	Structure	Mp or (bp) in °C	Yield (%)	Refs.
1	14	11	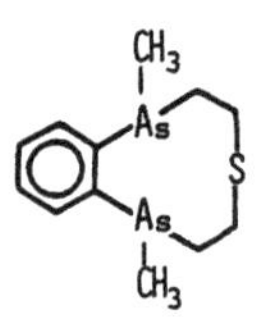	oil	nr	44

Table 6.2 (continued)

Cpd. No.	No. C's	Ring size	Structure		Mp or (bp) in °C	Yield (%)	Refs.
2	18	11	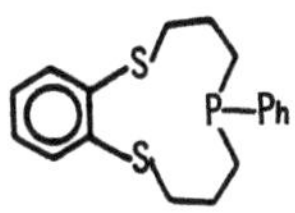		109.5–110 109.5–110	45 42	46 45
3	20	11	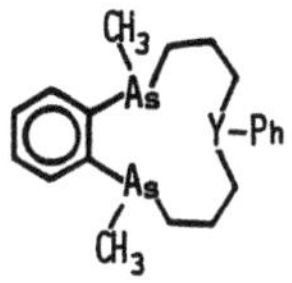	Y = As	oil	nr	44
4	20	11		Y = P	oil	nr	44
5	24	11	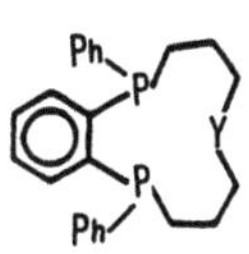	Y = O	glass	32	45
6	24	11		Y = S	112–112.5	51	45
7	25	11		Y = CH_2	120–121	33	45
8	25	11	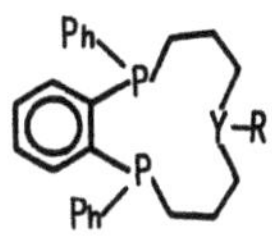	Y = N, R = CH_3	145–147	42	45
9	30	11		Y = N, R = Ph	147–149	56	45
10	30	11		Y = P, R = Ph	163–166	34	45
11	30	11	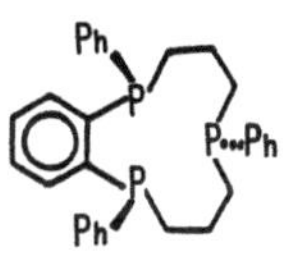		160.5–161.5	~ 5	46
12	30	14	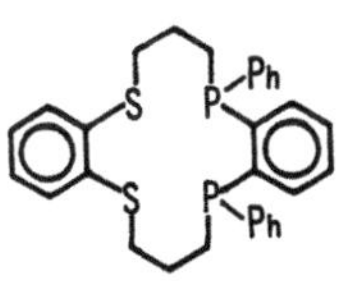		111–112	20	46
13	30	15	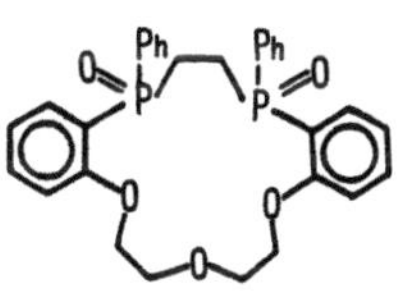		nr	nr	11
14	34	18	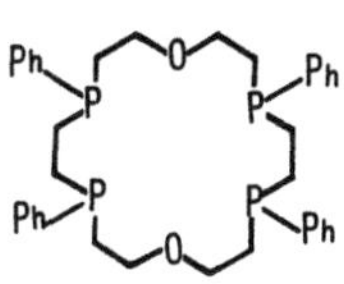	αα αβ	153–155 191–193	1.5 2	19 19

Table 6.2 (continued)

Cpd. No.	No. C's	Ring size	Structure		Mp or (bp) in °C	Yield (%)	Refs.
15	34	28	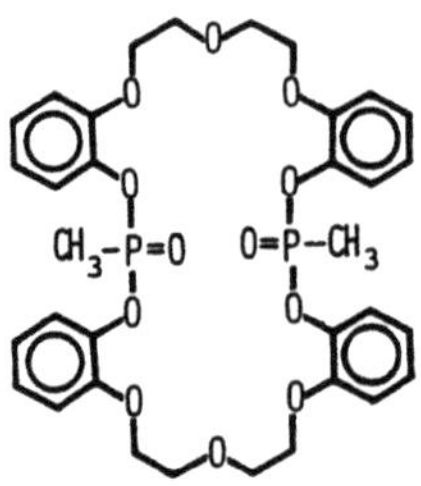		nr	11	43
16	36	16	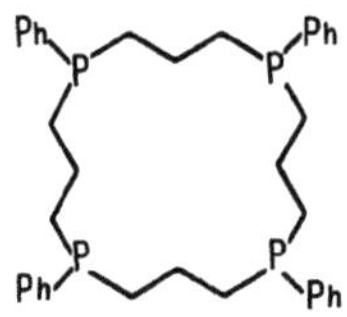		88–93	nr	32
17	36	16	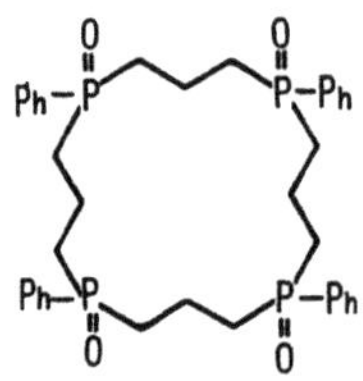		178–180	nr	32
18	39	15	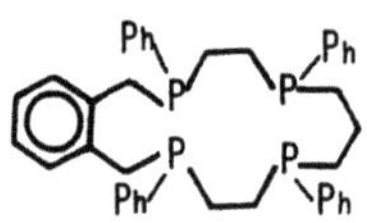		oil	10–11	27
19	42	14	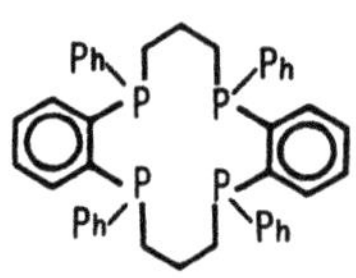	Isomer B	229–230 160–165	22 nr	46 26
20	44	20	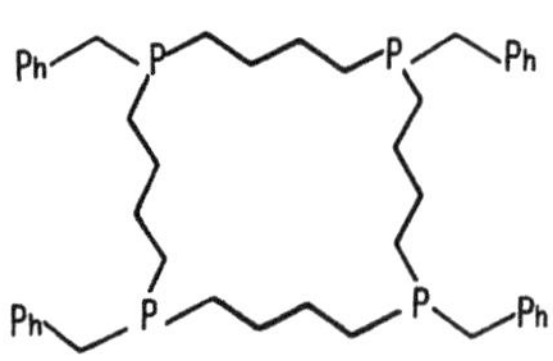		110	nr	32
21	44	20	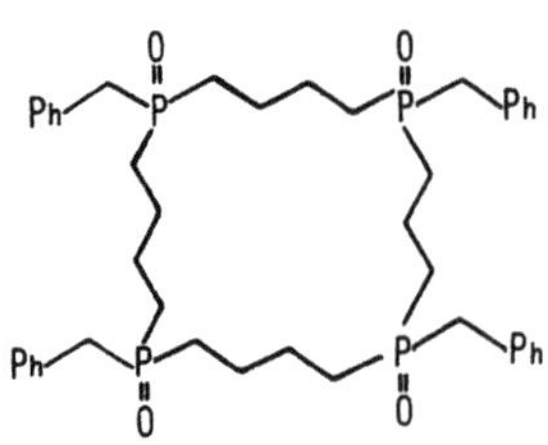		230–250	nr	32

Table 6.2 (continued)

Cpd. No.	No. C's	Ring size	Structure	Mp or (bp) in °C	Yield (%)	Refs.
			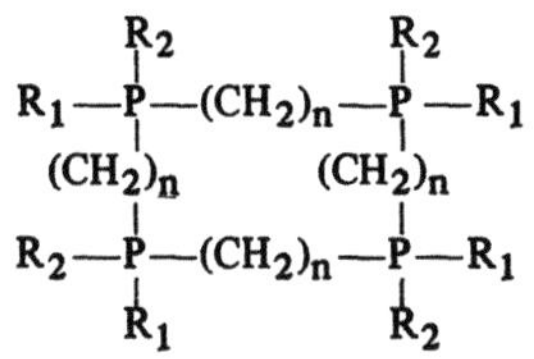			
22	64	16	n = 3, R_1 = Ph, R_2 = CH_2Ph	362	nr	32
23	72	20	n = 4, R_1 = R_2 = CH_2Ph	370	nr	32

Table 6.3. Macrocycles containing one or more sulfur atoms

Cpd. No.	No. C's	Ring size	Structure		Mp or (bp) in °C	Yield (%)	Refs.
1	6	9	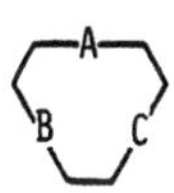	A = S, B = C = O	(40–48/0.15)	5	16
2	6	9		A = B = S, C = O	(62–64/0.1)	6	16
3	6	9		A = B = C = S	81–82 113	0.04 nr	29 64
4	6	10	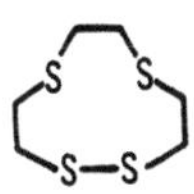	α β α	75 102–104 116	nr nr nr	64 64 64
5	8	11			(110/1.3)	1	57
6	8	11	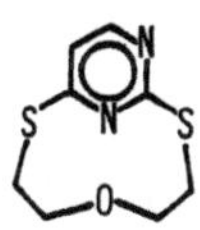		123–125	6	58
7	8	12	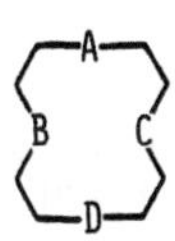	A = S B = C = D = O	(90–95/1) nr (80–87/1)	nr nr 41	18 25 16
8	8	12	A = B = S, C = D = O		(134–136/0.1)	19	16

Table 6.3 (continued)

Cpd. No.	No. C's	Ring size	Structure	Mp or (bp) in °C	Yield (%)	Refs.
9	8	12	A = C = S, B = D = O	(118–120/0.1)	12	16
				(118/1)	nr	18
				{22.5–23 (132–142/0.17)}	nr	25
10	8	12	A = B = C = S, D = O	89–91	26	16
11	8	12	A = B = C = D = S	{170–214 subl. 215–217}	1	68
				224–225	6.3	59
12	9	9	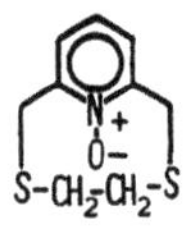	152–154	8	84
13	9	10	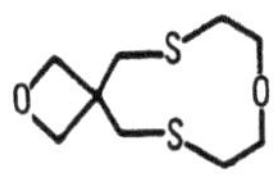	104–105	16	41
14	9	12		87–88	0.4	68
15	9	12	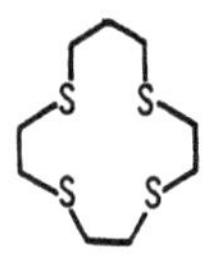	134–135	16	68
16	9	12		110–112	14	91
17	10	10		78–79	36	81
18	10	10		107–109	7	84
19	10	12	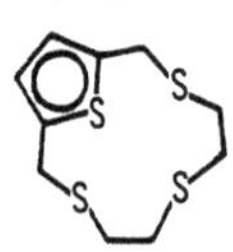	127–129	20	90
20	10	14		121–122.5	4.6	59
				119–120	nr	66, 78
				119–120	7.5	67

Table 6.3 (continued)

Cpd. No.	No. C's	Ring size	Structure	Mp or (bp) in °C	Yield (%)	Refs.
21	10	14	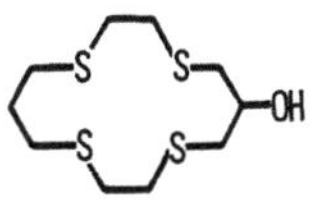	nr nr	55 nr	79 34
22	10	15	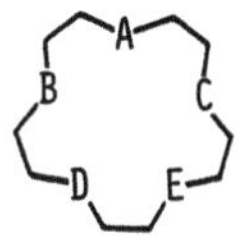 A = S, B = C = D = E = O	(123–124/0.1)	29	16
23	10	15	A = B = S, C = D = E = O	51–52	20	16
24	10	15	B = C = S, A = D = E = O	(150–151/0.1)	27	16
25	10	15	A = B = C = S, D = E = O	43–44	41	15
26	10	15	A = D = E = S, B = C = O	(180/1)	5	15
27	10	15	A = O, B = C = D = E = S	93–95	13	15
28	10	15	A = B = C = D = E = S	97.5–99	11	59
29	11	12		133–135 133–135	46 nr	91 86
30	11	12		162–163	50	90
31	11	12		151–152	18	92
32	11	12		171–174	26	90
33	11	13	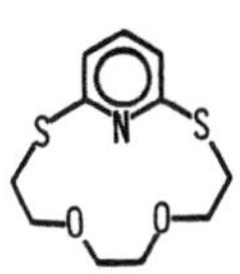	75–77	18	91

Table 6.3 (continued)

Cpd. No.	No. C's	Ring size	Structure	Mp or (bp) in °C	Yield (%)	Refs.
34	11	13		151–153 subl.	13	90
35	12	11	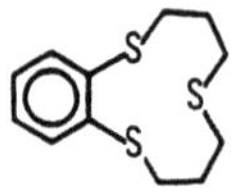	61–62	43	45
36	12	12	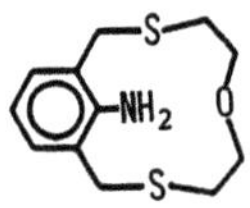	88–89	74	92
37	12	12	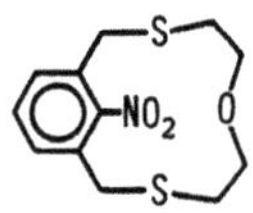	126–127	47	92
38	12	12	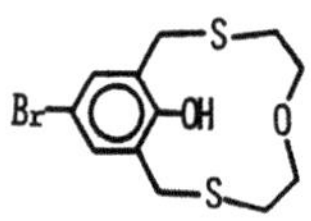	139–141	18	93
39	12	12	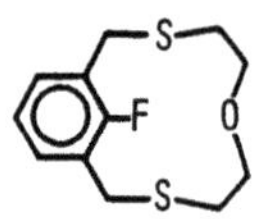	89–91	44	92
40	12	12	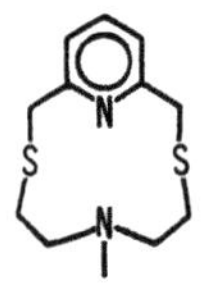	67–69	11	90
41	12	12		143–144 143–144	37 37	87 90
42	12	12	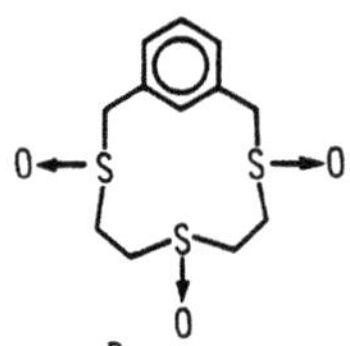	201–204 dec.	42	90
43	12	12	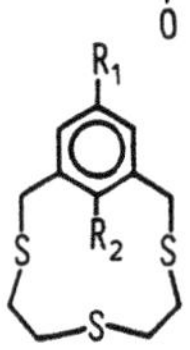 $R_1 = H$ $R_2 = NH_2$	138–140	77	93

Table 6.3 (continued)

Cpd. No.	No. C's	Ring size	Structure	Mp or (bp) in °C	Yield (%)	Refs.
44	12	12	R_1 = H, R_2 = Cl	130–132 nr	46 nr	87 90
45	12	12	R_1 = Br, R_2 = OH	172–174	7	93
46	12	12	R_1 = H, R_2 = F	113–115	51	92
47	12	12	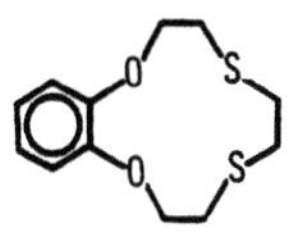	92	34	62
48	12	12	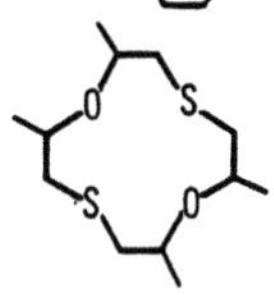	nr	nr	18
49	12	16	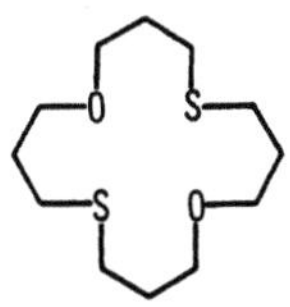	nr	nr	18
50	12	16		46 57.5–59	1 19.5	53 59
51	12	18	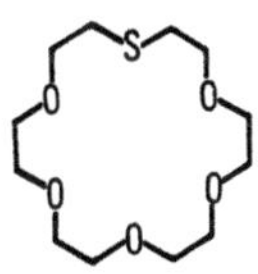	(164–170/0.1)	36	15
52	12	18		54–56	28	15
53	12	18	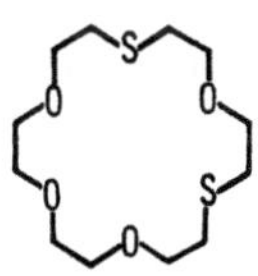	(174–179/1)	29	15
54	12	18	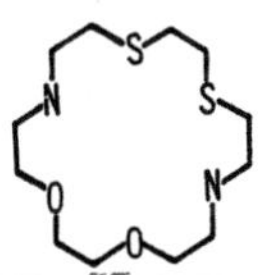	45	42	47a

Table 6.3 (continued)

Cpd. No.	No. C's	Ring size	Structure	Mp or (bp) in °C	Yield (%)	Refs.
55	12	18	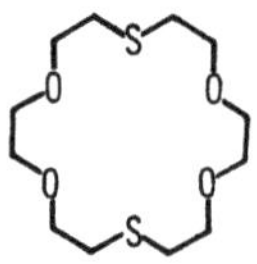	92–94 90–91	34 nr	18 25
56	12	18	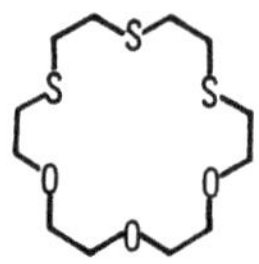	(200–206/2)	11	15
57	12	18	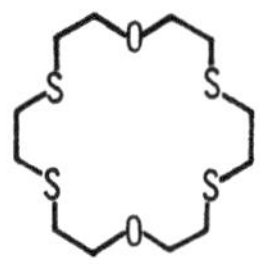	125–126 125 50	4 1.4 7	16 53 7
58	12	18	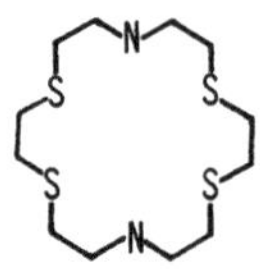	125	42	47a
59	12	18	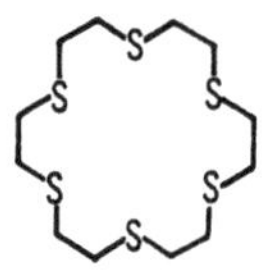	89.6–90 89–90 91–93	1.7 31 35	53 7 59
60	13	12		94–97 nr	15 nr	90 87
61	13	12	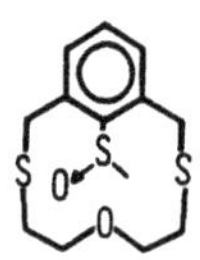	194–196 170–173	27 14	83 83
62	13	13	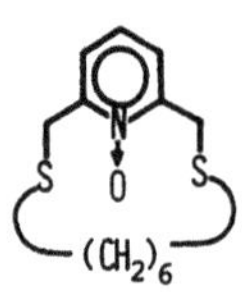	138–140	32	84
63	13	15		90–91 90–91	36 nr	91 86

Table 6.3 (continued)

Cpd. No.	No. C's	Ring size	Structure		Mp or (bp) in °C	Yield (%)	Refs.
64	13	15	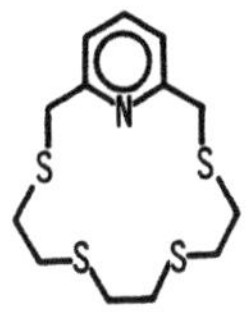		131–133	38	90
65	13	15	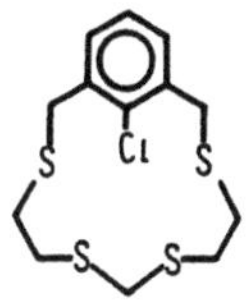		64–66	33	92
66	13	16			73–76	28	91
67	13	19	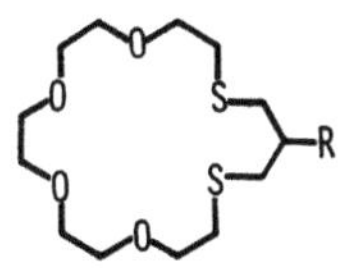	R = H	oil	24	17
68	13	19		R = OH	oil	8	17
69	14	12	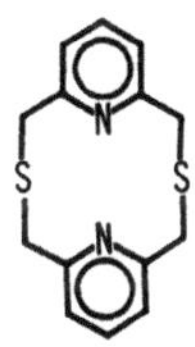		220–222	nr	85
70	14	12	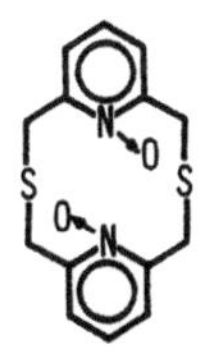		211 dec.	13	84
71	14	14	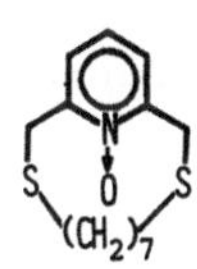		89	7	84
72	14	15	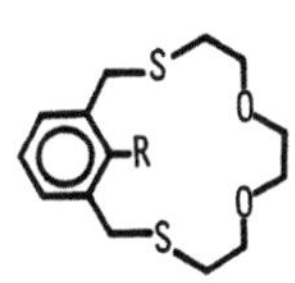	R = NH_2	73–74	40	92
73	14	15		R = NO_2	82–83	31	92
74	14	15		R = Cl	58–59	45	92

Table 6.3 (continued)

Cpd. No.	No. C's	Ring size	Structure		Mp or (bp) in °C	Yield (%)	Refs.
75	14	15	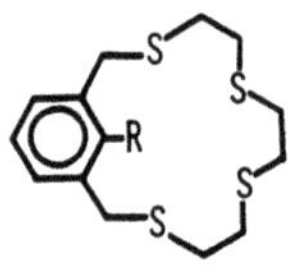	R = NO_2	118–120	55	92
76	14	15		R = Cl	67–69	38	92
77	14	16	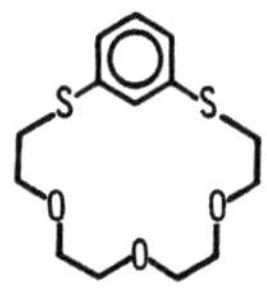		67–70	38	91
78	14	17	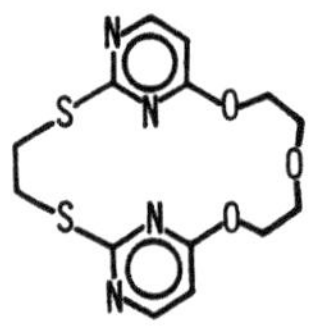		169–172	55	58
79	14	18	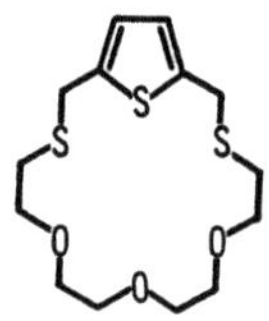		oil	nr	86
80	14	18			61	1.8	53
81	14	21	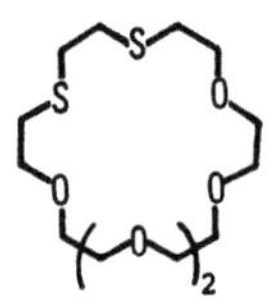		oil	10	17
82	14	21	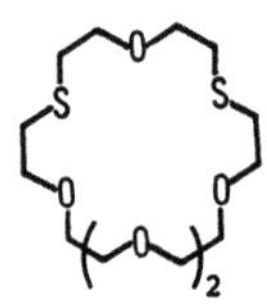		(180/1)	25	15
83	14	21	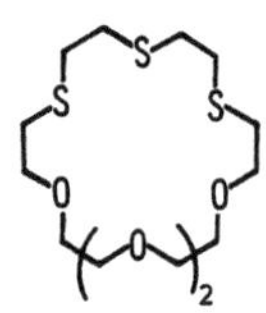		oil	11	15

Table 6.3 (continued)

Cpd. No.	No. C's	Ring size	Structure	Mp or (bp) in °C	Yield (%)	Refs.
84	15	12	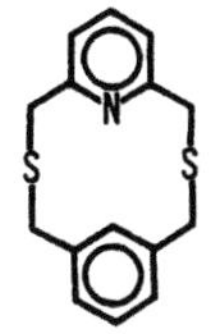	173–175	nr	85
85	15	12	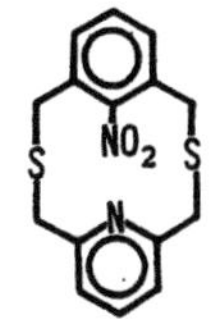	159–161	22	73
86	15	12	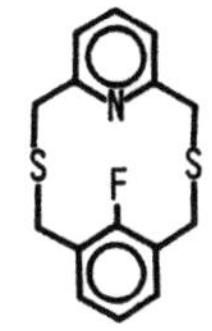	174–175	nr	85
87	15	14	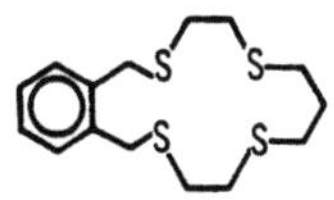	84–86	15	67
88	15	15	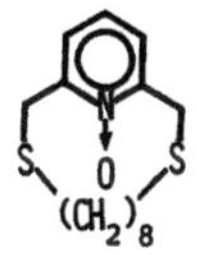	73–75	24	84
89	15	15		155–156	22	83
90	15	15	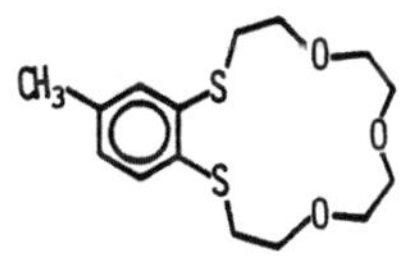	oil	30	61
91	15	18	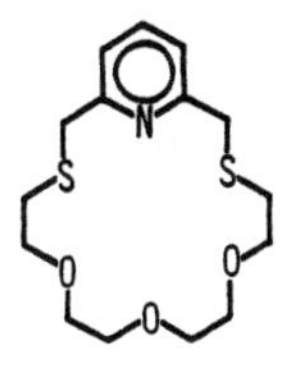	58–59 58–59	49 nr	91 86

Table 6.3 (continued)

Cpd. No.	No. C's	Ring size	Structure	Mp or (bp) in °C	Yield (%)	Refs.
92	15	18	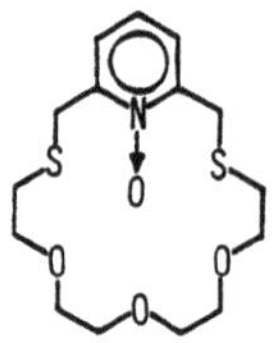	oil nr	47 nr	91 86
93	15	18	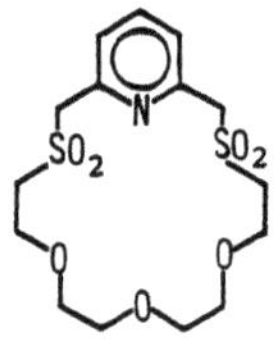	198–201	61	91
94	15	20	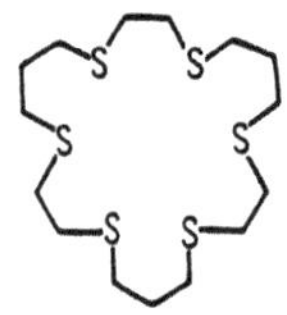	64–65	9.7	59
95	15	22	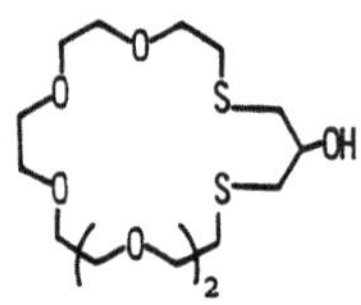	oil	15	17
96	16	11	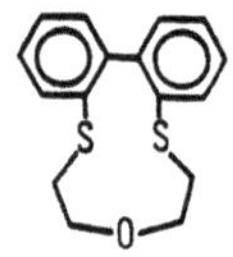	129	67	3
97	16	14	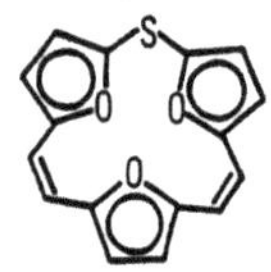	97–99	10.1	22
98	16	14	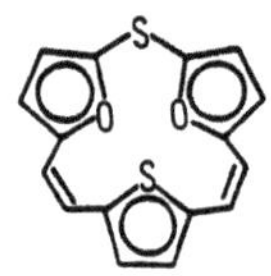	81–83	18.7	22
99	16	16	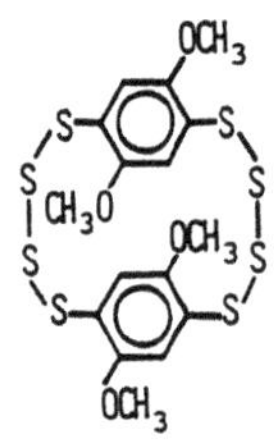	210–215	52	4

Table 6.3 (continued)

Cpd. No.	No. C's	Ring size	Structure	Mp or (bp) in °C	Yield (%)	Refs.
100	16	17	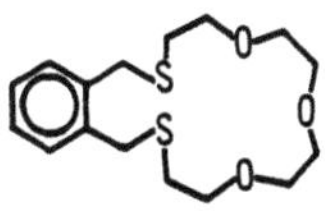	47–49 nr	53 nr	91 86
101	16	18	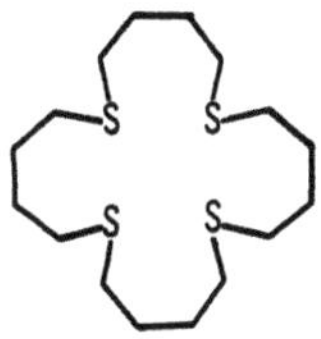	31–32	4	59
102	16	18	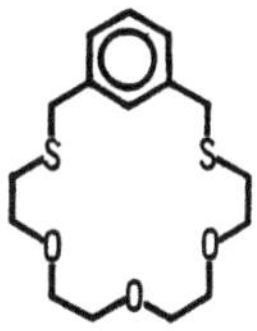	56–57 nr	24 nr	91 86
103	16	18	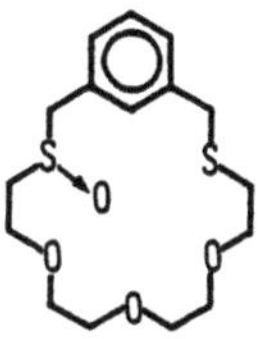	103–105	35	91
104	16	18	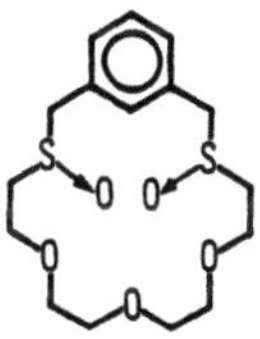	148–151	29	91
105	16	18	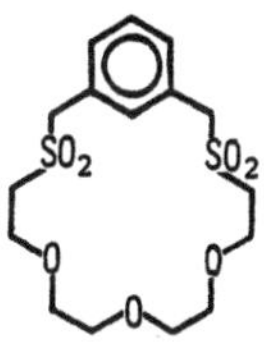	251–253	46	91
106	16	18	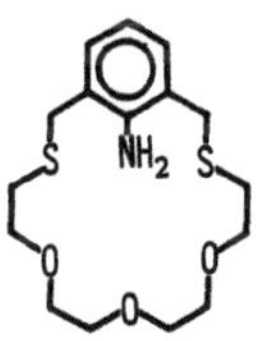	143–145	42	92
107	16	18	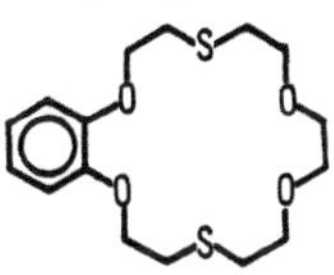	91	1	61, 62

Table 6.3 (continued)

Cpd. No.	No. C's	Ring size	Structure	Mp or (bp) in °C	Yield (%)	Refs.
108	16	18	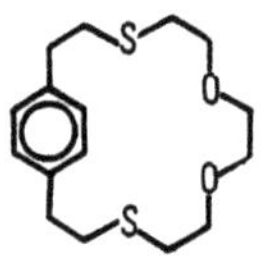	55.5–56	51	37
109	16	19		67–68	15	91
110	16	20	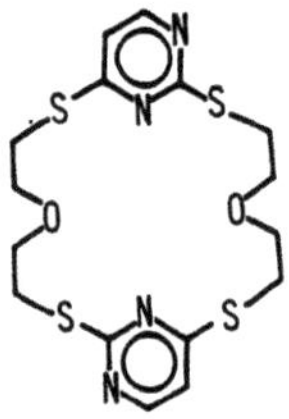	159–160	8	58
111	16	20	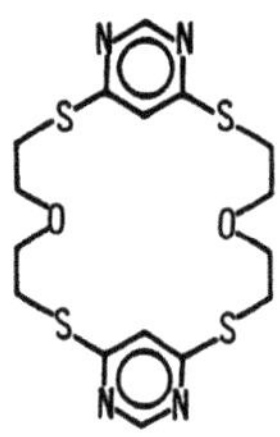	192–193	3	57
112	16	24	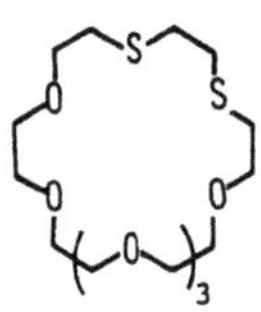	oil	5	17
113	16	24	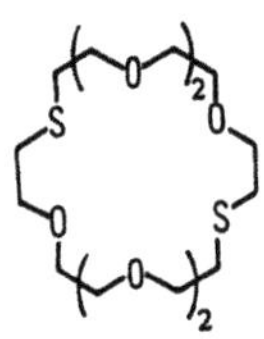	(220/0.01) nr	nr nr	18 25
114	17	14	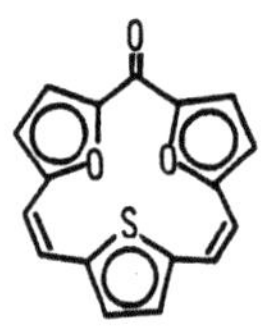	148–150	8.4	23

Table 6.3 (continued)

Cpd. No.	No. C's	Ring size	Structure		Mp or (bp) in °C	Yield (%)	Refs.
115	17	18	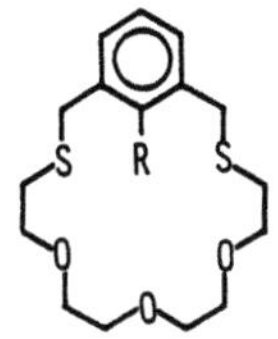	R = CN	72–74	46	91
116	17	18		R = OCH_3	oil	29	91
					nr	nr	86
117	17	18	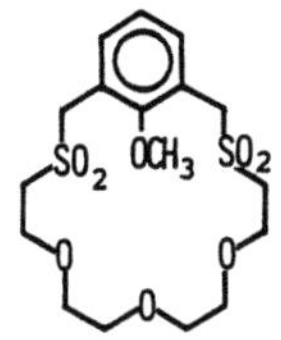		218–220	38	91
118	17	18	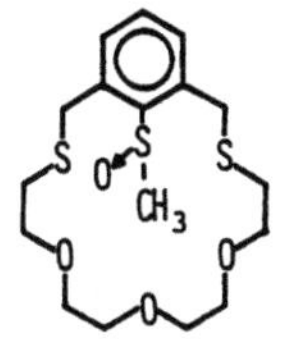		oil	48	91
119	17	18	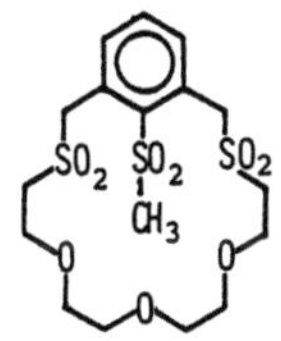		229–231	35	91
120	17	18	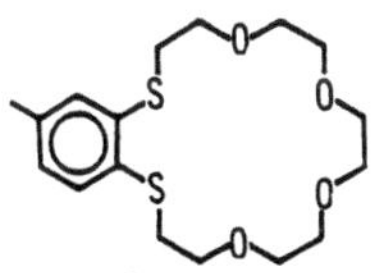		oil	56	61, 62
121	17	21	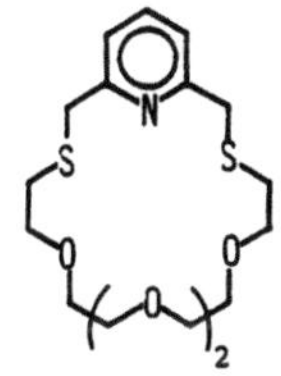		oil	nr	86
122	18	12	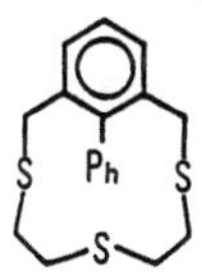		157–159	17	82
					nr	nr	87
					nr	nr	90
123	18	15			103–103.5	nr	5

Table 6.3 (continued)

Cpd. No.	No. C's	Ring size	Structure	Mp or (bp) in °C	Yield (%)	Refs.
124	18	16	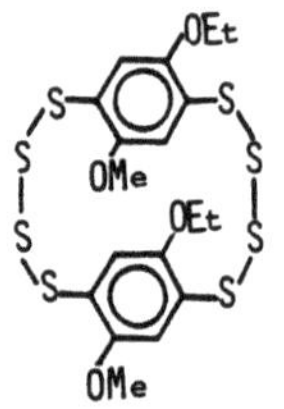	156–160	54	4
125	18	18	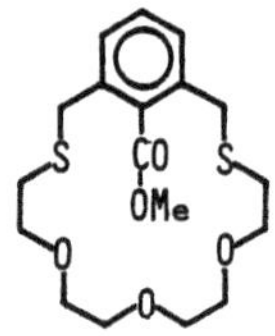	oil	34	91
126	18	18	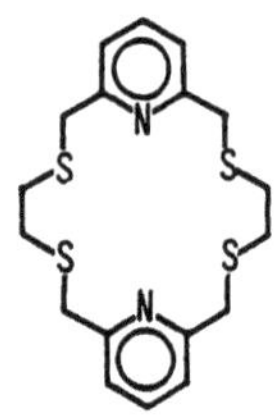	213–216	19	90
127	18	20	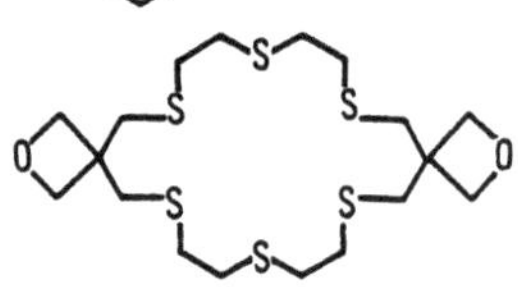	131–132	11	41
128	18	20	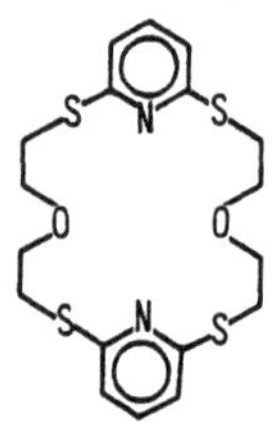	168–170	9	91
129	18	20	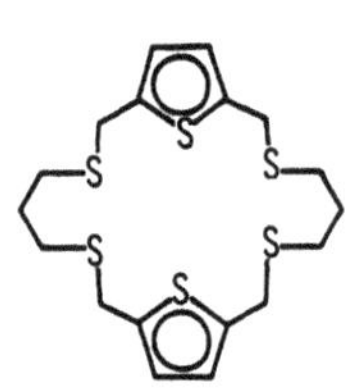	119.5–120.5	2.5	6
130	18	22	$S(CH_2)_6S$ $(CH_2)_3$ $(CH_2)_3$ $S(CH_2)_6S$	62	1.1	53

Table 6.3 (continued)

Cpd. No.	No. C's	Ring size	Structure		Mp or (bp) in °C	Yield (%)	Refs.
131	18	24	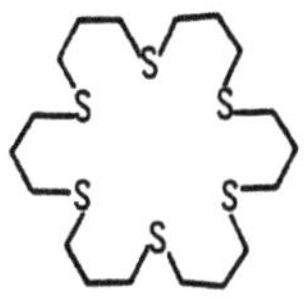		29–30	15.1	59
132	20	16	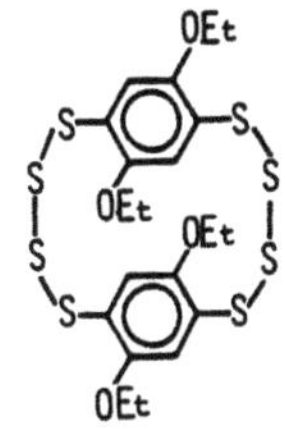		185–187	60	4
133	20	16	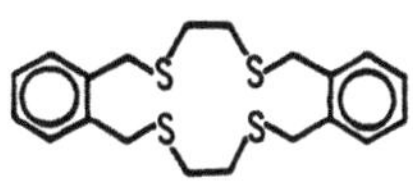		142–144	5	3
134	20	17	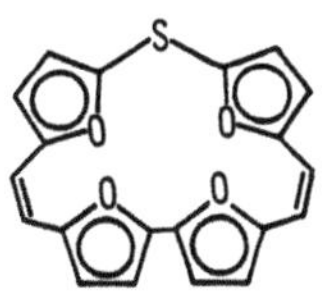		170–171	<1	22
135	20	17	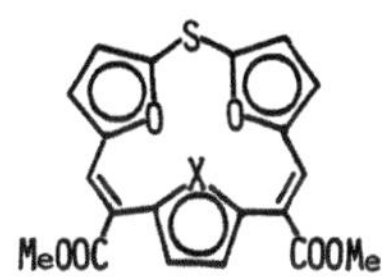	X = O	205–206	29	22
136	20	17		X = S	193–195	nr	22
137	20	17	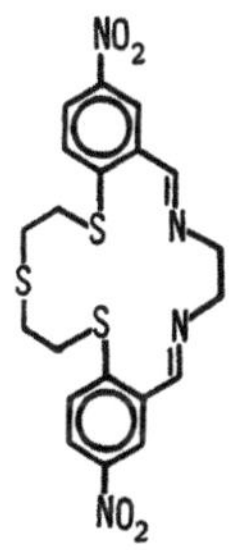		nr	nr	8
138	20	18	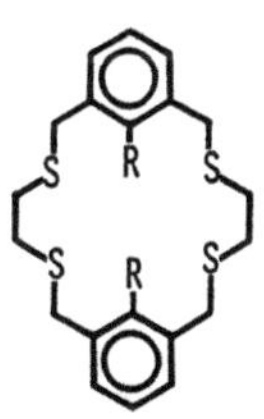	R = H	170–171	15	81
139	20	18		R = NH_2	215–217	49	93
140	20	18		R = NO_2	179–184	12	82

Table 6.3 (continued)

Cpd. No.	No. C's	Ring size	Structure	Mp or (bp) in °C	Yield (%)	Refs.
141	20	18	R = F	154–155	24	81
142	20	18	R = Cl	160–161	30	81
143	20	18	R = Br	176–177	22	81
144	20	18	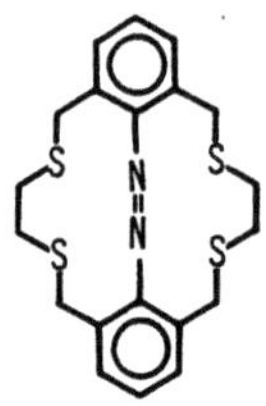	230–256 dec.	18	93
145	20	18	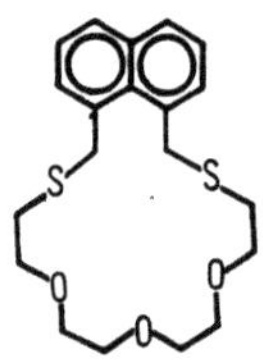	76–78	19	91
146	20	18	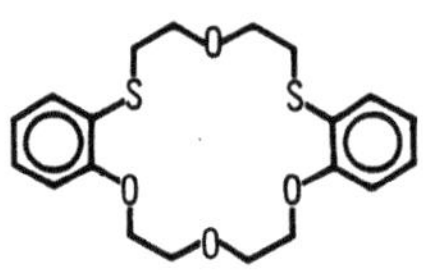	143–144	15	61, 62
147	20	18	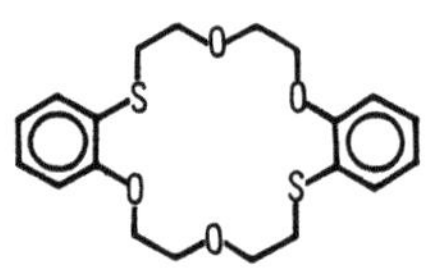	114–115	5	61, 62
148	20	18	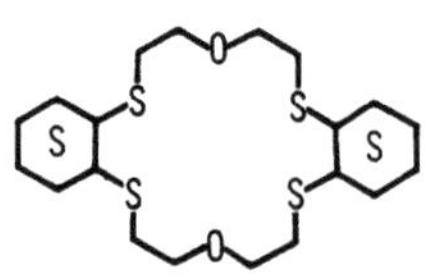	oil nr	24 30	61 62
149	20	20	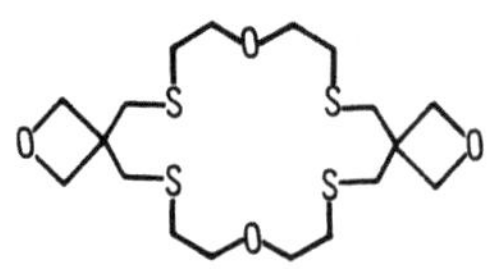	100.5–101.5	15	41
150	20	24	$[(CH_2)_5S]_4$	33–33.5	5.3	59
151	20	28	$[(CH_2)_3S(CH_2)_2S]_4$	66–67	25	78

Table 6.3 (continued)

Cpd. No.	No. C's	Ring size	Structure		Mp or (bp) in °C	Yield (%)	Refs.
152	22	18	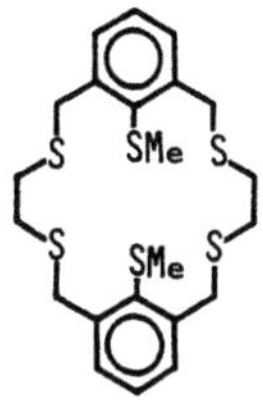		170–173	5	82
153	22	18	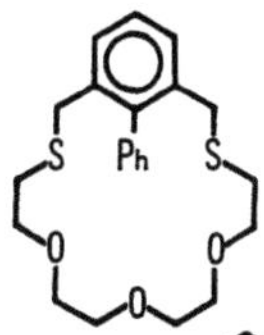		110–111	27	91
154	22	18	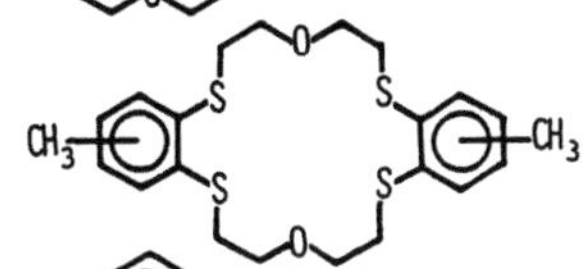		147	6	61, 62
155	22	18	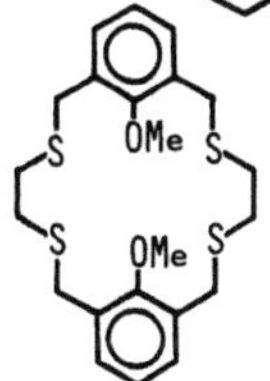		nr	nr	86
156	22	18	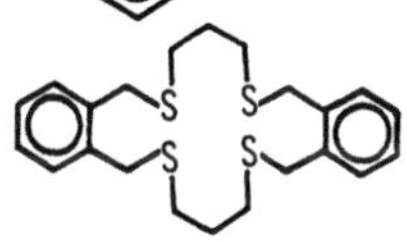		208–210	5	3
157	22	19	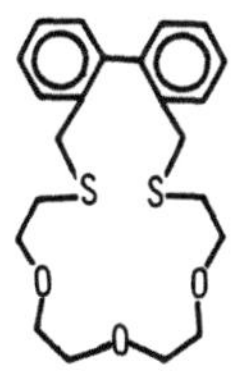		67–69	31	91
158	22	20	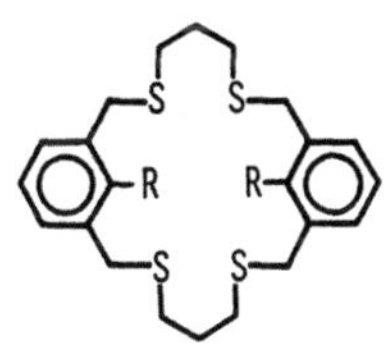	R = H	116–117	19	81
159	22	20		R = NH_2	113–114	88	93
160	22	20		R = NO_2	124	18	82
161	22	20		R = F	142–143	13	81
162	22	20		R = Cl	~101	11	81
163	22	20		R = Br	~165	14	81

Table 6.3 (continued)

Cpd. No.	No. C's	Ring size	Structure	Mp or (bp) in °C	Yield (%)	Refs.
164	22	21		50–52	36	91
165	23	19	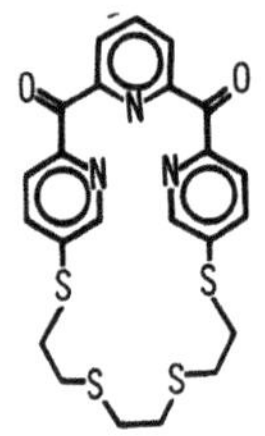	129–130	1–2	56
166	24	16	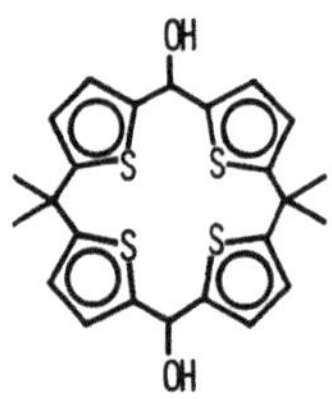	280 dec.	4	1
167	24	24	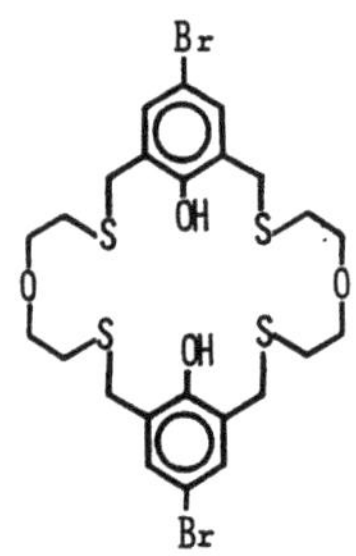	173–175	5	93
168	24	28	$[(CH_2)_6S]_4$	30–32	3.9	59
169	24	30	$[(CH_2)_4S]_6$	67–70	1.7	59
170	26	16	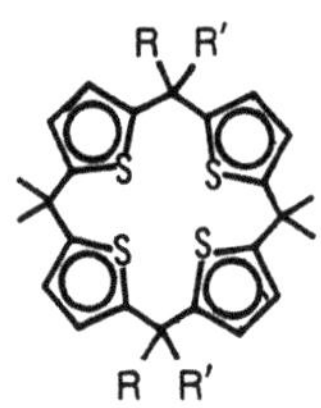 R = R′ = CH_2	250 dec.	0.8	1
171	26	16	R = CH_3, R′ = OH	280 dec.	2.4	1

Table 6.3 (continued)

Cpd. No.	No. C's	Ring size	Structure	Mp or (bp) in °C	Yield (%)	Refs.
172	28	17	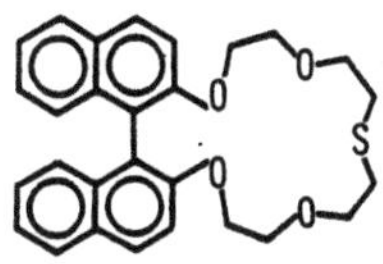	125–127	52	21, 47
173	30	20	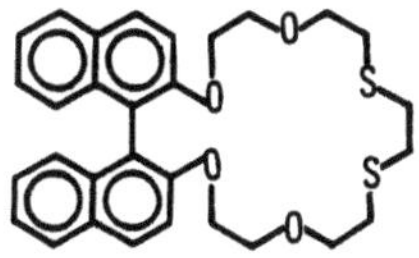	85–90	16	21
174	30	20	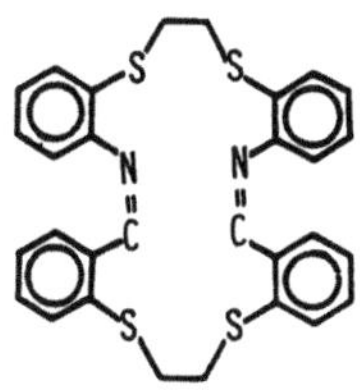	nr	nr	49
175	30	26	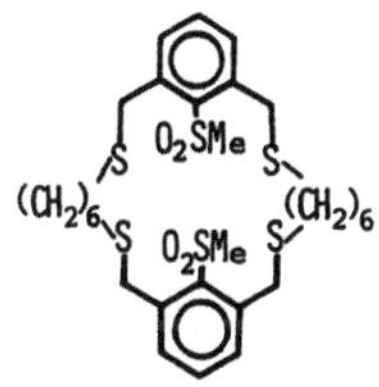	99–102	10	82
176	30	36	$[(CH_2)_5S]_6$	36.5–38	6.8	59
177	32	18	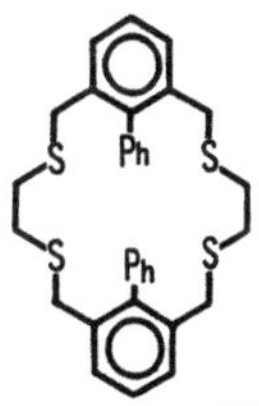	217–218	5	82
178	32	24	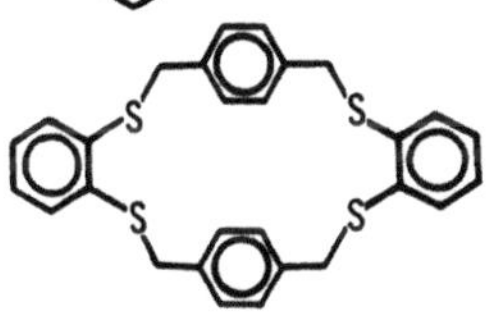	183–184	10	3
179	34	20	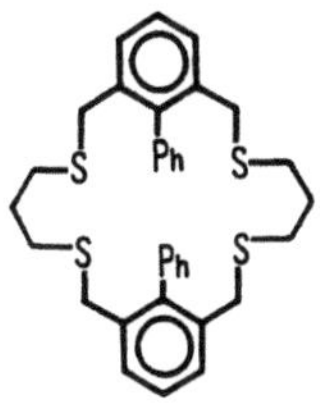	236–237	7	82

Table 6.3 (continued)

Cpd. No.	No. C's	Ring size	Structure	Mp or (bp) in °C	Yield (%)	Refs.
180	34	20	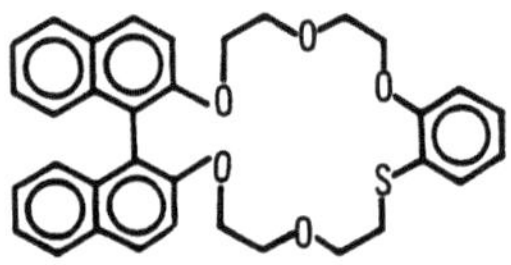	111–113 nr	38 nr	47 21
181	34	20	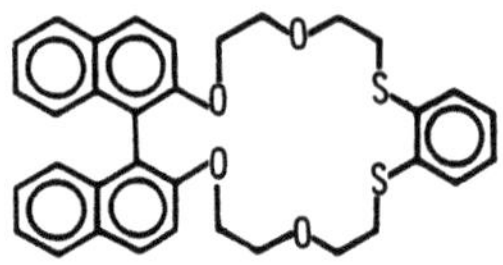	149–150 150–151	72 72	47 21
182	34	20	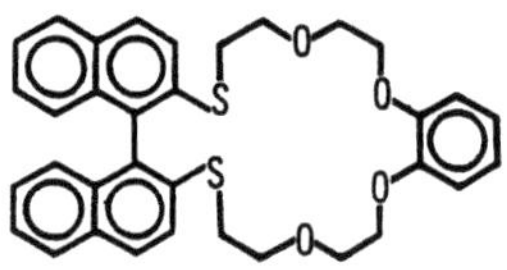	167–168	58	21
183	36	42	$[(CH_2)_6S]_6$	56–59.5	3.2	59
184	43	14	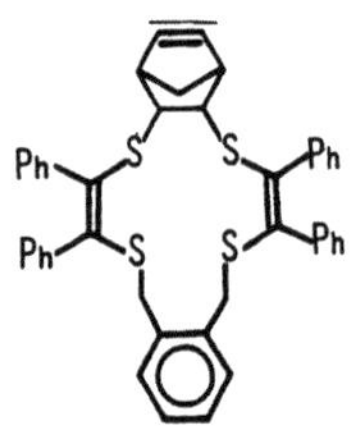	258	nr	70

Table 6.4. Crown-containing polymers and polycrowns

Cpd. No.	No. C's	Ring size	Structure	Mp or (bp) in °C	Yield (%)	Refs.
1	16	15	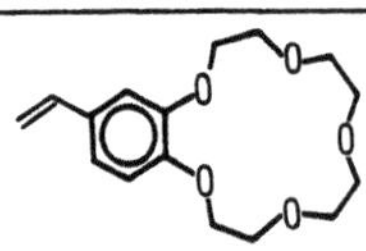	43–44	10	39
2	18	18	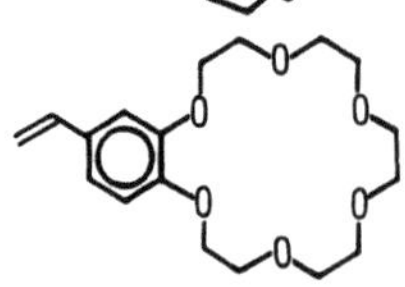	61–62	16	39
3	22	18	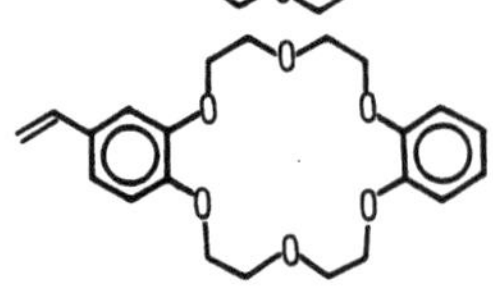	148–149	34	39

Table 6.4 (continued)

Cpd. No.	No. C's	Ring size	Structure		Mp or (bp) in °C	Yield (%)	Refs.
4	20	18	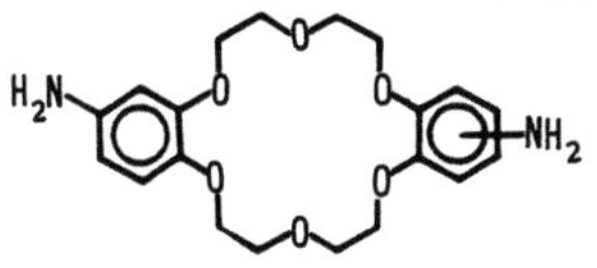	(4, 4′)	nr	nr	88
5	20	18		(4, 5′)	nr	nr	88
6	–	–	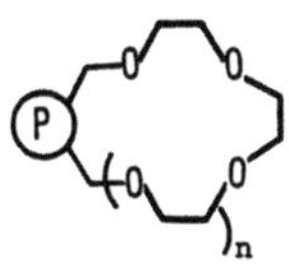	n = 2	nr	nr	89
7	–	–		n = 3	nr	nr	89
8	–	–		n = 4	nr	nr	89
9	–	–		n = 6	nr	nr	89
10	–	–		n = 8	nr	nr	89
11	–	–		n = 11	nr	nr	89
12	–	–		n = 13	nr	nr	89
13	–	–		n = 14	nr	nr	89
14	–	18	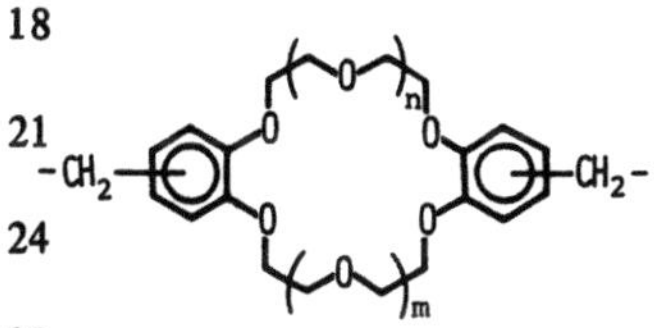	n = m = 1	nr	nr	9
15	–	21		n = 1, m = 2	nr	nr	9
16	–	24		n = 2, m = 2	nr	nr	9
17	–	30		n = 3, m = 3	nr	nr	9
18	–	18	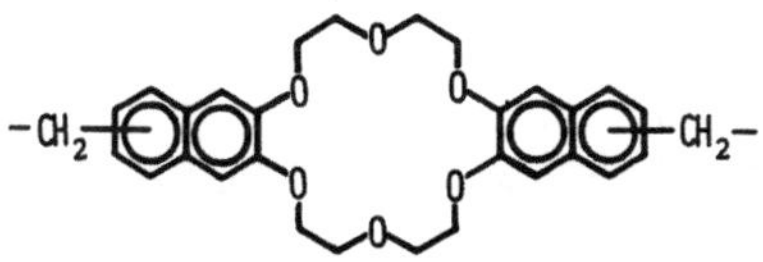		nr	nr	9
19	–	15	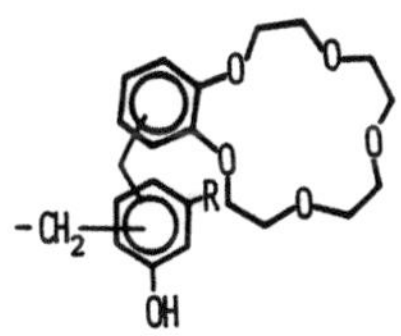	R = H	nr	nr	9
20	–	15		R = OH	nr	nr	9
21	–	18	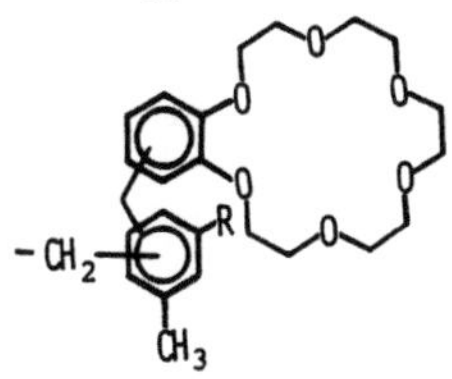	R = H	nr	nr	9
22	–	18		R = CH_3	nr	nr	9

Table 6.4 (continued)

Cpd. No.	No. C's	Ring size	Structure	Mp or (bp) in °C	Yield (%)	Refs.
23	–	18	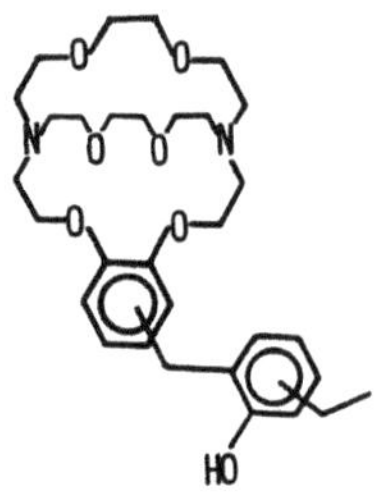	nr	nr	9
24	–	18	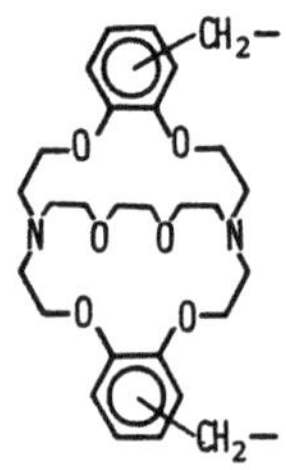	nr	nr	9
25	–	18	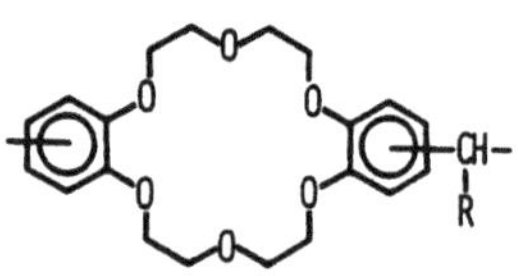 R = COOH	nr	96	77
26	–	18	R = CCl_3	nr	95	77
27	–	18	R = CCl_3, COOH, Ratio 1/1	nr	nr	77
28	–	18	R = CBr_3	nr	nr	77
			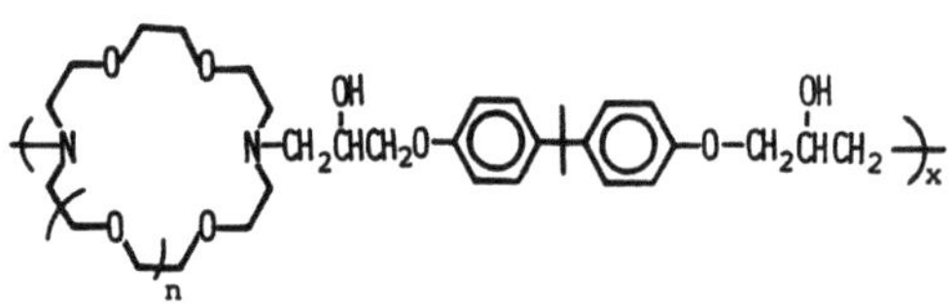			
29	–	15	n = 0	nr	nr	31
30	–	18	n = 1	nr	nr	31
31	–	15	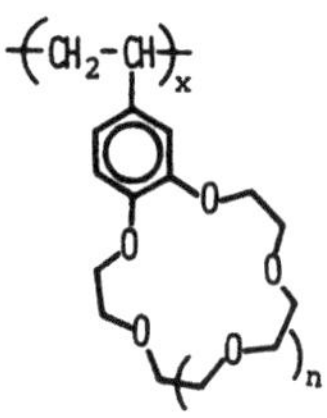 n = 1	nr	nr	38
32	–	18	n = 2	nr	nr	38

Table 6.4 (continued)

Cpd. No.	No. C's	Ring size	Structure	Mp or (bp) in °C	Yield (%)	Refs.
33	–	18	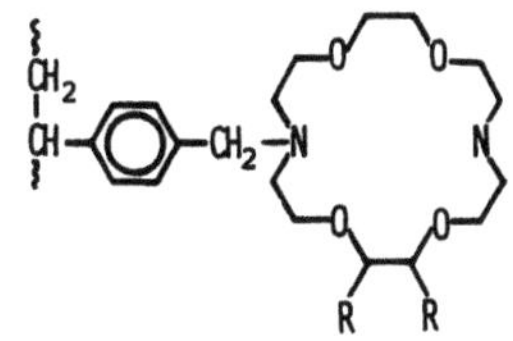 R = H	nr	nr	9
34	–	18	R = benzo	nr	nr	9
35	–	15	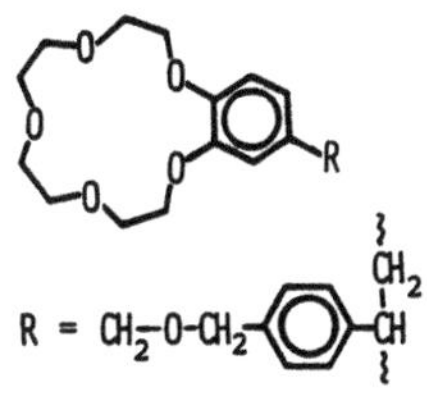	nr	nr	9
36	–	15	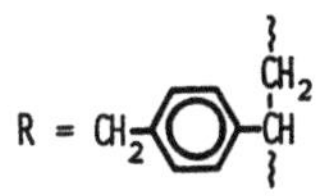	nr	nr	9
37	–	15	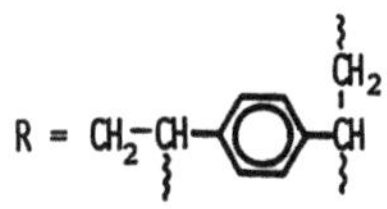	nr	nr	9
38	–	18	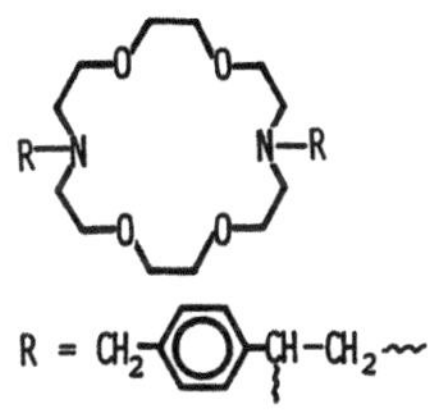	nr	nr	9
39	–	18	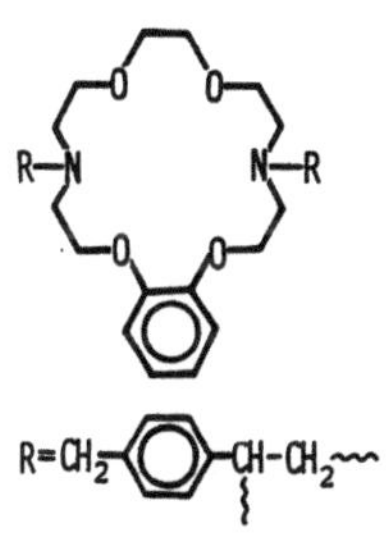	nr	nr	9
40	–	14	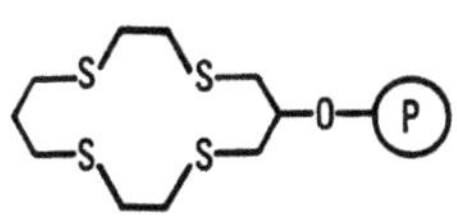	nr	nr	34

Table 6.4 (continued)

Cpd. No.	No. C's	Ring size	Structure	Mp or (bp) in °C	Yield (%)	Refs.
41	–	18	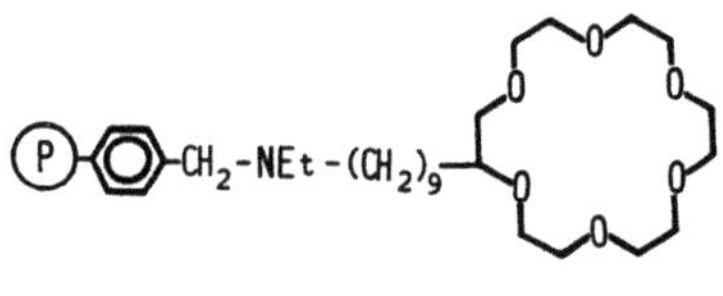	nr	nr	54
42	–	18	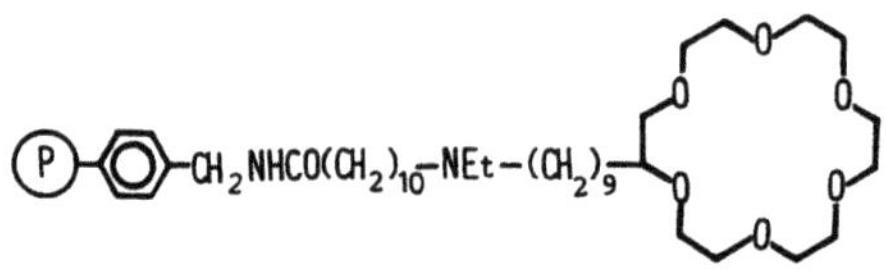	nr	nr	54
43	–	18	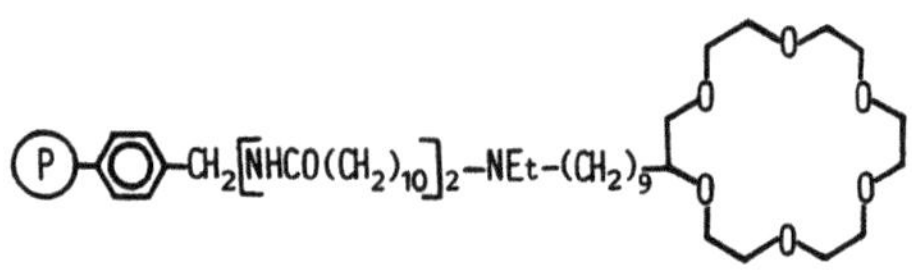	nr	nr	54
44	–	18	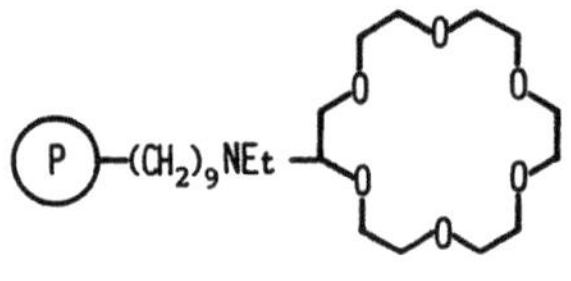	nr	nr	20
45	–	18	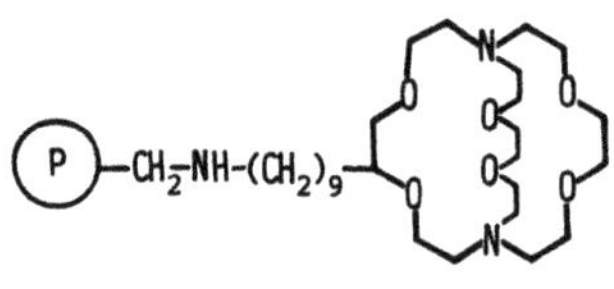	nr	nr	20

6.9 References

1. Ahmed, M., Meth-Cohn, O.: Tetrahedron Letters (1969) 1493
2. Albrecht, K., Fleischer, D., Rentsch, C., Yamaguchi, H., Schulz, R. C.: Makromol. Chem. (1977) *178*, 3191
3. Allen, D. W., Braunton, P. N., Millar, I. T., Tebby, J. C.: J. Chem. Soc. (C) (1971) 3454
4. Ariyan, Z. S., Martin, R. L.: J. C. S. Perkin Trans. I (1973) 1687
5. Badger, G. M., Elix, J. A., Lewis, G. E., Singh, V. P., Spotswood, T. M.: J. C. S. Chem. Commun. (1965) 269
6. Bhattacharjya, A., Hortmann, A. G.: J. Heterocyclic Chem. (1978) *15*, 1223

7. Black, D. St. C., McLean, I. A.: Tetrahedron Letters (1969) 3961
8. Black, D. St. C., McLean, I. A.: Inorg. Nucl. Chem. Lett. (1970) *6*, 675
9. Blasius, E., Janzen, K.-P., Adrian, W., Klautke, G., Lorscheider, R., Maurer, P.-G., Nguyen, V. P., Nguyen Tien, T., Scholten G., Stockemer, J.: Z. Anal. Chem. (1977) *284*, 337
10. Blasius, E., Janzen, K.-P., Newmann, W.: Mikrochimica Acta. (1977) *II*, 279
11. Bodrin, G. V., Polikarpov, Yu. M., Medved, T. Ya., Kabachnik, M. I.: Izv. Akad. Nauk SSSR, Ser. Khim. (1978) *8*, 1930
12. Bogatsky, A. V., Lvkyanenko, N. G., Kirichenko, T. I.: Tetrahedron Letters (1980) *21*, 313
13. Bonthrone, W., Cornforth, J.: J. Chem. Soc. (C) (1969) 1202
14. Bradshaw, J. S., Hui, J. Y. K.: J. Heterocyclic Chem. (1974) *11*, 649
15. Bradshaw, J. S., Hui, J. Y., Chan, Y., Haymore, B. L., Izatt, R. M., Christensen, J. J.: J. Heterocyclic Chem. (1974) *11*, 45
16. Bradshaw, J. S., Hui, J. Y., Haymore, B. L., Christensen, J. J., Izatt, R. M.: J. Heterocyclic Chem. (1973) *10*, 1
17. Bradshaw, J. S., Reeder, R. A., Thompson, M. D., Flanders, E. D., Carruth, R. L., Izatt, R. M., Christensen, J. J.: J. Org. Chem. (1976) *41*, 134
18. British Patent 942,868, 27 Nov. 1963
19. Ciampolini, M., Dapporto, P., Nardi, N., Zanobini, F.: J. C. S. Chem. Commun. (1980) 177
20. Cinquini, M., Colonna, S., Molinari, H., Montanari, F., Tundo, P.: J. C. S. Chem. Commun. (1976) 394
21. Cram, D. J.: Ger. Offen. 2,414,188, 21 Nov. 1974
22. Cresp, T. M., Sargent, M. V.: J. C. S. Chem. Commun. (1971) 1458
23. Cresp, T. M., Sargent, M. V.: J. C. S. Perkin Trans. I (1973) 2961
24. Dallacker, F., Mues, V.: Chem. Ber. (1975) *108*, 569
25. Dann, J. R., Chiesa, P. P., Gates, Jr., J. W.: J. Org. Chem. (1961) *26*, 1991
26. Davis, R. E., Hudson, C. W., Kyba, E. P.: J. Am. Chem. Soc. (1978) *100*, 3642
27. DelDonno, T. A., Rosen, W: J. Am. Chem. Soc. (1977) *99*, 8051
28. Diaddario, L. L., Zimmer, L. L., Jones, T. E., Sokol, L. S. W. L., Cruz, R. B., Yee, E. L., Ochrymowycz, L. A., Rorabacher, D. B.: J. Am. Chem. Soc. (1979) *101*, 3511
29. Gerber, D., Chongsawangrirod, P., Leung, A. K., Ochrymowycz, L. A.: J. Org. Chem. (1977) *42*, 2644
30. Gold, V., Sghibartz, C. M.: J. C. S. Chem. Commun. (1978) 507
31. Gramain, P., Frere, Y.: Macromolecules (1979) *12*, 1038
32. Horner, L., Kunz, H., Walach, P.: Phosphorus (1975) *6*, 63
33. Jones, T. E., Sokol, L. S. W. L., Rorabacher, D. B., Glick, M. D.: J. C. S. Chem. Commun. (1979) 140
34. Kakiuchi, H., Tomoi, M., Abe, O., Kihara, K.: Jpn. Kokai Tokkyo Koho (1980) *19*, 221
35. Kawakami, Y., Sugiura, T., Yamashita, Y.: Bull. Chem. Soc. Japan (1978) *51*, 3053
36. Kawakami, Y., Suzuki, J., Yamashita, Y.: Polymer Journal (1977) *9*(5), 519
37. Kawashima, N., Kawashima, T., Otsubo, T., Misumi, S.: Tetrahedron Letters (1978) 5025
38. Kopolow, S., Hogen Esch, T. E., Smid, J.: Macromolecules (1971) *4*, 359
39. Kopolow, S., Hogen Esch, T. E., Smid, J.: Macromolecules (1973) *6*, 133
40. Kopolow, S., Machacek, Z., Takaki, U., Smid, J.: J. Macromol. Sci., Chem. (1973) 1015
41. Krespan, C. G.: J. Org. Chem. (1974) *39*, 2351
42. Kudrya, T. N., Shtepanek, A. S., Kirsanov, A. V.: Zh. Obshch. Khim. (1978) *48*, 844
43. Kudrya, T. N., Shtepanek, A. S., Kirsanov, A. V.: Zh. Obshch. Khim. (1978) *48*, 927
44. Kyba, E. P., Chou, S.-S.P.: J. C. S. Chem. Commun. (1980) 449
45. Kyba, E. P., John, A. M., Brown, S. B., Hudson, C. W., McPhaul, M. J., Harding, A., Larsen, K., Niedzwiecki, S., Davis, R. E.: J. Am. Chem. Soc. (1980) *102*, 139
46. Kyba, E. P., Hudson, C. W., McPhaul, M. J., John, A. M.: J. Am. Chem. Soc. (1977) *99*, 8053
47. Kyba, E. P., Siegel, M. G., Sousa, L. R., Sogah, G. D. Y., Cram, D. J.: J. Amer. Chem. Soc. (1973) *95*, 2691
47a. Lehn, J.-M.: U.S. Patent 3,888,877, 10 June 1975
48. Lesiak, J., Bouchu, M. N., Descotes, G.: Synthesis (1980) 240
49. Lindoy, L. F., Busch, D. H.: J. C. S. Chem. Commun. (1968) 1589

50. Liptuga, N. I., Irodionova, L. F., Lozinskii, M. O.: Zh. Obshch. Khim. (1978) *48*, 1085
51. Magee, R. J., Mazurek, W., O'Connor, M. J., Phillip, A. T.: Aust. J. Chem. (1974) *27*, 1629
52. Manecke, G., Storck, W.: Angew. Chem. Int. Ed. Engl. (1978) *17*, 657
53. Meadow, J. R., Reid, E. E.: J. Am. Chem. Soc. (1934) *56*, 2177
54. Molinari, H., Montanari, F., Tundo, P.: J. C. S. Chem. Commun. (1977) 639
55. Moody, L. S.: U.S. Patent 3,155,683, 3 Nov. 1964
56. Newkome, G. R., Nayak, A., Sauer, J. D., Mattschei, P. K., Watkins, S. F., Fronczek, F., Benton, W. H.: J. Org. Chem. (1979) *44*, 3816
57. Newkome, G. R., Nayak, A., Sorci, M. G., Benton, W. H.: J. Org. Chem. (1979) *44*, 3812
58. Newkome, G. R., Nayak, A., Otemaa, J., Van, D. A., Benton, W. H.: J. Org. Chem. (1978) *43*, 3362
59. Ochrymowycz, L. A., Mak, C.-P., Michna, J. D.: J. Org. Chem. (1974) *39*, 2079
60. Pedersen, C. J.: J. Am. Chem. Soc. (1970) *92*, 391
61. Pedersen, C. J.: J. Org. Chem. (1971) *36*, 254
62. Pedersen, C. J.: U.S. Patent 3,856,813, 24 Dec. 1974
63. Pedersen, C. J.: U.S. Patent 3,873,569, 25 March 1975
64. Ray, P. C.: J. Chem. Soc. (1920) 1090
65. Rentsch, C., Schulz, R. C.: Makromol. Chem. (1977) *178*, 2535
66. Rosen, W., Busch, D. H.: Chem. Commun. (1969) 148
67. Rosen, W., Busch, D. H.: J. Am. Chem. Soc. (1969) *91*, 4694
68. Rosen, W., Busch, D. H.: Inorg. Chem. (1970) *9*, 262
69. Saegusa, T., Goethals, E., Eds.: ACS Symposium Series *59*, 1977; Yamashita, Y., Kawakami, Y.: p. 99
70. Schrauzer, G. N., Ho, R. K. Y., Murillo, R. P.: J. Am. Chem. Soc. (1970) *92*, 3508
71. Shchori, E., Jagur-Grodzinski, J.: J. Applied Polymer Sci. (1976) *20*, 773
72. Smid, J., Varma, A. J., Shah, S. C.: J. Am. Chem. Soc. (1979) *101*, 5764
73. Smith, G. F., Margesum, D. W.: J. C. S. Chem. Commun. (1975) 807
74. Staab, H. A., Herz, C. P., Doehling, A.: Chem. Ber. (1980) *113*, 233
75. Tabushi, I., Kobuke, Y., Nishiya, T.: Nature (1979) *280*, 665
76. Tabushi, I., Kobuke, Y., Nishiya, T.: Tetrahedron Letters (1979) 3515
77. Takekoshi, T., Scotia, Webb, J. L.: U.S. Patent 3,824,215, 16 July 1974
78. Travis, K., Busch, D. H.: J. C. S. Chem. Commun. (1970) 1041
79. Travis, K., Busch, D. H., Inorg. Chem. (1974) *13*, 2591
80. Vanlerberghe, G., Sebag, H.: Eur. Pat. Appl. 7,097, 23 Jan. 1980
81. Vögtle, F.: Tetrahedron (1969) *25*, 3231
82. Vögtle, F., Grütze, J., Nätscher, R., Wieder, W., Weber, E., Grün, R.: Chem. Ber. (1975) *108*, 1694
83. Vögtle, F., Nätscher, R.: Chem. Ber. (1976) *109*, 994
84. Vögtle, F., Risler, H.: Angew. Chem. Int. Ed. Engl. (1972) *11*, 727
85. Vögtle, F., Schunder, L.: Chem. Ber. (1969) *102*, 2677
86. Vögtle, F., Weber, E.: Angew. Chem. (1974) *86*, 126
87. Vögtle, F., Weber, E., Wehner, W., Nätscher, R., Grütze, J.: Chem. Zeitung (1974) *98*, 562
88. Volkora, M. S., Kiselena, T. M., Koton, M. M.: Vysokomol. Soedin, Ser. B (1977) *19*, 743
89. Warshawsky, A., Kalir, R., Deshe, A., Berkovitz, H., Patchornik, A.: J. Am. Chem. Soc. (1979) *101*, 4249
90. Weber, E., Vögtle, F.: Liebigs. Ann. Chem. (1976) 891
91. Weber, E., Vögtle, F.: Chem. Ber. (1976) *109*, 1803
92. Weber, E., Wieder, W., Vögtle, F.: Chem. Ber. (1976) *109*, 1002
93. Wieder, W., Nätscher, R., Vögtle, F.: Liebigs Ann. Chem. (1976) 924
94. Yatsimirskii, K. B., Sinyavskaya, E. I., Kudrya, T. N.: Dokl. Akad. Nauk SSSR (1978) *240*, 100

Chapter 7

Open-chained Equivalents of Crown Ethers

7.1 Introduction

With the discovery of the crowns[45] and related species, it was inevitable that a search would begin for simpler and simpler relatives which might be useful in similar applications. Perhaps these compounds would be easier and more economical to prepare and ultimately, of course, better in one respect or another than the molecules which inspired the research. In particular, the collateral developments of crown ether chemistry and phase transfer catalysis[65,76] fostered an interest in utilizing the readily available polyethylene glycol mono- or dimethyl ethers as catalysts for such reactions. Although there is considerable literature in this area, much of it relates to the use of simple polyethylene glycols in phase transfer processes. Since our main concern in this monograph is with novel structures, we will discuss these simple examples further only briefly, below.

A second source of inspiration for studying the open-chained equivalents of crown ethers was the observation that a number of naturally occurring antibiotics enhance cation transport and bear a structural similarity to open-chained crown ethers. A number of groups have examined neutral synthetic ionophores and a variety of novel cation carriers is now available. This is discussed in Sect. 7.4, below.

Although the return from the complex to the simple served as an important impetus for the study of open-chained crown ether equivalents, it should not be forgotten that Pedersen was originally attempting to prepare just such molecules when he made his fortuitous discovery of this class of compounds (see Sect. 1.2). Pedersen's original goal was the preparation of bis-phenol (*1*). In fact Pedersen holds a patent[46] which was filed in 1964 (three years before the crown ether disclosures) describing precisely the kind of compounds discussed in this chapter. A number of similar structures are shown in the tables at the end of this chapter. Further general information on this class of compounds can be found in a recent review by Vögtle and Weber[75]. These workers have suggested the term "podands" to indicate an open-chained equivalent of crown ether or cryptand compounds.

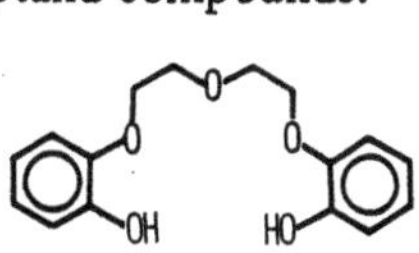

1

7.2 Uses of Open-chained Equivalents in PTC

It was noted early by Smid and his coworkers[66] that open-chained polyethylene glycol type compounds bind alkali metals much as the crowns do, but with considerably lower binding constants. This suggested that such materials could be substituted for crown ethers in phase transfer catalytic reactions where a larger amount of the more economical material could effect the transformation just as effectively as more expensive cyclic ethers. Knöchel and coworkers demonstrated the application of open-chained crown ether equivalents in 1975[30]. Recently, a number of applications have been published in which simple polyethylene glycols are substituted for crowns[5,19,25,32,33,39,58]. These include nucleophilic substitution reactions, as well as solubilization of arenediazonium cations[6–8]. Glymes have also been bound into polymer backbones for use as catalysts[54,64,83].

Chaput, Jeminet and Juillard[12] measured the association constants of several simple polyethylene glycols with Na^+, K^+, Cs^+, and Tl^+. Phase transfer catalytic processes and most biological processes are more likely to involve the first two cations rather than the latter two, so we will confine the discussion to these. Stability constants for the dimethyl ethers of tetra-, penta-, hexa-, and heptaethylene glycols were determined potentiometrically in anhydrous methanol solution and are shown in Table 7.I. In the third column of the table, the ratio of binding constants (Ks/K′s) is calculated. Note that Simon and his coworkers[3] have referred to this ratio as the "selectivity constant."

Since our concern here is primarily with syntheses and the variety of structures, we will comment only briefly on the information in this table. As expected, the more binding sites available, the greater is the binding constant, especially for the larger potassium cation. The binding constants for the last three compounds shown in the table suggest that there is a steric requirement. If the backbone prefers to adopt a geometry other than synclinal, the binding constant falls off considerably. This observation is identical to that made by Simon and his coworkers[2] at about the same time the data above appeared.

In specific applications to phase transfer catalysis, Knöchel and his coworkers[16,30] compared crown ethers, aminopolyethers, cryptands, octopus molecules ("krakenmoleküle", see below) and open-chained polyether compounds. They determined yields per unit time for reactions such as that between potassium acetate and benzyl chloride in acetonitrile solution. As expected, the open-chained polyethers were inferior to their cyclic counterparts, although a surprising finding was that certain aminopolyethers were superior to the corresponding crowns.

Pure, discrete diethylene, triethylene and tetraethylene glycols are all commercially available from a variety of sources. Pentaethylene glycol and longer polyoxyethylene glycols are generally prepared by condensation of two equivalents of a shorter glycol with a diol dichloride or ditosylate. Such methods have been reported by Pedersen[45], Cornforth[13], and Krespan[31]. The approach is illustrated in Eq. (7.1), below.

$$\underset{\text{(solvent)}}{HOCH_2CH_2OH} + ClCH_2(CH_2OCH_2)_nCH_2Cl \xrightarrow{KOH} HO(CH_2CH_2O)_{n+2}H \tag{7.1}$$

Table 7.I. Stability constants for several open-chained polyethers in anhydrous methanol at 25 °C[12]

Compound structure	Log Ks Na^+	Log Ks K^+	$\frac{Ks(K^+)}{Ks(Na^+)}$
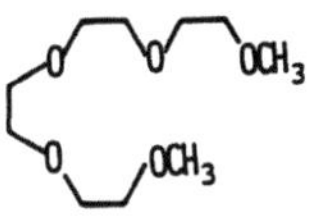	1.28	1.72	1.3
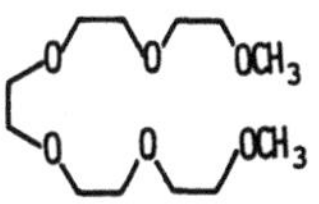	1.47	2.20	1.5
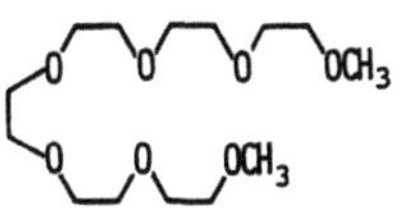	1.60	2.55	1.6
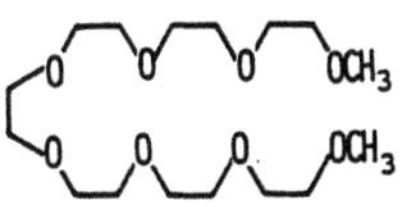	1.67	2.87	1.7
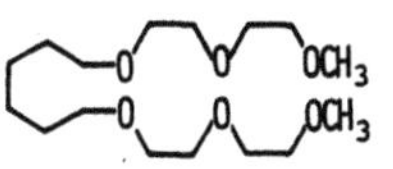	<0.1	<0.1	–
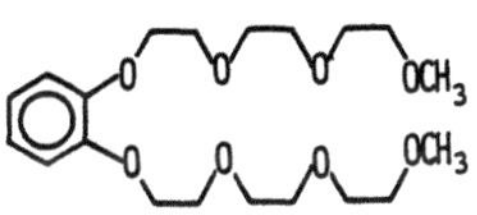	1.61	2.83	1.8
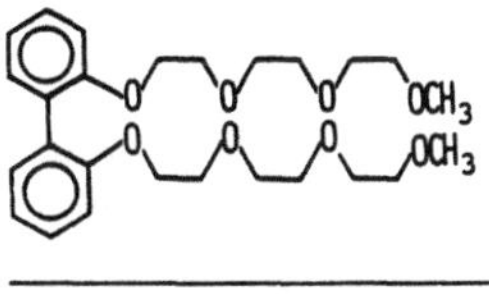	<0.1	1.45	–

A number of polyethylene glycols are also available in molecular weight ranges from commercial suppliers. These are made by initiating the polymerization of ethylene oxide by hydroxide ion. The corresponding monomethyl (or monoalkyl ethers) can be made in a similar fashion by initiating the polymerization with an alcohol. The general reaction is shown below in Eq. (7.2).

$$RO^- + n \cdot \overset{O}{\overbrace{CH_2-CH_2}} \longrightarrow RO-(CH_2CH_2O)_nH \qquad (7.2)$$

Obviously, these polymerized materials are sold as mixtures of compounds having an average molecular weight. For example, polyethylene glycol-400 (PEG-400) is a mixture of oligomers having molecular weights in the range 200–600[83]. A molecular weight of 400 corresponds to an average of between 8 and 9 ethyleneoxy units in the oligomer.

Montanari and his coworkers used the interesting "polypode" ligands derived from sym-trichlorotriazine as phase transfer catalysts for a variety of transformations[17]. These catalysts were quite successful and their formation is illustrated below in Eqs. (7.3)–(7.5). Comparisons were also made with certain pentaerythrityl derived polypodes as well. These latter compounds are listed in Table 7.1 as compounds 10–13.

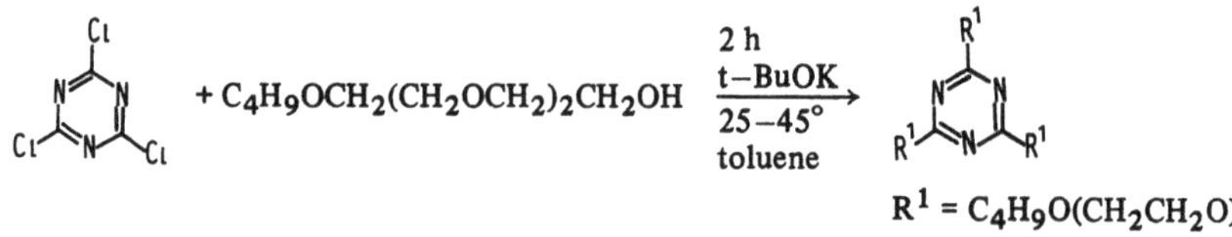

(7.3)

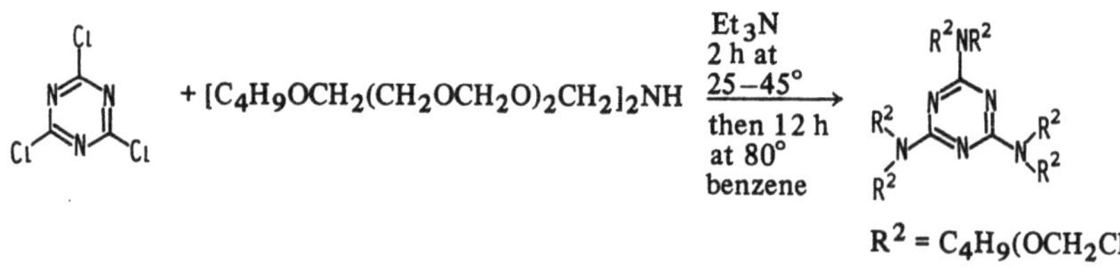

(7.4)

$$C_{10}H_{21}OCH_2\text{–}C(CH_2OH)_3 \xrightarrow[\text{2. LiAlH}_4]{\text{1. ClCH}_2\text{COOH}} \xrightarrow[\text{2. LiAlH}_4]{\text{1. ClCH}_2\text{ COOH}} C_{10}H_{21}OCH_2C[CH_2O(CH_2CH_2O)_2H]_3 \quad (7.5)$$

7.3 Octopus Molecules

In 1974, Vögtle and Weber[75] disclosed the preparation of a class of molecules which they reported showed "remarkable phenomenological parallelisms to the mode of food capture by an octopus using its suction pads". Such molecules have also been referred to as "hexa-hosts" (see below).

In the early work, benzene formed the basis of a variety of multi-armed structures. Analogs bearing from 2–6 "arms" were prepared and compared for cation binding ability. The only indication of mode of synthesis for the hexa-substituted benzene derivative is that it was obtained "on reaction of benzene-hexakis(methanethiol) and 1-bromo-3,6,9-trioxatridecane"[75]. The reaction is illustrated in Eq. (7.6), below, devoid of reaction conditions and yields which were not specified.

The authors report that the six-armed host is a "remarkably powerful complex ligand for cations". These compounds apparently can strongly bind alkali and alkaline earth metals including Li^+, Na^+, K^+, Rb^+, Cs^+, Mg^{2+}, Ca^{2+}, and Sr^{2+}. These ligands were relatively unsuccessful in binding several heavy metals which were tested.

As expected, the complexing tendency of the ligands was reduced in correspondence to the number of arms removed from the hexa-substituted system. Shortening

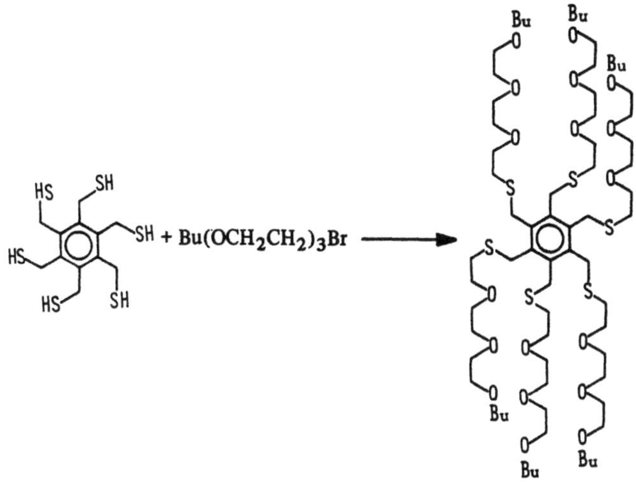

(7.6)

the length of the complexing arms with its concomitant reduction in the number of binding sites also reduced the complexing ability of these species.

The work of Hyatt on cyclotriveratrylene–derived octopus molecules contrasts with this[26]. Of course, these species have the advantage of ligand directionality absent in the benzene-derived octopus molecules. Except for the shortest-armed of the species (i.e., n = 1), all of the complexing agents (i.e., n = 2–4) were capable of complexing alkali metal cations. Synthesis of these species was accomplished as indicated below in Eq. (7.7). These variations of the original octopus molecules were also shown to catalyze the reaction between benzyl chloride and potassium acetate in acetonitrile solution and to effect the Wittig reaction between benzaldehyde and benzyltriphenylphosphonium chloride.

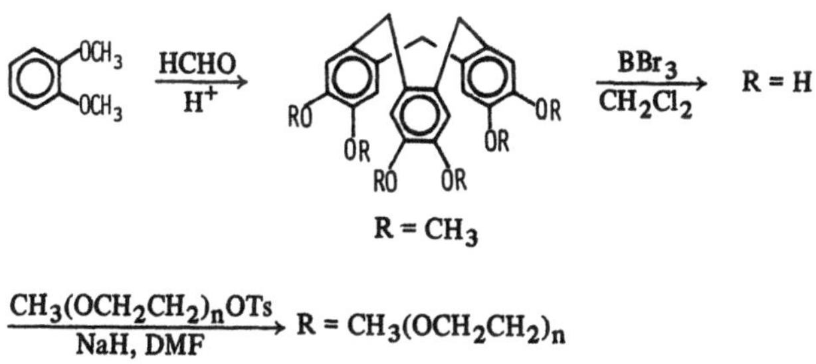

(7.7)

In further work from Vögtle's laboratory, a number of novel "krakenmoleculen" (compounds *2–4*) were reported to form complexes or inclusion compounds with a variety of molecular guests[77]. The guests included pyridine, DMF, chloroacetonitrile and a number of other substances. The general formula for these compounds is shown below. The substituents (R) reported in this paper were numerous and included 2-formylphenyl, 2-acetylphenyl, and 8-quinolinyl.

Molecular complexation or clathration was observed somewhat earlier using similar compounds referred to by MacNicol and Wilson as "hexa-hosts"[34]. These compounds formed clathrates with carbon tetrachloride, toluene, dioxane, bromoform and a num-

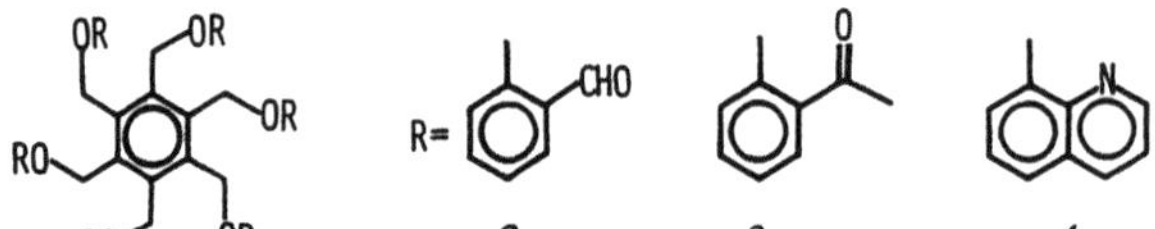

2 3 4

ber of other substances. As in the previously noted work on octopus type molecules, no detailed synthetic procedure was presented. The authors say only that the compound illustrated in Eq. (7.8) "was prepared by the reaction of hexachlorobenzene with PhSCu". Synthesis of other compounds reported in this paper were accomplished "by the action of the appropriate phenol, thiol, or selenol on hexakis(bromomethyl)benzene in the presence of base".

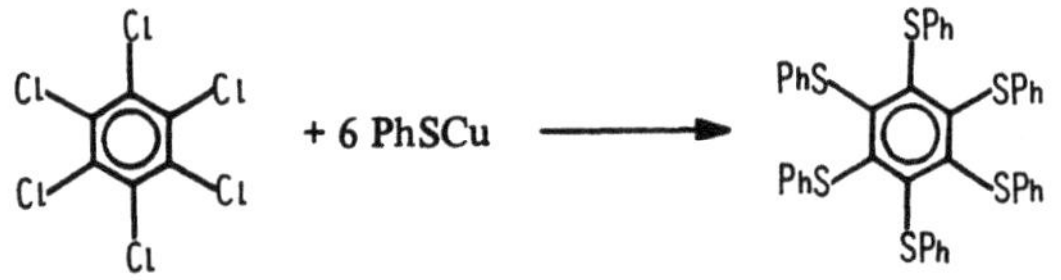

(7.8)

7.4 Substituted Podands

The bulk of the work which has been performed on open-chained crown ether and cryptand equivalents, especially for application to general cation binding studies has been accomplished by Vögtle and his coworkers. Vögtle has reviewed both his own and other work in this field[75].

Cram has published several studies indicating that complexation of ammonium salts was greatly enhanced by using cyclic polyethers over the corresponding acyclic ones. These molecules were analogs of the binaphthyl systems under study and were generally prepared by phenoxide substitution on the appropriate tosylate. The approach is illustrated below in Eq. (7.9)[67].

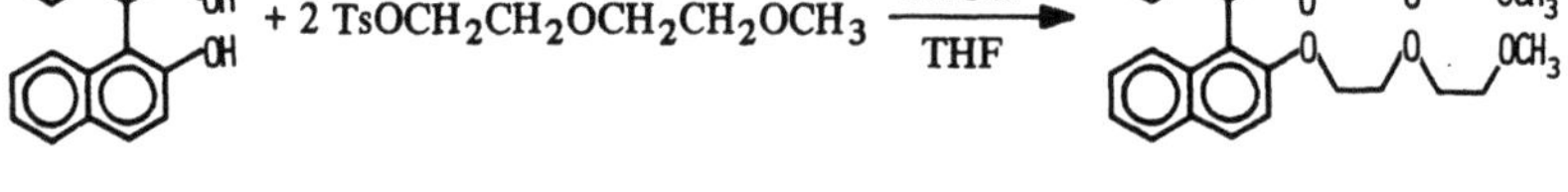

(7.9)

In 1975, Weber and Vögtle[78] showed that open-chained polyethers, 2,6-pyridinedimethanol, and ortho-xylene derivatives terminated in an 8-quinolinyl group could form stable, 1 : 1, crystalline complexes with a variety of metal salts. The "podands" were prepared from 8-hydroxyquinoline and the corresponding dihalides. A typical example is shown in Eq. (7.10).

In later work, Vögtle and his coworkers prepared analogs of both crown ethers[72] and cryptands[70]. These molecules are designed to have a "terminal donor group" which is capable of offering a complexed cation additional binding sites. Numerous

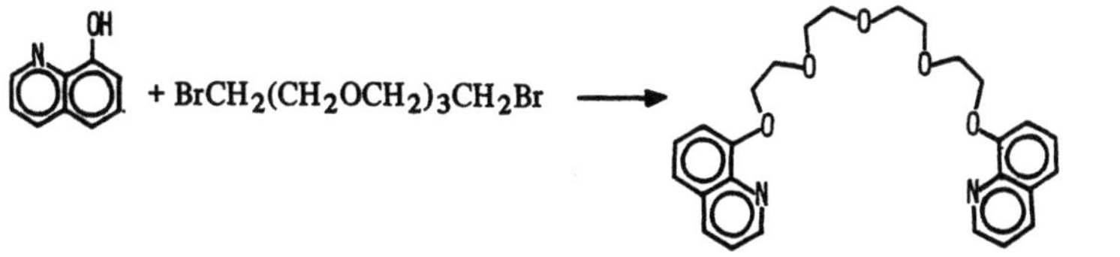

(7.10)

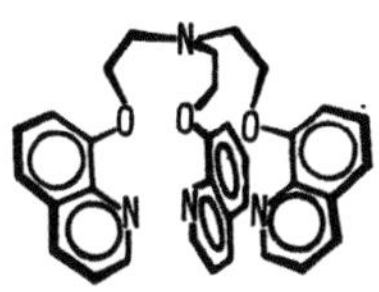

5

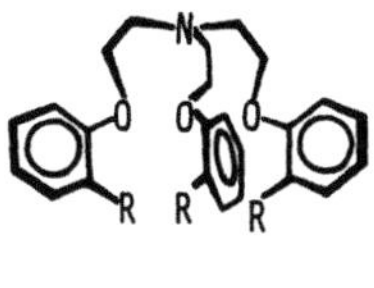

6

complexes were prepared between these hosts and metallic guests[50,70,72]. Examples of these structures are shown above as compounds *5* and *6*.

The materials shown and described above were generally prepared from the nucleophilic phenoxide or alkoxide and the appropriate bromide. The syntheses of a variety of such compounds were detailed in a report which appeared in 1977[69]. In the same report, complex stability and complexation kinetics are reported. Other, detailed studies, of a similar nature have recently appeared[49,68]. Vögtle and his collaborators have also demonstrated that solid complexes can be formed even from simple polyethylene glycol ethers[23,41,51,52,61,62]. Crystal structures of such species are also available[24].

A study of the complexation between $^{23}Na^{+}$ and the heptadentate amide shown in Eq. (7.11), led to the finding that in anhydrous pyridine, the enthalpy of binding was −17 kcal/mol and the entropy was the expectedly large value of −48 e.u.[20]

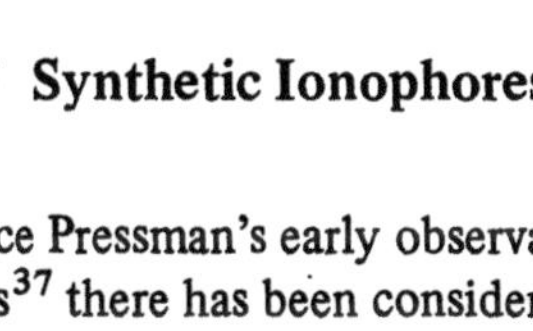

(7.11)

7.5 Synthetic Ionophores

Since Pressman's early observation that antibiotics of the nigericin type complexed cations[37] there has been considerable effort expended to mimic this action using simple, readily synthesized molecules. Simon and his collaborators have been especially active in this approach and have reviewed this area[38]. They have devised a number of struc-

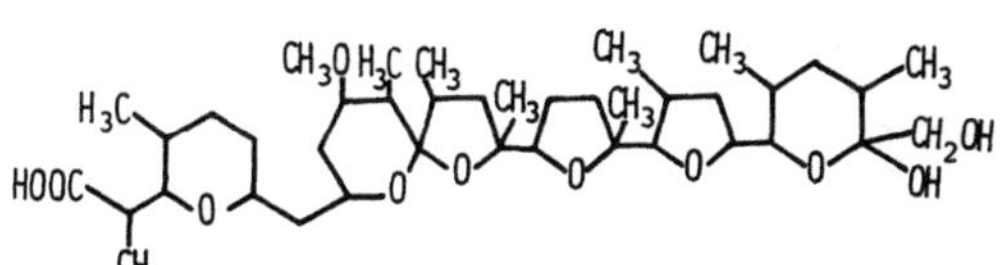

Nigericin

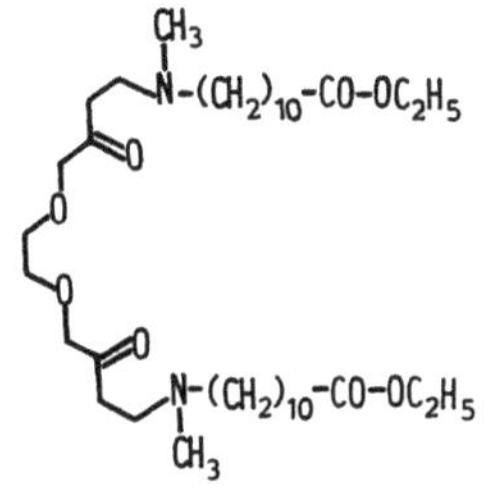

7

tures which bind cations, especially calcium cation[2–4,21,29]. They have used these in the preparation of a number of cation selective electrodes. A compound typical of Simon's cation carriers is 7, shown below. Filter paper was impregnated with this substance and the doped paper was used as the membrane in an electrochemical cell. The selectivity of the ligand was thus determined to be

Ca ~Rb > K ~Ba ~Cs ~Sr > Na ~Li ~Mg.

It has been observed that the best neutral cation carriers have the synclinal arrangement[2] of binding atoms in the backbone (see 7.2). Syntheses of both an aliphatic and aromatic ligand are shown below in Eqs. (7.12) and (7.13).

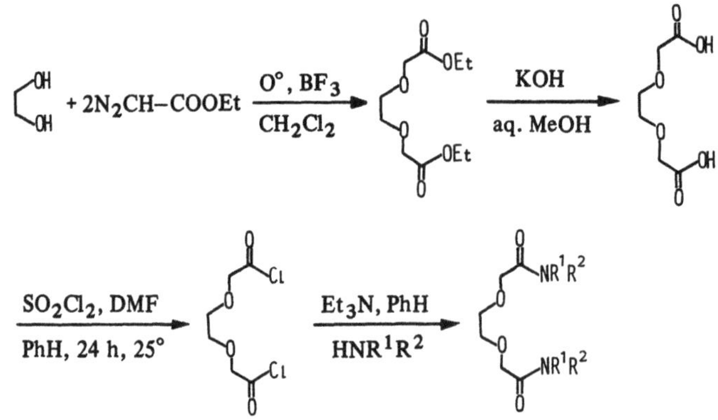

(7.12)

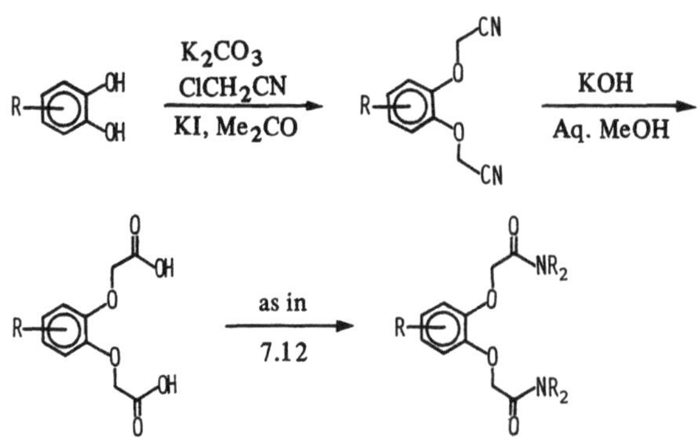

(7.13)

Recently, several other groups have contributed synthetic cation carriers. In particular, Vögtle has reported a large number of structures in this effort (see also Sect. 7.4)[56]. In addition, several ω-hydroxyacids have been prepared. The systems prepared by Yamazaki[81] (*8*) have been evaluated for ion transport action. Those prepared at Upjohn[79] have been reported to have Ca^{++} activity comparable to the natural antibiotic X-537A (*9*) and to be more active than crown ethers. The most active of their structures is shown as *10.*

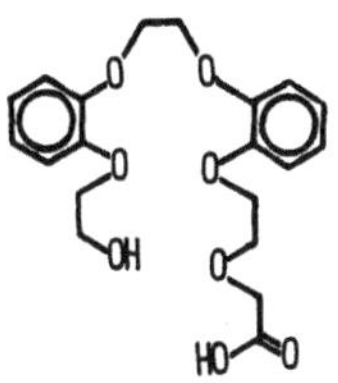

8

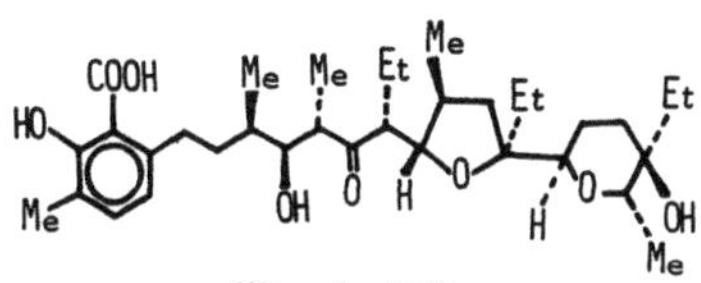

"Lasalocid"

9

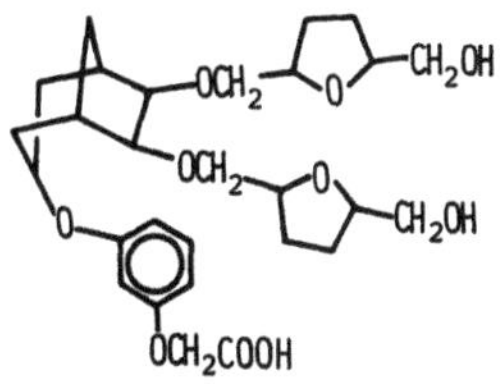

10

7.6 Listing of Tables

7.7 Tables 7.1–7.13

Table 7.1. Simple ethers

Cpd. No.	No. C's	Structure	Mp or (bp) in °C	Yield (%)	Refs.
		$R_1O(CH_2CH_2O)_nR_2$			
1	12	$n = 5, R_1 = R_2 = CH_3$	(108–113/0.3)	nr	12
2	12	$n = 5, R_1 = H, R_2 = CH_2CO_2H$	nr	nr	81
3	14	$n = 6, R_1 = R_2 = CH_3$	(145–149/0.6)	nr	12
4	16	$n = 7, R_1 = R_2 = CH_3$	(160–168/0.5)	nr	12
5	16	$(CH_2CH_2CH_2OCH_2CH_2OCH_2CH_2OCH_3)_2$	nr	nr	12
		$R_1O(CH_2CH_2O)_nR_2$			
6	18	$n = 4, R_1 = R_2 = CH_2CH_2\overset{O}{\overset{\|}{C}}OCH_2CH_3$	nr	nr	12
7	18	$n = 2, R_1 = R_2 =$ 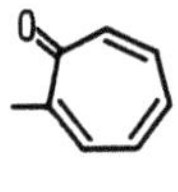	88–89	41	49
8	20	$n = 3, R_1 = R_2 =$ 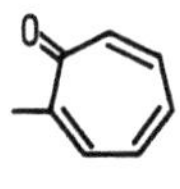	oil	51	49
9	22	$n = 4, R_1 = R_2 =$ 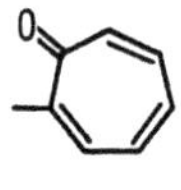	oil	55	49
		$CH_3(CH_2)_9OCH_2C(R)_3$			
10	27	$R = CH_2OCH_2CH_2OCH_2CH_2OH$	oil	nr	17
11	33	$R = CH_2(OCH_2CH_2)_3OH$	oil	nr	17
12	51	$R = CH_2(OCH_2CH_2)_2O(CH_2)_7CH_3$	oil	nr	17
13	57	$R = CH_2(OCH_2CH_2)_3O(CH_2)_7CH_3$	oil	nr	17

Table 7.2. Open-chained systems containing the disubstituted benzene subunit

Cpd. No.	No. C's	Structure		Mp or (bp) in °C	Yield (%)	Refs.
1	16	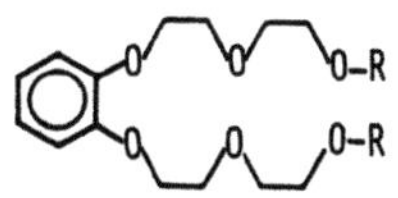	R = CH_3	oil	nr	12
2	18		R = CH_2CH_3	oil	nr	12
3	18	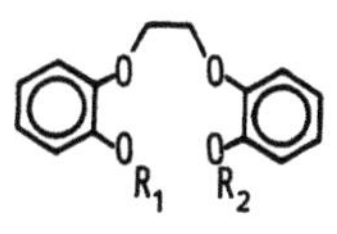	$R_1 = R_2 = CH_2CH_2OH$	99	nr	24
4	20		$R_1 = CH_2CH_2OH$ $R_2 = CH_2CH_2OCH_2CO_2H$	nr	nr	81
		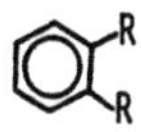				
5	20	R = $O(CH_2CH_2O)_3CH_3$		oil	nr	12
6	20	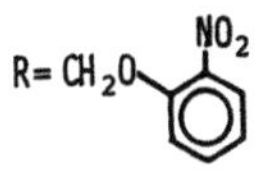		135	54	72
7	22	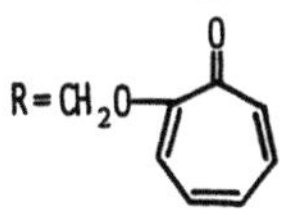		172–175	54	49
8	22	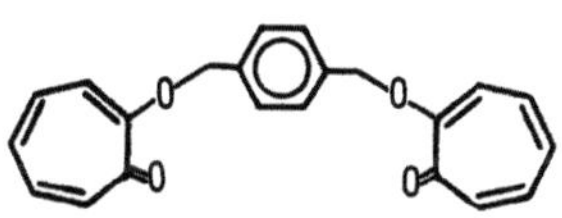		198–202	56	49
9	24	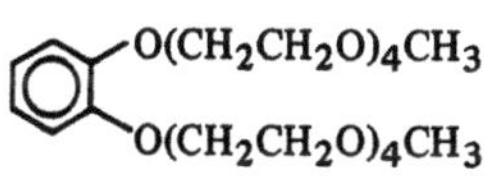		oil	nr	12
10	24	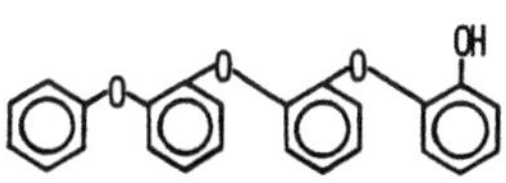		129	nr	28
11	26	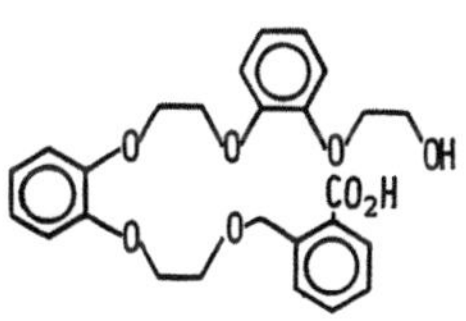		nr	nr	81
12	26	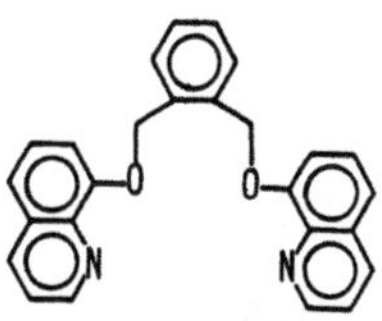		122–124	40	78

Table 7.2 (continued)

Cpd. No.	No. C's	Structure		Mp or (bp) in °C	Yield (%)	Refs.
13	26	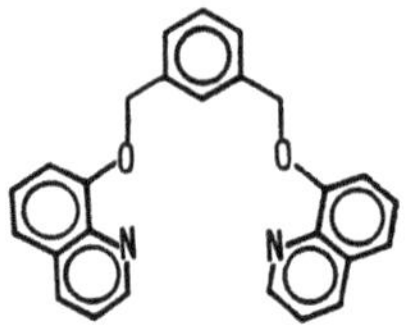		oil 228–230 picrate	36	78
14	28	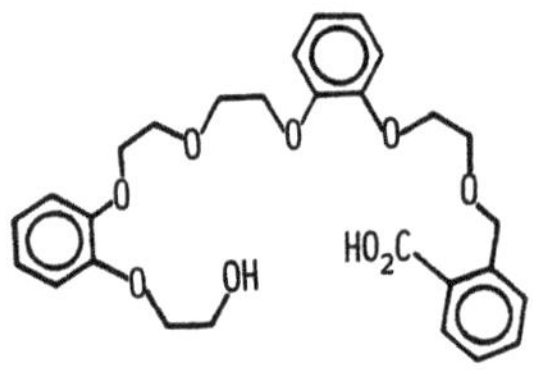		nr	nr	81
15	28	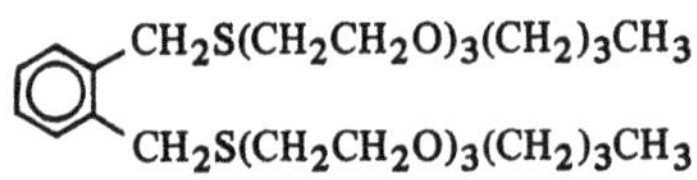			nr	74
16	30	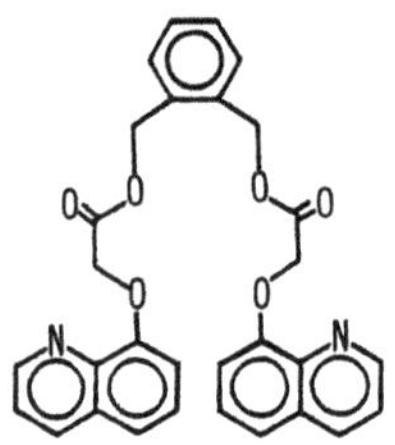		105–106	51	49
17	30	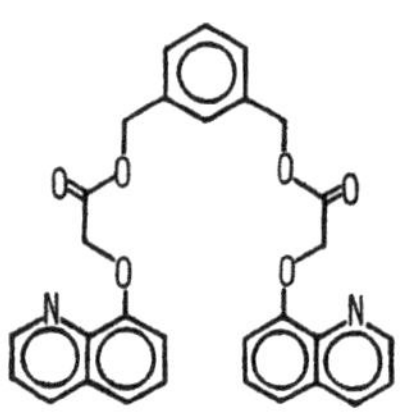		oil	56	49
18	30	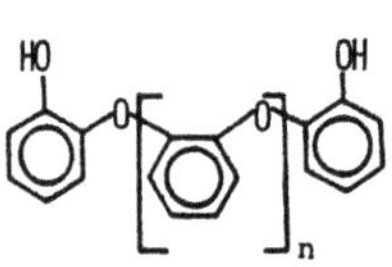	n = 3	69–70	nr	28
19	36		n = 4	152–153	nr	28
20	42		n = 5	91–93	nr	28

Table 7.3. Open-chained ligands having aromatic terminal groups

Cpd. No.	No. C's	Structure	Mp or (bp) in °C	Yield (%)	Refs.
1	10	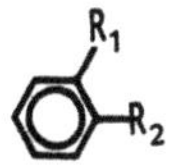 $R_1 = O(CH_2CH_2O)_2H$, $R_2 = NO_2$	oil	45	63
2	12	$R_1 = O(CH_2CH_2O)_3H$, $R_2 = NO_2$	oil	72	63
3	13	$R_1 = O(CH_2CH_2O)_3CH_3$, $R_2 = OH$	oil	nr	12
4	14	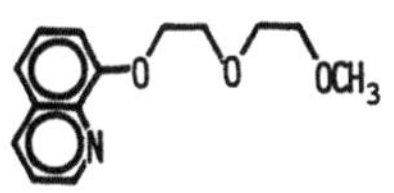	oil 131–132 picrate	53	78
5	15	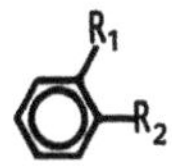 $R_1 = O(CH_2CH_2O)_4CH_3$, $R_2 = OH$	oil	nr	12
6	16	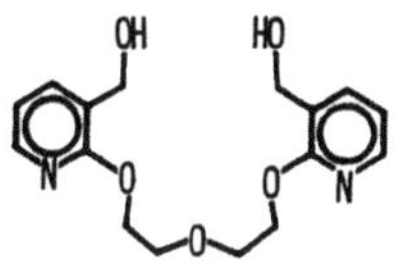	114.5–115.5	52	40
7	16	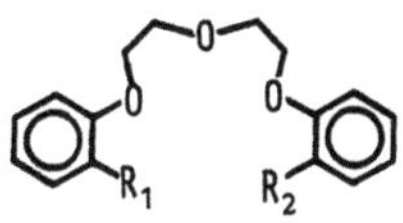 $R_1 = R_2 = OH$	86–88 · H_2O	23	41
8	17	$R_1 = NO_2$, $R_2 = CO_2H$	60–65	98	63
9	17	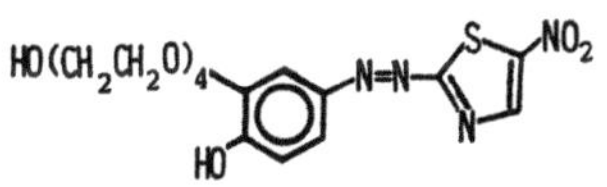	112–115	7	15
10	18	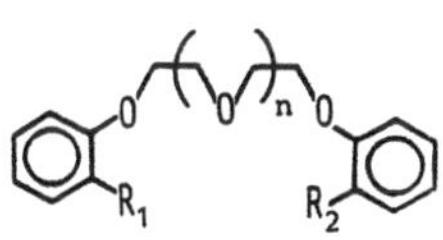 n = 2, $R_1 = R_2 = H$	65	75	72
11	18	n = 2, $R_1 = R_2 = OH$	oil 33–35 · H_2O	19	41
12	19	n = 2, $R_1 = NH_2$, $R_2 = CO_2H$	oil	88	63
13	19	n = 1, $R_1 = NO_2$, $R_2 = CO_2CH_2CH_3$	oil	78	63
14	19	n = 2, $R_1 = NO_2$, $R_2 = CO_2H$	76	97	63
15	20	n = 3, $R_1 = R_2 = H$	oil	59	72
16	20	n = 3, $R_1 = R_2 = OH$	oil	11	41
17	20	n = 2, $R_1 = R_2 = CO_2H$	116	91	63
18	20	n = 3, $R_1 = R_2 = NH_2$	oil oil	74 74	72 63
19	20	n = 3, $R_1 = R_2 = NO_2$	oil	47	63, 72

Table 7.3 (continued)

Cpd. No.	No. C's	Structure	Mp or (bp) in °C	Yield (%)	Refs.
20	20	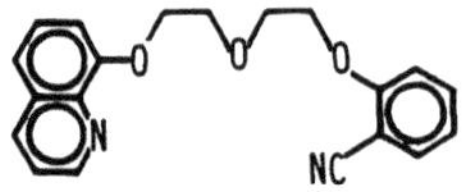	oil	78	72
21	21	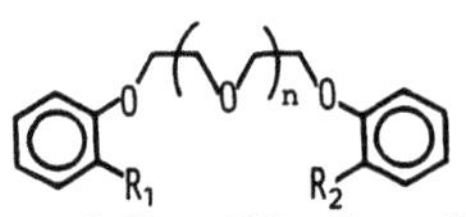 n = 2, $R_1 = NO_2$, $R_2 = CO_2CH_2CH_3$	oil	60	63
22	21	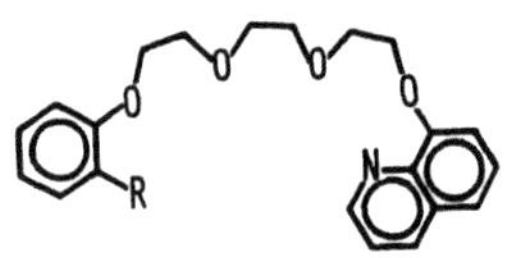 $R = NH_2$	oil	95	63
23	21	$R = NO_2$	oil	78	63
24	22	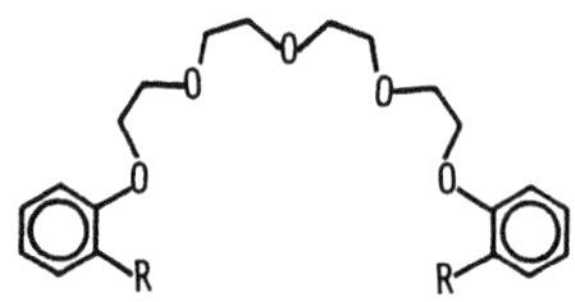 $R = OCH_3$	oil	88	72
25	22	R = CHO	oil	48	72
26	22	$R = CO_2H$	oil	89	63, 72
27	22	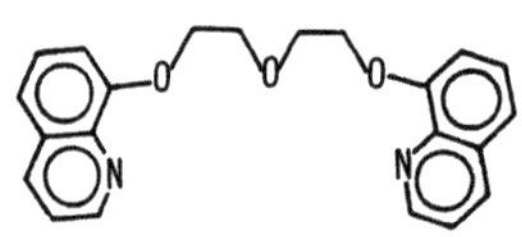	27	50	72
28	22	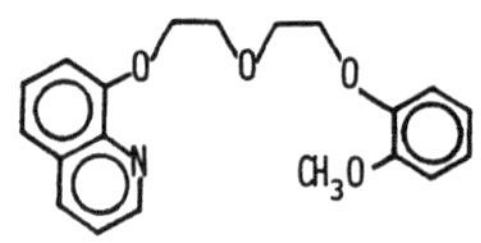	oil	70	72
29	23	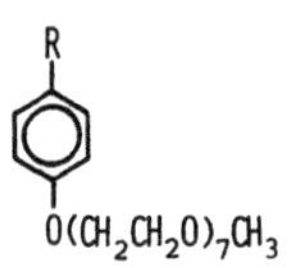 $R = CH{=}CH_2$	nr	70	64
30	23	$R = CHOHCH_3$	oil	85	64
31	23	$R = \overset{O}{\overset{\parallel}{C}}CH_3$	(190–194/4)	45	64
32	23	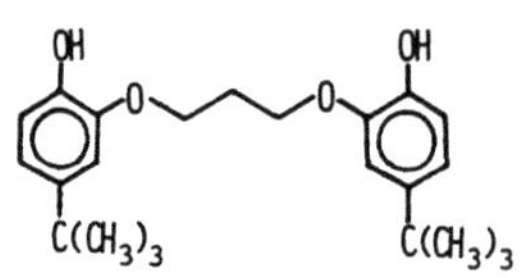	81–84	92	46
33	24	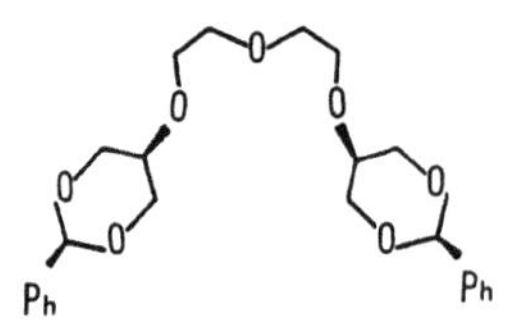	78–80	37	22

Table 7.3 (continued)

Cpd. No.	No. C's	Structure		Mp or (bp) in °C	Yield (%)	Refs.
34	24	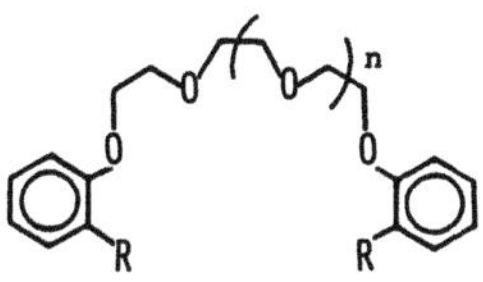	n = 1, R = $CO_2CH_2CH_3$	oil	65	63
35	24		n = 2, R = $NHCCH_3$ (C=O)	76–78	87	21, 63, 72
36	26		n = 2, R = $CO_2CH_2CH_3$	oil oil	47 67	72 63
37	24	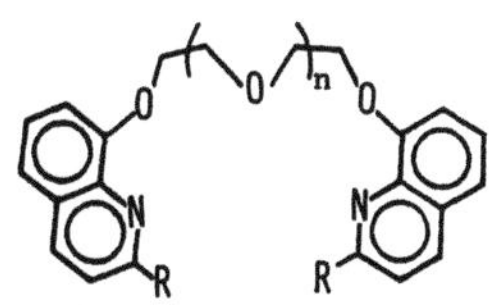	n = 2, R = H	103	43	72
38	26		n = 3, R = H	oil 153–155 · picrate	80 80	69 78
39	28		n = 3, R = CH_3	40	49	72
40	30		n = 5, R = H	oil	52	72
41	26	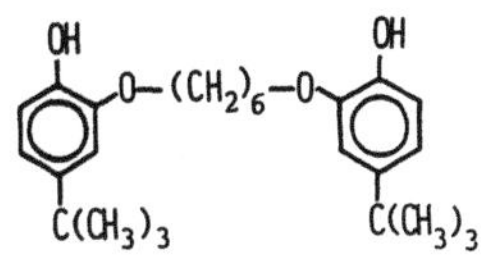		oil	99	46
42	26	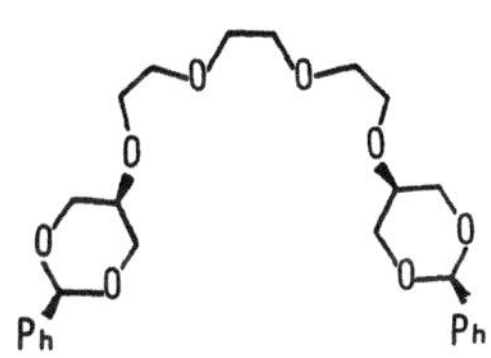		nr	nr	22
43	28	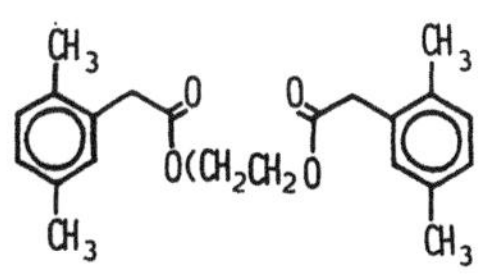	n = 4	oil	nr	57
44	30		n = 5	oil	nr	57
45	30	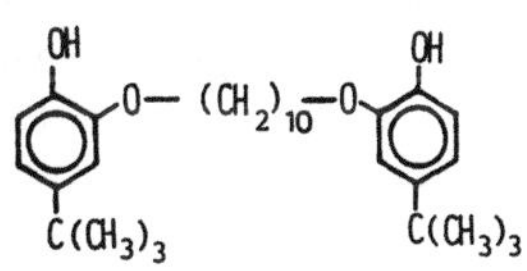		oil	~100	46
46	34	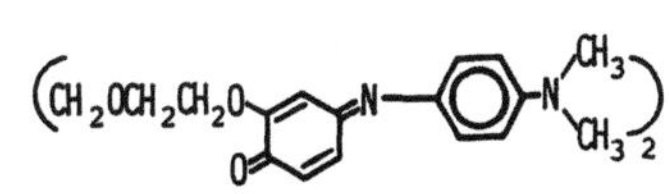		112–115	17	74
47	34	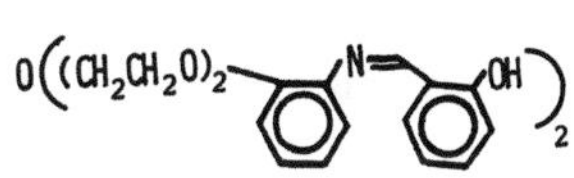		oil	88	72

Table 7.3 (continued)

Cpd. No.	No. C's	Structure		Mp or (bp) in °C	Yield (%)	Refs.
48	40	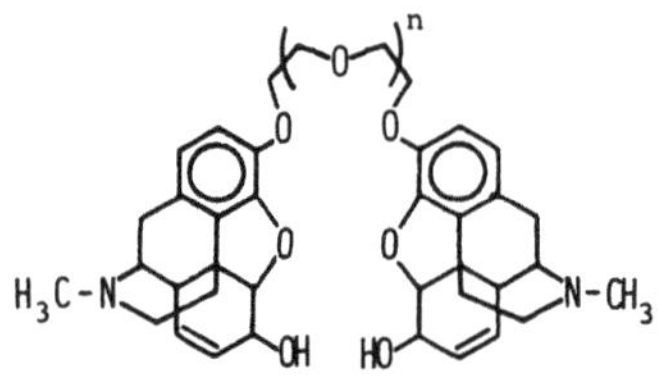	n = 2	69–70	8	18
49	42		n = 3	65–67	10	18

Table 7.4. Podands containing the biphenyl or binaphthyl subunit

Cpd. No.	No. C's	Structure		Mp or (bp) in °C	Yield (%)	Refs.
1	22	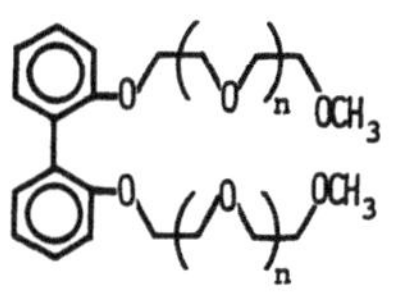	n = 1	oil	nr	12
2	26		n = 2	oil	nr	12
3	30	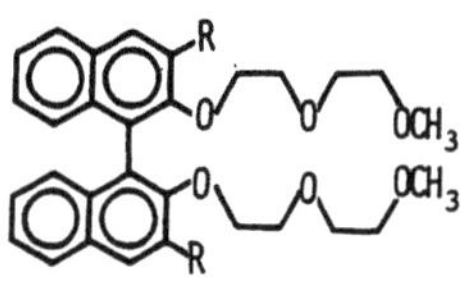	R = H	55.5–56.5	81	14
4	32		R = CH_2SH	nr	nr	11

Table 7.5. Open-chained esters, ketones, and carbamates

Cpd. No.	No. C's	Structure	Mp or (bp) in °C	Yield (%)	Refs.
1	6	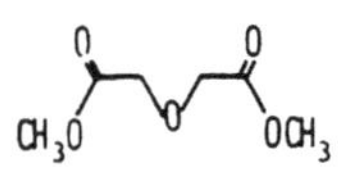	36–37	66	49
2	12	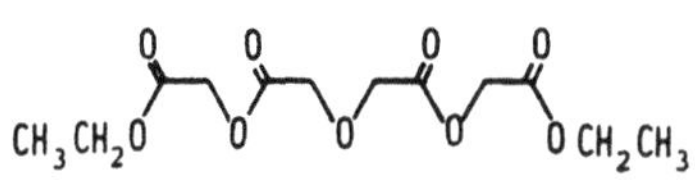	oil	49	49

Table 7.5 (continued)

Cpd. No.	No. C's	Structure	Mp or (bp) in °C	Yield (%)	Refs.
3	13	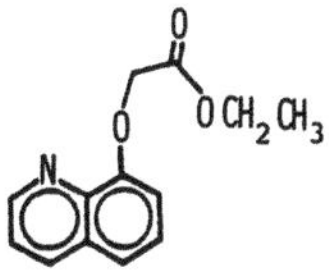	53	48	49
4	12	$RCO(OCH_2CH_2)_nOCOR$ 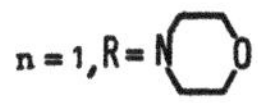	112	90	9
5	14	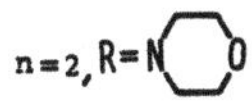	110	75	9
6	14	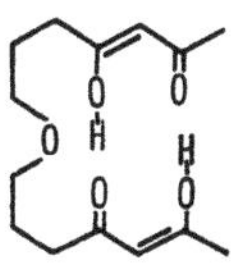	nr	70	1
7	14	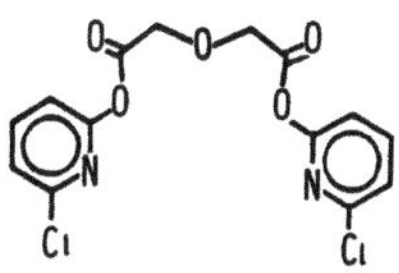	120–122	71	49
8	14	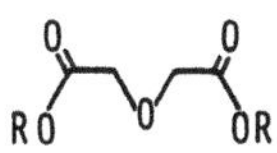 R = $(CH_2CH_2O)_2CH_3$	oil	59	49
9	16	R = Ph	73–75	69	64
10	16	R = 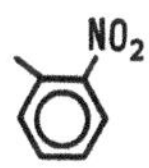	128	71	49
11	18	R = 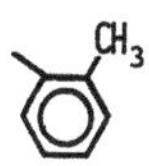	97	63	49
12	18	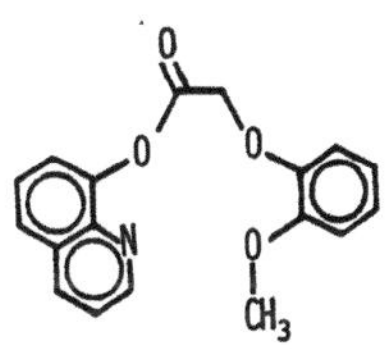	62–63	74	73
13	18	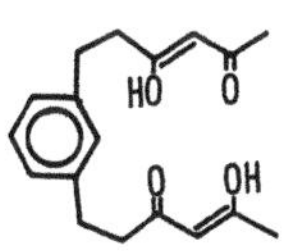	nr	95	1
		$RO(CH_2CH_2O)_nR$			
14	18	n = 4, R = $CH_2\overset{O}{\overset{\|}{C}}CH_2\overset{O}{\overset{\|}{C}}OCH_3$	nr	87	27

Table 7.5 (continued)

Cpd. No.	No. C's	Structure	Mp or (bp) in °C	Yield (%)	Refs.
15	20	n = 4, R = 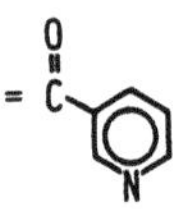	oil	90	47
16	20	n = 5, R = $CH_2\overset{O}{\overset{\|}{C}}CH_2\overset{O}{\overset{\|}{C}}OCH_3$	nr	84	27
17	22	n = 6, R = $CH_2\overset{O}{\overset{\|}{C}}CH_2\overset{O}{\overset{\|}{C}}OCH_3$	nr	75	27
18	24	n = 2, R = 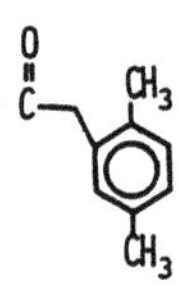	oil	nr	57
19	26	n = 3, R = 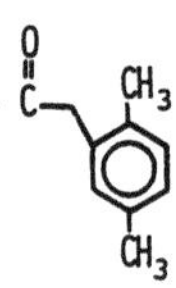	oil	nr	57
20	22	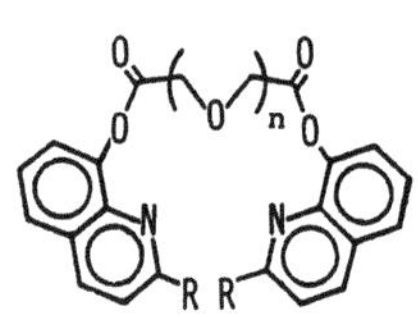 n = 1, R = H	123–125	67	49
21	24	n = 1, R = CH_3	118–120	63	49
22	26	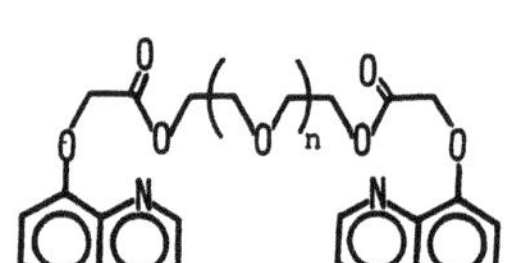 n = 1	oil	55	49
23	30	n = 3	oil	51	49
24	32	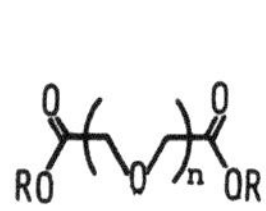 n = 1, R = 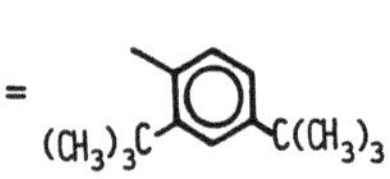	oil	68	49
25	32	n = 1, R = 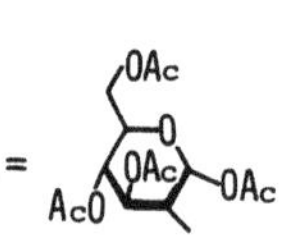	134–135	51	49
26	32	n = 1, R = 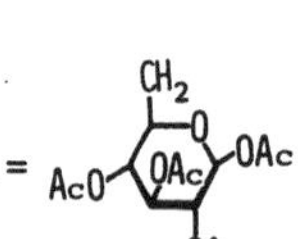	99–100	53	49

Table 7.5 (continued)

Cpd. No.	No. C's	Structure	Mp or (bp) in °C	Yield (%)	Refs.
27	32	n = 1, R = 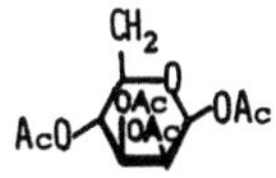	154–155	57	49
28	42	n = 2, R = 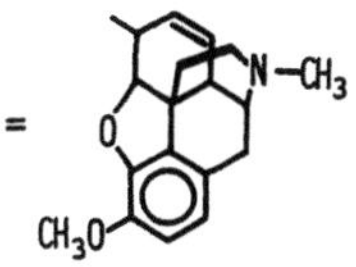	nr	nr	18

Table 7.6. Open-chained amides

Cpd. No.	No. C's	Structure	Mp or (bp) in °C	Yield (%)	Refs.
		$R_2NC(=O)(CH_2CH_2O)_nCH_2CH_2C(=O)NR_2$			
1	12	n = 4, R = H	94–95	nr	12
2	14	n = 5, R = H	82–83	nr	12
3	18	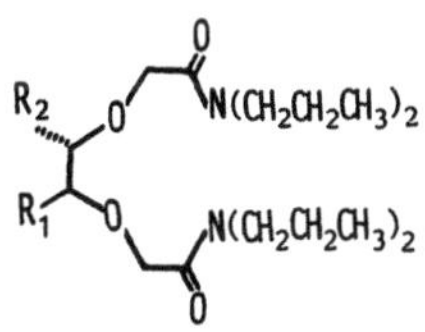 $R_1 = R_2 = H$	nr	85	2
4	19	$R_1 = CH_3, R_2 = H$	nr	22	2
5	20	$R_1 = R_2 = CH_3$ (R, R)	nr	51	2
		$(CH_2OCH_2C(=O)NHR)_2$			
6	20	$R = (CH_2)_3O(CH_2)_3CH_3$	nr	32	4
7	20	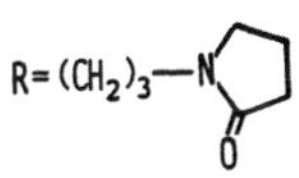	oil	9	4
8	22	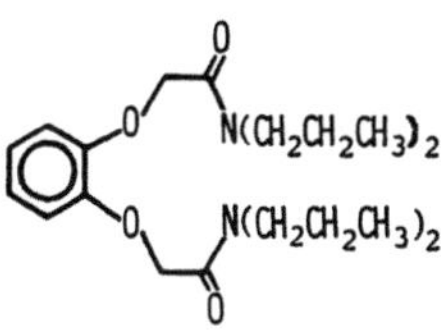	57.5–60	67	2

Table 7.6 (continued)

Cpd. No.	No. C's	Structure		Mp or (bp) in °C	Yield (%)	Refs.
9	22	$(CH_2OCH_2\overset{O}{\overset{\|}{C}}\overset{CH_3}{\overset{\|}{N}}(CH_2)_6CH_3)_2$		nr	42	4
10	22	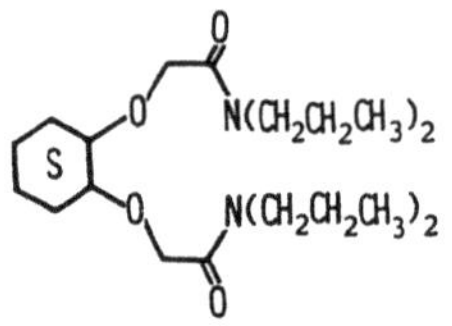		nr	63	2
11	22	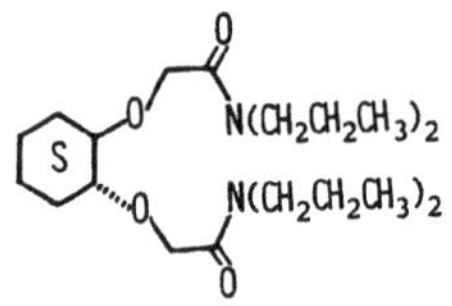		nr	47	2
12	23	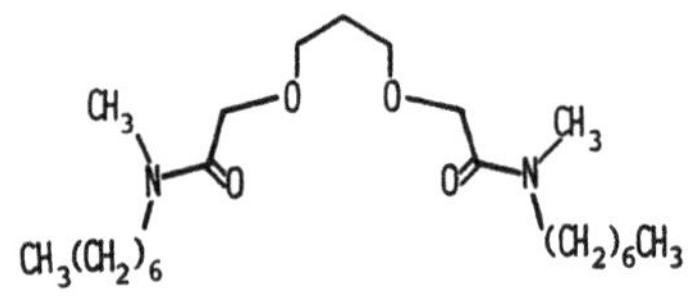		nr	nr	29
13	23	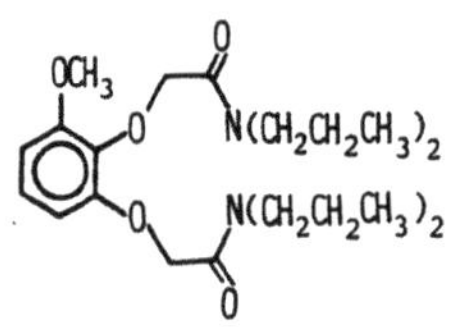		nr	61	2
14	25	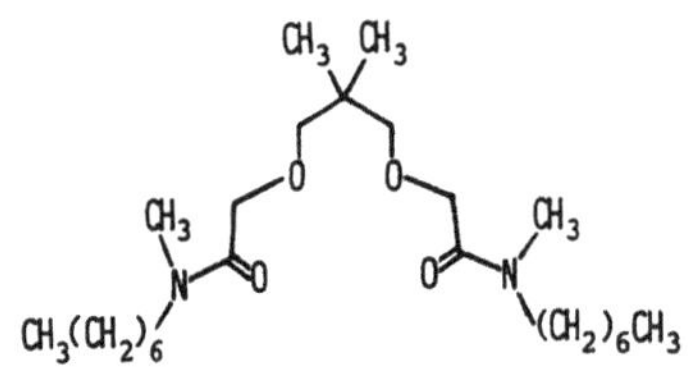		nr	52	29
15	26	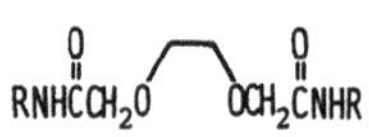	R = $(CH_2)_9CH_3$	nr	60	4
16	26		R = $(CH_2)_7CO_2CH_2CH_3$	nr	64	4
17	26	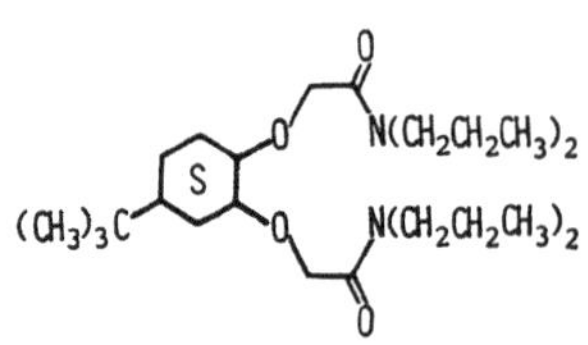		nr	26	2

Table 7.6 (continued)

Cpd. No.	No. C's	Structure	Mp or (bp) in °C	Yield (%)	Refs.
18	26	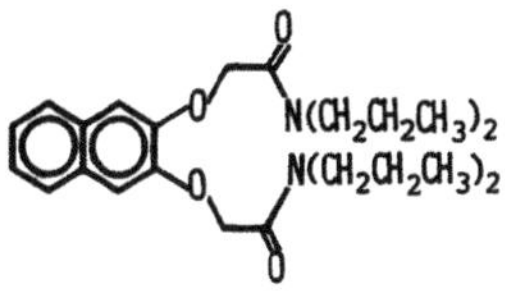	108.5–111.5	22	2
19	28	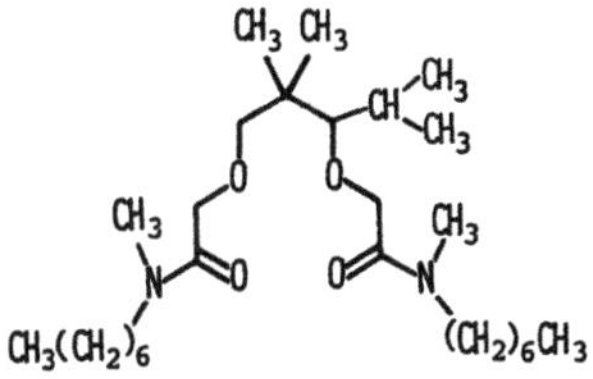	nr	27	29
20	29	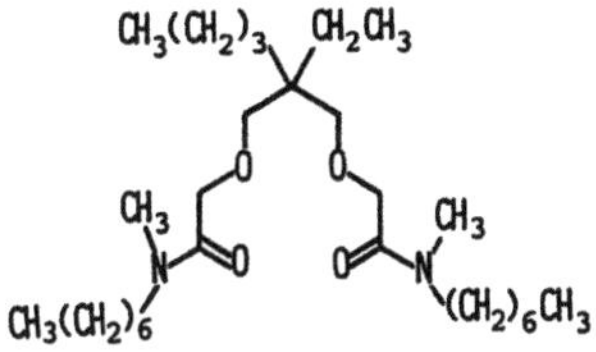	nr	34	29
		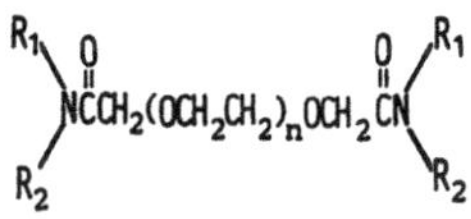			
21	30	n = 1, R_1 = R_2 = Ph	95–97	47	4
22	30	$Ph_2CHNH(CH_2CH_2O)_3CH_2CH_2NH\overset{O}{\overset{\Vert}{C}}CH=$ CHPh (E)	oil	25	73
23	30	$(CH_2OCH_2\overset{O}{\overset{\Vert}{C}}\overset{CH_3}{\overset{\vert}{N}}(CH_2)_{10}\overset{O}{\overset{\Vert}{C}}OCH_2Ph)_2$	nr	19	4
24	34	$O((CH_2CH_2O)_2$ 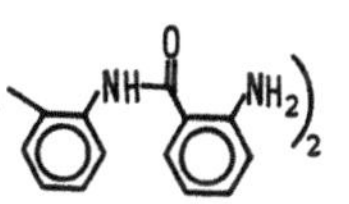	67–70	91	63
25	34	$O((CH_2CH_2O)_2$ 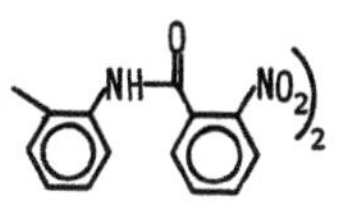	73	42	63
26	34	$(CH_2OCH_2\overset{O}{\overset{\Vert}{C}}\overset{CH_3}{\overset{\vert}{N}}(CH_2)_{10}CO_2CH_2CH_3)_2$	oil	50	3, 4

Table 7.6 (continued)

Cpd. No.	No. C's	Structure	Mp or (bp) in °C	Yield (%)	Refs.
27	36	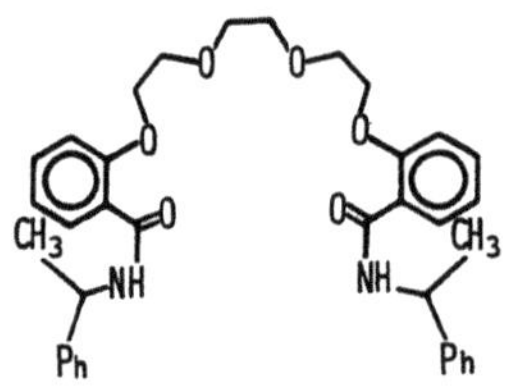 L-(+) D-(−)	oil oil	94 93	63 63
28	36	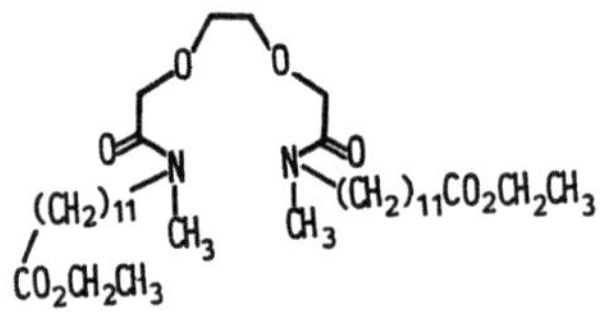	nr	30	2
29	36	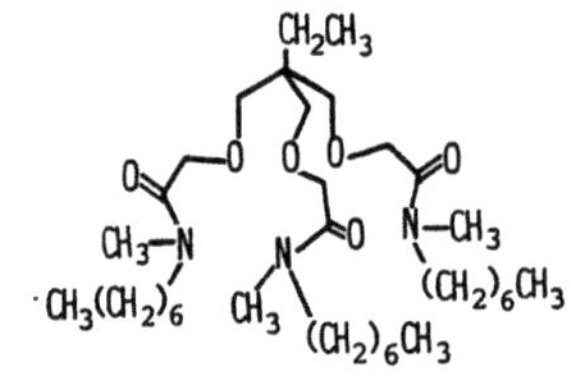	nr	50	21
30	37	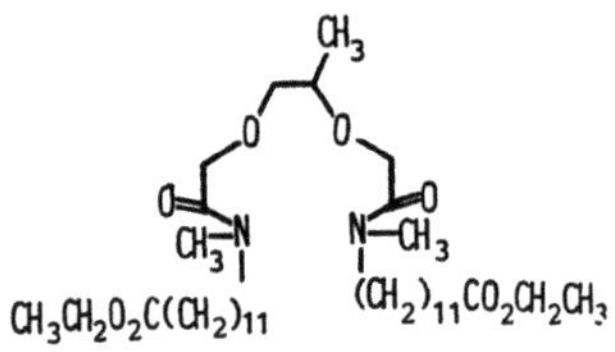	nr	21	2
31	38	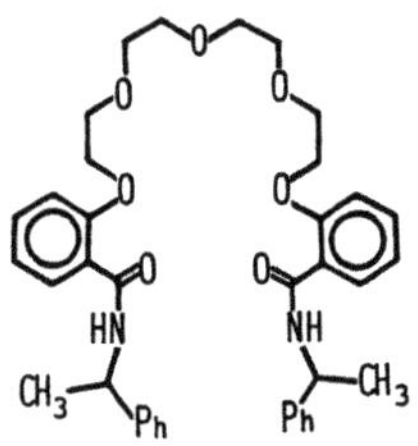 L-(+) D-(−)	oil oil	80 73	63 63
32	38	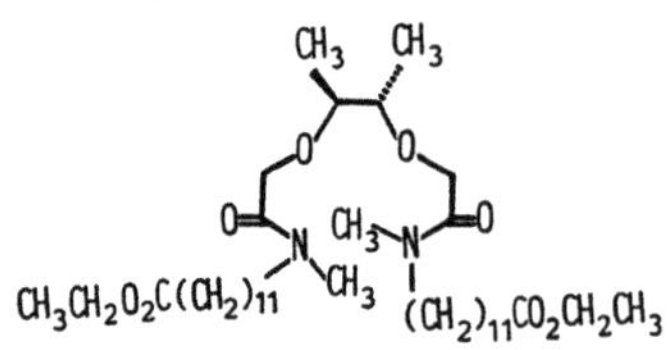	nr	47	2
33	38	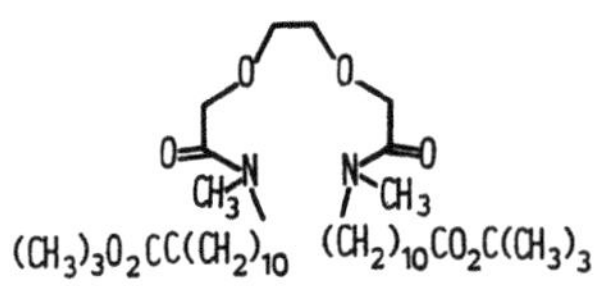	nr	5	4

Table 7.6 (continued)

Cpd. No.	No. C's	Structure	Mp or (bp) in °C	Yield (%)	Refs.
34	38	$(CH_2CH_2CH_2\overset{O}{\overset{\parallel}{C}}\overset{CH_3}{\overset{\mid}{N}}—(CH_2)_{11}—CO_2CH_2CH_3)_2$	nr	nr	4
35	40	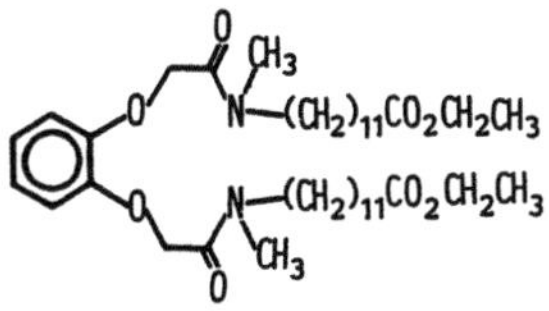	nr	67	2
36	40	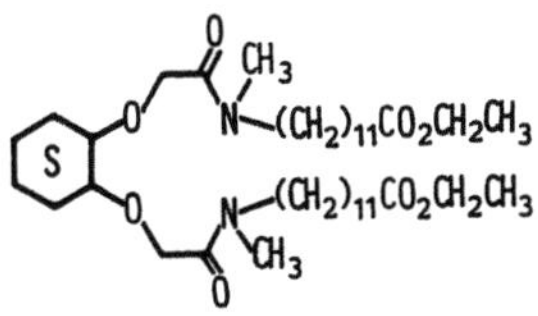 (E)	nr nr	59 45	2 2
37	41	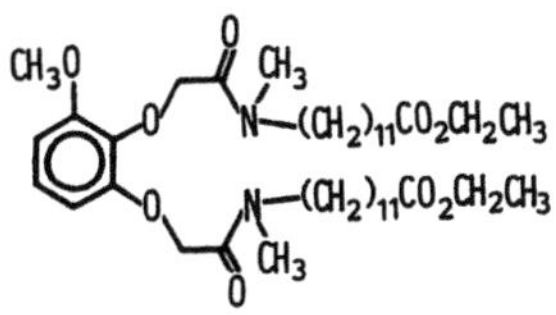	nr	63	2
38	44	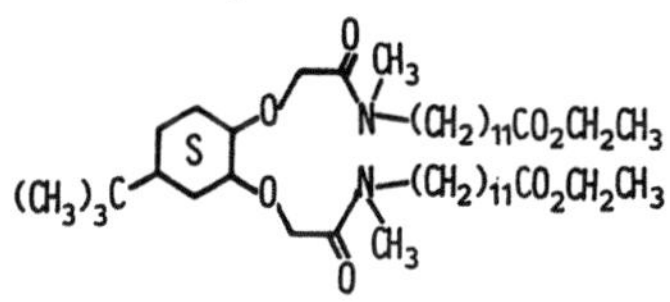	nr	22	2
39	44	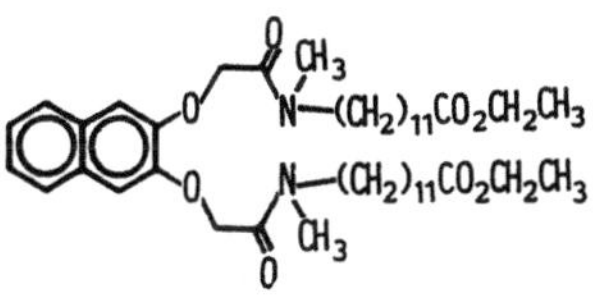	oil	37	2
		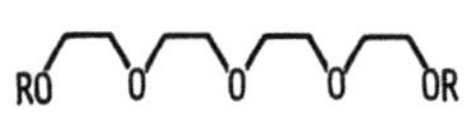			
40	48	R = 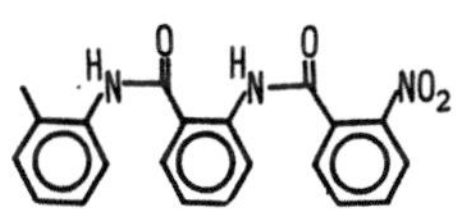	205	56	63
41	54	R = 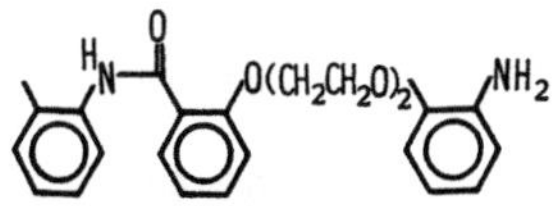	oil	95	63
42	54	R = 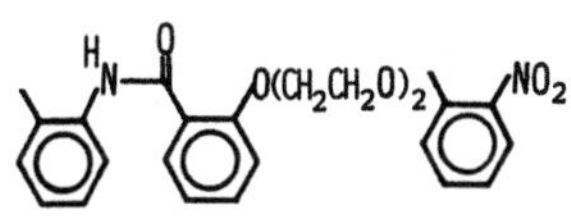	oil	67	63

Table 7.6 (continued)

Cpd. No.	No. C's	Structure	Mp or (bp) in °C	Yield (%)	Refs.
43	58	R= 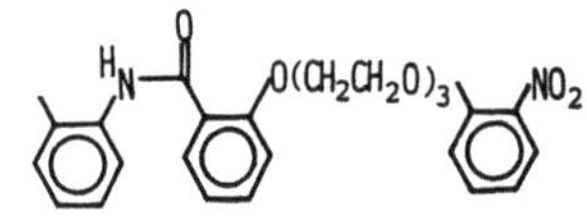	oil	65	63
44	88	R= 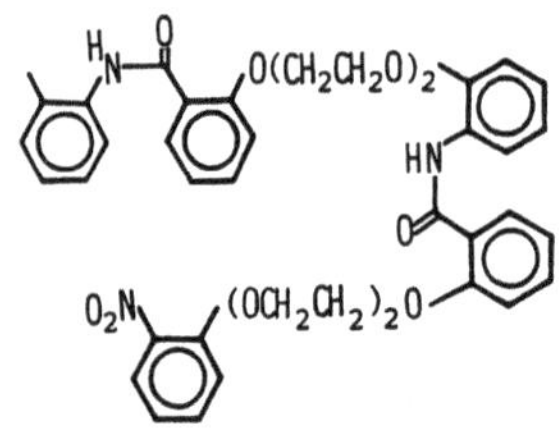	oil	60	63

Table 7.7. Multi-armed aromatic compounds

Cpd. No.	No. C's	Structure	Mp or (bp) in °C	Yield (%)	Refs.
1	11	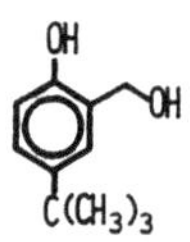	nr	99	46
2	11	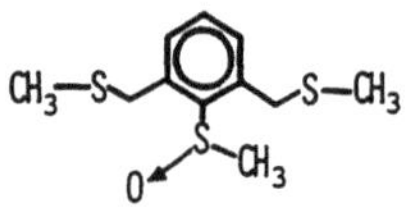	59–60	35	71
3	12	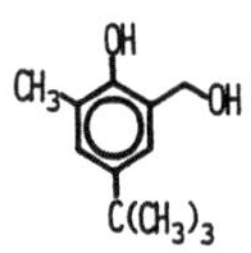	nr	97	46
4	16	R = NH_2	84–85	71	80
5	16	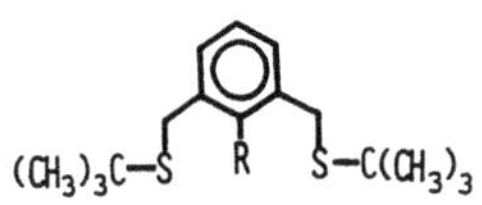 R = NO_2	65–67	87	80
6	17	R = S(→O)—CH_3	78–81	23	71

Table 7.7 (continued)

Cpd. No.	No. C's	Structure	Mp or (bp) in °C	Yield (%)	Refs.
		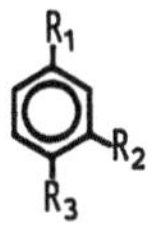			
7	17	$R_1 = CH_3$, $R_2 = R_3 = (OCH_2CH_2)_2OCH_3$	(176–178/1)	80	64
8	17	$R_1 = CHO$, $R_2 = R_3 = (OCH_2CH_2)_2OCH_3$	(196–197/1)	65	64
9	17	$R_1 = CH{=}CH_2$, $R_2 = R_3 = (OCH_2CH_2)_2OCH_3$	oil	67	64
10	18	$R_1 = CHOHCH_3$, $R_2 = R_3 = (OCH_2CH_2)_2OCH_3$	oil	70	64
		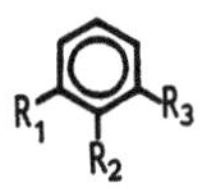			
11	18	$R_1 = R_3 = CH_2SC(CH_3)_3$, $R_2 = NC(=O)CH_3$	171	95	80
12	20	$R_1 = R_3 =$ 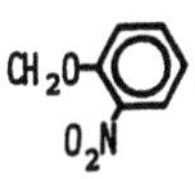, $R_2 = NO_2$	198	56	72
13	20	$R_1 = R_3 = CH_2SC(CH_3)_3$, $R_2 = N(Ac)_2$	93–94	78	80
14	21	$R_1 = R_3 = CH_2SPh$, $R_2 = \overset{O}{\overset{\uparrow}{S}}CH_3$	70–72	8	71
		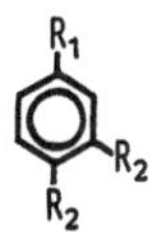			
15	21	$R_1 = CH_3$, $R_2 = (OCH_2CH_2)_3OCH_3$	(227–229/1)	70	64
16	21	$R_1 = CHO$, $R_2 = (OCH_2CH_2)_3OCH_3$	(234–235/1)	60	64
17	22	$R_1 = CH{=}CH_2$, $R_2 = (OCH_2CH_2)_3OCH_3$	oil	60	64
18	22	$R_1 = CHOHCH_3$, $R_2 = (OCH_2CH_2)_3OCH_3$	oil	70	64
19	26	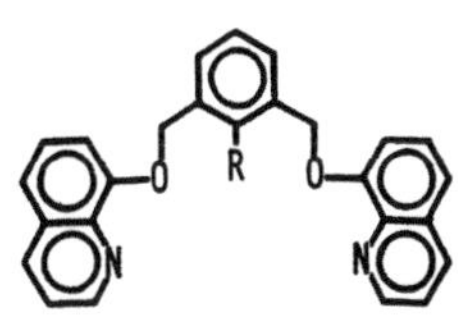 $R = NO_2$	208	48	72
20	27	$R = OCH_3$	226–227 · picrate	33	78

Table 7.7 (continued)

Cpd. No.	No. C's	Structure	Mp or (bp) in °C	Yield (%)	Refs.
21	33	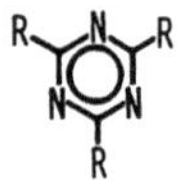 $R = (OCH_2CH_2)_3O(CH_2)_3CH_3$	oil	nr	17
22	36	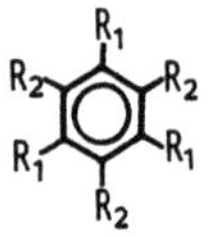 $R_1 = CH_3$, $R_2 =$ 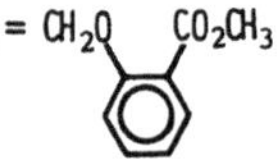	160–164	51	49
23	37	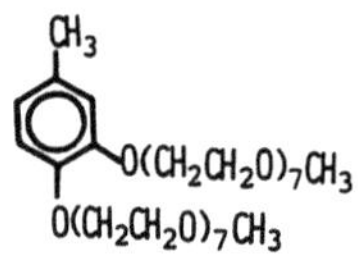	nr	40	64
		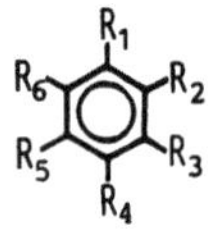			
24	39	$R_1 = R_3 = R_5 = CH_2S(CH_2CH_2O)_3(CH_2)_3CH_3$ $R_2 = R_4 = R_6 = H$	nr	nr	74
25	39	$R_1 = R_2 = R_4 = CH_2S(CH_2CH_2O)_3(CH_2)_3CH_3$ $R_3 = R_5 = R_6 = H$	nr	nr	74
26	39	$R_1 = R_2 = R_3 = CH_2S(CH_2CH_2O)_3(CH_2)_3CH_3$ $R_4 = R_5 = R_6 = H$	nr	nr	74
27	42	$R_1 = R_2 = R_3 = R_4 = R_5 = R_6 =$ $CH_2S(CH_2CH_2O)_2CH_3$	nr	nr	74
28	45	 $R = O(CH_2CH_2O)_3(CH_2)_7CH_3$	oil	nr	17
29	48	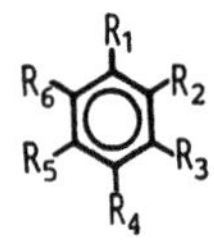 $R = CH_2OPh$	nr	nr	34
30	48	$R = CH_2SPh$	nr	nr	34
31	48	$R = CH_2SePh$	nr	nr	34
32	50	$R_1 = R_2 = R_3 = R_4 =$ $CH_2S(CH_2CH_2O)_3(CH_2)_3CH_3$ $R_5 = R_6 = H$	nr	nr	74
33	50	$R_1 = R_2 = R_3 = R_5 =$ $CH_2S(CH_2CH_2O)_3(CH_2)_3CH_3$ $R_4 = R_6 = H$	nr	nr	74

Table 7.7 (continued)

Cpd. No.	No. C's	Structure	Mp or (bp) in °C	Yield (%)	Refs.
34	50	$R_1 = R_2 = R_4 = R_5 = CH_2S(CH_2CH_2O)_3(CH_2)_3CH_3$; $R_3 = R_6 = H$	nr	nr	74
35	54	$R = CH_2SCH_2Ph$	nr	nr	34
36	61	$R_1 = R_2 = R_3 = R_4 = R_5 = CH_2S(CH_2CH_2O)_3(CH_2)_3CH_3$; $R_6 = H$	nr	nr	74
37	72	$R = CH_2S(CH_2CH_2O)_3(CH_2)_3CH_3$	nr	nr	74
38	72	R = 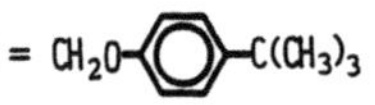	nr	nr	34
39	72	R = 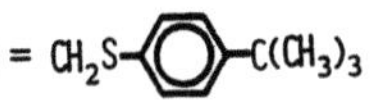	nr	nr	34
40	72	$R_1 = R_3 = R_5 =$; $R_2 = R_4 = R_6 = H$ 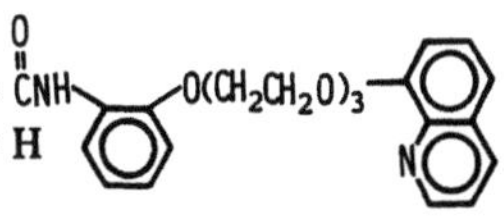	oil	65	63
41	75	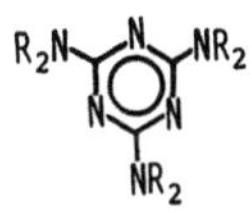 $R = (CH_2CH_2O)_4(CH_2)_3CH_3$	oil	nr	17
42	99	$R = (CH_2CH_2O)_4(CH_2)_7CH_3$	oil	nr	17

Table 7.8. Miscellaneous nitrogen-containing podands

Cpd. No.	No. C's	Structure	Mp or (bp) in °C	Yield (%)	Refs.
1	12	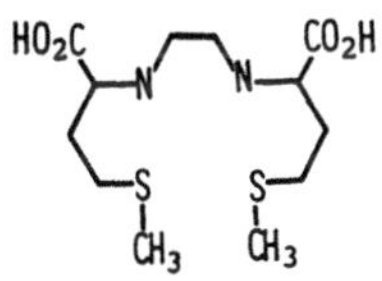	210–265 dec.	22	35
2	19	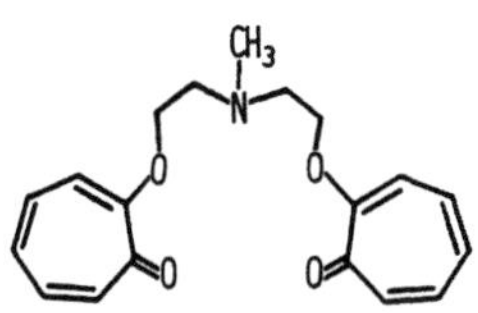	oil	48	49

Table 7.8 (continued)

Cpd. No.	No. C's	Structure	Mp or (bp) in °C	Yield (%)	Refs.
3	21	$Ph_2CHNH(CH_2CH_2O)_3CH_2CH_2NH_2$	oil	42–53	73
4	24	$(HOCH_2CH_2NCH_2CH_2OCH_2)_2$ Ts	86–88	30	53
5	24	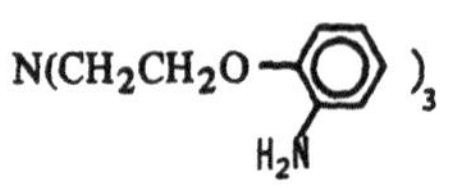	oil	84	63
6	26	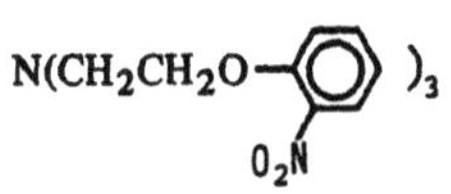	oil	63	63
7	30	$Ph_2CHNH(CH_2CH_2O)_3CH_2CH_2NHCH_2CH=$ CHPh (E)	oil	25	73
8	30	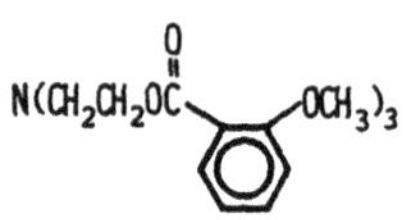	oil	71	49
9	33	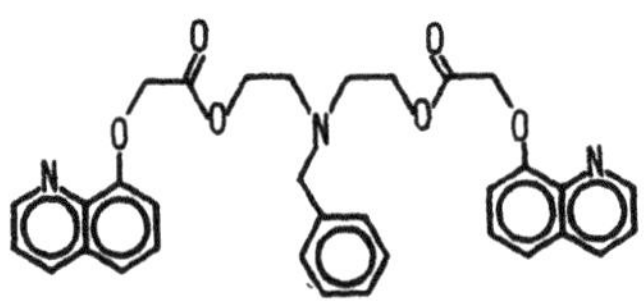	oil	57	49
10	81	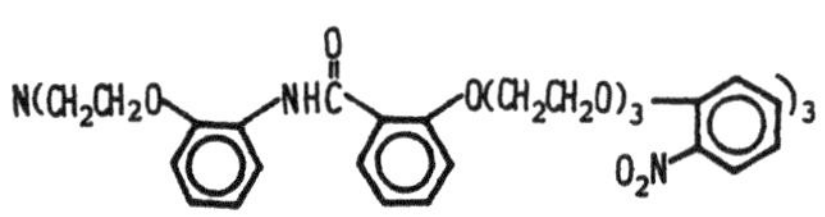	oil	63	63
11	–	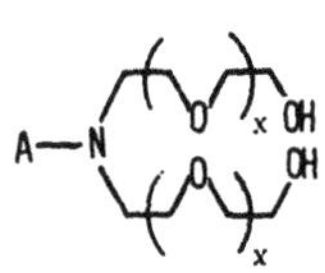 A = protecting groups x varies	nr	nr	43

Table 7.9. Open-chained imines

Cpd. No.	No. C's	Structure		Mp or (bp) in °C	Yield (%)	Refs.
1	12	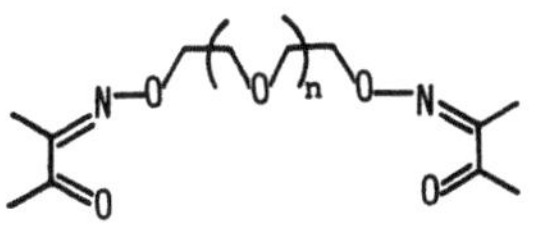	n = 1	oil	55	48
2	14		n = 2	oil	63	48
3	16		n = 3	oil	58	48
4	16	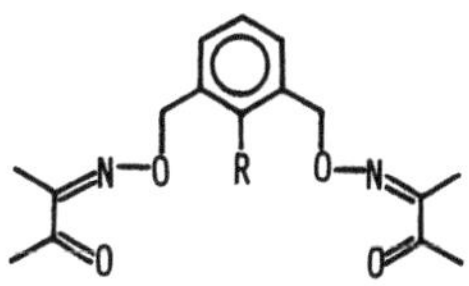	R = H	58–59	89	48
5	16		R = I	97	65	48
6	16		R = NO_2	101–103	73	48
7	16	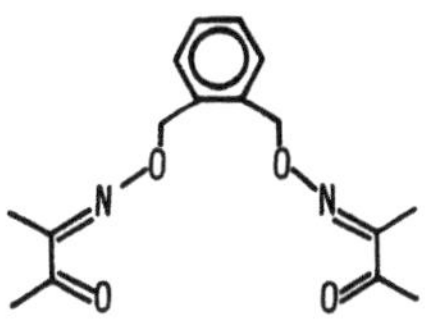		45–46	86	48
8	16	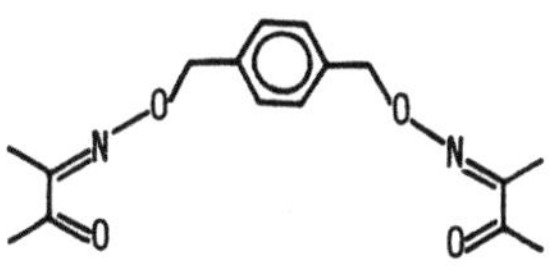		86–87	86	48
9	17	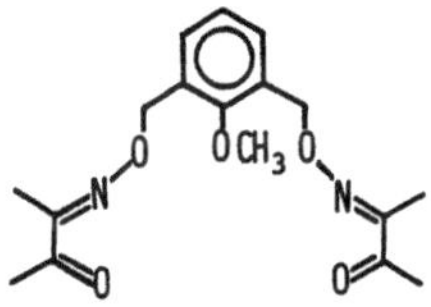		76–77	87	48
		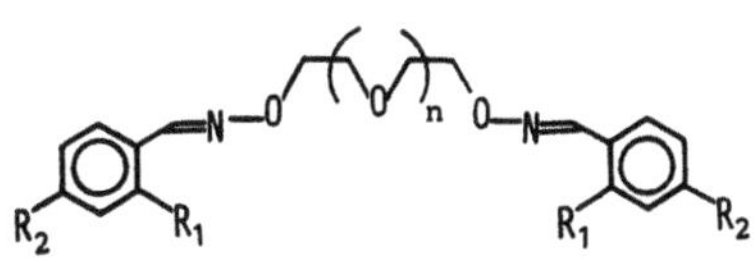				
10	18	n = 1, R_1 = OH, R_2 = H		oil	23	48
11	20	n = 1, R_1 = OCH_3, R_2 = H		oil	48	48
12	20	n = 2, R_1 = OH, R_2 = H		oil	28	48
13	22	n = 2, R_1 = H, R_2 = OCH_3		oil	70	48
14	22	n = 3, R_1 = OH, R_2 = H		oil	84	48
15	22	n = 2, R_1 = OCH_3, R_2 = H		156–162	51	48

Table 7.9 (continued)

Cpd. No.	No. C's	Structure	Mp or (bp) in °C	Yield (%)	Refs.
16	22	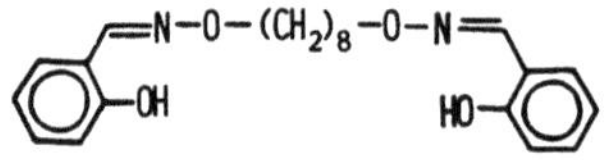	oil	89	48
17	24	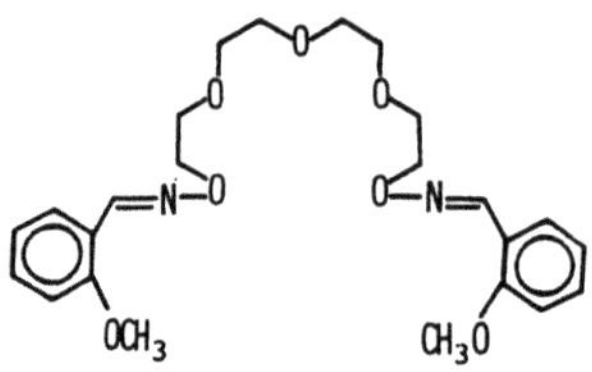	oil	54	48

Table 7.10. Sulfur-containing podands

Cpd. No.	No. C's	Structure	Mp or (bp) in °C	Yield (%)	Refs.
1	8	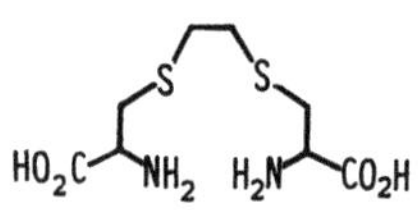	250–270 dec.	70	35
2	9	$CH_3S(CH_2)_2S(CH_2)_3S(CH_2)_2SCH_3$	(138–140/0.5)	22	55
3	10	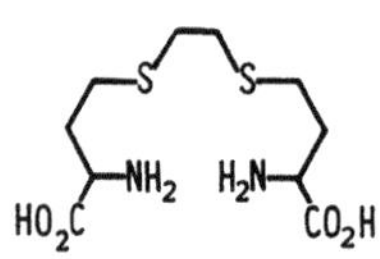	250–270 dec.	81	35
4	37	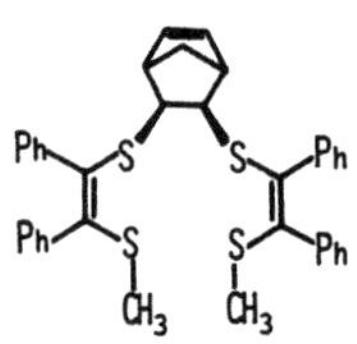	139	nr	59

Table 7.11. Open-chained ligands containing the 2,6-pyridyl subunit

Cpd. No.	No. C's	Structure	Mp or (bp) in °C	Yield (%)	Refs.
1	11	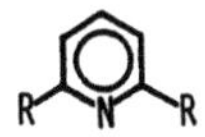 $R = CH_2OCCH_3$ (C=O)	oil	71	49
2	15	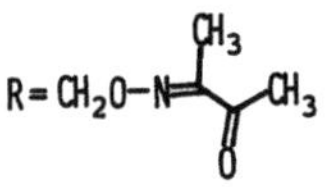	86–88	82	48
3	15	$R = CH_2N(CCH_3)_2$ (C=O)	158	85	10
4	17	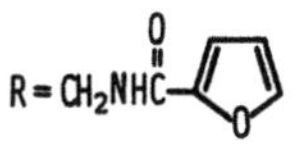	198–200	85	10
5	17	$R = CH_2S(CH_2CH_2O)_2CH_3$	nr	nr	74
6	21	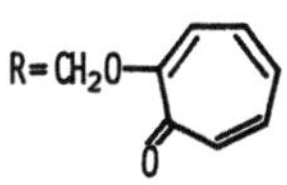	192–194	61	49
7	21	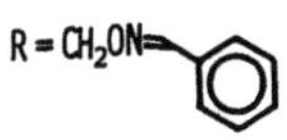	nr	nr	48
8	21	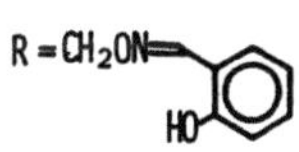	150–153	78	48
9	21	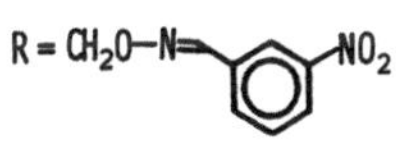	110	43	48
10	21	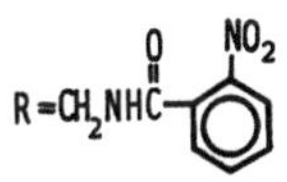	216–219	62	10
11	21	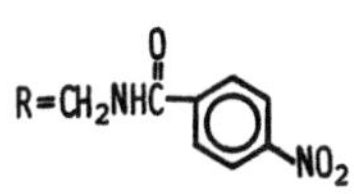	218	75	10
12	21	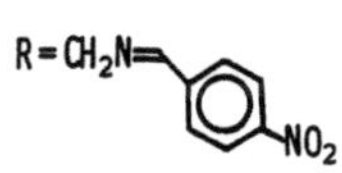	142	69	10
13	21	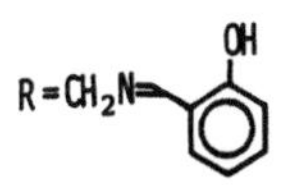	93–95	83	10
14	23	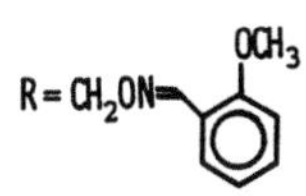	oil	85	48

Table 7.11 (continued)

Cpd. No.	No. C's	Structure	Mp or (bp) in °C	Yield (%)	Refs.
15	23	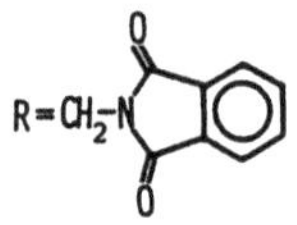	298–299	57	10
16	23	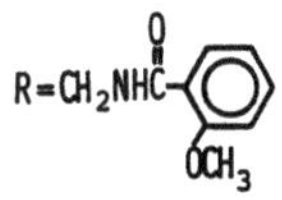	112	82	10
17	25	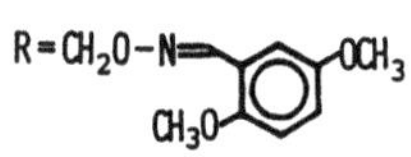	83–84	79	48
18	25	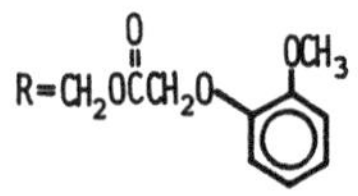	oil	59	49
19	25	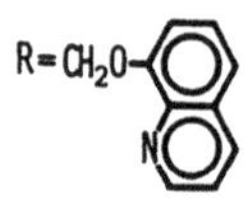	141–143 141–143	68 41	69 78
20	29	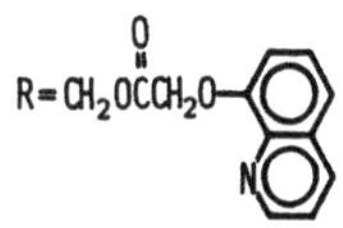	74–76	76	49
21	31	$R = CH_2S(CH_2CH_2O)_3CH_2CH_2CH_2CH_3$	nr	nr	74
22	37	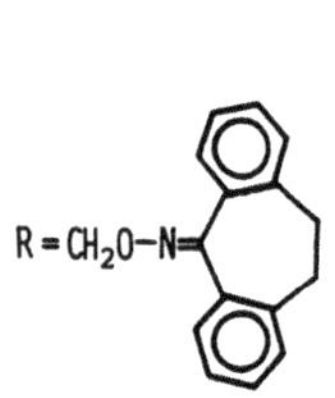	87–89	18	73
23	41	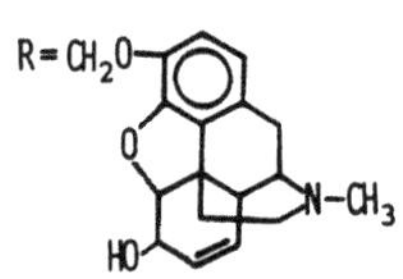	nr	nr	18

Table 7.12. Esters of pyridine-2,6-dicarboxylic acid

Cpd. No.	No. C's	Structure	Mp or (bp) in °C	Yield (%)	Refs.
1	9	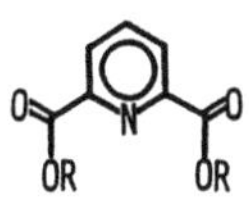 R = CH_3	121	80	49
2	11	R = CH_2CH_3	42–45	60	49
3	11	R = CH_2CH_2Cl	96	61	49
4	11	R = CH_2CF_3	110–111	65	49
5	11	R = CH_2CCl_3	222–223	67	49
6	11	R = CH_2CBr_3	245	71	49
7	13	R = $CH_2CH_2OCH_3$	oil	48	49
8	15	R = $CH_2CH_2OCH_2CH_2Cl$	oil	48	49
9	15	R = $CH_2CO_2CH_2CH_3$	62–63	44	49
10	17	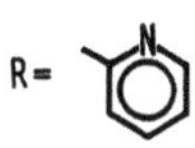	144	54	49
11	17	R = $(CH_2CH_2O)_2CH_3$	oil	51	49
12	19	R = Ph	175	79	49
13	19	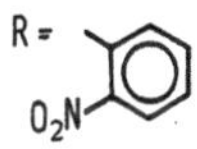	197–199	71	49
14	21	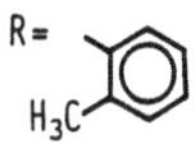	154–156	68	49
15	21	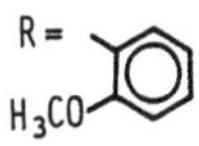	175	69	49
16	23	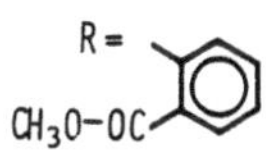	132–134	68	49
17	23	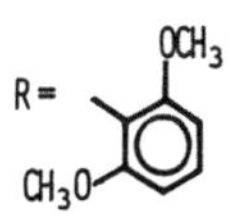	184	64	49
18	25	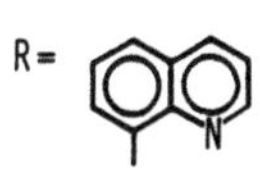	218–221	51	49

Table 7.12 (continued)

Cpd. No.	No. C's	Structure	Mp or (bp) in °C	Yield (%)	Refs.
19	33	R = 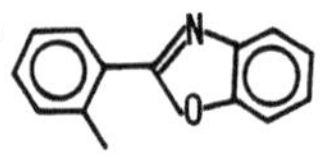	230–232	72	49
20	32	R = 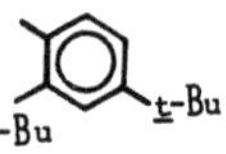	283–285	75	49
21	35	R = 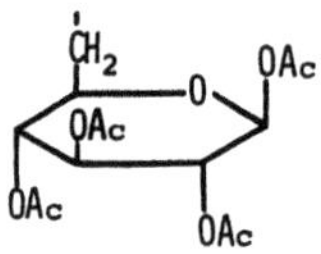	110–111	69	49

Table 7.13. Polymeric podands

Cpd. No.	Structure	Mp or (bp) in °C	Yield (%)	Refs.
	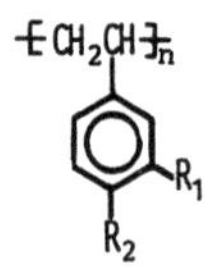			
1	$R_1 = R_2 = (OCH_2CH_2)_2OCH_3$	nr	nr	64
2	$R_1 = R_2 = (OCH_2CH_2)_3OCH_3$	nr	nr	64
3	$R_1 = H, R_2 = (OCH_2CH_2)_7OCH_3$	nr	nr	64
	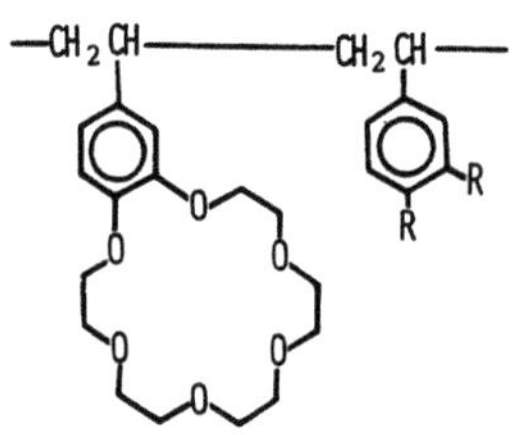			
4	$R = (OCH_2CH_2)_2OCH_3$	nr	nr	64
5	$R = (OCH_2CH_2)_3OCH_3$	nr	nr	64

7.8 References

1. Alberts, A. H., Cram, D. J.: J. Am. Chem. Soc. (1977) *99*, 3880
2. Ammann, D., Bissig, R., Guggi, M., Pretsch, E., Simon, W., Borowitz, I. J., Weiss, L.: Helv. Chim. Acta (1975) *58*, 1535
3. Ammann, D., Pretsch, E., Simon, W.. Tetrahedron Lett. (1972) 2473
4. Ammann, D., Pretsch, E., Simon, W.: Helv. Chim. Acta (1973) *56*, 1780
5. Balasubramanian, D., Sukumar, P., Chandani, R.: Tetrahedron Letters (1979) 3543
6. Bartsch, R. A., Juri, P. N.: Tetrahedron Lett. (1979) 407
7. Bartsch, R. A., Juri, P. N., Mills, M. A.: Tetrahedron Lett. (1979) 2944
8. Bartsch, R. A., Yang, I. W.: Tetrahedron Lett. (1979) 2503
9. Buhleier, E., Frensch, K., Luppertz, F., Vögtle, F.: Liebigs Ann. Chem. (1978) 1586
10. Buhleier, E., Wehner, W., Vögtle, F.: Liebigs Ann. Chem. (1978) 537
11. Chao, Y., Cram, D. J.: J. Am. Chem. Soc. (1976) *98*, 1015
12. Chaput, G., Jeminet, G., Juillard, J.: Can. J. Chem. (1975) *53*, 2240
13. Cornforth, J. W., Morgan, E. D., Potts, K. T. and Rees, R. J. W.: Tetrahedron (1973) *29*, 1659
14. Cram, D. J.: German Offen. 2,414,188, 21 Nov. 1974
15. Dix, J. P., Vögtle, F.: Chem. Ber. (1980) *113*, 457
16. Dorn, W. L., Knöchel, A., Oehler, J., Rudolph, G.: Z. Naturforsch. (1977) *326*, 776
17. Fornasier, R., Montanari, F., Podda, G. and Tundo, P.: Tetrahedron Letters (1976) 1381
18. Frensch, K., Vögtle, F.: Liebigs Ann. Chem. (1979) 2118
19. Gibson, T.: J. Org. Chem. (1980) *45*, 1095
20. Grandjean, J., Lazlo, P., Vögtle, F., Sieger, H.: Angew. Chem. Int. Ed. Engl. (1978) *17*, 856
21. Güggi, M., Oehme, M., Pretsch, E., Simon, W.: Helv. Chim. Acta (1976) *59*, 2417
22. Haines, A. H., Karntiang, P.: Carbohydr. Res. (1980) *78*, 205
23. Heimann, U. and Vögtle, F.: Angew. Chem. Int. Ed. Engl. (1978) *17*, 210
24. Hughes, O. L. and Wingfield, J. N.: J. C. S. Chem. Commun. (1978) 1001
25. Hünig, S., Öller, M., Wehner, G.: Liebigs Ann. Chem. (1979) 1925
26. Hyatt, J. A.: J. Org. Chem. (1978) *43*, 1808
27. Kellogg, R. M., Van Bergen, T. J., Van Doren, H., Hedstrand, D., Kool, J., Kruizinga, W. H., Troostwijk, C. B.: J. Org. Chem. (1980) *45*, 2854
28. Kime, D. E., Norymberski, J. K.: J. C. S. Perkin Trans. I. (1977) 1048
29. Kirsch, N. N. L., Funck, R. J. J., Pretsch, E., Simon, W.: Helv. Chim. Acta (1977) *60*, 2326
30. Knöchel, A., Oehler, J. and Rudolph, G.: Tetrahedron Letters (1975) 3167
31. Krespan, C. G.: J. Org. Chem. (1975) *40*, 1205
32. Lee, D. G., Chang, V. S.. J. Org. Chem. (1978) *43*, 1532
33. Lehmkuhl, H., Rabet, F., Hauschild, K.: Synthesis (1977) 184
34. MacNicol, D. D., Wilson, D. R.: J. C. S. Chem. Commun. (1976) 494
35. Magee, R. J., Mazurek, W., O'Connor, M. J., Phillip, A. T.: Aust. J. Chem. (1974) *27*, 1629
36. Montanari, F., Tundo, P.: Tetrahedron Letters (1979) 5055
37. Moore, C., Pressman, B. C.: Biochem. Biophys. Res. Commun. (1964) *15*, 562
38. Morf, W. E., Ammann, D., Bissig, R., Pretsch, E., Simon, W.: Progr. Macrocyclic Chem. (1978) *1*, 1
39. Movsumzade, M. M., Shabanov, A. L., Agaev, F. Kh., Abdullabekov, I. M.: Zh. Org. Chem. (1976) *12*, 2477
40. Newkome, G. R., Kawato, T., Benton, W. H.: J. Org. Chem. (1980) *45*, 626
41. Oepen, G., Dix, J. P., Vögtle, F.: Liebigs Ann. Chem. (1978) 1592
42. Ohmoto, H., Kai, Y., Yasuoko, N., Kasai, N., Yanagida, S., Okahara, M.: Bull. Chem. Soc. Jpn. (1979) *52*, 1209
43. Okahara, M.: Japan Kokai 78:33,319, 22 Mar. 1978
44. Patton, Jr., J. T.: Chem. Tech. (1976) 780
45. Pedersen, C. J.: J. Am. Chem. Soc. (1967) *89*, 7017
46. Pedersen, C. J.: U.S. Patent 3,361,778, 2 Jan. 1968
47. Piepers, O., Kellogg, R. M.: J. Chem. Soc. Chem. Commun. (1978) 383

48. Rasshofer, W., Müller, W. M., Oepen, G., Vögtle, F.: J. Chem. Res. (S) (1978) 72
49. Rasshofer, W., Müller, W. M. and Vögtle, F.: Chem. Ber. (1979) *112,* 2095
50. Rasshofer, W., Oepen, G., Müller, W. M., Vögtle, F.: Chem. Ber. (1978) *111,* 1108
51. Rasshofer, W., Oepen, G., Vögtle, F.: Chem. Ber. (1978) *111,* 419
52. Rasshofer, W., Vögtle, F.: Tetrahedron Letters (1978) 309
53. Rasshofer, W., Vögtle, F.: Liebigs Ann. Chem. (1978) 552
54. Regen, S. L., Heh, J. C. K., McLick, J.: J. Org. Chem. (1979) *44,* 1961
55. Rosen, W., Busch, D. H.: J. Am. Chem. Soc. (1969) *91,* 4694
56. Saenger, W., Brand, H., Vögtle, F., Weber, E., in Pullman, B. and Goldblum, N. (eds.): Metal-Ligand Interactions in Organic Chemistry and Biochemistry, Part 1, 1977. D. Reidel Publ. Co., p. 363
57. Sakamoto, K., Oki, M.: Bull. Chem. Soc. Jpn. (1976) *49,* 3159
58. Santiello, E., Manzocchi, A., Sozzani, P.: Tetrahedron Letters (1979) 4581
59. Schrauzer, G. N., Ho, R. K. Y., Murillo, R. P.: J. Am. Chem. Soc. (1970) *92,* 3508
60. Shinkai, S., Ogawa, T., Manabe, O.: J. C. S. Chem. Commun. (1980) 375
61. Sieger, H., Vögtle, F.: Angew. Chem. Int. Ed. Engl. (1978) *17,* 198
62. Sieger, H., Vögtle, F.: Tetrahedron Letters (1978) 2709
63. Sieger, H., Vögtle, F.: Liebigs Ann. Chem. (1980) 425
64. Sinta, R. and Smid, J.: Macromolecules (1980) *13,* 339
65. Starks, C. M.: J. Am. Chem. Soc. (1971) *93,* 195
66. Takaki, U., Hogen Esch, T. E. and Smid, J.: J. Phys. Chem. (1972) *76,* 2152
67. Timko, J. M., Helgeson, R. C., Newcomb, M., Gokel, G. W., Cram, D. J.: J. Am. Chem. Soc. (1974) *96,* 7097
68. Tümmler, B., Maass, G., Vögtle, F., Sieger, H., Heimann, U., Weber, E.: J. Am. Chem. Soc. (1979) *101,* 2588
69. Tümmler, B., Maass, G., Weber, E., Wehner, W., Vögtle, F.: J. Am. Chem. Soc. (1977) *99,* 4683
70. Vögtle, F., Müller, W. M., Wehner, W., Buhleier, E.: Angew. Chem. Int. Ed. Engl. (1977) *16,* 548
71. Vögtle, F., Nätscher, R.: Chem. Ber. (1976) *109,* 994
72. Vögtle, F. and Sieger, H.: Angew. Chem. Int. Ed. Engl. (1977) *16,* 396
73. Vögtle, F., Ulrich, E.: Chem. Ber. (1978) *111,* 1434
74. Vögtle, F. and Weber, E.: Angew. Chem. Int. Ed. Engl. (1974) *13,* 814
75. Vögtle, F. and Weber, E.: Angew. Chem. Int. Ed. Engl. (1979) *18,* 753
76. Weber, W. P. and Gokel, G. W.: Phase Transfer Catalysis in Organic Synthesis. Springer Verlag, Berlin, 1977
77. Weber, E., Müller, W. M. and Vögtle, F.: Tet. Lett. (1979) 2335
78. Weber, E. and Vögtle, F.: Tet. Lett. (1975) 2415
79. Wierenga, W., Evans, B. R., Woltersom, J. A.: J. Am. Chem. Soc. (1979) *101,* 1334
80. Wieder, W., Nätscher, R., Vögtle, F.: Liebigs Ann. Chem. (1976) 924
81. Yamazaki, N., Nakahama, S., Hirao, A., Negi, S.: Tet. Lett. (1978) 2429
82. Yanagida, S., Noji, G., Okahara, M.: Tetrahedron Letters (1977) 2893
83. Yanagida, S., Takahashi, K., Okahara, M.: Bull. Chem. Soc. Jpn. (1977) *50,* 1386
84. Yanagida, S., Takahashi, K., Okahara, M.: J. Org. Chem. (1979) *44,* 1099

Chapter 8

Cryptands and Related Polycyclic Systems

8.1 Inception and General Syntheses

In 1969, Dietrich, Lehn and Sauvage[22] published the first report of a type of molecule which was to engender tremendous interest in the chemical community. These compounds were macrobicyclic polyethers having nitrogen pivot atoms and exhibiting the ability to bind a variety of metal cations. The syntheses of the first examples of macrobicyclic amines were carried out as illustrated in Eq. (8.1). The appropriate diaminopolyether and polyether diacyl chloride were condensed under high dilution conditions to give diamide *1* in 80% yield. Reduction was then effected using $LiAlH_4$ to tetrahydrofuran to yield diamine *2* in 75% yield. A second equivalent of the diacyl chloride used in the first step was allowed to react with *2* (again, high dilution) affording the macrobicyclic diamide *3* in 45% yield. Reduction using diborane afforded the compound called [2.2.2]-cryptand (*4*) as its bis-borane adduct. Removal of the borane residues could be effected by treatment with 6 N HCl affording *4* in more than 90% yield from *3*. The overall yield for the process was about 25%.

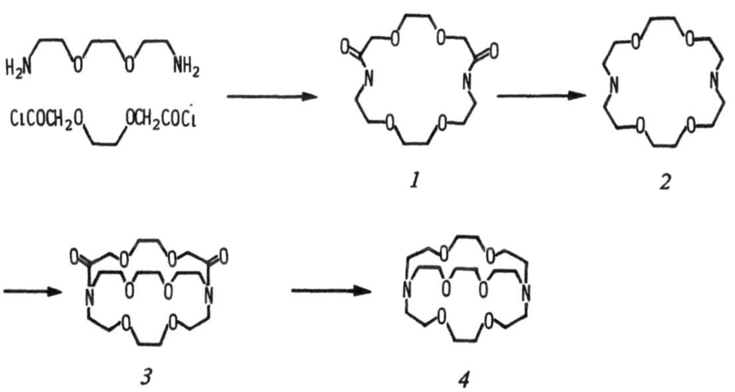

(8.1)

In his original paper, Lehn recognized the interesting possibility of "exo-endo" isomerism[22]. This kind of isomerism is shown in Eq. (8.2) and is similar to that referred to by Simmons and Park as "in-out" isomerism. Lehn and his coworkers recognized from the very beginning that this new class of molecules had the potential for binding metal cations of many types[23]. Since the molecules were designed to have molecular cavities, or holes, in them, Lehn suggested the term "cryptate" to describe such complexes. The origin of the term is the Greek *kruptos* meaning hidden. Indeed, a cation within the cryptand hole is certainly shielded or hidden from the outside environment. It has become general practice to refer to any complex involving a macrobi- or polycyclic system as a cryptate in accordance with the original suggestion[23] and to refer to the ligand itself as a cryptand (see Chap. 1).

(8.2)

"in-in" (endo-endo) "in-out" (exo-endo) "out-out" (exo-exo)

Dietrich, Lehn and Sauvage recognized not only the possibility of enclosing a cation completely in a lipophilic shell, but they also recognized the potential for using such systems for activating associated anions[23]. This is made particularly clear in a paper which appeared some years later: "One of the original motivations for our work on cryptates rested on their potential use for *salt solubilization, anion activation* and *phase transfer catalysis*"[16]. This particular application is discussed below in Sect. 8.3.

Although Lehn and his coworkers prepared a large number of cryptands and derived complexes over the years, the synthetic approach to these compounds remained essentially similar for most of them. Details are presented for a number of such compounds in ref. 26. The essential features of these syntheses were use of amide-forming reactions in the absence of templating ions with reliance on a high dilution step to form the second ring. An alternative approach for the synthesis of cryptands was developed by Dye and his coworkers[29]. Their approach involved the use of a flow synthesis to replace the high dilution step.

The procedure which had originally been used by Lehn et al.[22] involved slow addition (over a period of ca. 8 h) of ca. 0.1 M solutions of diamine and diacyl halide in benzene. Dye et al. found that the reactions could be conducted more rapidly as long as stirring was kept efficient[29]. This observation suggested the use of a mixing chamber of the type normally used for stopped-flow kinetic studies. Utilizing this type of set-up, the latter authors were able to obtain a 70% yield for *1*, slightly inferior to the yield reported by Lehn, but a "similar" yield of *3* which is better than that previously obtained[22]. Note that the chemical features of this synthetic method are essentially identical to the approach shown in Eq. (8.1) and differ primarily in the mechanics.

It should also be noted that 1,10-diaza-18-crown-6 is a key intermediate in the preparation of numerous cryptands. Considerable attention has been given to the synthesis of this molecule and is noted in Sect. 4.4 to which the reader is referred.

Alder and Sessions[2] have reported that reductive cleavage of hydrazinium dications is a useful approach to macrobicyclic amines of the type developed by Park and Simmons[57,62], but the method does not appear to have been successfully applied to any cryptand syntheses.

It should be noted here that the brevity of this section is due in part to the fact that essentially this single method is utilized in the synthesis of most cryptands which have thus far been prepared. For a detailed and fascinating discussion of design strategies, topological and synthetic considerations, the reader is referred to Lehn's article in *Structure and Bonding*, Vol. 16, p. 1.

8.2 Cryptands Containing Sulfur and Nitrogen in the Bridges

Shortly after their first report of all-oxygen bridged cryptands[22], Dietrich, Lehn and Sauvage reported incorporation of sulfur in the strands[24]. The experimental methods used were essentially similar to those applied in the syntheses of the parent cryptands. As in previous cases, a diacyl chloride was condensed with a diamine under high dilution conditions. In this case, however, the diamine contained sulfur atoms rather than oxygen. The synthesis of compound *5* was accomplished in two stages as illustrated below in Eq. (8.3). The first cyclization step affords the macrocyclic amine in 55% yield. The macrobicyclic product (*5*) is formed in 25% yield from the monocyclic diamine and the acid chloride.

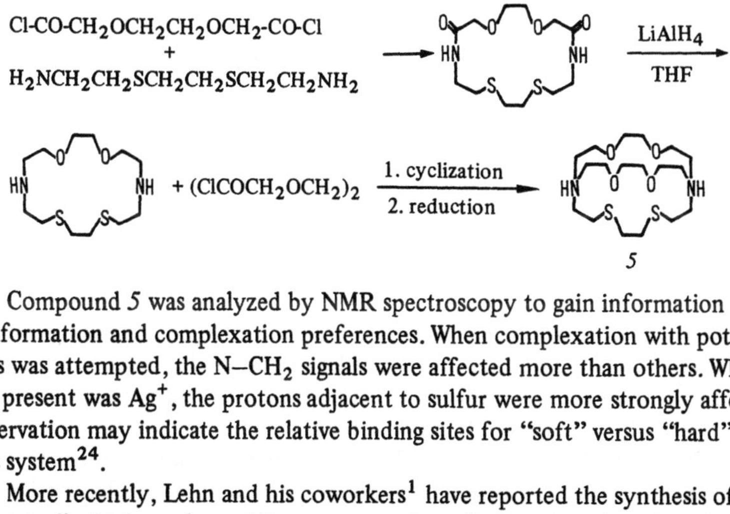

(8.3)

Compound *5* was analyzed by NMR spectroscopy to gain information relative to conformation and complexation preferences. When complexation with potassium cations was attempted, the N–CH_2 signals were affected more than others. When the cation present was Ag^+, the protons adjacent to sulfur were more strongly affected. This observation may indicate the relative binding sites for "soft" versus "hard" cations in this system[24].

More recently, Lehn and his coworkers[1] have reported the synthesis of "binuclear cryptates" which are formed from macropolycyclic ligands rather than the simple bicycles illustrated above. Formation of these compounds involves high dilution condensation of preformed sulfur-containing macrocycles using diacid chlorides as bridge precursors. In this sense, the strategy is similar to that which has been utilized in previous cryptand syntheses. After the appropriate bridges are formed, the amides are reduced to the corresponding amines as before. The process is illustrated in Eq. (8.4). Note that the presence of two macrocyclic systems in the same molecule allows for the possibility of binding two cations in the essentially cylindrical system. It is apparently envisioned that two cations may fit in the rings and leave enough room between them for the insertion of oxygen or nitrogen. Such complexes would be of obvious interest.

Synthesis of cryptands which contain nitrogen atoms in the bridges has also attracted Lehn's attention[43]. In general, a similar synthetic strategy was utilized, but ni-

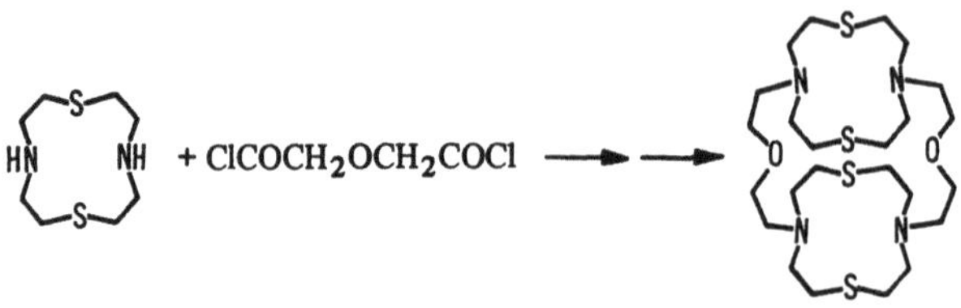

(8.4)

trogen, being tri-, rather than divalent, poses an additional problem. As noted in Sect. 4.2, when nitrogen is present, there is the possibility of forming more than one bond in certain situations. Lehn and Montavon[43] have circumvented this problem by protecting the nitrogen atom as the tosylate. Removal of the sulfonamide residues was then accomplished reductively using alkali metal in ammonia. The alkali metal was chosen so that the ligand cavity would not match it. That is, lithium in ammonia would be used for a larger cryptand and potassium in ammonia would be applied to the smaller cavity system. This choice was intended to preclude complications associated with complexation. After the tosyl residues were removed, the nitrogen atoms were methylated by either the Eschweiler-Clarke method (formic acid and formaldehyde) or using methyl chloroformate followed by reduction of the amide. The latter sequence is illustrated in Eq. (8.5).

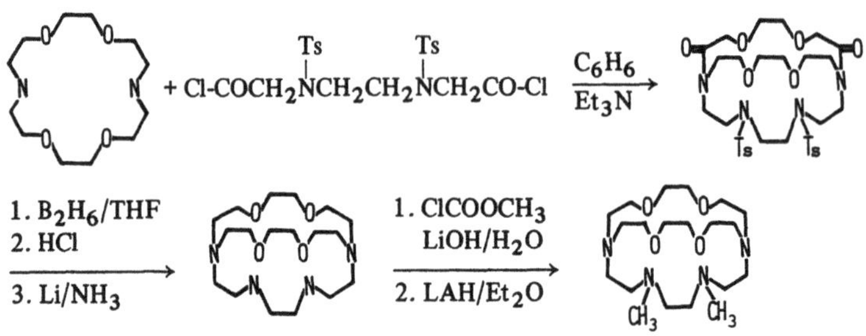

(8.5)

Lehn and his coworkers have further extended both the nitrogen containing cryptand and the binuclear cryptand notions into the realm of "nonbiological molecular receptor" molecules[42,44]. Again, the synthetic techniques are variations on the themes previously described. The important difference in these structures arises not merely from substitution of nitrogen in the bridges, but in locating them strategically so that they may be protonated and serve as hydrogen bond donors to encased molecular species. An example of the concept is illustrated below as structure *6*, the cryptate complex of azide anion. In this case, there appears to be almost perfect complimentarity between the ellipsoidal cavity of the hexaprotonated ligand and the linear or elliptical anion, azide[42]. It should be obvious that cation inclusion is also possible with such structures in the non-protonated form[44].

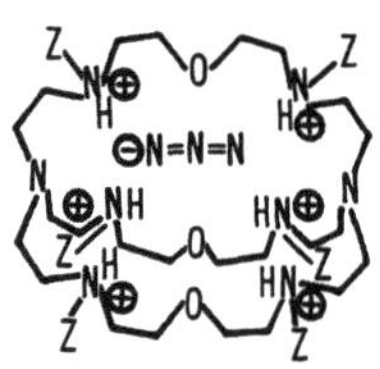

6

An interesting alternative approach to the synthesis of a cryptand having nitrogen atoms in the bridges was presented by Newkome and coworkers[56]. This group condensed triethanolamine with 2,6-dichloropyridine in a relatively straightforward but low yield (5%) nucleophilic aromatic substitution to form 7, illustrated below in Eq. (8.6). The identity of the compound was confirmed by X-ray structural analysis.

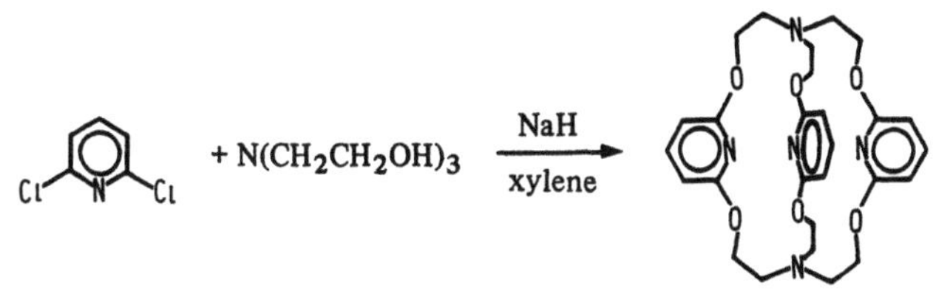

7

(8.6)

8.3 Lipophilic Cryptands

Although the cryptands are powerful cation complexing agents, there has been a need felt for increasing the lipophilicity of these materials. In particular, Montanari and his coworkers have utilized the lipophilic cryptands in phase transfer catalytic processes[15,40] as have Lehn and his group[16]. In all of this work, the principal structural variation has been in appending a hydrocarbon chain to the cryptand molecule. The approaches taken by Lehn and Montanari differ somewhat in the execution but are quite similar in principle. Both involve oxidation of an alkene to a diol followed by chain elongation to ultimately afford a bis-acyl halide which forms the third strand upon reaction with a macrocyclic diamine. The process is illustrated below as Eq. (8.7).

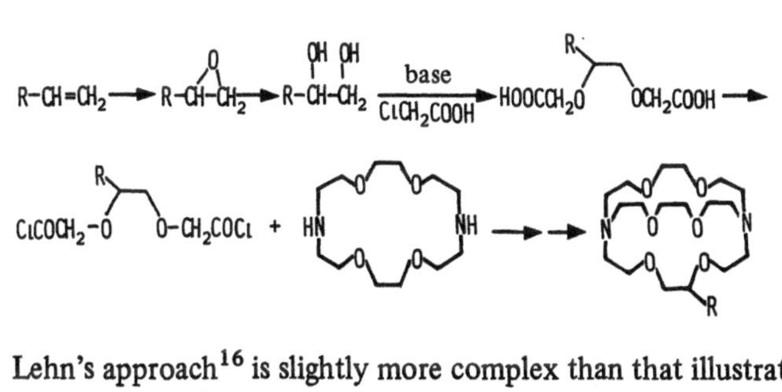

(8.7)

Lehn's approach[16] is slightly more complex than that illustrated above in that the diol is chloromethylated and then treated with cyanide. Hydrolysis then affords the diacid which may be carried through as shown. It should also be noted that once the bis-acyl halide is in hand, it may be treated directly with an open-chained amine to yield a lipophilic diazacrown, after reduction[15].

Landini, Montanari and Rolla[40] were able to incorporate cyclohexyl residues in the strands and thereby increase lipophilicity. As with the compounds referred to above, these were effective phase transfer catalysts.

Vögtle and his coworkers have reported the synthesis of cryptands containing lipophilic structural elements like *o*-, *m*- and *p*-phenylene, biphenyl and pyridine nuclei[7].

By incorporating trisubstituted aromatic rings, Vögtle has prepared what he calls "endo-lipophilic" cryptands[68]. Such a compound is illustrated below as *8*. The synthetic approach to these systems parallels those outlined in previous discussion[68].

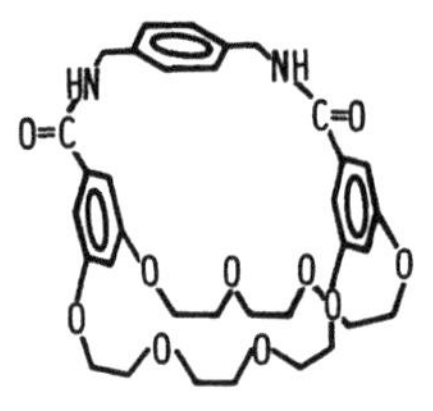

8

8.4 Cryptands Containing a Carbon Bridgehead

The cryptands are generally characterized by three heteroatom-containing strands which meet at a common point to which we have generally referred as "bridgehead". For most of the compounds which have been reported, this bridgehead is a nitrogen atom and conformational flexibility is attributed to inversion at this atom. In this section we consider other bridgehead units, specifically those in which carbon is the bridgehead atom. In fact in these cases, the flexibility associated with nitrogen inversion is absent and it is the flexibility of the entire group which permits whatever conformational mobility there is.

We have noted above (Sect. 8.3) that certain substituted benzoic acids have been used in the synthesis of "endolipophilic" cryptands[68] but it is the glycerol and pentaerythritol units which have been used most commonly in this application. These are discussed separately, below.

There are essentially three methods which have been used for incorporation of the glycerol unit in macrobicyclic species. In all of these the pervasive problem is the tendency of glycerol to form five-membered acetals involving the primary and secondary hydroxyls rather than the two primary hydroxyls and the tendency of the glycerol equivalent 1,3-dichloro-2-propanol to form epichlorohydrin.

Despite the latter problem, Parsons[58] treated glycerin dichloride with 2-benzyloxyphenol and formed the alcohol illustrated below as compound *9*. Hydrogenolysis of *9* to remove the benzyl protecting groups affords triol-*10*. This can be treated with another equivalent of glycerin dichloride and the final bridge can be added in the conventional way using tri- or tetraethylene glycol ditosylate. The overall approach is shown in Eq. (8.8)[58].

An alternative method for incorporating the glycerol unit is to form the acetal or ketal. Gregory, Haines and Karntiang[32] have formed the five-membered dimethyl ketal of glycerol and attached a strand to the remaining primary hydroxyl group of glycerol. After hydrolysis of the ketal, the primary hydroxyl could be selectively protected as its trityl ether. The second strand could then be added and after hydrolysis of the trityl ether, the macrocyclic diol could be converted into the macrobicycle in the usual fashion[32].

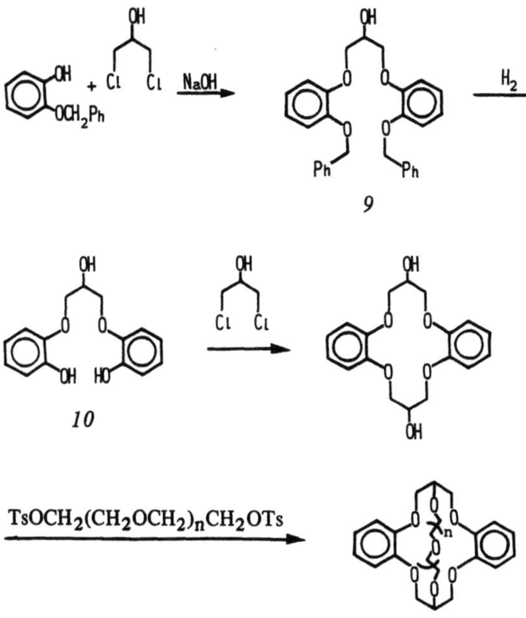

(8.8)

Haines and Karntiang later adopted a somewhat more symmetrical approach in the synthesis of glycerol-based cryptands when they utilized the six-membered benzylidene derivative of glycerol[33]. Details of the synthesis described in Eq. (8.8), above, are also presented in this paper. The "symmetrical" approach is quite similar to that described in Eq. (8.8), and is outlined below in Eq. (8.9).

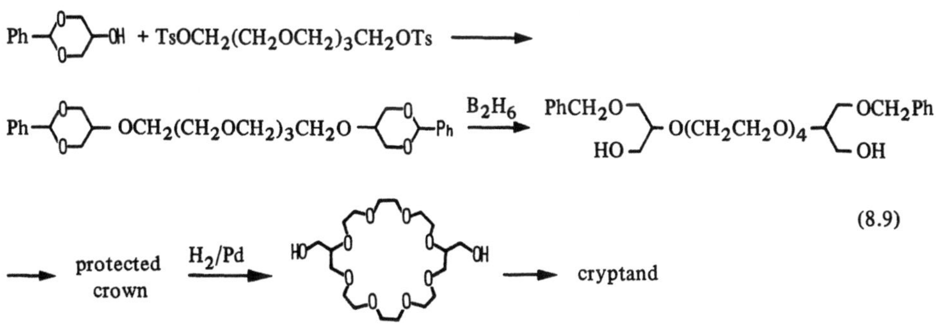

(8.9)

Coxon and Stoddart[17,18] have directed their attention to the formation of pentaerythritol-derived cryptands. With these molecules, the strategy was to block one pair of hydroxyl groups as an acetal and form a crown from the remaining diol. In the first of the two reports cited above, this was accomplished by treating the 0-benzylidine derivative of pentaerythritol with base and diethylene glycol ditosylate. The crown was then treated with a mixture of $LiAlH_4$ and BF_3 which gives partial reduction of the acetal as shown in (8.9), above. The monoprotected diol could now be treated in a fashion similar to that previously described and the benzyloxy cryptand (*11*) would result. The scheme is illustrated below as Eq. (8.10).

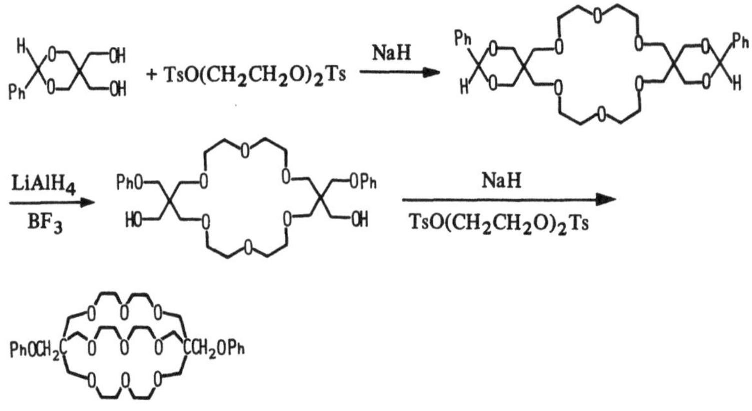

11

(8.10)

Similar pentaerythritol cryptands have been prepared by a slight modification of the above approach. Although the general strategy is as illustrated in Eq. (8.11), acetal formation is accomplished by reaction of pentaerythritol with paraformaldehyde. This reaction leads to diacetal *12* which is hydrolyzed in dilute H_2SO_4 to yield the monoacetal, *13*. The latter is then used in a fashion similar to that described in Eq. 8.10, above.

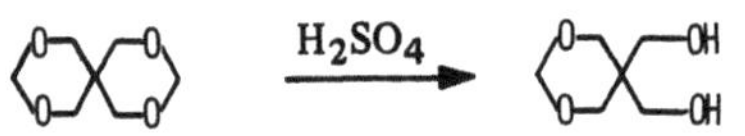

12 13

(8.11)

An electrophilic equivalent of *13*, above, is 3,3-bis(halomethyl)oxetane, *14*. Krespan has utilized these substances in the preparation of a variety of azacrowns and cryptands[37,38]. Note that this material is also mentioned by Coxon and Stoddart[18]. The strategy and techniques utilized with *14* are similar to those discussed above and are not noted here in detail. An overall transformation is illustrated in Eq. (8.12).

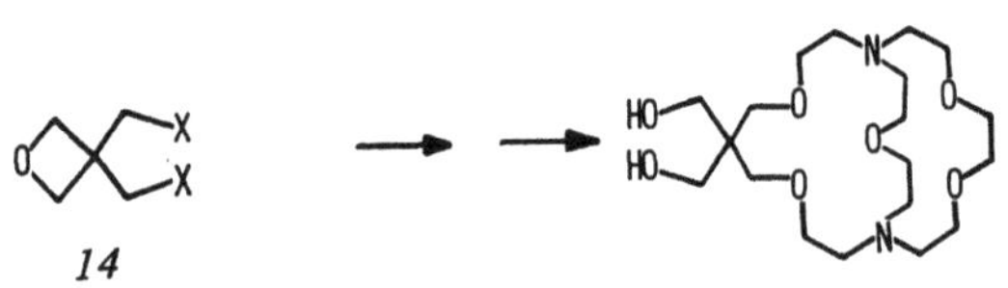

14

(8.12)

8.5 Chiral Cryptands

Lehn and his coworkers have prepared a number of chiral cryptands based upon the 2,2′-binaphthyl unit[27,28,46]. In a typical preparation, the binaphthyl units are treated with bromoacetic acid to form the phenoxyacetic acid derivatives which are then converted into the corresponding diacyl chlorides (*15*). Reaction of *15* with 1,10-diaza-18-

crown-6 yields a chiral cryptand diamide. Reduction in the usual fashion yields the cryptand, *16*. The approach is illustrated in Eq. (8.13).

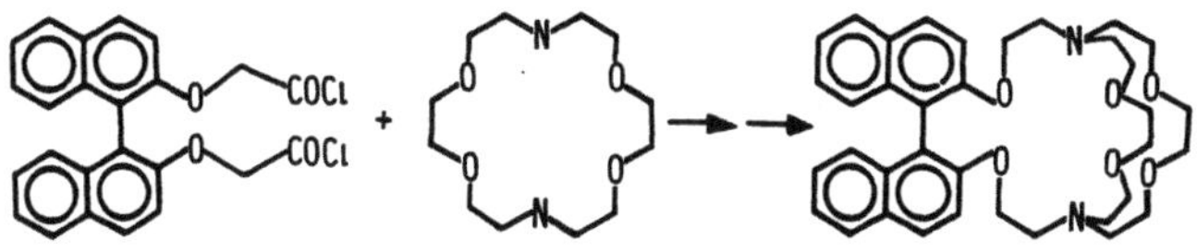

(8.13)

15 *16*

If a bridged bis-crown is used instead of diaza-18-crown-6, the cryptand contains two macrorings facing each other (see Table 8.5). Note also that the 2,2′-binaphthyl unit has been used extensively by Cram and his coworkers to provide chirality to monocyclic systems as well (see Sect. 3.13).

8.6 Miscellaneous Cryptands

A number of bi- or polycyclic structures have been prepared over the years which are somewhat difficult to classify within the narrow confines of crown and cryptand nomenclature. Nevertheless, these molecules deserve mention and are noted here.

The similarity between the cryptands and the first of these molecules is obvious[5]. Compound *17* is a urethane equivalent of [2.2.2]-cryptand. The synthesis of *17* was accomplished using a diacyl halide and 1,10-diaza-18-crown-6 (shown in Eq. 8.13). Since amidic nitrogen inverts less rapidly than a tertiary amine nitrogen, Vögtle and his coworkers who prepared *17*[5], analyzed the proton and carbon magnetic resonance spectra to discern differences in conformational preferences. Compound *17* was found to form a lithium perchlorate complex.

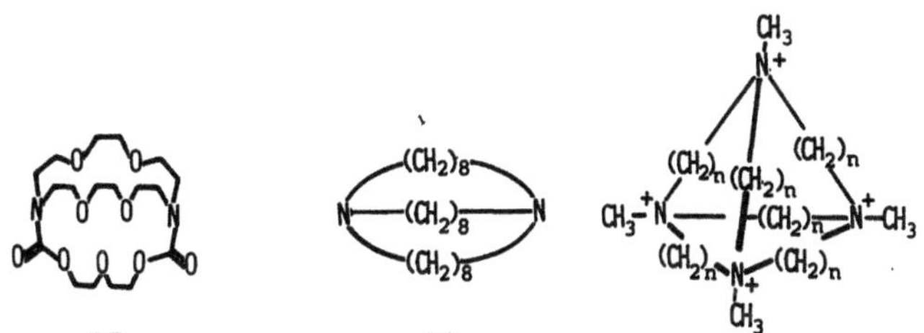

17 *18* *19*

The "in-out" bicyclic amines prepared by Simmons and Park[57,62] bear a remarkable semblance to the cryptands but lack the binding sites in the bridges. As a result, these molecules interact with electrophiles in a fashion similar to other tertiary amines and generally do not exhibit strong interactions with alkali or alkaline earth metal ions. The "in-out" bicyclic amines are prepared by reaction of the appropriate acid chlorides and amines in two stages to yield the macrobicyclic amine after reduction of the amidic linkages. A typical amine is shown above as compound *18*.

A relative of the latter class of compounds are the macrotricyclic quaternary ammonium salts which have been reported by Schmidtchen[61]. The bridges may contain either methylenes or ethyleneoxy units and the nitrogens are quaternarized. The underlying principle is to provide a cavity suitable for solvating or at least trapping anions. Schmidtchen presents evidence which suggests the formation of halide inclusion complexes. The synthesis of these molecules is accomplished along more or less traditional lines[61]. Such a species is illustrated above as compound *19*.

Finally, Vögtle and his coworkers[9] have prepared a number of "cascade" molecules which are structurally related to the aforementioned systems. These are repeating ring units of increasingly large cavity size and are prepared by repetitive synthetic procedures. Typically, an amine is cyanoethylated, the nitrile reduced to an amine which may then be further cyanoethylated and reduced or cyclized with a diacid halide. The rather elaborate scheme is illustrated in ref. 61 and examples of the structural type are shown in Table 8.4.

8.7 Macropolycyclic Cryptands

In other sections in this chapter, we have referred to a variety of macropolycyclic structures which are more elaborate than the simple three-stranded bicyclic cryptands. This includes bridged double-macrocycles[1,11,13,27,36,47,50,51], in-out bicyclic amines[57,62] and the macrotricyclic quaternary ammonium salts of Schmidtchen[61]. In addition to these, there are two other types of compounds which deserve special note. The first of these is a stacked twin-ring cryptand, but it is a hybrid molecule rather than a double-cryptand[10]. The species shown below as *20* is a "crowned" porphyrin, and was designed to provide a pair of metal cation binding sites similar to those which might be available in natural biological systems[10].

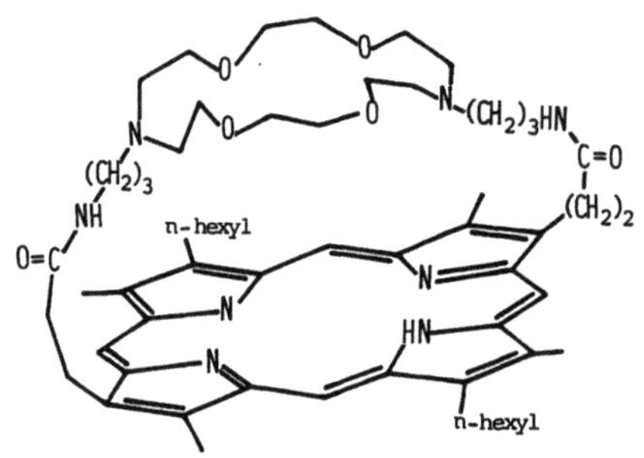

20

A somewhat different approach to providing tailored cavities for metal cations was taken by the groups of Cram[19] and Lehn[30]. Graf and Lehn prepared the spheroidal molecule *21* which has an interesting molecular architecture. The molecule has ten coordination sites within it, six which form an octahedral array and four which are in a tetrahedral arrangement. This remarkable compound "is soluble in all solvents from petroleum ether to water"[30] and forms complexes in solvents as diverse as water and chloroform with cations as varied at NH_4^+ and Cs^+.

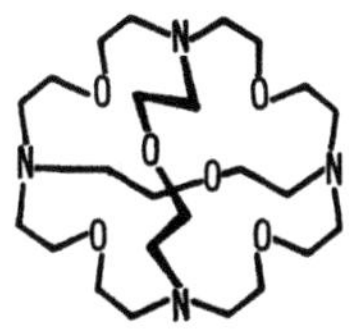

21

The "spherand" prepared by Cram and coworkers[19] was designed to have a relatively small molecular cavity and appeared to prefer complexation with Li^+ and Na^+ over larger cations like K^+, Rb^+, etc[19]. The spheroidal cryptand prepared by Lehn[30] involved strategy employed previously but the spherand *24* was prepared by quite a different approach.

Oxidation of *p*-cresol afforded a triphenol *22* which is approximately half the molecule. The central hydroxyl of the triphenol could be selectively methylated and then the compound was *ortho*-brominated and bridged using 1,3-dibromopropane to give *23*. Metallation with butyllithium followed by iron catalyzed coupling afforded the macrocycle as indicated.

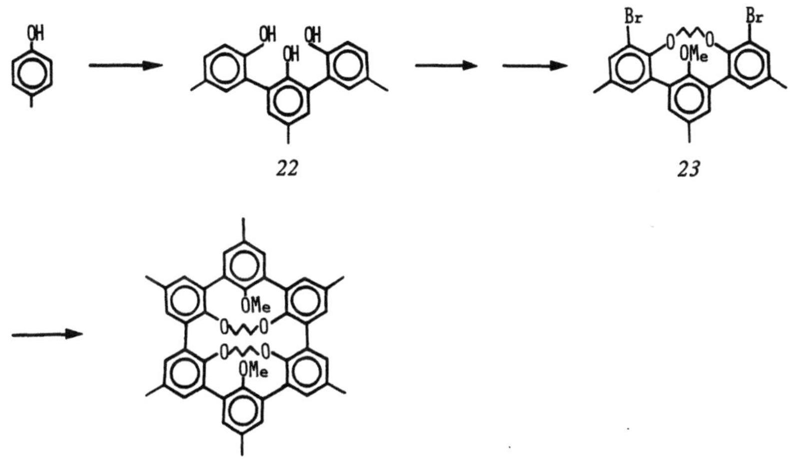

22 *23* *24*

(8.14)

8.8 Polymer-bound Cryptands

Two groups have bound cryptands to polymeric backbones. Although the syntheses of the ligands *per se* generally parallel the methods previously described, the approach to anchoring differs. Blasius and Maurer[4] have prepared cryptands with benzo groups in the bridges and then utilized phenol-formaldehyde type condensations to give a crosslinked resin as illustrated in Eq. (8.15)[4].

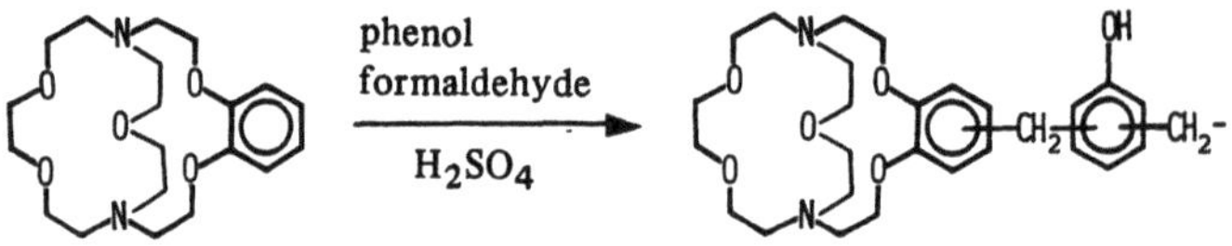

(8.15)

Tomoi, Kihara and Kakiuchi[64] have prepared a cryptand with a hydroxymethyl group in one of the strands and then utilized this as nucleophile in a reaction with chloromethylated polystyrene. The type of compound prepared by this method is illustrated in Eq. (8.16).

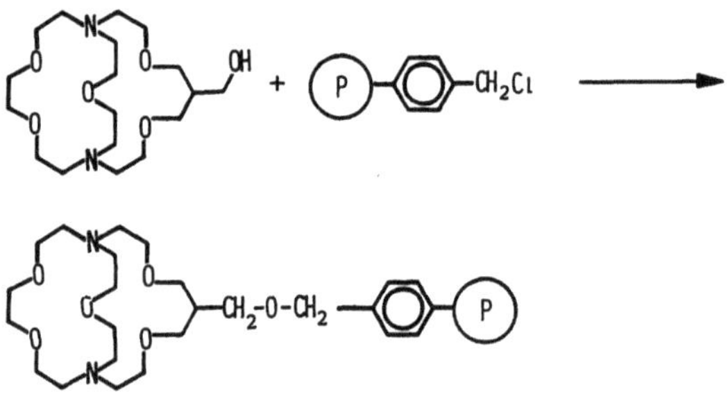

(8.16)

8.9 Listing of Tables

Note on Table Usage in Chapter 8

In an effort to maximize economy of production and still retain clarity in presentation we have resorted to the extensive use of abbreviations in these tables. For the most part, these abbreviations should be self-explanatory and are decoded at the bottom of each page. Two less obvious abbreviations deserve special note.

First: An asterisk (*) placed in the table indicates that the information can be gleaned only from an examination of the structure which is shown.

Second: Where one strand of a bicyclic contains no ligating elements or it is not clear to what extent these atoms may be involved in binding, it is impossible to present a meaningful "molecular class" designation. In such cases we have used "z" to indicate that a designation is unclear and the reader is directed to the structure shown. The first such case is compound 34 in Table 8.3. Two strands are bis-ethyleneoxy (EOEOE) chains and the third is an octyl chain. Since the former have two heteroatoms each and the latter none, the molecular class would be designated "2.2.z." The alternative designation "2.2.0" might lead one to conclude that a simple ethylene linkage is present.

8.10 Tables 8.1–8.5

Table 8.1. In-out Bicyclic amines

Cpd. No.	No. C's	Pivot Atom(s) or Unit	Strand A	Strand B	Strand C	MP or (BP) °C	Yld. (%)	Refs.
1	20	N, N	C_6	C_6	C_8	45	40	62
2	21	N, N	C_7	C_7	C_7	35	20	62
3	24	N, N	C_8	C_8	C_8	120	75	62
4	24	N, N	C_6	C_8	C_{10}	<0	60	62
5	26	N, N	C_8	C_8	C_8	55	70	62
6	27	N, N	C_9	C_9	C_9	32	50	62
7	30	N, N	C_{10}	C_{10}	C_{10}	114	75	62

Abbreviations:

M = CH_2; E = CH_2CH_2; P = $CH_2CH_2CH_2$; C_n = $(CH_2)_n$; CO = $\overset{O}{\overset{\|}{C}}$; Me = CH_3; Et = CH_2CH_3; Py = pyridyl;

2B = ; 3B = ; 4B = 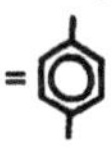; Ph = C_6H_5

Table 8.2. Monocyclic cryptand precursors

Cpd. No.	No. C's	No. Atoms Lgst. Ring	Molecular Class	Pivot Atom(s) or Unit	Strand A	Strand B	MP or (BP) °C	Yld. (%)	Refs.
1	8	12	1.1	N,N	EOE	EOE	83–84	46	41
2	8	13	2.1	N,N	COOEOCO	EOE	185	37	5
3	10	15	2.1	N,N	EOEOE	EOE	oil	67	41
4	10	16	2.2	N,N	COOEOCO	EOEOE	122	84	5
5	10	16	3.1	N,N	COOEOEOCO	EOE	126	89	5
6	12	18	2.2	N,N	EOEOE	EOEOE	115–116	60	22
7	12	18	2.2	N,N	EOEOE	ESESE	45	42	41
8	12	18	2.2	N,N	ESESE	ESESE	125	42	41
9	12	19	3.2	N,N	EOEOEOE	COOEOCO	88	84	5
10	12	19	3.2	N,N	COOEOEOCO	EOEOE	112	77	5
11	13	16	*	*	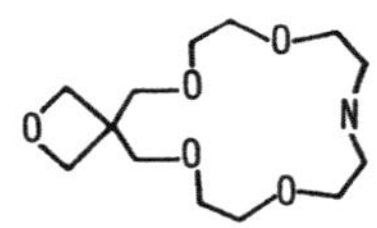		79–81	67	36
12	13	16	*	*	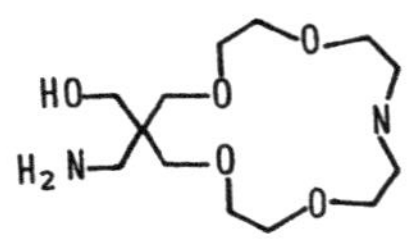		$(105/10^{-6})$	35	36

Table 8.2 (continued)

Cpd. No.	No. C's	No. Atoms Lgst. Ring	Molecular Class	Pivot Atom(s) or Unit	Strand A	Strand B	MP or (BP) °C	Yld. (%)	Refs.
13	13	16	*	*	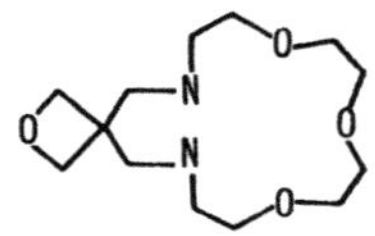		(108–110/10^{-7})	8	36
14	13	16	3.1	N,N	COOEOEOCO	M-2,6-Py-M	135	59	5
15	14	16	2.2	N,N	$EOCH_2C(=CH_2)CH_2OE$	EOEOE	nr	nr	53
16	14	16	3.z	N,N	COOEOEOCO	M3BM	145–147	66	5
17	14	21	3.2	N,N	EOEOEOE	EOEOE	15[a] 0[a]	nr 52	22 41
18	14	22	3.3	N,N	COOEOEOCO	EOEOEOE	62	90	5
19	15	16	2.2	N,N	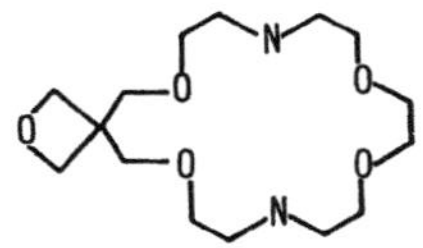		oil	71	36
20	15	16	*	*	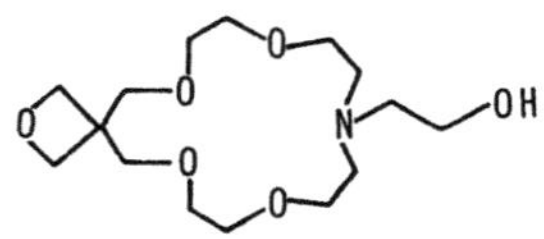		oil	58	36

Table 8.2 (continued)

Cpd. No.	No. C's	No. Atoms Lgst. Ring	Molecular Class	Pivot Atom(s) or Unit	Strand A	Strand B	MP or (BP) °C	Yld. (%)	Refs.
21	16	9	*	*	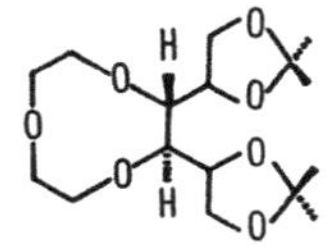	(DD)	oil	0.2	20
22	16	18	*	*	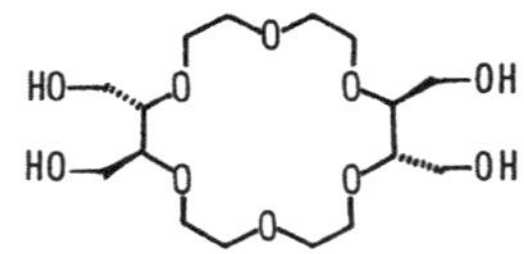	(LL)	nr	4	20
23	16	21	3.2	N,N	COOEOEOCO		89	59	5
24	16	24	3.3	N,N	EOEOEOE	EOEOEOE	15 liq	40 nr	41 22
25	18	20	2.2	N,N	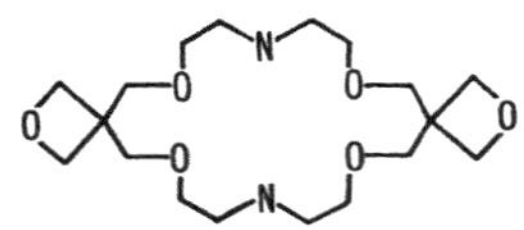		118.5–119	17	36
26	18	20	3.3	*	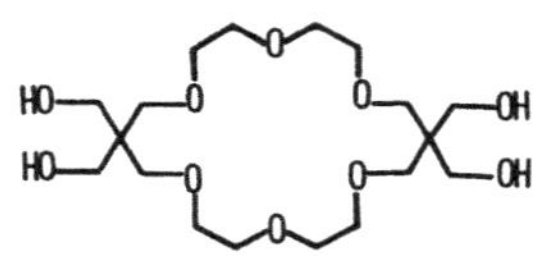		80–81	16	38

Table 8.2 (continued)

Cpd. No.	No. C's	No. Atoms Lgst. Ring	Molecular Class	Pivot Atom(s) or Unit	Strand A	Strand B	MP or (BP) °C	Yld. (%)	Refs.
27	18	20	3.3	*	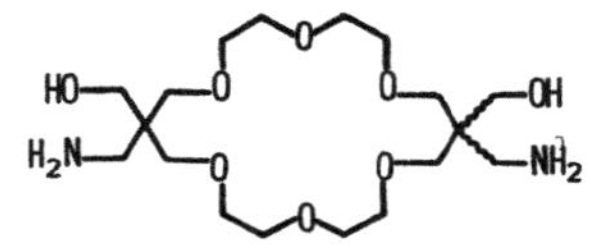		118–127	66	38
28	20	18	3.3	*	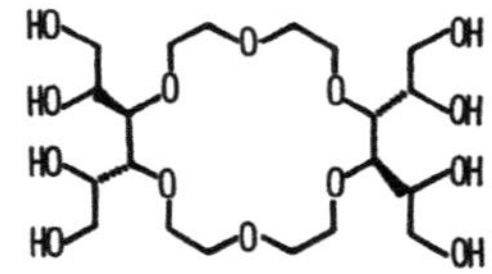		58–60	5	20
29	20	26	3.3	*	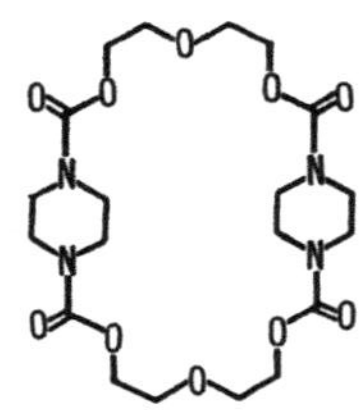		108–112	35	5
30	24	18	3.3	*	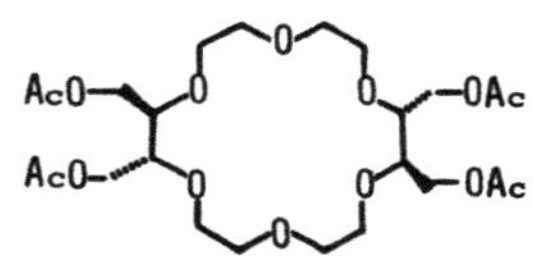		nr	nr	20

Table 8.2 (continued)

Cpd. No.	No. C's	No. Atoms Lgst. Ring	Molecular Class	Pivot Atom(s) or Unit	Strand A	Strand B	MP or (BP) °C	Yld. (%)	Refs.
31	24	18	3.3	*	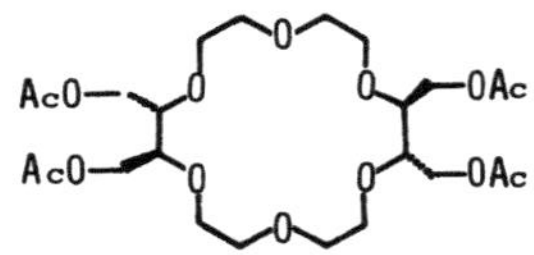	(LL)	69–74	nr	20
32	28	18	3.3	*	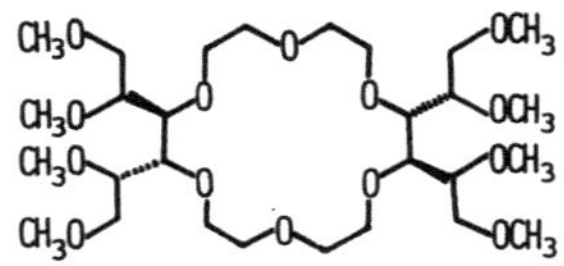	(DD)	nr	nr	20
33	32	18	3.3	*	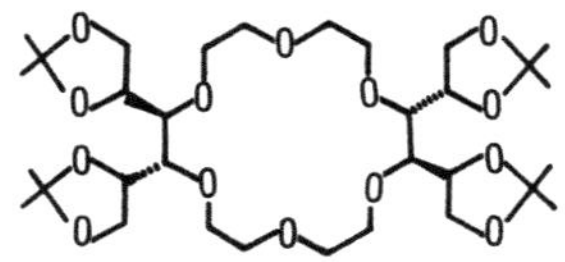	(DD)	oil	5	20
34	36	18	3.3	*	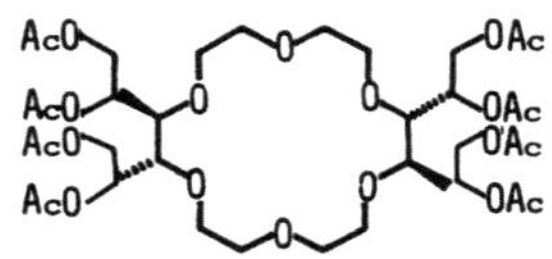	(DD)	nr	nr	20

Table 8.2 (continued)

Cpd. No.	No. C's	No. Atoms Lgst. Ring	Molecular Class	Pivot Atom(s) or Unit	Strand A	Strand B	MP or (BP) °C	Yld. (%)	Refs.
35	42	18	3.3	*	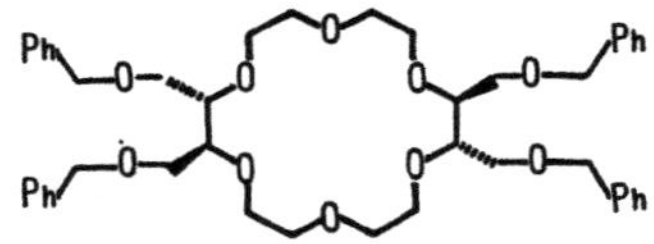	(LL)	oil	6	20
36	92	18	3.3	*	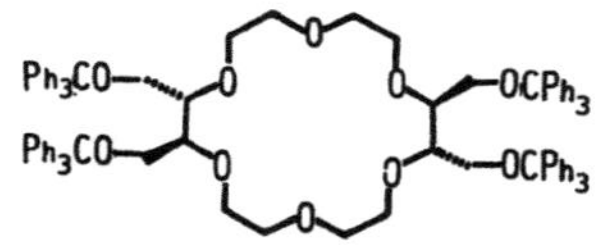	(LL)	nr	nr	20

Abbreviations:

M = CH_2; E = CH_2CH_2; P = $CH_2CH_2CH_2$; C_n = $(CH_2)_n$; CO = C(=O); Me = CH_3; Et = CH_2CH_3; Py = pyridyl; 2B = 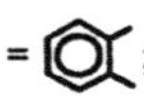; 3B = 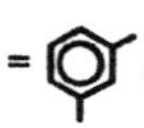; 4B = 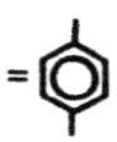; Ph = C_6H_5

[a] approximate

Table 8.3. Cryptands

Cpd. No.	No. C's	No. Atoms Lgst. Ring	Molecular Class	Pivot Atom(s) or Unit
1	12	12	1.1.1	N,N
2	14	15	2.1.1	N,N
3	14	15	2.1.1	N,N
4	15	19	*	*
5	16	15	2.1.1	N,N
6	16	18	2.2.2	N,N
7	16	18	2.2.1	N,N
8	17	15	2.1.1	N,N
9	17	15	2.2.2	N,[a]
10	18	15	2.1.1	N,N
11	18	16	2.2.1	N,N
12	18	18	2.2.2	N,N
13	18	18	2.2.2	N,N
14	18	18	2.2.2	N,N
15	18	18	2.2.2	N,N
16	18	18	2.2.2	N,N
17	18	18	2.2.2	N,N
18	18	18	2.2.2	N,N

Strand A	Strand B	Strand C	MP or (BP) °C	Yld. (%)	Refs.
EOE	EOE	EOE	82–83	≤4	11, 12
EOEOE	EOE	EOE	oil (130/0.002)	19 15	41 26
ENENE	EOE	EOE	nr	nr	41, 43
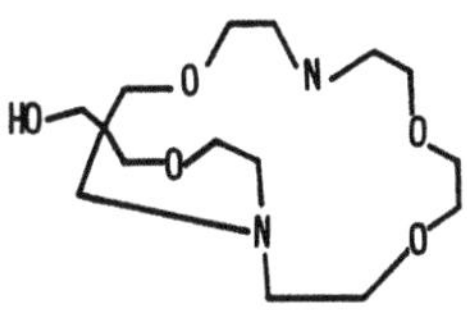			oil	27	36
ENMeENMeN	EOE	EOE	oil	25	43
COOEOCO	EOEOE	EOEOE	106	40	5
EOEOE	EOEOE	EOE	$(175/10^{-3})$ oil	30 29	26 41
CO-2,6-Py-CO	EOEOE	EOE	275–276	29	66
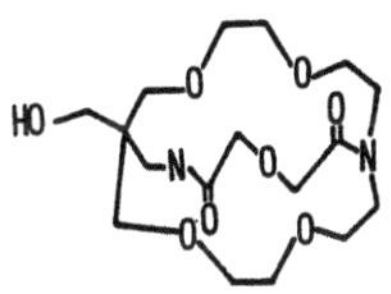			186–187.5	12	36
EO2BOE	EOE	EOE	119–120	nr	41
$EOCH_2CH(OH)CH_2OE$	EOEOE	EOE	oil	nr	64
EOEOE	EOEOE	EOEOE	68–69 68–69	24 48	22, 26 41
ENEOE	EOEOE	EOEOE	53	58	43
ENENE	EOEOE	EOEOE	69	43	41
ESESE	EOEOE	EOEOE	78–80 78–80	25 20	24 41
ESESE	ESEOE	EOEOE	oil	48	41
ENENE	ENENE	EOEOE	75	24	41
ESESE	ENENE	EOEOE	nr	nr	41

Table 8.3 (continued)

Cpd. No.	No. C's	No. Atoms Lgst. Ring	Molecular Class	Pivot Atom(s) or Unit
19	18	18	2.2.2	N,N
20	18	18	2.2.2	N,N
21	18	18	3.2.2	N,N
22	18	20	*	*
23	18	20	3.3.1	HOMC 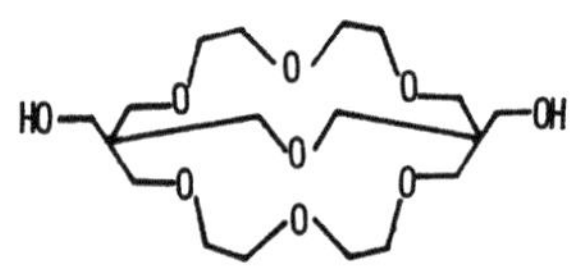
24	18	20	3.3.1	HOMC, 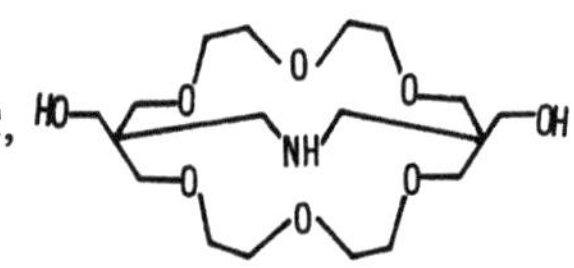
25	18	21	3.2.1	N,N
26	18	21	4.1.1	N,N
27	19	16	2.2.1	N,N
28	19	16	2.2.1	N,N
29	19	18	2.2.2	N,N
30	19	18	2.2.1	N,N
31	19	18	2.2.1	N,N

Strand A	Strand B	Strand C	MP or (BP) °C	Yld. (%)	Refs.
ESESE	ESESE	ENENE	86–87 86–87	16 20	41 24
ESESE	ESESE	ENENE	172 172	7 7	41 24
COOEOEOCO	EOEOE	EOEOE	104–105	36	5
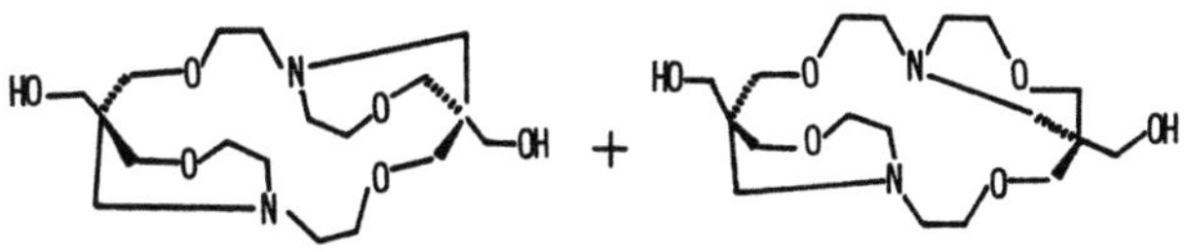			177.5–178	38	36
MOEOEOM	MOEOEOM	MOM	123–124	14	38
MOEOEOM	MOEOEOM	MNM	nr	49	38
EOEOEOE	ENENE	EOE	nr	nr	41
EOEOEOEOE	EOE	EOE	nr	nr	41
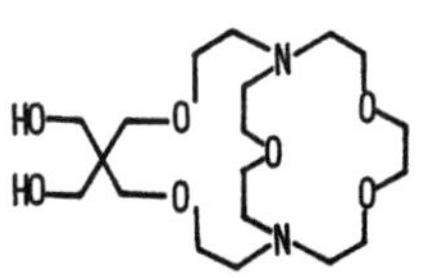	EOEOE	EOE	oil	⩽14	38
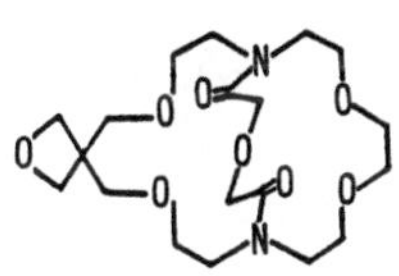	EOEOE	COMOMCO	180.5–182	21	36
ENMeEOE	EOEOE	EOEOE	oil	44	43
M-2,6-Py-M	EOEOE	EOEOE	95–96	63	66
CO-2,6-Py-CO	EOEOE	EOEOE	185–186	74	66

Table 8.3 (continued)

Cpd. No.	No. C's	No. Atoms Lgst. Ring	Molecular Class	Pivot Atom(s) or Unit
32	19	19	2.2.2	N,N
33	19	19	2.2.2	N,N
34	20	18	2.2.z	N,N
35	20	18	2.2.z	N,N
36	20	18	2.2.z	N,N
37	20	18	2.2.z	N,N
38	20	18	2.2.2	N,N
39	20	19	2.2.2	N,N
40	20	20	2.2.2	N,N
41	20	20	2.2.2	N,N
42	20	20	3.3.1	HOMC, 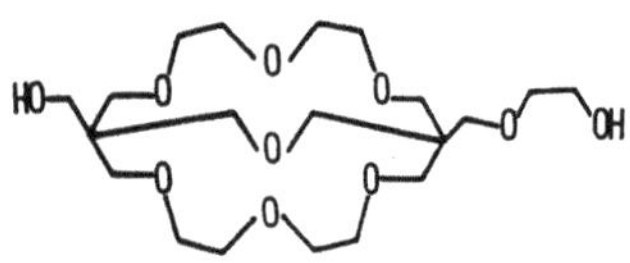
43	20	21	3.2.2	N,N
44	20	23	3.3.3	HC, HC 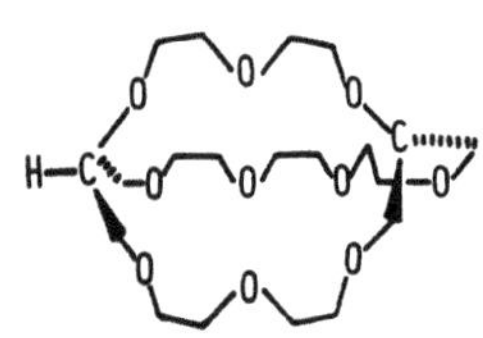
45	20	24	4.2.1	N,N
46	21	21	3.2.1	N,N
47	21	21	2.2.2	N,N
48	21	21	2.2.2	N,N

Strand A	Strand B	Strand C	MP or (BP) °C	Yld. (%)	Refs.
ENPNE	EOEOE	EOEOE	nr	nr	41
ENPNE	ESESE	ESESE	nr	nr	41
C_8	EOEOE	EOEOE	90–91	38	41
M3BM	EOEOE	EOEOE	100–101 oil	41 40	66 7
CO3BCO	EOEOE	EOEOE	224–225	72	7, 66
C_8	EOEOE	EOEOE	90–91	<48	26
ENMeENMeE	EOEOE	EOEOE	liq	38	41, 42, 43
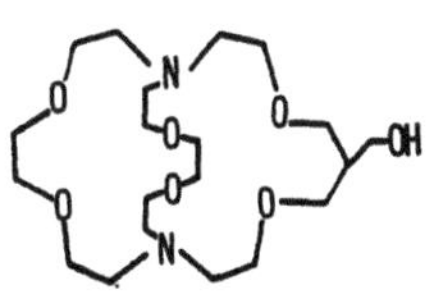	EOEOE	EOEOE	oil	nr	64
ENC_4NE	EOEOE	EOEOE	nr	nr	41
ENC_4NE	ESESE	ESESE	nr	nr	41
MOM	MOEOEM	MOEOEM	97–98	15	41
EOEOEOE	EOEOE	EOEOE	liq liq	<29 27	41 22, 26
MOEOEOM	MOEOEOM	OEOEO	oil	1.4	33
EOEOEOEOE	EOEOE	EOE	nr	nr	41
CO-2,6-Py-CO	EOEOEOE	EOEOE	oil	53	7
ENC_5NE	EOEOE	EOEOE	nr	nr	41
ENC_5NE	ESESE	ESESE	nr	nr	41

Table 8.3 (continued)

Cpd. No.	No. C's	No. Atoms Lgst. Ring	Molecular Class	Pivot Atom(s) or Unit
49	22	16	2.2.1	*
50	22	18	2.z.z	N,N
51	22	18	2.2.z	N,N
52	22	18	2.2.z	N,N
53	22	18	2.2.2	N,N
54	22	18	2.2.2	N,N
55	22	20	3.3.3	CMe; CMe
56	22	20	3.3.3	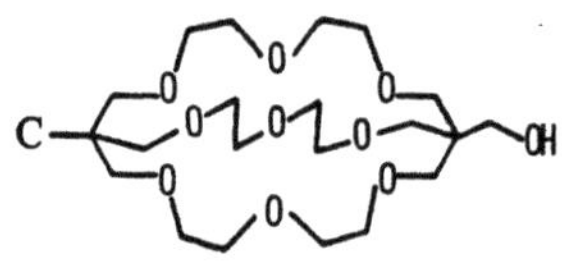
57	22	20	3.3.3	HOMC, HOMC
58	22	20	3.3.1	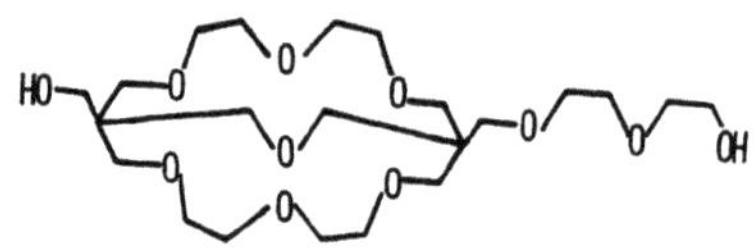
59	22	21	3.2.z	N,N
60	22	21	3.3.2	N,N
61	22	22	2.2.2	N,N
62	22	22	2.2.2	N,N
63	22	23	4.4.3	HC, HC, 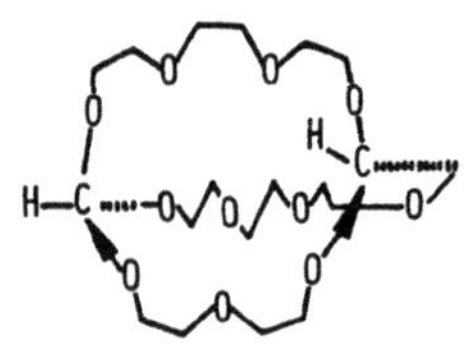
64	22	24	3.3.1	N,N
65	22	24	3.3.2	N,N

Strand A	Strand B	Strand C	MP or (BP) °C	Yld. (%)	Refs.
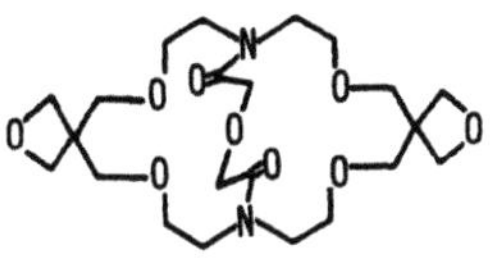			190–192	7	37
E4BE	COC_4CO	EOEOE	168	34	7
E4BE	EOEOE	EOEOE	122	89	7
COM4BMCO	EOEOE	EOEOE	208	47	7
EO2BOE	EOEOE	EOEOE	oil	40	41
ENMeENMeE	ENMeENMeE	EOEOE	32–33	18	41, 42, 43
MOEOEOM	MOEOEOM	MOEOEOM	59–60	nr	18
MOEOEOM	MOEOEOM	MOEOEOM	nr	nr	18
MOEOEOM	MOEOEOM	MOEOEOM	64–65	nr	17, 18
MOEOEOM	MOEOEOM	MOM	81.5–83	25	38
CO3BCO	EOEOEOE	EOEOE	oil	58	7
EOEOEOE	EOEOEOE	EOEOE	liq	25	41
ENC_6NE	EOEOE	EOEOE	nr	nr	41
ENC_6NE	ESESE	ESESE	nr	nr	41
MOEOEOEOM	OEOEOEO	MOEOEOM	oil	1.5	32, 33
EOENMeEOE	EOENMeEOE	EOE	oil	nr	31
EOEOEOE	EOEOEOE	EOEOE	liq	26	22, 26

Table 8.3 (continued)

Cpd. No.	No. C's	No. Atoms Lgst. Ring	Molecular Class	Pivot Atom(s) or Unit
66	22	24	4.2.2	N,N
67	22	24	4.2.2	N,N
68	22	24	4.2.2	N,N
69	22	26	4.4.3	HC, HC 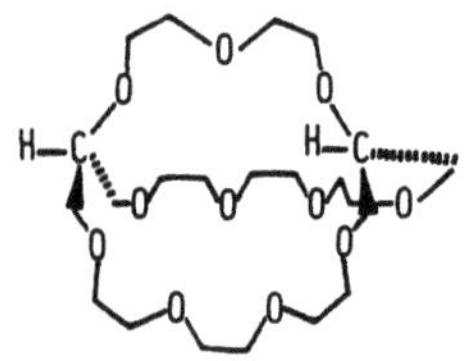
70	22	27	4.3.1	N,N
71	23	18	2.2.1	N,N
72	23	24	3.3.1	N,N
73	24	18	4.2.2	HC, HC 
74	24	18	2.2.z	N,N
75	24	18	2.2.z	N,N
76	24	18	2.2.z	N,N
77	24	20	2.2.2	N,N
78	24	20	2.2.2	N,N
79	24	20	2.2.2	N,N
80	24	20	3.3.3	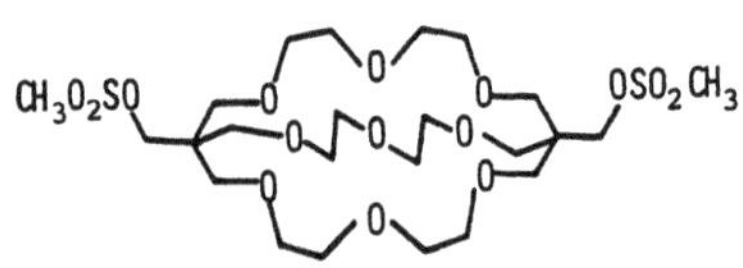
81	24	24	3.3.z	N,N

Strand A	Strand B	Strand C	MP or (BP) °C	Yld. (%)	Refs.
EOEOEOEOE	ESESE	EOEOE	nr	nr	41
EOEOEOEOE	ESESE	ENENE	nr	nr	41
EOEOEOEOE	ESESE	ESESE	nr	nr	41
MOEOEOEOM	MOEOEOEOM	OEOEO	oil	<1	33
EOEOEOEOE	EOEOEO	EOE	nr	nr	41
CO-2,6-Py-CO	EO2BOE	EOEOE	244	47	7
CO-2,6-Py-CO	EOEOEOE	EOEOEOE	oil	61	7
OEOEOEO	MO2BOM	MO2BOM	154	nr	58
EOEOE	EO2BOE	M3BM	oil	40	7
EOEOE	COMO2BOMCO	M3BM	253	41	7
EOEOE	EO2BOE	CO3BCO	259	61	7
EOEOE	EOEOE	EOM2BMOE	nr	nr	41
EOEOE	ESESE	EOM2BMOE	nr	nr	41
ESESE	ESESE	EOM2BMOE	nr	nr	41
MOEOEOM	MOEOEOM	MOEOEOM	90–91	nr	17, 18
EOEOEOE	EOEOEOE	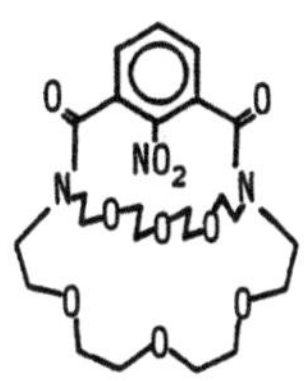	176–177	26	7

Table 8.3 (continued)

Cpd. No.	No. C's	No. Atoms Lgst. Ring	Molecular Class	Pivot Atom(s) or Unit
82	24	24	3.3.3	N,N
83	24	24	4.2.2	N,N
84	24	24	3.3.3	N,N
85	24	24	3.3.3	N,N
86	24	26	4.4.4	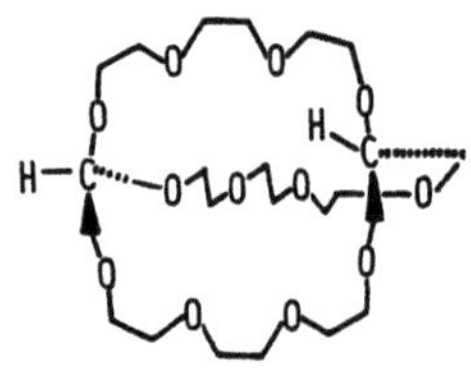
87	24	27	4.3.2	N,N
88	26	18	2.2.1	N,N
89	26	18	2.z.z	N,N
90	26	18	2.z.z	N,N
91	26	18	2.2.z	N,N
92	26	18	2.2.2	N,N
93	26	18	2.2.z	N,N
94	26	18	2.2.z	N,N
95	26	18	2.2.z	N,N
96	26	20	2.2.2	N,N
97	26	21	5.2.2	HC, HC
98	26	27	4.3.3	N,N
99	26	30	4.4.2	N,N

Strand A	Strand B	Strand C	MP or (BP) °C	Yld. (%)	Refs.
EOEOEOE	EOEOEOE	EOEOEOE	liq	52	13, 22
			oil	24	41
EOEOEOEOE	EOEOE	ENMeENMeE	nr	nr	41
ENEOENE	ENEOENE	ENEOENE	oil	38	44
ESESESE	ESESESE	ESESESE	nr	nr	41
MOEOEOEOM	MOEOEOEOM	OEOEOEO	nr	1.3	32
EOEOEOEOE	EOEOEOE	EOEOE	nr	nr	41
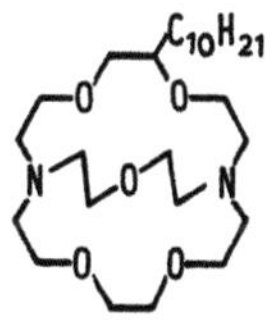	EOEOE	EOE	nr	nr	15
EOEOE	E4BE	E4BE	182	42	7
EOEOE	E4BE	COM4BMCO	258	21	7
EOEOE	EO2BOE	E4BE	oil	51	7
EOEOE	EO2BOE	EO2BOE	oil	nr	41
EOEOE	EO2BOE	COM4BMCO	151	29	7
EOEOE	COMO2BOMCO	E4BE	221–222	41	7
EOEOE	EOEOE	CO2B2BCO	271	51	7
EOEOE	EOM2BMOE	ENMeENMeE	nr	nr	41
MOEOEOEOEOM	MO2BOM	MO2BOM	117	nr	58
EOEOEOEOE	EOEOEOE	EOEOEOE	nr	nr	41
EOEOEOEOE	EOEOEOEOE	EOEOE	nr	nr	41

Table 8.3 (continued)

Cpd. No.	No. C's	No. Atoms Lgst. Ring	Molecular Class	Pivot Atom(s) or Unit
100	26	30	4.4.2	N,N
101	27	20	3.3.3	N,N
102	27	24	2.2.2	N,N
103	28	18	2.2.2	N,N
104	28	18	4.2.2	HC, HC
105	28	24	2.2.2	N,N
106	28	26	4.4.2	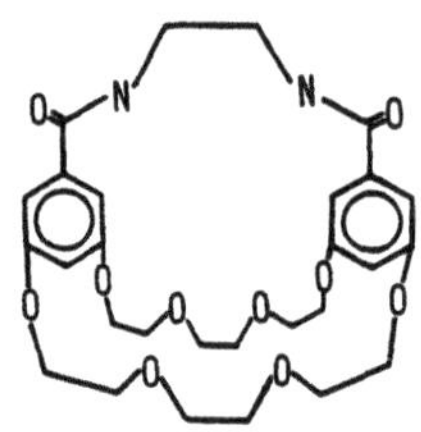
107	29	18	2.2.2	N,N
108	29	25	2.2.2	N,N
109	30	18	2.2.2	N,N
110	30	18	2.2.z	N,N

Strand A	Strand B	Strand C	MP or (BP) °C	Yld. (%)	Refs.
EOEOEOEOE	EOEOEOEOE	ESESE	nr	nr	41
EO-2,6-Py-OE	EO-2,6-Py-OE	EO-2,6-Py-OE	228	5	56
POPOP	POPOP	POPOP	nr	nr	41
EOEOE	EOEOE	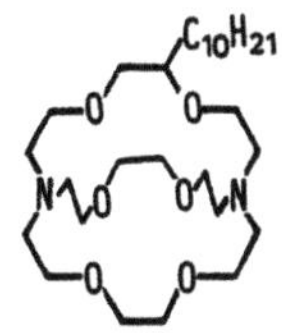	~-5	49	15
OEO2BOEO	MO2BOM	MO2BOM	nr	nr	34
EOEOE	EOEOE	M4BOEO4BM	101	nr	67
OEOEOEO	OEOEOEO	CONENCO	239	35	68
EOEOE	EOEOE	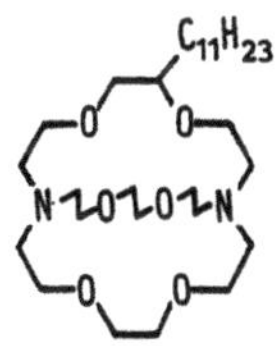	94.5–95.5	nr	15
EOEOE	EOEOE	M4BOPO4BM	84	nr	67
			oil	≤30	40
EOEOE	EO2BOE	CO2B2BCO	285	43	7

Table 8.3 (continued)

Cpd. No.	No. C's	No. Atoms Lgst. Ring	Molecular Class	Pivot Atom(s) or Unit
111	30	24	3.3.3	N,N
112	30	26	2.2.2	N,N
113	30	30	4.4.4	N,N
114	30	30	4.4.4	N,N
115	31	26	2.2.2	N,N
116	32	18	4.2.2	HC, HC
117	32	18	2.2.2	N,N
118	32	26	4.4.2	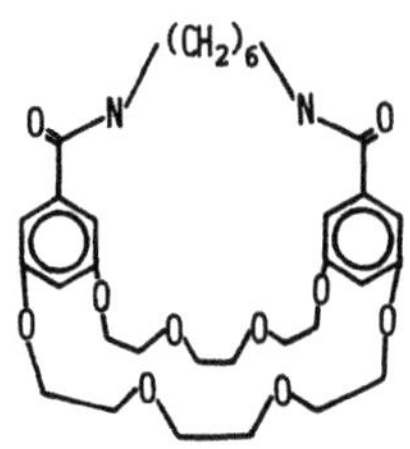
119	32	26	4.4.4	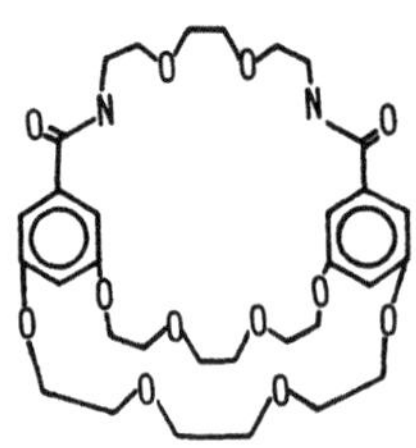
120	32	28	2.2.2	N,N

Strand A	Strand B	Strand C	MP or (BP) °C	Yld. (%)	Refs.
ENMeEOENMeE	ENMeEOENMeE	ENMeEOENMeE	oil	34	44
EOEOE	EOEOE	$M4BOC_4O4BM$	91	21–27	67
EOEOEOEOE	EOEOEOEOE	EOEOEOEOE	nr	nr	41
ESESESESE	ESESESESE	ESESESESE	nr	nr	41
EOEOE	EOEOE	$M4BOC_5O4BM$	101	nr	67
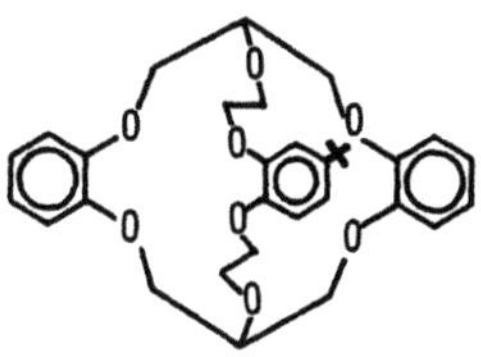	MO2BOM	MO2BOM	nr	nr	34
EOEOE	EOEOE	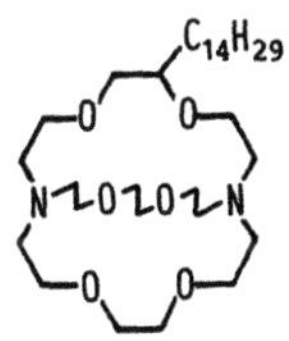	91.5–92.5[b]	nr	15
OEOEOEO	OEOEOEO	$CONC_6NCO$	158	32	68
OEOEOEO	OEOEOEO	CONEOEONCO	143	54	68
EOEOE	EOEOE	$M4BOC_6O4BM$	liq	nr	67

Table 8.3 (continued)

Cpd. No.	No. C's	No. Atoms Lgst. Ring	Molecular Class	Pivot Atom(s) or Unit
121	34	26	4.4.2	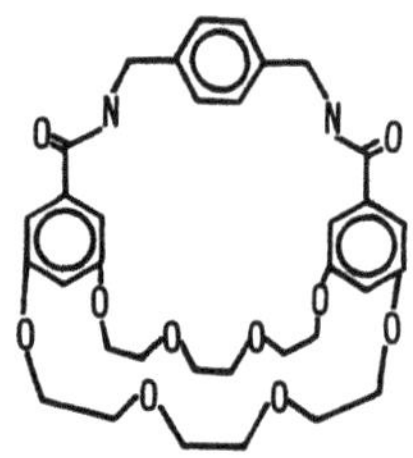
122	35	35	4.4.2	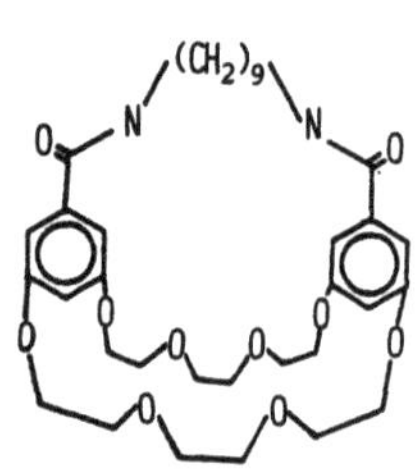
123	36	20	3.3.3	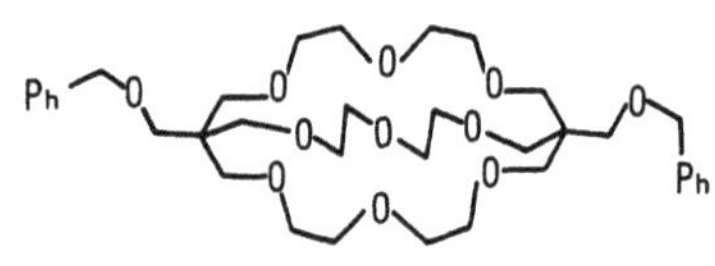
124	36	20	2.2.2	N,N
125	36	32	3.3.3	N,N
126	36	36	5.5.5	N,N
127	38	18	2.2.2	N,N

Strand A	Strand B	Strand C	MP or (BP) °C	Yld. (%)	Refs.
OEOEOEO	OEOEOEO	CONM4BMNCO	279	33	68
OEOEOEO	OEOEOEO	$CONC_9NCO$	98	47	68
MOEOEOM	MOEOEOM	MOEOEOM	48–49	nr	17, 18
EOEOE	EOEOE	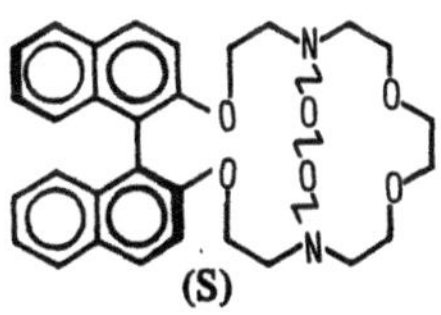 (S)	oil	51	27
POPOPOP	POPOPOP	POPOPOP	nr	nr	41
EOEOEOEOEOE	EOEOEOEOEOE	EOEOEOEOEOE	nr	nr	41
EOEOE	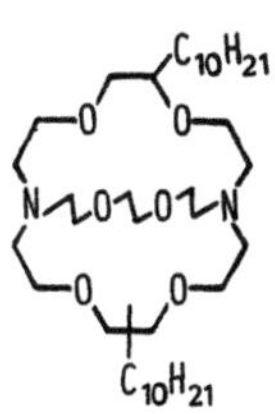		nr	nr	15

Table 8.3 (continued)

Cpd. No.	No. C's	No. Atoms Lgst. Ring	Molecular Class	Pivot Atom(s) or Unit
128	38	18	2.2.2	N,N
129	42	42	6.6.6	N,N
130	46	18	2.2.2	N,N
131	60	40	2.2.2	N,N
132	60	40	2.2.2	N,N
133	63	56	6.6.6	N,N

Abbreviations:

M = CH_2; E = CH_2CH_2; P = $CH_2CH_2CH_2$; C_n = $(CH_2)_n$; CO = $\overset{O}{\overset{\|}{C}}$; Me = CH_3; Et = CH_2CH_3; Py = pyridyl;

2B =  ; 3B = 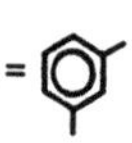 ; 4B = 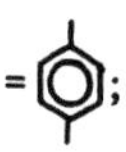 ; Ph = C_6H_5

[a] = pentaerythrityl

[b] as $NaBF_4$ complex

Strand A	Strand B	Strand C	MP or (BP) °C	Yld. (%)	Refs.
EOEOE	EOEOE	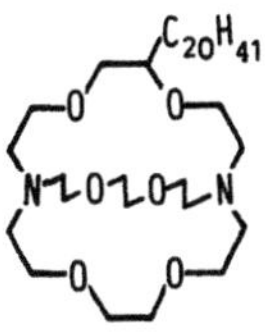	73–75[b]	nr	15
EOEOEOEOEOEOE	EOEOEOEOEOEOE	EOEOEOEOEOE	nr	nr	41
EOEOE	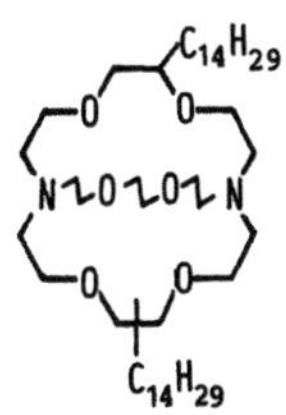		nr	nr	15
M4BOC$_6$O4BM	M4BOC$_6$O4BM	M4BOC$_6$O4BM	179	41	69
M4BOC$_6$O4BM	M4BOC$_6$O4BM	CO4BOC$_6$O4BCO	217	49	69
POPOPOPOPOPOP	POPOPOPOPOPOP	POPOPOPOPOPOP	nr	nr	41

Table 8.4. Cascade precursors

Cpd. No.	No. C's	Structure		Mp or (bp) in °C	Yield (%)	Refs.
1	12	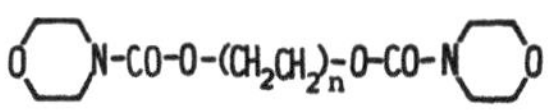	n = 1	112	90	7
2	14		n = 2	110	75	7
3	14	$[CH_2N(CH_2CH_2CH_2NH_2)_2]_2$		oil	24	9
4	16	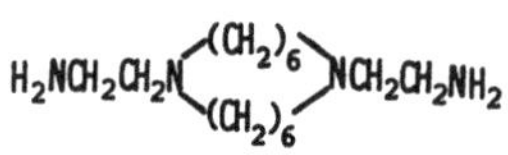		oil	83	9
5	16	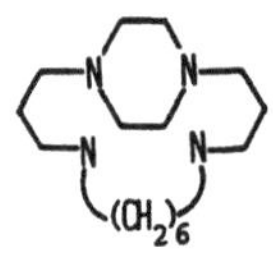		oil	84	9
6	18	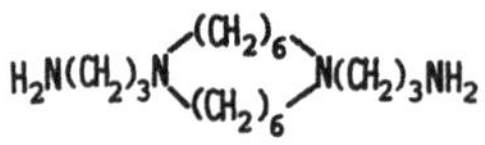		oil	58	9
7	19	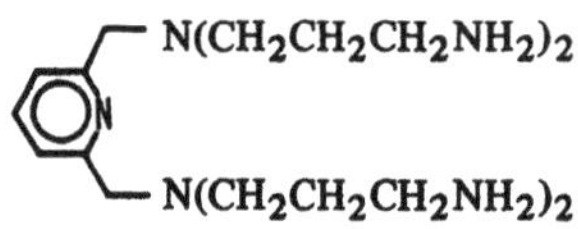		oil	63	9
8	20	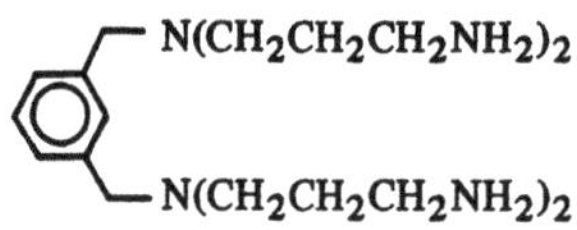		oil	57	9
9	22	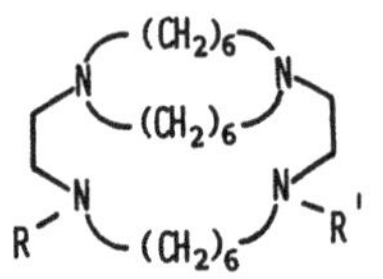	R = R′ = H	oil	79	9
10	25	$PhCH_2N[CH_2CH_2CH_2N(CH_2CH_2CH_2NH_2)_2]_2$		oil	35	9
11	28	See structure 9, R = R′ = $(CH_2)_3NH_2$		oil	61	9
12	34	See structure 9, R = R′ = $-(CH_2)_3N\text{-}CO\text{-}(CH_2)_4\text{-}CO\text{-}N(CH_2)_3-$		oil	12	9
13	38	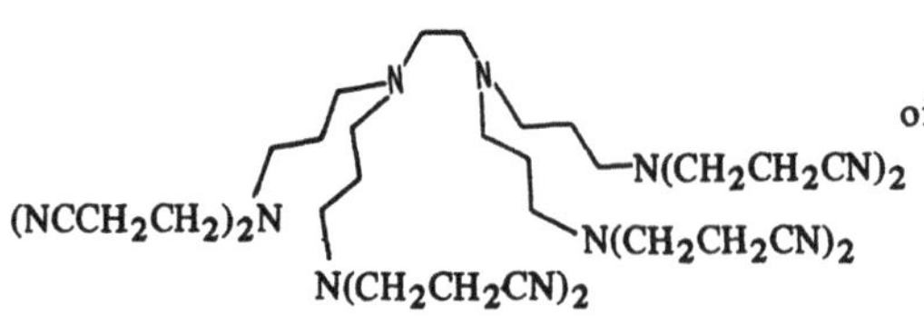		oil	64	9

Table 8.4 (continued)

Cpd. No.	No. C's	Structure	Mp or (bp) in °C	Yield (%)	Refs.
14	43	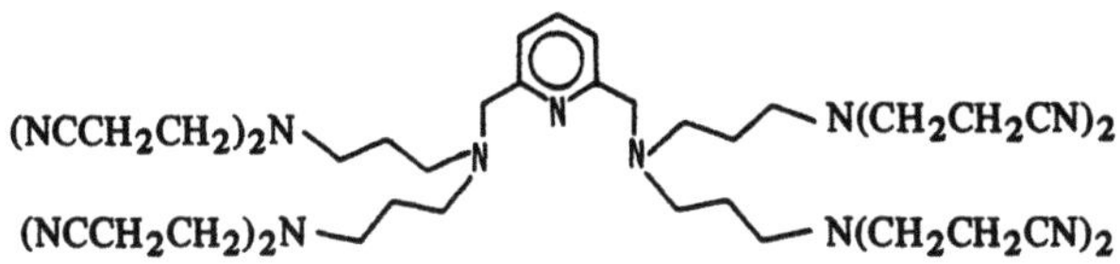	oil	42	9

Table 8.5. Polycyclic cryptands and spherands

Cpd. No.	No. C's	Ring size	Structure	Mp or (bp) in °C	Yield (%)	Refs.
1	24	18		196–200	31	30
2	24	24	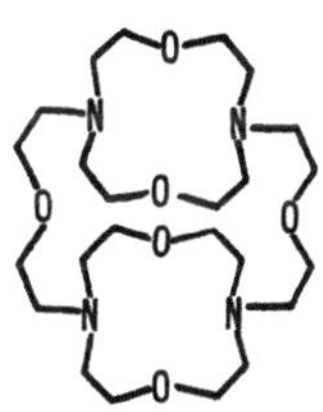	52	24	11,13
3	24	24	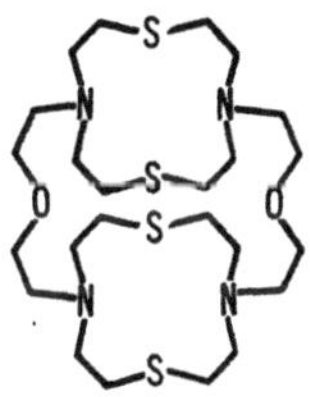	102–103	7.5–22.5	1
4	25	18	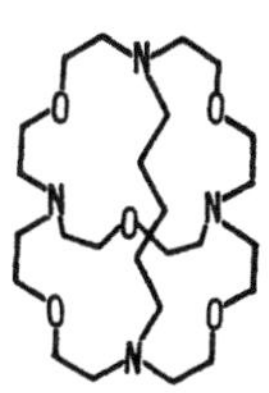	214–216	nr	31

Table 8.5 (continued)

Cpd. No.	No. C's	Ring size	Structure	Mp or (bp) in °C	Yield (%)	Refs.
5	28	18	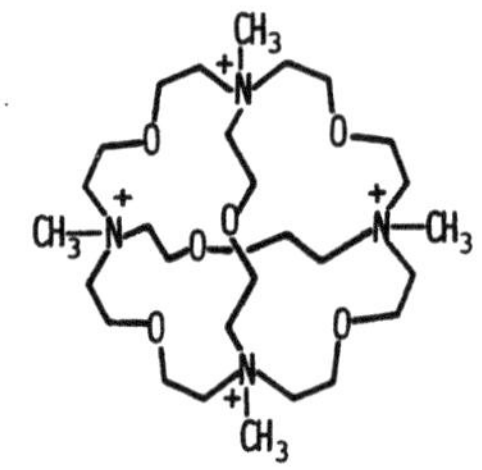	nr	nr	61
6	28	27	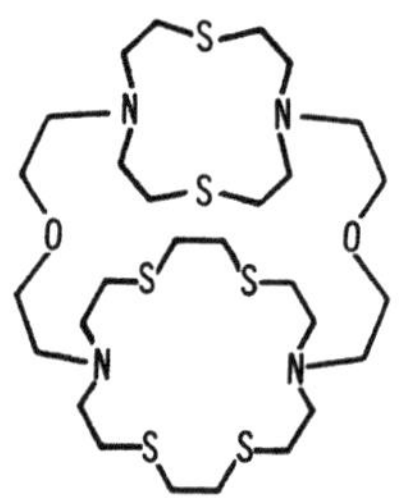	135–136	40	1
7	32	30	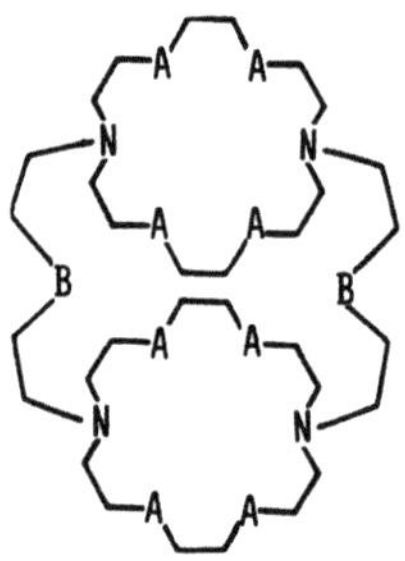 A = B = O	64	28	47
8	32	30	A = O, B = N	92–94	16	47
9	32	30	A = S, B = O	130–131	48	1
10	34	30	A = O, B = CH_2	45–46	30	47
11	38	28	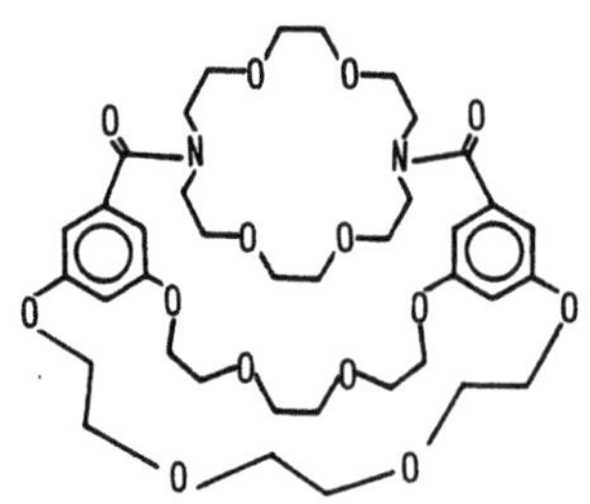	75	35	68
12	32	26	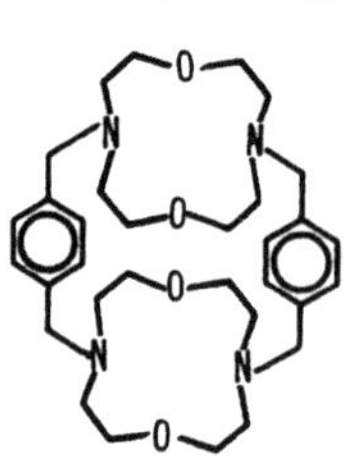	nr	nr	51

Table 8.5 (continued)

Cpd. No.	No. C's	Ring size	Structure	Mp or (bp) in °C	Yield (%)	Refs.
13	40	24	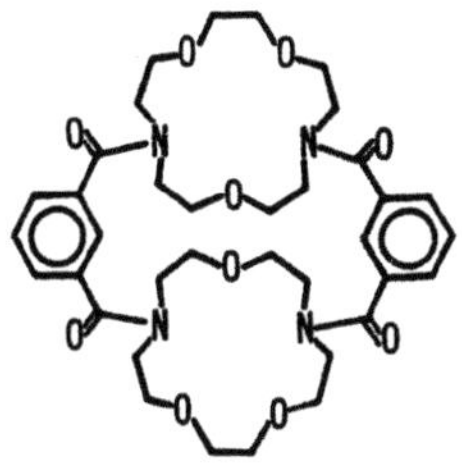	257–259	30	66
14	40	28	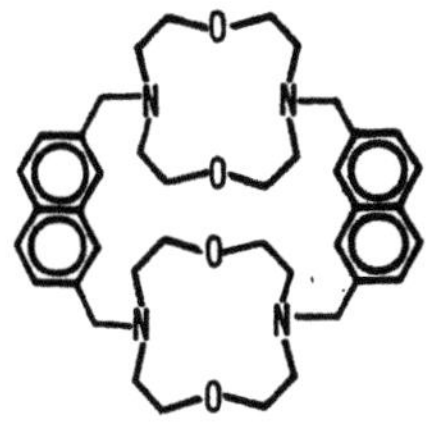	nr	nr	51
15	40	30	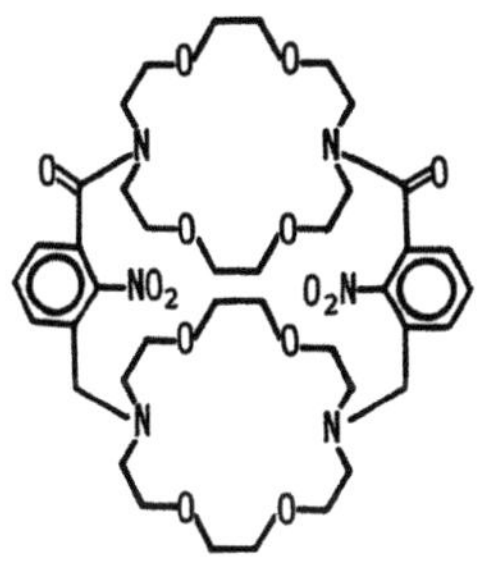	295–296	5.2	7
16	40	32	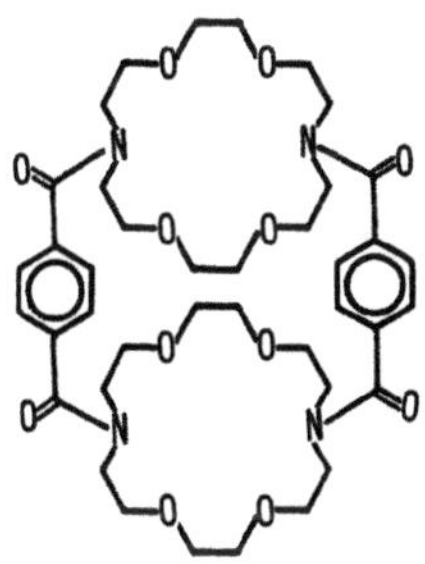	190	38	7
17	40	36	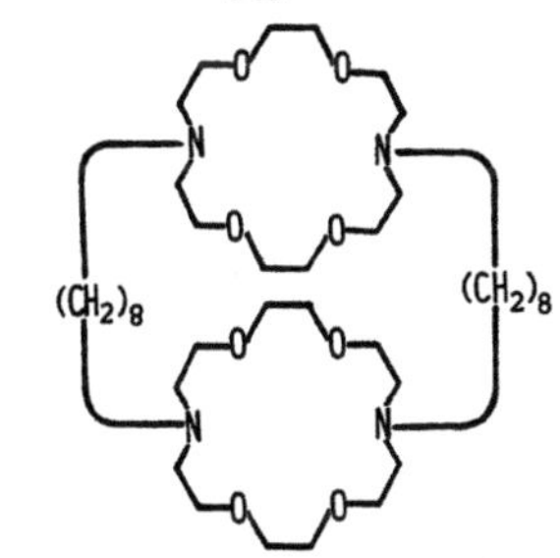	oil	58	36

Table 8.5 (continued)

Cpd. No.	No. C's	Ring size	Structure	Mp or (bp) in °C	Yield (%)	Refs.
18	44	27	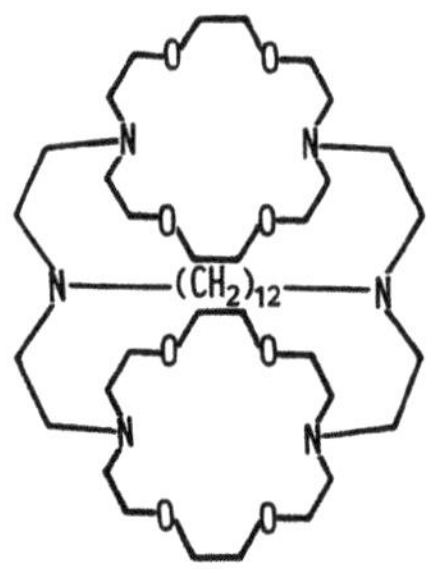	oil	6	47
19	44	34	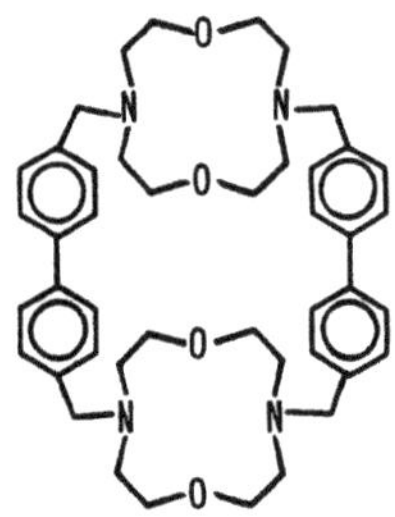	nr	nr	51
20	48	36	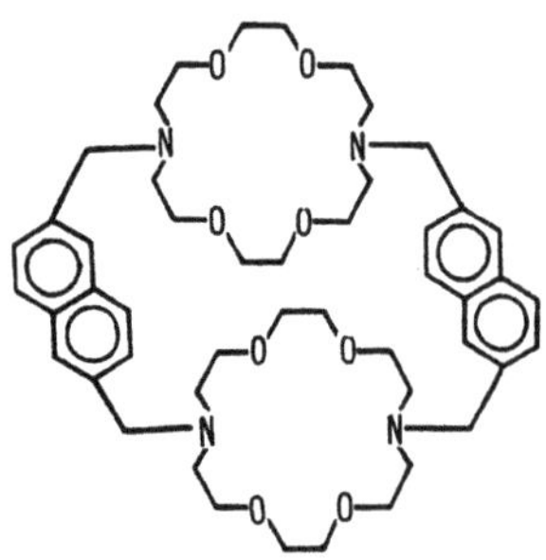	210	65	35
21	50	18	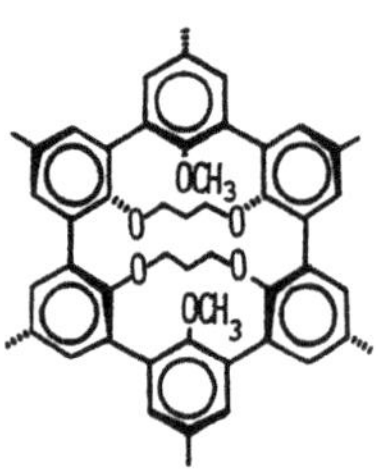	ca. 360° d.	1.2	19

Table 8.5 (continued)

Cpd. No.	No. C's	Ring size	Structure	Mp or (bp) in °C	Yield (%)	Refs.
22	52	35	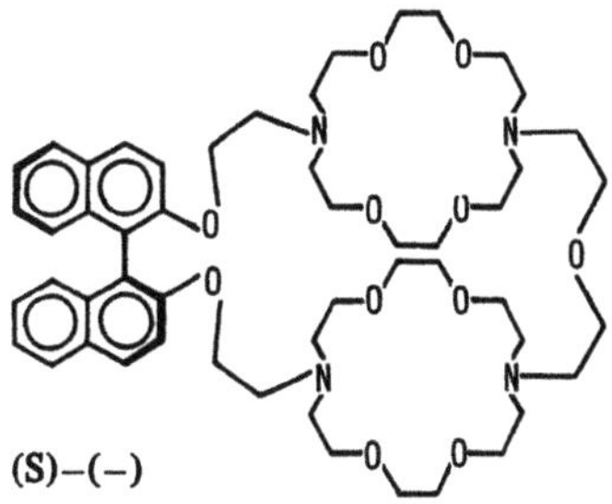 (S)–(–)	oil	43	27
23	52	35	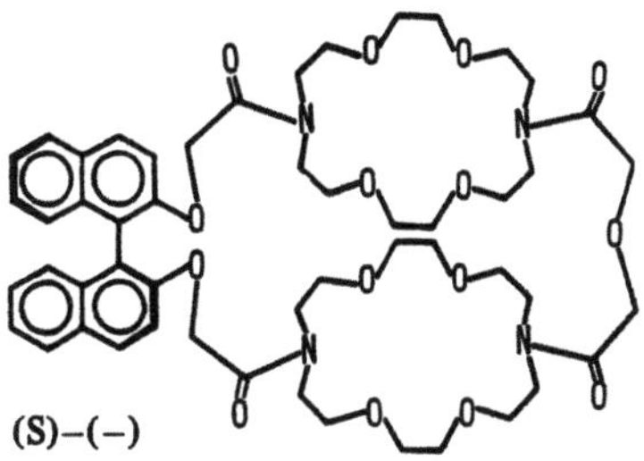 (S)–(–)	nr	50	46
24	52	35	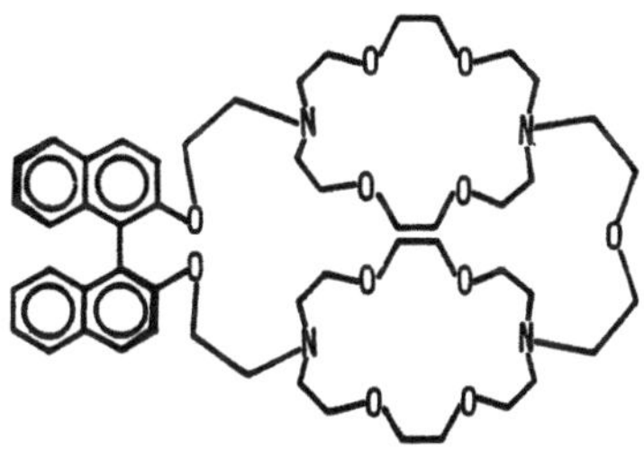	nr	45	46
25	52	39	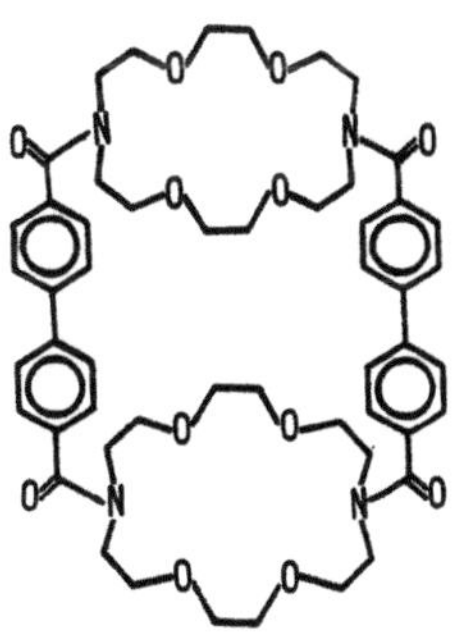	180	69	7

Table 8.5 (continued)

Cpd. No.	No. C's	Ring size	Structure	Mp or (bp) in °C	Yield (%)	Refs.
26	52	40	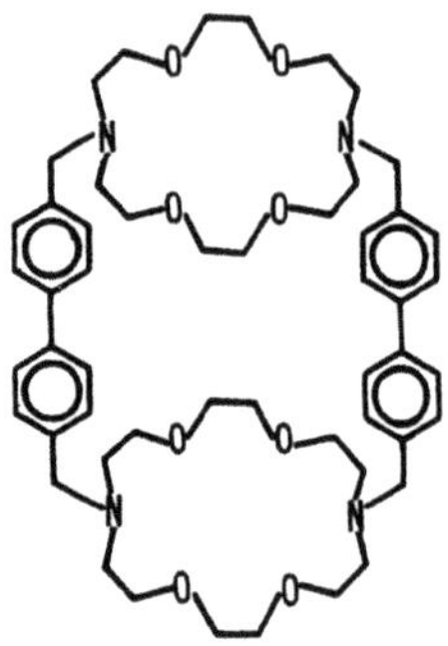	174	78	36
27	60		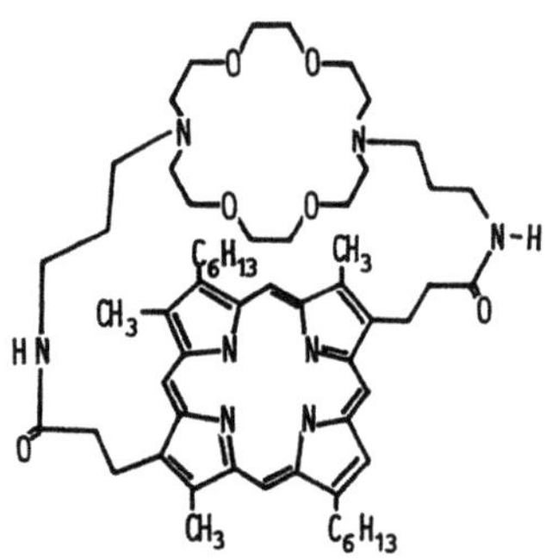	nr	65	10
28	78		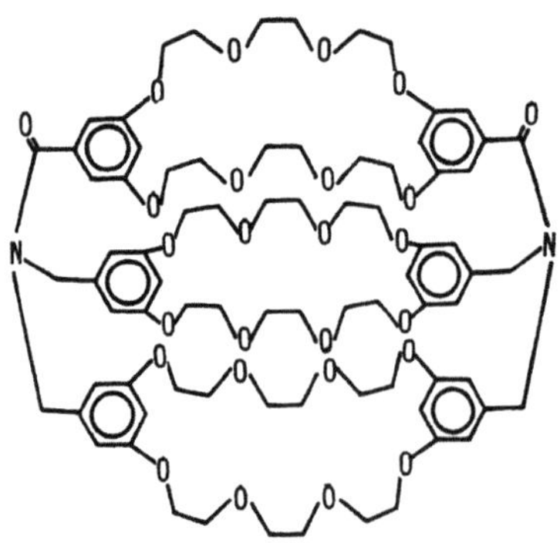	190	5	69

8.11 References

1. Alberts, A. H., Annunziata, R., Lehn, J. M.: J. Am. Chem. Soc. (1977) *99,* 8502
2. Alder, R. W., Sessions, R. B., Mellor, J. M., Rowlins, M. F.: J. C. S. Chem. Commun. (1977) 747
3. Bell, A. P., Hall, C. D.: J. C. S. Chem. Commun. (1980) 163
4. Blasius, E., Maurer, P.-G.: Makromol. Chem. (1977) *178,* 649
5. Buhleier, E., Frensch, K., Luppertz, F., Vögtle, F.: Liebigs. Ann. Chem. (1978) 1586
6. Buhleier, E., Wehner, W., Vögtle, F.: Chem. Ber. (1978) *111,* 200
7. Buhleier, E., Wehner, W., Vögtle, F.: Chem. Ber. (1979) *112,* 546
8. Buhleier, E., Wehner, W., Vögtle, F.: Chem. Ber. (1979) *112,* 559
9. Buhleier, E., Wehner, W., Vögtle, F.: Synthesis (1978) 155
10. Chang, C. K.: J. Am. Chem. Soc. (1977) *99,* 2820
11. Cheney, J., Kintzinger, J. P., Lehn, J. M.: Nouveau. J. Chem. (1978) *2,* 411
12. Cheney, J., Lehn, J. M.: J. C. S. Chem. Commun. (1972) 487
13. Cheney, J., Lehn, J. M., Sauvage, J. P., Stubbs, M. E.: J. C. S. Chem. Commun. (1972) 1100
14. Cinquini, M., Montanari, F., Tundo, P.: J. C. S. Chem. Commun. (1975) 393
15. Cinquini, M., Montanari, F., Tundo, P.: Gazz. Chim. Ital. (1977) *107,* 11
16. Clement, D., Damm, F., Lehn, J. M.: Heterocycles (1976) *5,* 477
17. Coxon, A. C., Stoddart, J. F.. J. C. S. Chem. Commun. (1974) 537
18. Coxon, A. C., Stoddart, J. F.: J. C. S. Perkin Trans. I (1977) 767
19. Cram, D. J., Kaneda, T., Lein, G. M., Helgeson, R. C.: J. C. S. Chem. Commun. (1979) 948
20. Curtis, W. D., Laidler, D. A., Stoddart, J. F., Jones, G. H.: J. C. S. Chem. Commun. (1975) 833
21. Curtis, W. D., Laidler, D. A., Stoddart, J. F., Jones, G. H.: J. C. S. Chem. Commun. (1975) 835
22. Dietrich, B., Lehn, J. M., Sauvage, J. P.: Tetrahedron Letters (1969) 2885
23. Dietrich, B., Lehn, J. M., Sauvage, J. P.: Tetrahedron Letters (1969) 2889
24. Dietrich, B., Lehn, J. M., Sauvage, J. P.: Chem. Commun. (1970) 1055
25. Dietrich, B., Lehn, J. M., Sauvage, J. P.: J. C. S. Chem. Commun. (1973) 15
26. Dietrich, B., Lehn, J. M., Sauvage, J. P., Blanzat, J.: Tetrahedron (1973) *29,* 1629
27. Dietrich, B., Lehn, J. M., Simon, J.: Angew. Chem. Int. Ed. Engl. (1974) *13,* 406
28. Dietrich, B., Lehn, J. M., Simon, J.: Angew. Chem. (1974) *86,* 443
29. Dye, J. L., Lok, M. T., Tehan, F. J., Ceraso, J. M., Voorhees, K. J.: J. Org. Chem. (1974) *38,* 1773
30. Graf, E., Lehn, J. M.: J. Am. Chem. Soc. (1975) *97,* 5022
31. Graf, E., Lehn, J. M.: J. Am. Chem. Soc. (1976) *98,* 6403
32. Gregory, B. J., Haines, A. H., Karntiang, P.: J. C. S. Chem. Commun. (1977) 918
33. Haines, A. H., Karntiang, P.: J. C. S., Perkin Trans. I (1979) 2577
34. Hanson, I. R., Parsons, D. G., Truter, M. R.: J. C. S. Chem. Commun. (1979) 486
35. Johnson, M. R., Sutherland, I. O., Newton, R. F.: J. C. S. Chem. Commun. (1979) 309
36. Kotzyba-Hibert, F., Lehn, J. M., Vierling, P.: Tetrahedron Letters (1980) *21,* 941
37. Krespan, C. G.: J. Org. Chem. (1975) *40,* 1205
38. Krespan, C. G.: J. Org. Chem. (1980) *45,* 1177
39. Kulstad, S., Malmsten, L. A.: Tetrahedron Letters (1980) *21,* 643
40. Landini, D., Montanari, F., Rolla, F.: Synthesis (1978) 223
41. Lehn, J. M.: U.S. Patent 3,888,877, 10 June 1975
42. Lehn, J. M., Montavon, F.: Tetrahedron Letters (1972) 4557
43. Lehn, J. M., Montavon, F.: Helv. Chim. Acta (1976) *59,* 1566
44. Lehn, J. M., Pine, S. H., Watanabe, E., Willard, A. K.: J. Am. Chem. Soc. (1977) *99,* 6766
45. Lehn, J. M., Sauvage, J. P.: J. Am. Chem. Soc. (1975) *97,* 6700
46. Lehn, J. M., Simon, J., Moradpour, A.: Helv. Chim. Acta (1978) *61,* 2407
47. Lehn, J. M., Simon, J., Wagner, J.. Angew. Chem. Int. Ed. Engl. (1973) *12,* 578
48. Lehn, J. M., Simon, J., Wagner, J.: Angew. Chem. Int. Ed. Engl. (1973) *12,* 579
49. Lehn, J. M., Sonveaux, E., Willard, A. K.: J. Am. Chem. Soc. (1978) *100,* 4914
50. Louis, R., Agnus, Y., Weiss, R.: J. Am. Chem. Soc. (1978) *100,* 3604

51. Mageswaran, R., Mageswaran, S., Sutherland, I. O.: J. C. S. Chem. Commun. (1979) 722
52. Metz, B., Boras, D., Weiss, R.: J. C. S. Perkin Trans. II (1976) 423
53. Mitsubishi Chem. Ind., Jpn. Kokai 78:91,328, 9 Feb. 1980
54. Mitsubishi Chem. Ind., Jpn. Kokai 78:91,329, 9 Feb. 1980
55. Montanari, F., Tundo, P.: Tetrahedron Letters (1979) 5055
56. Newkome, G. R., Majestic, V., Fronczek, F., Atwood, J. C.: J. Am. Chem. Soc. (1979) *101,* 1047
57. Park, C. H., Simmons, H. E.: J. Am. Chem. Soc. (1968) *90,* 2429
58. Parsons, D. G.: J. C. S. Perkin Trans. I. (1978) 451
59. Pedersen, C. J., Bromels, M. J.: U.S. Patent 3,847,949, 12 Nov. 1974
60. Pedersen, C. J., Bromels, M. J.: U.S. Patent 4,031,111, 21 June 1977
61. Schmidtchen, F. P.: Angew. Chem. Int. Ed. Engl. (1977) *16,* 720
62. Simmons, H. E., Park, C. H.: J. Am. Chem. Soc. (1968) *90,* 2428
63. Tehan, F. J., Barnett, B. L., Dye, J. L.: J. Am. Chem. Soc. (1974) *96,* 7203
64. Tomoi, M., Kihara, K., Kakiuchi, H.: Tetrahedron Letters (1979) 3485
65. Vögtle, F., Dix, P.: Liebigs Ann. Chem. (1977) 1698
66. Wehner, W., Vögtle, F.: Tetrahedron Letters (1976) 2603
67. Wester, N., Vögtle, F.: J. Chem. Res. (S) (1978) 400, J. Chem. Res. (M) (1978) 4856
68. Wester, N., Vögtle, F.: Chem. Ber. (1979) *112,* 3723
69. Wester, N., Vögtle, F.: Chem. Ber. (1980) *113,* 1487

Author Index

Subject Index

MIX
Papier aus verantwortungsvollen Quellen
Paper from responsible sources
FSC® C105338

If you have any concerns about our products,
you can contact us on
ProductSafety@springernature.com

In case Publisher is established outside the EU,
the EU authorized representative is:
Springer Nature Customer Service Center GmbH
Europaplatz 3, 69115 Heidelberg, Germany

Printed by Libri Plureos GmbH
in Hamburg, Germany